Companion to microbiology
Selected topics for further study

Edited by

Alan T. Bull
Professor of Fermentation Microbiology, University of Wales Institute of
Science and Technology, Cardiff

Pauline M. Meadow
Reader in Biochemistry, University College London

Longman
London and New York

Longman Group Limited London

*Associated companies, branches and representatives
throughout the world*

*Published in the United States of America
by Longman Inc., New York*

© Longman Group Limited 1978

First published 1978

Library of Congress Cataloging in Publication Data

Main entry under title:

Companion to microbiology.

    Includes index.
    1. Microbiology—Addresses, essays, lectures.
I. Bull, Alan T. II. Meadow, Pauline.
QR58.C63   576   77-13967
ISBN 0-582-46067-0

Printed in Great Britain by
Fletcher & Son Ltd, Norwich

Contents

Preface   ix

1   DNA tumour viruses   *B. E. Griffin*   1
      Introduction.   Histories of polyoma virus and SV40.   Physical description of polyoma virus and SV40 and their DNAs.   DNA replication.   Transcription in permissive cells.   Proteins of polyoma virus and SV40. Genetics.   Transformation.   Comparison of polyoma virus and SV40. Conclusions.

2   RNA tumour viruses   *R. A. Weiss*   27
      Introduction.   Assay methods for oncornaviruses.   Structure and replication. Transmission of oncornaviruses.   Genetics of oncornaviruses. Oncogenesis.   Human RNA tumour viruses.

3   Control of DNA synthesis and cell division in bacteria   *R. James*   59
      A system for studying differentiation.   The *E. coli* cell-division cycle. Regulation of DNA initiation.   Regulation of cell septation.   Reflection.

4   The biology of plasmids   *P. A. Williams*   77
      Introduction.   Characterisation of a function as plasmid-coded.   Classes of plasmid.   Some important features of plasmid biology.   Concluding remarks.

5   Bacteriocins   *K. G. Hardy*   109
      Introduction.   Classification and chemistry of colicins.   Colicin synthesis. Col plasmids.   Effects of colicins on sensitive bacteria.   Ecology of colicins. Other bacteriocins: the problem of defining a bacteriocin.   Uses of bacteriocins.   Summary.

6   Molecular basis of toxin action   *J. P. Arbuthnott*   127
      Scope of chapter.   Toxins and pathogenicity. Diversity and special features of toxins. Approaches to the study of mechanism of action.   Diphtheria

vi

toxin. The enterotoxins. The neurotoxins. The cytolytic (membrane-damaging) toxins. Staphylococcal epidermolytic toxin. Lipopolysaccharide endotoxin.

7 Efficiency of microbial growth   *D. E. F. Harrison*   155
   Introduction. Expression and measurement of growth efficiency. Maintenance energy and 'true yield'. Factors affecting growth efficiency. The regulation of growth efficiency. The importance of growth efficiency.

8 Interactions between microbial populations   *J. H. Slater and A. T. Bull*   181
   Introduction. Basic types of interaction. The kinetics of competition. Competition as a mechanism in enzyme evolution. Microbial communities. Conclusions.

9 Microbial hormones   *G. W. Gooday*   207
   Introduction. Autotactic attractants. Sperm attractants. Fungal, algal and protozoal sex hormones. Conclusions.

10 Morphogenesis in bacteria   *R. Whittenbury and C. S. Dow*   221
   Introduction. The aquatic environment and the morphology of microbes. The colonial microbes. Categorisation of cell cycles and related events. 'Budding' bacteria. Varieties of budding bacteria. *Escherichia coli*—molecular biology of the cell cycle. *Arthrobacter crystallopoietes*—nutritionally-controlled morphogenesis. *Bacillus* species—endospore morphogenesis and differentiation; non-obligate differentiation. *Caulobacter* species—obligate dimorphism. *Rhodomicrobium vannielii*—summation of microbial diversity. Filamentous cyanobacteria—(multicelled prokaryotes?). Myxobacteria.

11 Membrane fluidity in microorganisms   *D. J. Ellar*   265
   Introduction. Membrane isolation and analysis. Membranes as an asymmetric fluid mosaic. Membrane variability and the role of fluidity. Membrane biosynthesis and the fluid mosaic model.

12 Transport of organic solutes by bacteria   *H. Tristram*   297
   Introduction. Transport of β-galactosides in *Escherichia coli*. Group translocation. Isolation and characterisation of transport proteins. Transport of amino acids and peptides. Transport of galactose in *Escherichia coli*. Transport of hexose phosphates. The problem of energy-coupling to transport. Concluding remarks.

13 Bacterial motility and chemotaxis   *D. G. Smith*   321
   Introduction. Flagellar movement. Spirochaetal movement. Gliding movement. Bacterial chemotaxis. Chemotaxis in non-flagellate bacteria. Summary and prospects.

14   Biological nitrogen fixation   *J. R. Postgate*   343
      Introduction. Biochemistry. Physiology. Ecology. Genetics. Some
      future prospects.

15   Bioenergetics of chemolithotrophic bacteria   *D. P. Kelly*   363
      Introduction. The organisms and their substrates. Energy requirements
      for autotrophic growth and the nature of chemolithotrophic energy genera-
      tion. The energetic problems facing the chemolithotrophs. Pathways of
      oxidations in chemolithotrophs. Continuous flow culture as a tool in
      elucidating chemolithotrophic energy-coupling mechanisms. Concluding
      remarks.

16   Photosynthetic endosymbionts of invertebrates   *D. C. Smith*   387
      Introduction. Symbiotic dinoflagellates. Symbiotic *Chlorella*. *Convoluta
      roscoffensis*. Chloroplast symbiosis. Cyanellae. Discussion.

17   Current problems and developments in medical mycology   *D. Kerridge*   415
      Introduction. Detection. The role of exocellular products in fungal
      infections. Structural dimorphism. Chemotherapy of fungal infections.
      Prospects.

18   Epidemiology and control of protozoal pathogens of man   *J. R. Baker*   431
      Introduction. Classification. Parasites   acquired   orally. Parasites
      acquired via the blood. Parasites acquired congenitally. Parasite acquired
      venereally—*Trichomonas vaginalis*. Conclusions.

19   Microbial control of pest insects   *J. R. Norris*   459
      Introduction. History. *Bacillus thuringiensis*. *Bacillus popilliae/Bacillus
      lentimorbus*. Viruses. Safety of microbial insecticides. The develop-
      ment of resistance to microbial control agents. Future prospects.

Index   481

Acknowledgements

We are grateful to the following for permission to reproduce copyright material:

Authors and Academic Press Inc. for figs. by M. Fried and B. Griffin from *Advances in Cancer Research*, Vol. 24 (1977), Eds. Klein and Weinhouse and a table by R. N. McElhaney from *The Biological Significance of Alterations in the Fatty Acid Compositions of Microbial Membranes in Response to Changes in Environmental Temperature Extreme Environments*, Ed. M. R. Heinrich; Authors and Academic Press Inc. (London) Ltd. for modified fig. from *Journal of Molecular Biology*, Vol. 78 (1973), Eds. Danna, Sack and Nathans, Fig. 11 b,c,d from *Journal of Molecular Biology*, Vol. 70 (1972), Eds. O'Brien and Bennett and a table by L. A. Falcon from *Microbial Control of Insects and Mites*, Eds. Burgess and Hussey; American Society for Microbiology for fig. and tables from *Bacteriological Review*, Vol. 35 (1971), 'Physical Properties of Membrane Lipids' in *Bacteriological Review*, Vol. 39 (1975), Eds Gronan and Gelmann, and *Bacteriological Review*, Vol. 26 (1962), Eds. Mortenson, Mower and Carnahan; Blackwell Scientific Publications Ltd. for a fig. from *Cell and Tissue Kinetics*, Vol. 9 (1976), Eds. Botchen *et al.*; Authors and Cold Spring Harbor Laboratory for figs. from *Symposium of Quantitative Biology*, Vol. 39 (1974), Ed. Fried *et al.*, *Symposium of Quantitative Biology*, Vol. 28 (1963), Ed. Jacob *et al.* and a table by E. P. Kennedy from *The Lactose Operon*, Eds. J. R. Beckwith and D. Zipser; Elsevier/North–Holland Biomedical Press for a table by Haest, De Gier and van Deenen from *Chem. Phys. Lipids*, Vol. 3 (1969), modified Figs. from *Biochem. Biophys. Acta*, Vol. 4 (1971), Eds. Wong and Wilson and *Biochem. Biophys. Acta*, Vol. 457 (1976), Eds. Postma and Roseman; W. H. Freeman & Co. for a fig. from *Scientific American* after Luria, Vol. 233 (1975); Authors for figs. by Fried, Griffin and Cowie from *Proceedings of the National Academy of Sciences*, Vol. 71 (1974); Authors for a fig. by Drs. Graessmann and Graessmann from *Proceedings of the National Academy of Sciences*, Vol. 71 (1974); Macmillan Journals Ltd. for figs. by Teich *et al.* from *Nature*, Vol. 256 (1975), and by Donachie from *Nature*, Vol. 219 (1968); Author for table by M. Sinesky from *Proceedings of the National Academy of Sciences*, Vol. 71 (1974); Springer-Verlag for a fig. by Kepes from *Journal of Membrane Biology*, Vol. 4 (1971).

Whilst every effort has been made to trace the owners of copyright, in a few cases this has proved impossible and we take this opportunity to offer our apologies to any author whose rights may have been infringed.

Preface

Over the last ten years or so there has been a steady and welcome flow of textbooks of general microbiology including some with a medical, ecological or biochemical slant. There have also been some excellent specialist texts and reviews devoted to narrow aspects of the subject. In this *Companion to Microbiology* we have selected a few topics for discussion in broader perspective than is usual in review articles and specialist texts and in more depth than is possible in a general textbook. Inevitably, the selection reflects the personal bias of the editors, but we have tried to rationalise it in terms of the important growth points of Microbiology and those topics which are rarely discussed outside review articles and research Journals. We invited our contributors to write for students and to assume that they would have a basic background of microbiology and biochemistry. We have not tried to make the chapters uniform and have been glad to accommodate a variety of styles so that authors could discuss their subject as they thought appropriate. The referencing too has largely been a matter of personal choice. Some chapters required extensive bibliographies to enable the reader to follow up points of interest. Other subjects could adequately be signposted by a few key review articles.

We hope that *Companion to Microbiology* will prove useful to advanced undergraduate and postgraduate students not only to supplement lecture courses but also to encourage them to explore facets of microbiology outside their curricula. Above all we hope the Companion will provide interesting and stimulating reading. We would welcome comments from students and teachers on ways of improving the book and suggestions for further topics which might be included in new editions.

A. T. B.
P. M. M.

1
DNA tumour viruses

B. E. Griffin

Imperial Cancer Research Fund, Lincoln's Inn Fields, London

1 Introduction

Animal viruses with DNA genomes have been classified into five groups — parvoviruses, papovaviruses, adenoviruses, herpes viruses and poxviruses — in order of increasing complexity. Of these, the papovaviruses, especially polyoma virus and Simian virus 40 (SV40) have received particular attention in the past decade. The reason is undoubtedly two-fold. On the one hand, the very small size of the viral genomes suggests that they might play a useful role as probes for studying the much more complicated genomes of eukaryotic cells, and might provide a means of unravelling some of the molecular biology of the cell. On the other hand, these viruses are known to be capable of producing tumours in a variety of animals. The possibility that they might therefore be used for studying the mechanism by which a normal cell is converted into a malignant cell makes them of great practical significance.

So far, studies on polyoma virus and SV40 have been more successful in defining the molecular biology of the two viruses themselves than in answering broader questions about the role of the viruses in de-regulating cells — the carcinogenic role of viruses. It still seems reasonable to believe that cancer is a disease that may be initiated at the cellular level, that viruses may act directly by altering the genetic make up of one or more of the chromosomes of a cell, and that once this genetic change has occurred, it is transferred to the progeny of that cell. When a newly-created cancer cell divides, both its descendants presumably carry the cancer properties and they in turn pass them on to their progeny. If this is a reasonable premise, one returns again and again to the original question: can the initial genetic change be attributed to the action of the virus on the cell? If so, what *was* the change and how was it effected?

The question — do human tumour viruses exist? — is still unanswerable. Within the past few years viruses with morphologies similar to polyoma virus and SV40 have been isolated from human sources. A virus isolated in 1972 from brain tissue of patients with progressive multifocal leukoencephalopathy (PML), a rapidly progressive neurological disorder found in association with carcinomas, appears to be antigenically related to SV40. It can be grown in culture in primary monkey cells as well as in human foetal brain cells. Its DNA has been shown to have about 30 per cent homology with SV40 DNA. Whether these human papova viruses cause PML or any other disease in man remains to be seen. Papova viruses have also been found in the urine of patients who have been immunosuppressed in conjunction with kidney transplants. The reason for their presence is still obscure.

A number of other DNA viruses have been implicated in tumour formation, but their relevance to human cancer stands on no stronger ground than that of the papova viruses. The discovery in the 1960s that certain human adenoviruses could induce tumours in newborn hamsters initially caused much excitement, since adenovirus infections are very common in human populations and most people contain antibodies against one or more adenovirus. Adenoviruses are much larger and more complex that papovaviruses, but have much in common with them, namely:

a Both groups of viruses replicate and are assembled in the nuclei of permissive cells.

b Both cause tumours in a variety of newborn rodents, and transform cells in culture.

The genetic information for adenoviruses resides in their DNAs, which are double-stranded linear molecules with molecular weights about 23×10^6 daltons (or 6–7 times the size of polyoma virus and SV40 DNAs). Like the papova viruses, the molecular biology of the adenoviruses is slowly being understood, and studies with viral DNA fragments made by cleavage with restriction enzymes show that only a small fragment from the adenovirus genome is essential for cellular transformation. Extensive studies on a large number of human cancers have, however, failed to show the presence of any transcription products of adenoviruses in human cancer.

Herpes viruses have also been associated with cancer in five sorts of animals, in frogs, domestic fowls, guinea pigs, monkeys and man. There are many herpes viruses, and it is not clear at present how they relate one to another. They are defined as large viruses with complex DNA genomes (molecular weight about 100×10^6 daltons), a capsid which contains 162 capsomers (or structurally equivalent subunits) arranged in the form of an icosahedron, and a lipid–glycoprotein envelope. Exposure of these viruses to lipid solvents or to lipases destroys their infectivity. Very little is known about the mechanism of viral DNA synthesis, although it takes place in the nucleus and appears to require protein synthesis.

Herpes viruses can transform cells in culture, but, as with the other DNA viruses, very little is known at present about the interactions between herpes viruses and their non-permissive or semi-permissive host cells. What little is known suggests that the herpes genomes may establish stable associations with host cell genomes by integration in much the same way that polyoma virus and SV40 DNAs do. Whether viral products are needed for the maintenance of the transformed state is not known.

Of all the human herpes or herpes-like viruses, the most likely candidate for a cancer-causative agent at present is the Epstein–Barr virus (EB virus). This virus was first noted in cultured human lymphoblast cells from a patient with Burkitt's malignant lymphoma. It fulfils several of the criteria expected of a human tumour virus:

a Virtually every biopsy of Burkitt's lymphoma and of cell lines that have been established from tumour tissue have proved to contain at least one copy, and often many more, of EB virus DNA per cell.

b Antigens on the surface of Burkitt's lymphoma cells are also present on the envelopes of EB virus particles.

c The epidemiologies of Burkitt's lymphoma and EB virus suggest that the tumour might develop in people suffering simultaneously from both lymphocyte depletion and virus infection.

Although this evidence appears more persuasive than most of the evidence which links DNA viruses to cancer, there remains the overriding question of why Burkitt's lymphoma (and post-nasal carcinoma, which has also been linked with EB virus) is so common in certain parts of the world (notably East Africa for Burkitt's lymphoma) and virtually absent in other parts of the world, whereas the virus itself is ubiquitous. If EB virus has a causative role, perhaps it only induces cancer in the presence of other accessory factors. And what then are these factors?

It is not clear what experiments should or could be carried out to indicate more clearly whether the DNA viruses play a significant role in human cancer. It is clear however that they may ultimately be of considerable use in unravelling the complicated machinery of the cell.

Because of their small size and the relative ease with which they can be handled, much progress has been made toward understanding the two tumour viruses, polyoma and SV40. This chapter will deal largely with these two viruses.

2 Histories of polyoma virus and SV40

The principal viruses of the papova group are the papilloma (wart) viruses, polyoma virus (a mouse virus) and SV40 (a vacuolating virus of monkeys). The name papova is derived from the first two letters of the original names for each virus, *pa*pilloma, *po*lyoma and *va*cuolating virus. The papilloma viruses have been much less well defined than polyoma virus and SV40, at least in molecular terms, mainly because no one has yet succeeded in growing them in the laboratory in cell culture systems.

The histories of the discoveries of both polyoma virus and SV40 are interesting. Polyoma virus was first isolated by Gross in the early 1950s during a study of leukemia in mice. In attempting to isolate a mouse leukemia agent, he obtained a filterable extract which was capable of producing salivary gland carcinomas when inoculated into newborn mice. As it seemed unlikely that a single agent would be responsible for both events, he concluded that the mice might be carrying *two* distinct oncogenic agents, one producing leukemia and the other tumours. Subsequently he found that normal healthy mice contained the virus, polyoma virus, which was responsible for the tumours. Polyoma virus has now been found in a large proportion of both wild and laboratory mice, and in most cases is probably a harmless passenger of the species. It can be grown in culture in a number of mouse cell lines, called *permissive* lines. The word permissive means that the cells support the complete replication and growth of the virus and ultimately die, liberating virus particles into the surrounding media. This is referred to as *lytic* infection, or the lytic cycle. Polyoma virus has also been found to infect other animal cells (such as hamster and rat cells) in culture without concomitant production of large amounts of virus. These cells are *transformed* by the virus into cells with some of the characteristics of tumour cells (Fig. 1.1). Cells that can be infected but do not liberate virus particles are called *non-permissive* cells and infection is termed *non-lytic*. Polyoma virus ultimately received its name from the wide spectrum of tumours which it could be shown to induce.

In the early 1960s it was found by Eddy and others that malignant tumours were induced in newborn hamsters when they were injected with extracts from rhesus monkey kidney cells. The agent responsible for tumour formation was shown to be a virus, but the tumours were somewhat different from those induced by polyoma virus. This virus, carried by monkeys, was identical to a vacuolating virus identified earlier in monkey cells by Sweet and Hilleman and called SV40. The discovery that SV40 could induce tumours was doubly disturbing since it had been found as a contaminant, often in extensive quantities, in the live attenuated vaccines being used against poliomyelitis. The question then arose of the safety of giving such a vaccine to human subjects, particularly infants, and subsequent batches of vaccine were presumably carefully screened for traces of SV40.

SV40 can be grown in culture in permissive monkey cells. It transforms mouse and hamster cells. Human cells, in general, have been found to be *semi-permissive* for SV40, i.e. some of the cells support the replication of the virus and ultimately die, whereas others become trans-

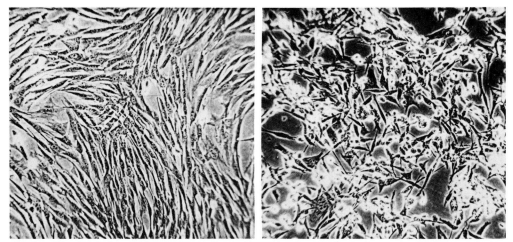

Fig. 1.1 Transformation of BHK 21 (hamster) cells by polyoma virus. (*Left*) Colony which illustrates the regular parallel arrangement of the non-transformed elongated BHK 21 fibroblastic cells. (*Right*) BHK 21 cells transformed by polyoma virus, illustrating the random orientation and more rounded nature of the transformed cells. (Photograph by courtesy of Dr M. G. P. Stoker.)

formed. Human cells are currently thought to be non-permissive for polyoma virus.

The DNAs from polyoma virus and SV40 have been found to have very little homology in terms of their nucleic acid sequences. Moreover, polyoma virus and SV40 do not appear to share common antigens. In terms of size and overall genome organisation, however, the two viruses appear to be remarkably similar.

3 Physical description of polyoma virus and SV40 and their DNAs

3.1 Physical properties

Polyoma virus and SV40 are very similar in size and gross structures. The particles are about 45 nm in diameter and have a sedimentation coefficient, S_w^{20}, of 240. They have icosahedral symmetry, each with 72 capsomers and a triangulation number, t, of 7. Figure 1.2 shows an electron micrograph of polyoma virus. The virus particles contain only protein and DNA. Since they lack lipids, they can be extracted with lipophilic solvents without the loss of infectivity. They are relatively stable to heat but can be inactivated with formaldehyde.

The genetic information for both viruses resides within their DNAs. The molecular weights of the DNAs are $3-3.5 \times 10^6$ daltons (about 5 000 nucleotide pairs, or enough genetic information to code for about 200 000 daltons of protein). In both viruses, the DNA exists in *native* form as covalently-closed, circular superhelices (supercoils) with a sedimentation co-efficient of 21S. Supercoiled DNA is known as Form I DNA. When this is nicked at a single site in one strand (by cleavage of a phosphodiester bond) the superhelical form relaxes to a circular form (Form II), with sedimentation coefficient 16S (see Fig. 1.2). SV40 DNA has a guanine + cytosine (G + C) content of about 41 per cent; polyoma DNA has a G + C content of about 47–48 per cent. In an analysis of the frequencies of nearest-neighbour nucleotide sequences in DNA, it has been observed that the CpG doublet is very rare in mammalian cell DNA. The CpG doublet is also found with low frequency in the DNAs of both polyoma virus and SV40. The argument has been put forward that the virus can only use efficiently the translation apparatus

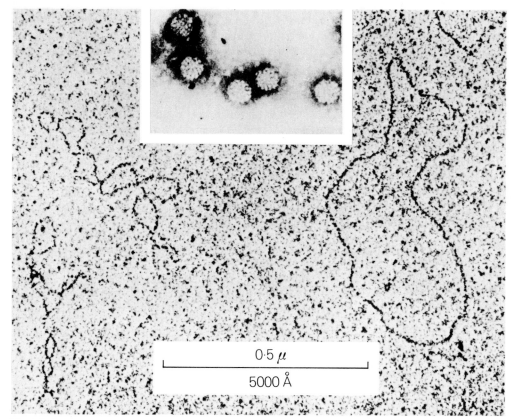

Fig. 1.2 Electron micrograph of polyoma virus particles and DNA. *Top* (*centre*), polyoma virus. *Left,* polyoma superhelical (Form I) DNA. *Right,* polyoma circular (Form II) DNA. A single nick in the DNA converts Form I to Form II DNA. In this photograph, the same scale is used for virus particles and DNAs, which points to the very interesting question of how viral DNA is packaged. (Photograph, copyright Drs L. V. Crawford and E. Follett.)

of a host cell whose nearest-neighbour pattern closely resembles that of the virus. Whether such an hypothesis has any validity will only become evident as primary sequence information becomes available both for viral and host cell DNA.

3.2 DNA physical maps

Following the early physical description of the viruses and their DNAs, further work of this nature was considerably hampered by the absence of physical and/or biological markers on the genomes. The DNAs do not apparently even contain any minor bases among their nucleotides. This situation was dramatically altered by the discovery of bacterial restriction enzymes (endonucleases). These are enzymes which recognise specific nucleotide sequences in double-stranded DNA and cleave both strands of duplex DNA. They have now been isolated from a large number of different microorganisms. At present, nearly 100 specific endonucleases have been discovered, and they recognise more than 40 different specific sequences, most of which appear to be tetra- or hexa-nucleotide pairs with characteristic symmetries. Many enzymes have been discovered from different sources which recognise the same nucleotide sequences. The term *isoschizomer* has been given to these enzymes. Moreover, some microorganisms have been found to contain more than one restriction enzyme. In any particular microorganism, the restriction endonucleases are part of a restriction–modification system, modification (methylation) being used by the organism as a means of protecting its own DNA against cleavage by the enzyme. Foreign DNA, on the other hand, is unprotected and susceptible to cleavage. Thus, many microorganisms possess in their restriction enzymes a means of self-protection against invading species and in their modification systems a protection against self-destruction. To date, no such restriction–modification system has been reported for mammalian cells – presumably, the immune surveillance system in higher organisms makes it unnecessary. Although several different types of restriction endonucleases have been reported, the most useful are enzymes which recognise and cleave the same set of nucleotide sequences. A specific and reproducible set of DNA fragments can be made from a species of DNA which contains sequences corresponding to those required by the restriction enzyme (see Fig. 1.3). Table 1.1 gives a list of some of the restriction enzymes known to cleave polyoma virus and SV40 DNAs, together with the sequences at which cleavage occurs. Restriction enzymes can even be seen to provide

Table 1.1   Cleavage of polyoma virus and SV40 DNAs by restriction endonucleases

Microorganism	*Abbreviation*	*Sequence* $5' \rightarrow 3'$ $3' \leftarrow 5'$	*Number of sites*	
			Polyoma	SV40
Brevibacterium albidum	Bal I	CGG↓CCG GCC↑GGC	0	0
Haemophilus parainfluenzae	Hpa I	GTT↓AAC CAA↑TTG	0	4
Bacillus amyloliquefaciens	Bam I	G↓GATCC CCTAG↑G	1	1
Escherichia coli RI	Eco RI	G↓AATTC CTTAA↑G	1	1
H. aegyptius	Hae II	RGCGC↓Y Y↑CGCGR	1	1
H. influenzae-d	Hind II	GTY↓RAC CAR↑YTG	2	7
H. influenzae-d	Hind III	A↓AGCTT TTCGA↑A	2	6
H. haemolyticus	Hha I	GCG↓C C↑GCG	3	2
H. parainfluenzae	Hpa II	C↓CGG GGC↑C	8	1
H. aegyptius	Hae III	GG↓CC CC↑GG	24	16–18

Enzyme nomenclature uses a three-letter abbreviation for the host enzyme followed by a strain designation where required, and a roman number when more than one enzyme is obtained from the same organism. For example, Hind III is one of the restriction enzymes obtained from *Haemophilus influenzae*, serotype *d*.

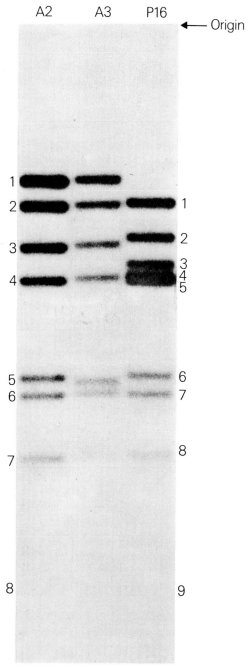

Fig. 1.3 An autoradiogram of fragments obtained when ³²P-labelled polyoma DNAs from three different strains, strains A2, A3 (large plaque morphology) and P16 (small plaque morphology) are cleaved with a restriction enzyme from *Haemophilus parainfluenzae*, Hpa II. Fragments are separated by electrophoresis on an acrylamide gel. A2 and A3 polyoma DNAs are each cleaved to eight fragments, but size differences can be seen. The P16 polyoma DNA is cleaved into nine fragments. Restriction enzymes have proved very useful for looking at minor differences between strains of both polyoma virus and SV40. (Taken from Fried *et al.* 1974, *Cold Spring Harbor Symp. Quant. Biol.,* **39**, 45–52.)

limited information about primary sequences present in DNAs. For example, one of the enzymes from *Haemophilus influenzae* (Hind II) recognises the four different hexanucleotide sequences specified by 5′–GTYRAC–3′, where Y is any pyrimidine and R is any purine. This enzyme cleaves polyoma DNA twice and SV40 DNA seven times. An enzyme from *H. parainfluenzae* (Hpa I) recognises the specific sequence 5′–GTTAAC–3′, which is one of the four possible Hind II recognition sites. Hpa I cleaves SV40 DNA four times and does not cleave polyoma DNA at all. Thus, the sequence GTTAAC must be absent in polyoma DNA, and must account for four of the seven Hind II recognition sites in SV40 DNA.

Physical maps of both polyoma virus and SV40 DNAs have been determined from cleavage of the DNAs into discrete fragments with restriction enzymes, and subsequent arrangement of the fragments in topographical order. A number of methods have been used to order the DNA fragments. One includes a limited digest with a particular restriction enzyme; DNA fragments are then separated, and further cleaved with the same restriction enzyme to show which fragments are normally joined by the enzyme site. For example, a product of partial digestion of polyoma DNA with *H. parainfluenzae* (Hpa II) would have fragments 3 and 5 (Figs. 1.3 and 1.4) still covalently linked. Further treatment with Hpa II would produce fragments 3 and 5 as individual fragments. Analysis of all the products which on further digestion would yield two fragments ultimately leads to a physical map. Pulse-chase experiments with isotopes followed by enzymatic cleavage and simultaneous cleavage with two or more restriction enzymes have also been used to locate particular cleavage sites. The reference map now most frequently used for SV40 DNA is that obtained by cleavage with a mixture of the two restriction endonucleases from *H. influenzae* (Hind II and III, 13 sites) and for polyoma DNA that obtained by cleavage with one of the restriction endonucleases from *H. parainfluenzae* (Hpa II, 8 sites). Figure 1.3 shows the Hpa II cleavage patterns of three strains of polyoma viruses. Physical maps for polyoma virus and SV40 are shown together in Fig. 1.4.

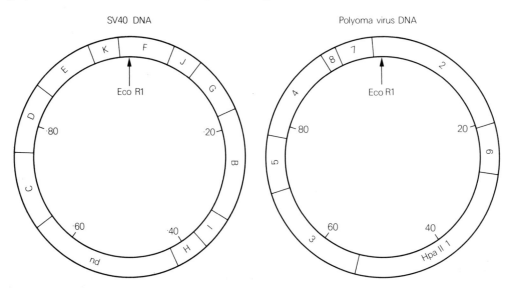

Fig. 1.4 Physical (restriction enzyme) maps of polyoma virus and SV40 DNAs. The SV40 *Haemophilus influenzae* (Hind) map is divided into a total 1·0 map unit. The polyoma virus *Haemophilus parainfluenzae* (Hpa II) map is divided into 100 map units. For both maps, the single *E. coli* RI (Eco RI) restriction enzyme cleavage site is designated to lie at zero map unit. (The SV40 physical map is modified from Danna, Sack and Nathans (1973), *J. Molec. Biol.,* **78**, 363, and 11 of the 13 fragments are shown. The polyoma virus map is from Griffin, Fried and Cowie (1974), *Proc. Natl. Acad. Sci. USA,* **71**, 2077.)

3.3 Primary sequence studies

The simplest method of examining the primary nucleotide sequences of DNA involves the chemical depurination reactions first described by Burton and Petersen. When DNA is treated with formic acid the purines (adenine and guanine) are cleaved from the DNA chain leaving so-called 'apurinic acids'. Subsequent treatment with a base (usually diphenylamine) results in the further cleavage of phosphodiester bonds at sites which formerly contained purines. The result leaves DNA fragments which contain only pyrimidines (thymine and cytosine), or so-called 'tracts' of pyrimidines. With double-stranded DNA the tracts of pyrimidines represent a picture of the entire molecule since a tract of pyrimidines on one strand of the DNA will have a complementary tract of purines on the other strand. Fingerprints made of the depurination products from any particular DNA or segment of DNA provide an 'identi-kit' of that DNA. The depurination products of polyoma virus and SV40 DNAs are shown in Fig. 1.5. A comparison of the two fingerprints, either from a qualitative or a quantitative point of view, suggests that there is very little obvious similarity between the sequences of the two DNAs.

Direct DNA-sequencing procedures have now advanced to such a state that within the next few years the primary sequences of large portions, if not all, of both viral DNAs should be available for direct comparison.

3.4 Nucleoprotein complexes

DNA in eukaryotic cells associates with histones to form compact structures called nucleosomes or v-bodies. Each nucleosome contains about 200 base pairs of DNA and eight histone molecules, adjacent nucleosomes being linked by short (enzyme-susceptible) bridges of DNA. Nucleosome structures may therefore be used within the cell to expose or protect certain sites on the DNA. Preliminary data only are available on the cleavage of nucleoprotein complexes of polyoma virus and SV40 DNAs with restriction endonuclease. It suggests that although nucleosomes do not occupy unique positions, the nucleosome arrangement in any species is probably not entirely random. This is clearly a topic which requires further study.

4 DNA replication

The genomes of both polyoma virus and SV40 are replicated in the nuclei of permissive cells. Replicating intermediates (see Figs. 1.6 and 1.7) of both viral DNAs have been isolated from permissive host cells by a variety of techniques. They sediment faster in neutral sucrose gradients than Form I viral DNA ($S_W^{20} = 25$), presumably as a result of their increased mass and more compact structure. It is not yet entirely clear whether viral DNA becomes integrated into host cell DNA in the lytic infection, and whether viral replication occurs *autonomously* or from an integrated molecule.

A study of replicating molecules by electron microscopy showed species with two untwisted loops of about equal size (the presumed replicating region) attached to a superhelical tail (presumed unreplicated region). The template for replication should be the continuously-closed circles of DNA, while the newly-replicated part of the molecules (relaxed loops) should also contain single strands of polynucleotides, hydrogen-bonded to separated template strands. Biochemical studies supported this model. Additional studies by electron microscopy showed single-stranded regions of DNA at the position of the replicating forks in SV40 DNA.

(a)

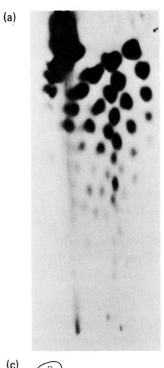

(b)

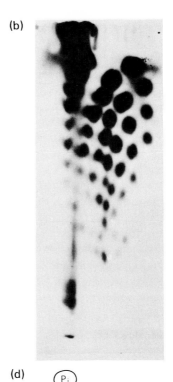

(c)

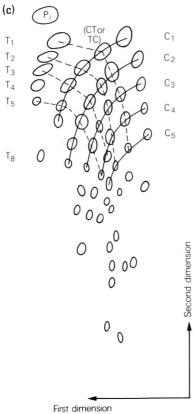

(d)

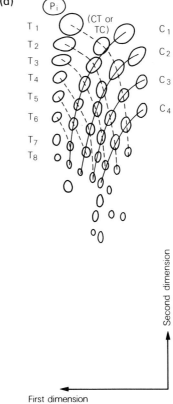

First dimension

Second dimension

First dimension

Second dimension

A model for the autonomous replication of both polyoma virus and SV40 involves an overall mechanism that can be divided into four distinct stages: (1) initiation; (2) chain elongation; (3) segregation of the progeny molecules; and (4) maturation of newly-made daughter molecules. This is summarised in Fig. 1.6. Some aspects of the detailed mechanisms within each stage are known, but many still remain to be defined.

1 Initiation. Both polyoma virus and SV40 have been shown to have a unique origin for viral DNA replication, and both DNAs have been shown to replicate bidirectionally at about the same rate in each direction. Analyses of replicating intermediates with restriction enzymes have

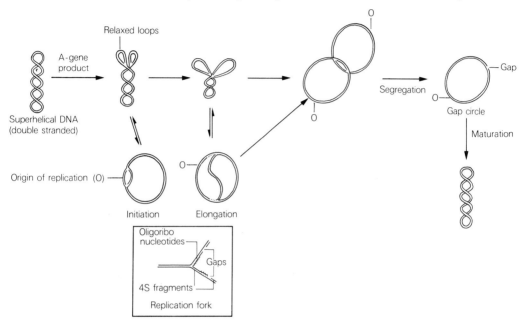

Fig. 1.6 A model depicting stages in the replication of polyoma virus or SV40 DNA. For SV40, initiation occurs at a unique site (0·67 map units, see Figs. 1.8 and 1.12) on the DNA and requires the A gene product. For polyoma virus DNA, the origin lies at 71 map units (see Figs. 1.8 and 1.12). Polynucleotide chain elongation proceeds from this origin site bidirectionally by addition of small DNA fragments (about 4S in size) to the replication fork. The closed circular and superhelical replicative intermediates are in equilibrium with the active relaxed replicative form. When replication proceeds about 180° around the circles, segregation of the two daughter molecules takes place resulting in a circular molecule which contains a single-stranded region or gap at the terminus (about 0·15 map units for SV40 DNA) of replication. Maturation of this DNA requires a DNA polymerase and polynucleotide ligase activity to produce a closed circular mature viral DNA molecule. (Modified from Levine, Van der Vliet and Sussenbach (1976), *Current Topics in Microbiology and Immunology,* **73,** 67.)

Fig. 1.5 Fingerprints of the depurination products of ^{32}P-labelled polyoma virus DNA (a) and SV40 DNA (b). The tracts of pyrimidines were separated in the first dimension by electrophoresis at pH 3·5 and in the second dimension by chromatography. (Separation in the first dimension is based on charge, and in the second on size.) Schematic diagrams of the polyoma fingerprint (c) and the SV40 fingerprint (d) are shown. C_1 means cytidylic acid, C_2 cytidine dinucleotide, T_1, thymidylic acid, etc. By following the lines between pyrimidine tracts which have only C or T, tracts containing both nucleotides can be found, and their compositions (but not sequences) predicted (see TC, for example). The fingerprints of the two viral DNAs show little similarity at the level of complexity where individual tracts of pyrimidines begin to be significant (octanucleotide and greater). Polyoma DNA has nine oligopyrimidines greater than ten units long whereas SV40 DNA has only three. SV40 DNA has many more tracts containing thymidine only than does polyoma DNA.

shown the origin of replication to lie in SV40 DNA in the third largest Hind II + III restriction fragment (called Hind-C) at about 0·67 map units on the SV40 physical map, and in polyoma virus DNA near the junction of the third and fifth largest Hpa II restriction fragments (called Hpa II 3–5 junction) at about 71 map units on the polyoma physical map, see Figs. 1.7 and 1.8. DNAs with multiple repeats of the region containing the viral origin of replication are frequently found in stocks of both polyoma virus and SV40, which suggests that the origin is an essential site and required for DNA replication. It seems clear from several lines of evidence that each round of viral DNA replication requires protein synthesis. It also appears that at least one of the proteins required for initiation is a viral gene product (see section 7). There is some evidence that a globular protein may be involved in the initiation of DNA replication since such a protein has been detected in association with SV40 DNA, and the complex has been shown to involve the region around the origin of replication. It is not at present clear whether this is the same viral protein required for replication or a cellular protein. The primary nucleotide sequences from the origin region show symmetrical, repeated sequences that are rich in adenine (A) and thymine (T) residues. These may ultimately prove to be important, either in specific protein binding and unwinding of the DNA prior to replication, or in binding of primer oligoribonucleotides (see p. 13), or as signals for transcription, or in other as yet unknown functions.

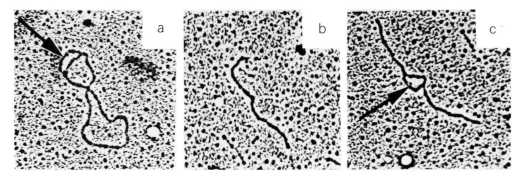

Fig. 1.7 Mapping the viral origin of replication. Electron micrograph of replicating DNA. Replicating bubbles are indicated by arrows, × 45 000. (A): Polyoma DNA replicating molecule, less than 50 per cent replicated. (B and C): a replicating molecule cleaved with restriction enzyme Hind III. Hind III cleaves polyoma DNA into two unequally-sized fragments. The larger of the two fragments has the replicating bubble. The midpoints of the bubbles in molecules represented by (C) were measured to be 25 ± 3 per cent from one end, and 30 ± 3 per cent from the other. (Taken from Griffin, Fried and Cowie (1974) *Proc. Natl. Acad. Sci. USA*, 71, 2077.)

2 Elongation. Once DNA synthesis has begun, it proceeds bidirectionally from the unique origin. Elongation appears to proceed discontinuously, since replicative intermediates have been found to contain small nascent DNA chains (4S in size or about 150 nucleotides long) which appear to represent true intermediates in the elongation of the progeny strands of DNA. Short stretches of oligoribonucleotides have been found at the 5'-ends of many of the small DNA pieces. This suggests that growth of each small DNA fragment is initiated by a short (about decanucleotide length) piece of RNA, which satisfies the primer requirements of DNA polymerase. The discontinuous mode of synthesis can be exaggerated and studied by the addition of inhibitors of DNA synthesis. Single-stranded gaps have been shown to exist between the short DNA segments and the progeny strand of DNA. These may be generated when the initiator oligoribonucleotides are removed. For completion of the progeny strand, enzymes such as DNA polymerase and ligase are needed to fill the gaps and make the appropriate phosphodiester bonds.

DNA polymerases require a primer that contains a 3'-OH group to which nucleoside triphosphates can be added. The presence of an initiator RNA molecule and a discontinuous mode of replication easily accommodates synthesis ($5' \rightarrow 3'$) on one side of a replication fork. Whether it is the mode of synthesis from the other ($3' \rightarrow 5'$) strand is not clear. Experiments from some laboratories support a discontinuous mode of synthesis from one strand and a continous mode of synthesis from the other. The question must at the moment remain open.

The unwinding of superhelical strands is a problem that must also be considered, since elongation, particularly in its final stages, must involve either some form of unwinding or strand interruption and resynthesis. Pools of replicating molecules are known to contain a larger proportion of late-replicating (almost completed) forms than early-replicating forms. Moreover, a single round of SV40 DNA replication appears to require between 5 and 15 min. to complete *in vivo*. Since the rate of polynucleotide chain elongation of mammalian cell DNA is about 2×10^{-6} cm min.$^{-1}$, and since SV40 DNA has a length of $1 \cdot 66 \times 10^{-6}$ cm and replication is bidirectional, in the absence of any rate-limiting step, one round of replication should be completed in 25 s. The relaxation of the late-replicative intermediates may therefore be introducing a slow, rate-determining step into DNA replication of polyoma virus and SV40.

3 Segregation of replicated molecules. Chain elongation appears to proceed until the two replicating folks approach each other at a terminus region. Studies with deletion mutants of SV40 suggest that the termination of replication is probably not sequence dependent, but occurs about 180° from the origin. At the termination site the two interlocking replicating circles, each of which contains a circular parental strand and a linear progeny strand, separate. Segregation appears to occur before the progeny strand is full length. It is not clear however whether the degree of single-strandedness is the same in both replicated species. It is also not clear whether protein synthesis is required for segregation.

4 Maturation. Maturation of the viral DNA requires the completed synthesis of the progeny strand, ligation to complete the circle, and the winding of the DNA strands to produce the supercoiled form. Presumably proteins are involved in the winding process. Histones may be attached to produce nucleoprotein complexes and may even be important in the winding step. At present very little is known about this final stage in the overall process of DNA replication.

It appears likely that a viral function is required for initiation of viral DNA replication but that cellular functions are involved in elongation, termination and maturation events for both polyoma virus and SV40. Viral DNA replication appears to be accompanied by an increase in replication of cellular DNA, suggesting that the viral event, as it becomes better understood, can be used to study cellular replication.

5 Transcription in permissive cells

Polyoma virus and SV40 specific RNAs are made in small amounts in permissive cells *early* (within 12 h) after infection. Following the onset of viral DNA replication (12–18 h), considerably larger quantities of viral RNAs (about 100-fold more) are made. According to current usage, viral RNA made *before* DNA replication is called *early* RNA and RNA made after DNA replication is called *late* RNA. (The DNA strand from which early RNA is made is called early strand DNA). The terms have very little real meaning since small amounts of early RNA continue to be made even after DNA replication begins. The viruses do not inhibit cellular RNA production and may even stimulate it.

Experiments have been carried out with both polyoma virus and SV40 that show that early RNA contains sequences complementary only to one of the two DNA strands, while late RNA contains sequences complementary only to the other. It appears therefore that for both viruses, a *strand switch* may occur in DNA transcription during the lytic cycle. Questions subsequently raised and answered (at least in part) about viral transcription include: How much of each DNA strand is transcribed? How many viral messenger RNAs are there? What size are they? Where do they come from on the physical maps of the DNAs? For what proteins do they code? What is the direction of transcription?

Three specific viral messenger RNAs have been isolated from polyoma virus and SV40 infected cells. By definition, these are RNAs which are released from cytoplasmic polyribosomes. With both viruses, only one early messenger RNA has been found which sediments at 19–20S. The molecular weight of this is 7 to 8×10^5 daltons, or about half the size of one strand of viral DNA. Two late messenger RNAs have been identified, one which sediments at 19S and the other at 16S. Cytoplasmic RNA has been found to be derived from only about one-half of each of the two strands of DNA. These data lead to the obvious conclusions that if the assigned sizes are correct, there must be only one early mRNA, if all of the 19–20S *early* mRNA is in fact coded for by the virus. Both 19S and 16S *late* mRNAs must be derived from about 50 per cent of late strand DNA, and from a consideration of their sizes must be transcribed at least in part from the *same* region of the DNA. It is still not clear whether late 19S RNA is a precursor of 16S RNA or whether the two species are transcribed independently.

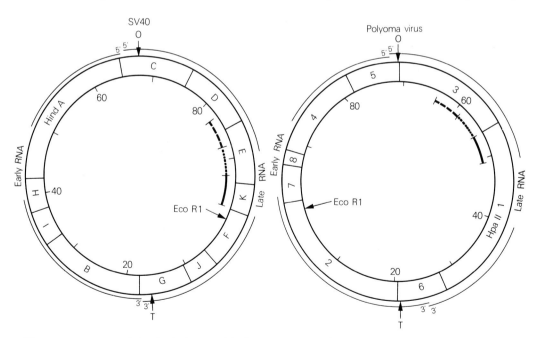

Fig. 1.8 Comparison of the physical maps of SV40 and polyoma DNAs, aligned with the viral origins of DNA replication (O) at the top of each map. The stable early and late mRNA transcripts have been mapped as shown, with the 5′-ends of the RNA near the origin and the 3′-ends near the termination (T) of DNA replication. The homologous regions between the two DNAs are indicated inside the map: solid line (strong homology), dotted line (weak homology), dashed line (intermediate homology). (Taken from Fried *et al.* (1974). *Cold Spring Harbor Symp. Quant. Biol.*, **39**, 45.)

In order to locate the topographical positions of the early and late gene sequences, hybridisation studies have been carried out between viral mRNAs and restriction enzyme fragments of the appropriate viral DNAs. For polyoma virus, the early transcription product was found to be derived from DNA contained in contiguous restriction fragments Hpa II 5, 4, 8, 7, 2 and (in part) 6 and late transcription products from contiguous fragments Hpa II 3, 1 and (in part) 6. For SV40, the early transcription product came mainly from contiguous Hind II + III fragments A, H, I and B and late transcription products from contiguous Hind II + III C, D, E, K, F, J and G (Fig. 1.8).

It appears to be a general rule that transcription of DNA proceeds in a $5' \rightarrow 3'$ direction with respect to the synthesis of the messenger RNA molecules, or $3' \rightarrow 5'$ along the DNA template strand. Since early and late polyoma virus and SV40 mRNAs are transcribed from different DNA strands and since strands of DNA in duplex molecules have opposite polarities, it becomes apparent that transcription of early and late viral sequences occurs in opposite directions. In order to determine direction, a unique population of linear DNA molecules was obtained (using restriction enzymes) and the $3'$ and $5'$ ends of each DNA strand were distinguished. Experiments equivalent to strand separation were carried out and from the knowledge of the polarity of each DNA strand, and the mRNA to which it could be hybridised, the transcriptional direction for both viruses has been deduced (Fig. 1.8).

The $3'$ ends of mRNAs from both viruses are polyadenylated, although the length of the polyadenylic acid is unknown. It is not coded for by the virus and is presumably added as part of a post-transcriptional processing mechanism. In SV40 mRNAs (and probably also in polyoma virus mRNAs), the $5'$ triphosphate end is capped with 7-methylguanosine. In primary sequence studies on SV40 DNA it has been found that the $5'$ ends of the early and late 19S mRNAs overlap, i.e. in a short segment of DNA which lies near the origin of DNA replication, both DNA strands appear to be transcribed. As yet, the promoter sites are not known for either virus, nor indeed is it known whether transcription occurs *in vivo* from a free or from an integrated form of the viral DNA. If from an integrated form, the virus might even use cellular promoters. The virus appears to use cellular enzymes for its transcription. Little is known about enzymes used for processing or folding viral RNAs or promoting the strand-switch during transcription (if one occurs), nor is it clear what controls termination of transcription or whether DNA sequences in the termination region are transcribed.

In the nucleus of the cell, viral RNAs have been found which are larger than the full-length viral genome and which appear to hybridise to all of the sequences of both early and late strand DNA. How RNA is selected, processed and transported to the cytoplasm is one of the many unsolved problems in cell biology, and some insight may eventually be obtained by studying these same processes in the greatly simplified case of viral RNAs (see Ch. 2). Some evidence from the current literature suggests that two control mechanisms regulate viral gene expression in SV40 infected cells. One is transcriptional and determines the frequencies of transcription from the late and early strands of DNA, and the other is post-transcriptional and determines the stability of the primary gene products and the fragments of the total transcripts which ultimately reach the cytoplasm.

6 Proteins of polyoma virus and SV40

6.1 Virion proteins

The best-defined and most fully characterised virus-specified proteins of polyoma virus and

SV40 are the structural (capsid) proteins. For SV40, two capsid proteins only have been identified. The major capsid protein (called VP1) has a molecular weight of about 45 000 daltons and the minor capsid protein (VP2) has a molecular weight of about 30 000 daltons. SV40 may also have a third structural protein (VP 3), but the data are not yet consistent. For polyoma virus, three capsid proteins have been observed with molecular weights of about 45 000 (VP1), 36 000 (VP2) and 23 000 (VP3). Fingerprints made of tryptic peptide digests of purified proteins show that there are no common sequences between the major (VP1) and minor (VP2/VP3) capsid proteins, but common sequences are observed in VP2 and VP3. It is therefore not clear whether VP2 acts as precursor of VP3, the latter being the functional protein, or whether VP3 is a specific breakdown product of a functional VP2, or whether indeed both proteins are functional and essentially translated independently. For both polyoma virus and SV40, it has been shown that the late 16S messenger RNA contains the coding information for the major capsid protein VP1 and the late 19S mRNA contains the coding information for the minor capsid protein(s) VP2, and possibly VP3. The DNA content of both viruses allows only for about 200 000 daltons of protein. It seems clear from the data on capsid proteins that the functions of both of the known late mRNAs have been accounted for as well as about half the total coding capacity of the DNA. For SV40, the N-terminal amino acid sequences of VP1 have been found to be Ala–Pro–Thr–Lys–Arg–Lys–Gly–. Primary sequence studies on restriction enzyme fragments of SV40 DNA have shown that the DNA sequences G–C–C, C–C–A, A–C–A, A–A–A, A–G–A, A–A–A, and G–G–A–, which correspond to the N-terminal amino acid sequence of VP1, lie on the SV40 physical map in fragment Hind K, close to the Hind E–K junction (Fig. 1.8). In this virus then, the precise location of one protein with respect to the DNA is known.

Whereas viral DNA is itself infectious, infection with intact virions is much more effective than with DNA alone. The capsid proteins appear to play a valuable role in reproduction of the viruses, possibly by protecting the DNA. They may also increase the efficiency of viral absorption by the cell.

6.2 Virus-induced proteins and cellular proteins

One of the most intriguing questions still unanswered for polyoma virus and SV40 is whether there is only one (or more than one) early gene product, and what it is. A number of virus-specific antigens are known to be synthesised in the cell early after infection, but they have as yet only been superficially characterised. These antigens have been detected primarily by immunological methods. One of them T antigen (or tumour-antigen) can be detected in the cell nucleus very early after infection, and before DNA replication begins. It persists in the cell throughout the lytic cycle and is also found in transformed cells. T antigen from SV40 does not cross-react with T antigen from polyoma virus and is the best candidate for an early viral gene product. Evidence for it being a virally-coded protein comes from the fact that different species of cells infected (or transformed) with the same virus give rise to T antigens which appear to be immunologically cross-reacting. UV-irradiated virus or temperature sensitive (ts) mutants which are defective in an early function either fail to make T antigen, make reduced amounts of it, or make an apparently altered product. Interferon (considered to be a specific inhibitor of viral protein synthesis but not of cellular protein synthesis) inhibits the synthesis of T antigen. More direct evidence comes from the use of a micro-injection technique which allows nucleic acids to be injected directly into mammalian cells in culture (Fig. 1.9). SV40 specific RNA complementary to the early viral DNA strand was synthesised *in vitro* by *E. coli* DNA-dependent RNA polymerase, isolated and injected into epitheloid cells in a confluent mouse cell culture. After 4

(a)   (b)

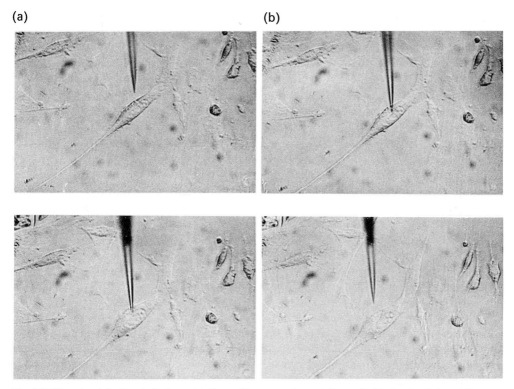

Fig. 1.9 The microinjection technique: (a) the capillary near the recipient cell; (b) the tip of the capillary inside the cell; (c) the cell immediately after injection; (d) recipient cell after injection, with the capillary outside the cell. The volume transferred into the cytoplasm of the cells in a typical experiment is $1-2 \times 10^{-8}$ μl per cell. In this experiment, RNA complementary to SV40 early strand DNA was injected into cells in order to look for expression of T-antigen. Twenty-four hours after injection, about 70 per cent of the cells were T-antigen positive. (Photograph, courtesy of Prof. A. Graessmann, taken from Graessmann and Graessmann (1976), *Proc. Natl. Acad. Sci. USA*, 73, 366.)

to 5 h, a large number of cells were found by immunofluorescence to contain T antigen. The proportion of T antigen-positive cells increased with time and reached a maximum (60–80 per cent) at 24 to 26 h. Similar experiments have been carried out with polyoma virus-specific RNA (in this case complementary to both strands of polyoma virus DNA) and similar results were obtained. These experiments strongly suggest that the T antigens are virus-coded. There remains the possibility however that they are made from a messenger RNA species transcribed from host DNA containing integrated viral DNA (or that they are host proteins induced by viral DNA). If this were the case, T antigen could be, at least in part, coded for by the cell. This possibility is raised because some of the existing genetic evidence, particularly with polyoma virus, suggests that there should be at least one other early gene product. Although the size of T antigen still remains a subject for much controversy, it appears that the protein has a molecular weight somewhere between 80 000 and 100 000 daltons. A product of this size would require almost all of the coding capacity currently assigned to the early strand of viral DNA. Assuming all viral proteins to be made from the same triplet code frame, there is simply no room left for any but a minute protein on early strand DNA. Until the size of T antigen has been established with certainty, or a second early messenger RNA identified, the question of the early gene product (or products) of polyoma virus and SV40 remains a topic for speculation and research. The broader questions of translation have scarcely been raised.

Both polyoma virus and SV40 have been shown to have *a tumour-specific transplantation antigen* (TSTA) which appears to be localised at the cell surface. TSTA has been shown to be synthesised during the lytic cycle, but whether it is a virus-coded or a virally-induced cellular protein is not known. It is also present on transformed cells but has not been purified from either polyoma virus or SV40. A surface antigen (S-antigen) has also been identified by immunofluorescence. For SV40, another antigen called U antigen has been found by immuno-fluorescence at the nuclear membrane. U antigen closely resembles T antigen in the kinetics of its synthesis and its morphological appearance in SV40 infected and transformed cells, but whereas T antigen is heat labile, U antigen is heat stable. Whether U antigen is a variant or subunit of T antigen is not known.

In the virion, viral DNA is associated with four histones which appear to correspond to cellular histones. Furthermore, the synthesis of a number of enzymes is known to be stimulated shortly after viral infection. Most of these enzymes seem to play a role in DNA synthesis, and resemble cellular enzymes. All these proteins are probably virally induced rather than virally coded.

7 Genetics

Since polyoma virus and SV40 are able to cause cancer in animals and to change (transform) the growth properties of infected cells in culture (Fig. 1.1), genetic studies on these viruses have been carried out with the aim of identifying the viral genes and understanding how they affect cell growth and regulation. To date, three genes have been identified and mapped for both viruses. One of these is an early gene which appears to code for a protein of about 90 000–100 000 daltons. This gene, called the A gene, appears to code for a protein which is identical to T antigen. The second gene, called the B/C gene in SV40, codes for the major viral capsid protein, VP1. The third gene, called the D gene in SV40, codes for a minor viral capsid protein, VP2 or VP3.

Over 300 independently-isolated temperature-sensitive (ts) mutants* have been made and characterised for polyoma virus and SV40. In studies on the function affected by the mutation which creates the temperature-sensitivity, the A gene appears to be pleiotropic, in that it has several phenotypes. Four distinct functions appear to be defective in cells infected or trans-formed with A mutants at non-permissive temperatures. These are: (1) initiation of viral DNA synthesis; (2) stimulation of cellular thymidine kinase activity; (3) initiation of transcription of late genes, and (4) establishment (and possibly maintenance) of transformation. There is some evidence that the stimulation of cellular DNA synthesis is also related to the function of the A gene.

Temperature-sensitive B and C mutants complement each other, but are now thought to be examples of intragenic complementation, and in fact to contain mutations in a single gene, the late B/C gene which codes for the major viral capsid protein (VP1). B/C mutants synthesise viral DNA at the non-permissive temperature but do not produce infectious virus. They are capable

---

*Temperature-sensitive mutants are conditional lethal mutants. They are generally thought to be variants of the DNA which contain a base-pair substitution. This changes the primary structure of the protein specified by that part of the DNA which contains the mutation in such a way that it is only able to take on and retain a functional structure at the *permissive* temperature. At *non-permissive* temperatures, the protein is denatured and non-functional. The corresponding wild-type protein is of course functional at both temperatures. Analysis of the mutation in terms of functional changes allows the DNA to be related to the protein.

of transforming cells at the non-permissive temperature so that it seems clear that this gene is not involved in cellular transformation.

The temperature-sensitive D mutants are defective for *uncoating*. Viruses containing DNA with a mutation in the D gene are not infectious at non-permissive temperatures. If, at the non-permissive temperature, the protein (minor capsid protein VP2 or VP3) is not released from the DNA, none of the viral functions could be expressed. Although the D gene is a late gene in terms of the protein for which it codes, it also affects early functions. There are some reports that D mutants are involved in the maintenance of the phenotype of transformed cells.

A large number of temperature-sensitive mutants has been analysed and their mutations assigned to regions on the restriction enzyme maps of polyoma virus and SV40 by means of a technique called *marker rescue*, which is illustrated in Fig. 1.10. It was first used for the assignment of genes to the bacterial virus, ϕX 174, and has proved useful for studying genes of polyoma virus and SV40. The data obtained by marker rescue have been somewhat unexpected. For both viruses, all the temperature-sensitive A mutants map in only about half of the region of the DNA assigned to expression of an early function (or functions) (Fig. 1.12). Most of the late mutants also map in about only half of the region of the DNA assigned to expression of late functions. For polyoma virus, only one mutant has so far been assigned to the D class of temperature-sensitive mutants, all others apparently being B/C types. Whether the limited portions of the genome in which mutations (leading to temperature-sensitivity) have been found has any functional significance or merely reflects a selection process is not clear.

In addition to temperature-sensitive mutants, a large number of viral deletion mutants have

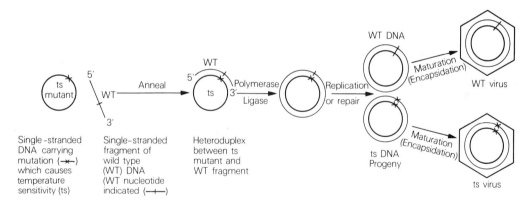

Fig. 1.10 A schematic diagram of the marker rescue method which allows a mutation (X) to be mapped and correlated with the region of the DNA which contains the mutation and thereby prescribes the functional change. A complex (heteroduplex) is made by hydrogen-bonding the single-stranded closed circular form of the DNA which contains the mutation (X) with a single-stranded fragment of wild type (WT) DNA. Cells infected with this complex presumably have the enzymatic machinery (polymerase/ligase) to extend the fragment primer as well as join the extended portion to form full-length closed-circular duplex DNA. By processes as yet ill-defined but presumably involving replication and/or repair, the duplex is replicated and encapsidated to give in part wild type virus, which now grows at the non-permissive temperature. Thus the marker (temperature sensitivity) in a number of molecules has been rescued. The original marker rescue studies were carried out on mutants of bacterial viruses using heterogeneous fragments of DNA made by non-specific fragmentation methods. Subsequent studies have used specific fragments made by cleaving DNA with restriction enzymes followed by denaturation to convert them to single-stranded pieces. A correlation between the restriction fragment that successfully carries out the marker rescue (i.e. covers the site of the mutation) and the position of this fragment on a physical map of the DNA allows the original mutation, and the protein affected by it, to be mapped within a region of the DNA. Marker rescue has also been used to map deletion mutants.

been made, mainly by excising segments from DNA using restriction enzymes, and infecting cells directly with this DNA. The cellular mechanisms involved in the creation of deletion mutants by this procedure are not understood, although cell-mediated circularisation may play a role. Many of the deletion mutants made in this way have however been found to have deletions which extend beyond the DNA originally excised by the restriction enzyme. It has been suggested that the mutants may arise as a result of intramolecular recombination events which occur prior to replication. These mutants have been especially useful for studying regions of the DNA not covered by the temperature-sensitive mutants.

Mutants of SV40 with deletions in the early or late regions of the genome have been tested for biological activity. Some deletion mutants lacking portions of both late genes (B/C and D) but with an intact early region have been shown to induce T antigen in infected cells, to replicate their DNA in the absence of helper virus, to stimulate cellular DNA and to transform non-permissive cells. Deletion mutants lacking substantial portions of early DNA did not exhibit all these activities. Small deletions around the viral origin of replication do not seem to interfere with infectivity of DNA replication, and these mutants have normal transformation properties. Deletion mutants have been used to confirm the conclusions that the product of the B/C gene is the major capsid protein, VP1. Since a deletion mutant of SV40 made by the excision of the restriction fragment Hind E (Fig. 1.4) does not complement any of the late ts-mutants, it has been suggested that the DNA deleted from this region may not ultimately be translated but may contain a signal required for expression of the B/C gene – such as a promoter, a processing signal for forming late 16S mRNA or a ribosomal binding site.

One of the apparent major differences between the polyoma virus and SV40 genomes has been uncovered by the use of mutants. A mutant in which 10 per cent of the early region of SV40 DNA has been deleted (from 0·54 to 0·59 on the SV40 physical map, or about 0·13 to 0·08 map units from the origin of viral replication (Fig. 1.12)) has been found to be viable and is reported to transform cells. Mutants with deletions in a similar region of polyoma DNA have a number of interesting properties. They are defective in their ability to grow in permissive cells, and they lack the ability to transform cells in culture. They will grow in cells already transformed by polyoma virus, and the initial assumption was that the transformed cell DNA supplied a function missing in the mutant DNA. Since they will also grow in cells transformed by agents other than polyoma virus, this may be too facile an explanation. Translation studies with this class of mutants, known as *host-range mutants*, or hr-t mutants, show that a protein is made which is the same size as polyoma virus T antigen, and appears by immunological studies to be T antigen. In marker rescue experiments (Fig. 1.10), a small restriction fragment, Hpa II-4 (Fig. 1.8) can restore both normal host range and transforming properties to these mutants.

Studies on the genetics of polyoma virus and SV40 point to the presence of at least three viral genes, one containing coding information for an early protein, probably T antigen, and two coding for late viral capsid proteins. Making the assumption that for such a small virus most of the DNA is ultimately translated into functional proteins, host-range mutants of polyoma DNA suggest that a second early viral protein may exist. It is not even clear from the study of mutants that there are only two late genes. Using restriction enzymes to create mutants is a technique of such wide application that it seems only a matter of time before these unsolved questions of genes and gene products will be answered.

8 Transformation

Although it is still not clear whether the viral DNA becomes integrated into host DNA in the infectious process that leads to cell lysis, in the process that leads to cellular transformation, viral DNA becomes covalently bound to cellular DNA, and in this integrated form is called the *provirus*. With a cyclic DNA it is easy to see how integration can occur without loss of viral genetic information. In the transformed state, viral DNA is retained and replicated along with cellular DNA. It has also been shown that provirus is transcribed into messenger RNA many generations after the initial event which led to cellular transformation, suggesting that this state is stable. The gene that codes for T antigen is expressed in the transformed state, but genes that code for viral capsid proteins are apparently not expressed.

Studies with viral mutants suggest that the A gene is required for the initiation of the transformed state. As discussed in section 7, the A gene is an early gene and may code for T antigen, which may therefore play a role in cellular transformation, possibly by binding to cellular DNA. Purified T antigen has been shown to have a high affinity for binding to double-stranded DNA. Expression of a viral gene also seems to be required for the maintenance of the transformed state, but it is still not clear which viral gene this is, nor whether it is translated into a functional protein.

Certain cells, known as abortively-transformed cells, behave as transformed cells for a few generations after viral infection and then return to normal. When they are normal again they no longer contain any detectable viral DNA. It can be postulated that following the initial (abortive) transformation events, the gene needed for maintenance of transformation may be excised. Alternatively, it may be integrated into the cellular DNA in such a way that it is subsequently not expressed or expressed poorly. There are no data yet to support any such hypotheses about abortive transformation.

Cells transformed in culture have certain characteristics that distinguish them from non-transformed cells:

a They grow to higher densities, having apparently overcome the factors which normally control the growth of cells. Normal cells are spoken of as being *contact-inhibited* whereas transformed cells are not contact-inhibited.

b They grow even in the presence of very low concentrations of serum. Growth of normal cells is easily arrested by serum starvation.

c They grow in soft agar or methocel suspension, media which do not support the growth of normal cells.

d They show random orientation during growth, whereas the growth of normal cells is more regular and orientated (see Fig. 1.1).

e They induce tumours in susceptible animals.

f They exhibit a number of biochemical changes at their cell surfaces.

Clearly, none of these characteristics is well-defined, and the observations lack meaning unless normal and transformed properties of the same types of cell are being compared. Nonetheless, it is useful to identify simple changes by which transformation can be monitored.

Transformation may not be a single event, but rather may proceed through a series of steps which involve functional changes and mutations within the cell. None of the cellular events has yet been defined. Viral proteins may replace normal regulatory proteins in the cell, and thus interfere with the control and regulation of the growth of the cell. Results from studies with polyoma virus and SV40 suggest that for the generation of the stably-transformed cell, cellular mutations may be required. Some of these mutations may even be virus-induced, since during early stages of transformation by papova viruses cells have frequent chromatid breaks.

It should be obvious that changes in cellular functions during (and following) transformation are an important field for study, but a difficult one. Until more is known about growth factors and control within a normal cell, irregular growth and regulation will probably not be understood, although it can be seen that viral interference may be a useful probe for defining normal events.

To return to the viruses themselves, it has been shown for SV40 that only the early region of DNA is required for transformation. Restriction fragments which contain all the early sequences and the origin of DNA replication (about 60 per cent of total SV40 DNA) have been shown to be capable of stably transforming non-permissive cells. In some interesting recent studies, a number of SV40 transformed cell lines (i.e. cells that have been transformed by SV40 virus, cloned and carried in culture for a number of generations) have been investigated in order to study integration of viral DNA into host DNA. Since integration presumably requires some sort of recombination between viral and host cell DNA, the simplest model would predict a linear insertion of the viral DNA into cellular DNA. With this model, the question would still arise as to whether integration involved a site-specific mechanism or a more or less random recombination event. In order to study this, two groups of investigators looked at SV40 viral sequences in different transformed cell lines. In both cases, the DNA from the transformed cells was cleaved with restriction enzymes and, using a very sensitive technique (see Fig. 1.11), the fragments were examined to see how much viral DNA was present and whether the same viral DNA was present in each cell line. The results suggested that different amounts of viral DNA were present in different cell lines and that recombination sites mapped at different places on the viral DNA. They suggested that integration was not site-specific but could not exclude the possibility that it might be occurring at specific nucleotide sequences if such sequences were located at different sites on the host and/or viral DNA. In all the transformed cells, the early region of the viral genome has been found. This confirms other data which suggest that this region is essential for the initiation of transformation, and possibly for its maintenance. An important consequence of these studies is that a number of integration sites for SV40 DNA have been mapped, and these may be useful in studying the ways in which the expression of viral information is controlled in transformed cells.

9 Comparison of polyoma virus and SV40

As more becomes known about polyoma virus and SV40 it becomes clear that in the overall organisation of their genetic information, these two viruses are remarkably similar. This is illustrated in Fig. 1.12. Nonetheless, attempts to find sequence homologies between the DNAs of the two viruses have shown that there is only a very limited region of homology, which maps in the late portion of the DNA. Even in this region, the sequences are only in part identical.

In spite of similarities at the genetic level, cells susceptible either to lytic infection or transformation by the one virus are not generally susceptible to the other.

10 Conclusions

One of the reasons for presenting the detailed molecular biology of polyoma virus and SV40, as it is now understood, is to point to the remarkable similarity of the two viruses (Fig. 1.12). The mechanisms of DNA replication, the areas of the DNA transcribed *early* and *late*, the proteins coded for, the genetics — all the data obtained from one of the viruses fit that from the other to an almost uncanny extent. And yet, very little homology has been found between the viral DNAs, and the host specificities of the two viruses are almost entirely different. What factors make up the differences and can they be seen in the genome?

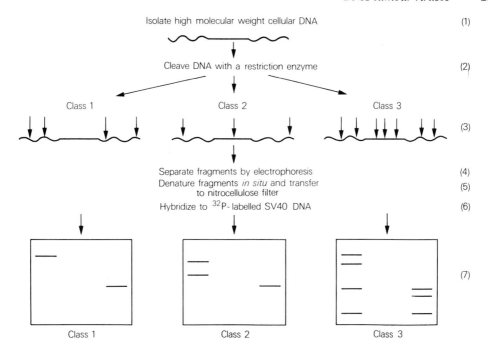

Fig. 1.11 Scheme for determining the arrangement of SV40 sequences in the DNA of transformed cells. (1 to 3) DNA from transformed cell lines is isolated and cleaved with three different kinds of restriction enzymes: those which do not cleave SV40 DNA (Class 1), those which cleave viral DNA only once (Class 2), and those which cleave it more than once (Class 3). (SV40 DNA is indicated by a straight line and host cell DNA by curved lines.) (4) The DNA fragments are fractionated according to size by electrophoresis on an agarose gel. (Large fragments move slowly in the electrical field and smaller fragments move more rapidly.) (5) The fragments are denatured *in situ* by treatment with alkali and then transferred to a nitrocellulose filter, and (6) hybridised to SV40 DNA ^{32}P-labelled to high specific activity. In this technique the labelled viral DNA is present in vast excess and provides the driving force for the reaction. (Hybridisation is a technique which allows homologous DNA to be detected. Thus, 'probe' ^{32}P-labelled SV40 DNA can be used to locate SV40 DNA sequences in cellular DNA.) (7) The distribution of viral sequences among the different sized fragments of transformed cellular DNA is determined by autoradiography. Fragments that contain as little as 10^{-13} g of viral DNA can be detected using high specific activity 'probe' SV40 DNA. For an indication of the size of the fragments obtained from cellular DNA, a control experiment shown on the right of each block in (7) is carried out using SV40 DNA alone. Full-length SV40 DNA is used as control for the Class 1 and 2 experiments, and SV40 DNA cleaved with the appropriate restriction enzyme is used as control in the Class 3 experiment.

With enzymes that do not cleave SV40 DNA (Class 1), all fragments of DNA must come from the cleavage of the host cell DNA, and the number of radioactive bands from any one cell line provides an estimate of the number of separate sites at which SV40 is integrated.

Enzymes that cleave SV40 DNA once (Class 2) can be used to detect the presence in transformed cells of partially duplicated or tandem copies of SV40 DNA. These enzymes should cleave within repetitious DNA to generate a whole copy of viral DNA that should migrate together with linear SV40 DNA.

Enzymes that cleave SV40 DNA at more than one site (Class 3) can be used to catalogue viral DNA sequences within the transformed cell by comparing fragment sizes with those obtained by cleaving viral DNA alone.

Results of these and similar experiments carried out on a number of established SV40-transformed cell lines have shown that:

a   In any particular cloned transformed cell line, the location of viral sequences within the cellular DNA remain stable.

b   There can be multiple copies of intact or partial viral genomes integrated at different regions within the host DNA.

c   The integration site between viral and host DNA maps at different positions on the viral DNA.

(Adapted from Botchan *et al.* (1976), *Cell*, **9**, 269.)

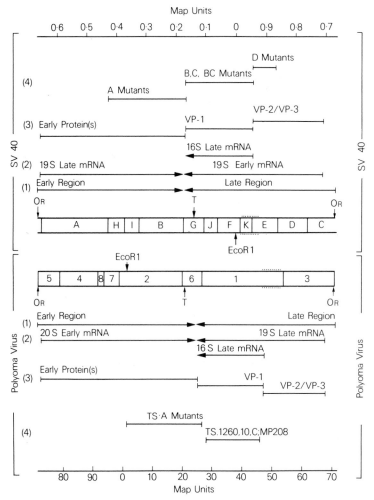

Fig. 1.12 Comparison of SV40 and polyoma virus. SV40 (*top half of figure*): the circular Hind physical map which consists of 13 fragments (11 of which are shown) (see Fig. 1.8) is shown as linear by breaking the molecule at 0·67 map units where the origin of DNA replication (OR) has been mapped. The termination of DNA replication (T) is also indicated. The single Eco R1 restriction enzyme cleavage site is shown at zero map unit. The region of Hind K and E indicated by the dotted lines is the portion of the SV40 genome that shows the strongest homology with the polyoma virus genome. Above the linear DNA map, the lines from bottom to top represent: (1) the early and late regions of the DNA; (2) the regions of the DNA within which the viral mRNAs (early 19 S, late 19 S, and late 16 S) have been mapped; (3) the regions of the DNA within which the viral capsid proteins (VP-1 and VP-2/VP-3) and the early protein (s) have been mapped; (4) the regions of the genome in which the early (A mutants) and late (B/C and D) temperature-sensitive mutants have been mapped.

Polyoma virus (*bottom half of figure*): the circular Hpa II physical map which consists of 8 restriction fragments (see Fig. 1.8) has been linearised by breaking the molecule at 70·8 map units, where the origin of DNA replication (OR) has been mapped. The termination of DNA replication (T) is also indicated. The single Eco R1 restriction enzyme cleavage site is shown at zero map unit. The region in Hpa II-1 indicated by the dotted lines is the portion of the polyoma virus genome which shows the strongest homology with the SV40 genome. Below the linear DNA map, the lines from top to bottom represent: (1) the early and late regions of the DNA; (2) the region of the DNA within which the viral mRNAs (early 19–20 S, late 19 S and late 16 S) have been mapped; (3) the regions of the DNA within which the viral capsid proteins (VP-1 and VP-2/VP-3) and the early viral protein(s) have been mapped; (4) the regions of the genome in which the early (TS-A type) and late (TS-1260, -10, and -C) temperature-sensitive mutants and the minute-plaque morphology mutant (208) have been mapped. (Taken from Fried and Griffin (1977), *Adv. Cancer Res.*, **24**, 67–113.)

Considerable progress has already been made in determining the primary sequence of the SV40 genome. Sequence studies on polyoma DNA are also being undertaken. The primary sequence of one genome alone may provide little information, but it will be disappointing if a great deal of information is not forthcoming from a comparison of the DNA sequences of the two viruses. In a matter of only a few years, we should know.

Addendum

Recent studies with adenovirus have shown that certain of the viral messenger RNAs consist of sequences which are complementary to *non-contiguous* portions of the viral DNA. Short segments (leader sequences) appear to be *spliced* to the main body of the message. These results suggest that some sort of *post-transcriptional* control may be important in eukaryotic cells. (BERGET, S. M., MOORE, C. and SHARP, P. A. (1977) 'Spliced segments at the 5'terminus of adenovirus 2 late mRNA', *Proc. Natl. Acad. Sci. USA,* **74**, 3171–5.) Similar results have now been observed for polyoma virus and SV40 messenger RNAs.

Bibliography

FENNER, F., McAUSLAN, B. R., MIMS, C. A., SAMBROOK, J. and WHITE, D. O. (1974) *The Biology of Animal Viruses* (2nd edn.), Academic Press, New York and London.
TOOZE, J. (Ed.) (1973) *The Molecular Biology of Tumor Viruses*, Cold Spring Harbor Laboratory. *Cold Spring Harb. Symp. Quant. Biol.* (1974), **39** (1 and 2).

Review articles or papers dealing with special aspects

ACHESON, N. H. (1976) 'Transcription during productive infection with polyoma virus and simian virus 40', *Cell,* **8**, 1–2.
ARBER, W. (1974) 'DNA modification and restriction', *Prog. Nucl. Acid Res. Molec. Biol.,* **14**, 1–37.
CAMPBELL, A. M. (1976) 'How viruses insert their DNA into the DNA of the host cell', *Scient. Am.,* **235**, 103–12.
DULBECCO, R. (1976) 'From the molecular biology of oncogenic DNA viruses to cancer', *Science,* **192**, 437–40.
ECKHART, W. (1974) 'Genetics of DNA tumour viruses', *Ann. Rev. Genetics,* **8**, 301–17.
FRIED, M. and GRIFFIN, B. E. (1977) 'Organization of the genomes of polyoma virus and SV40', *Adv. Cancer Res.,* **24**, 67–113.
KELLY, T. J. and NATHANS, D. (1977) 'The genome of simian virus 40', *Adv. Virus Res.,* **21**, 85–173.
LEVINE, A. J. (1976) 'SV40 and adenovirus early functions involved in DNA replication and transformation', *Biochem. et Biophys. Acta,* **458**, 213–41.
LEVINE, A. J., van der VLIET, P. C. and SUSSENBACH, J. S. (1976) 'The replication of papovavirus and adenovirus DNA', *Current Topics in Microbiology and Immunology* (Springer Verlag, Berlin), **73**, 67–124.
NATHANS, D. and SMITH, H. O. (1975) 'Restriction endonucleases in the analysis and restructuring of DNA molecules', *Ann. Rev. Biochem.,* **44**, 273–93.

26

NEWTON, A. A. (1974) 'Viruses'. In: *Companion to Biochemistry,* (Eds. Bull, A. T., Lagnado, J. R., Thomas, J. O. and Tipton, K. F.) pp. 277–306. Longman, London.

ROBERTS, R. J. (1976) 'Restriction endonucleases', *Crit. Rev. Biochem.,* **4**, 123–64.

SALZMAN, N. P. and KHOURY, G. (1974) 'Reproduction of papovaviruses'. In: *Comprehensive Virology* (Eds. Fraenkel-Conrat, H. and Wagner, R. R.), pp. 63–141. Plenum Press, New York.

2
RNA tumour viruses

R. A. Weiss

Imperial Cancer Research Fund Laboratories, Lincoln's Inn Fields, London

1 Introduction

RNA tumour viruses are classified as retroviruses because they replicate by forming a proviral DNA copy of the viral genome in the infected cell. Being oncogenic (cancer-causing) and containing RNA, they are commonly called *oncornaviruses*. Many species of vertebrate are hosts to oncornaviruses, which have been isolated from bony fish, reptiles, birds and mammals including primates. Whether oncornaviruses occur as natural infections of humans is not yet clear, although the evidence is growing and will be discussed in section 7. Oncornaviruses cause malignant diseases in the host, such as leukaemia and sarcoma in chickens, lymphoma and mammary carcinoma in mice. Particular strains of virus are associated with different types of disease, and with different host species.

Oncornaviruses were first studied 70 years ago. Leukaemia in chickens was recognised as a viral disease by Ellerman and Bang in 1908, several years before the discovery of bacteriophages, and in 1911 Rous first isolated the chicken sarcoma virus that bears his name. For many years medical opinion was reluctant to believe that viruses, or filterable agents as they were then usually called, could be a genuine cause of cancer, even in animals, and Rous received little support for his pioneering studies. Doubt was expressed as to whether the tumours induced by Rous sarcoma virus were really malignant, and when Rous shows they were, it was argued that malignant tumours in chickens were irrelevant to mammalian cancer. In 1936, however, Bittner showed that mammary cancer in mice, which had been thought to be an inherited disease, was in fact caused by a virus secreted in the milk. Similarly, in 1951 Gross demonstrated that the hereditary lymphoma of AKR mice was transmissible by cell-free extracts inoculated into newborn mice of a low-incidence strain.

More recently, many strains of oncornavirus have been isolated, and the leukaemias of cats and cattle have been recognised as infectious, epidemic diseases caused by these viruses. Not all oncornaviruses cause cancer; some strains may not be pathogenic at all, while others are associated with autoimmune disease. The oncogenic properties of oncornaviruses have recently become amenable to genetic and molecular analysis; for instance, sarcoma viruses appear to carry a specific cancer-transforming gene, or *oncogene*, and these viruses provide fascinating tools for studying fundamental processes in neoplasia (see section 5).

One of the most interesting aspects of oncornaviruses is that they exist in two distinct forms, as infectious virus particles containing RNA genes, and as DNA genes integrated into the host chromosomal DNA. The idea that the virus could exist as a DNA copy in the infected cell, and that this form was indeed an obligatory replicative intermediate, was first postulated by Temin in 1964, and became known as the *DNA provirus hypothesis*. It was generally regarded as heresy by most molecular biologists until Mizutani and Temin, and Baltimore in 1970 discovered the enzyme responsible for producing the DNA copy, *reverse transcriptase,* inside oncornavirus particles. Since then, this enzyme has become popular with molecular biologists as it can be used to synthesise complementary DNA from RNA, thus providing a useful tool for probing gene expression, and also for cloning genes, by reverse transcription of purified messenger RNA.

The alternation of generations of oncornaviruses between infectious RNA-containing particles and host-integrated DNA genes has been adapted to different modes of natural transmission during their evolution. Whereas some strains depend on infectious transmission from one host to another, other strains are inherited by their host as proviruses passed from parent to offspring. In other words, the viral genes masquerade as *host Mendelian traits* because they have become integrated into the chromosomes of germ cells. Thus the virus causing lymphoma in AKR mice is transmitted in that host as an endogenous, genetic trait, but it becomes activated into infectious form. As an inherited genome, the oncornavirus may remain latent throughout the life of the host, perhaps for numerous host generations. Furthermore, some hosts have developed resistance to reinfection by their own inherited viruses, which when activated may only be infectious for other host species, a phenomenon called xenotropism. The various modes of virus transmission will be discussed in more detail in section 4.

The transmission of oncornaviruses is further complicated by their frequent existence as polymorphic populations, both in nature and in the laboratory. Many of the more oncogenic strains are defective and can only be maintained in infectious form by the presence of related *helper* viruses. These helper viruses provide the functions and structural components lacking in the highly oncogenic virus but by themselves have a different effect on the host. In time, the virus population may change as one helper virus complementing a defective virus becomes substituted by another. Since the helper virus provides some or all of the antigens of the virus particles, the serological classification of the virus complex therefore may not be correlated with its pathogenicity. Furthermore, related oncornaviruses frequently give rise to *genetic recombinants* during mixed infection and thus the properties of the viruses may change rapidly. These phenotypic and genetic interactions between oncornaviruses will be disussed in section 5.

Table 2.1.   Types of tumour virus infection

Productive	*Non-productive − DNA and RNA viruses*
1. Lytic − DNA viruses 2. Non-lytic − RNA viruses	1. Abortive − viral genome lost 2. Defective − part of viral genome missing 3. Non-permissive − host factors missing 4. Inactive − part or all of viral genome not transcribed

Tumour viruses interact with host cells in several different ways, as summarised in Table 2.1. In *productive infection*, progeny virus is synthesised and released. Productive infection by DNA viruses, such as the papovaviruses discussed in Ch. 1 or the herpes viruses, kills the host cell, and hence is referred to as lytic infection. RNA tumour viruses, on the other hand, are not usually lytic, for the cell does not burst open to release thousands of progeny; instead, the virus particles bud from the cell membrane without noticeable cytopathic effect and the cells become chronic virus producers. Whereas productive infection by DNA tumour viruses is not oncogenic (dead cells, after all, cannot become cancer cells), productive infection by oncorna-viruses may well result in malignant transformation. Nevertheless, *non-productive infection* is frequently encountered with RNA tumour viruses as with DNA tumour viruses, and may conveniently be divided into four classes:

a In *abortive* infection, the viral genome following initial infection is lost from the cell or from daughter cells of subsequent divisions. The virus may exert a transient effect on the host cell, including changes in growth properties resembling malignancy, but these are not maintained when the viral genome is lost.

b If the virus is *defective,* that is some part of the genome necessary for complete replication is missing, no infectious progeny is synthesised, but the virus can persist in the cell and transform it. As mentioned already, the defective virus may be rescued by a helper virus.

c A competent virus may infect a cell which is not able to support the expression of all viral functions; such cells are called *non-permissive.* Whether non-permissive cells lack certain essential factors for viral replication or whether they specifically repress viral functions is not clear; the former is more likely the case when non-permissive infection of unnatural host species is encountered, as with avian sarcoma virus in mammalian cells. Virus may be rescued from non-permissive cells by fusion with permissive cells.

d A viral genome may remain *inactive* in the host cell even when the virus is not defective and the cell is normally permissive. This happens with inherited viral genomes which frequently remain unexpressed or only partially expressed. However, these viruses may on occasion become activated by inducing agents, for example, ionising radiations.

The varied modes of transmission and interactions with the host or with other viruses are manifest in the diverse forms and behaviour of RNA tumour viruses. In the remaining sections of this chapter, some of these phenomena will be described in more detail. The oncornaviruses are not only fascinating cancer-inducing agents, but represent ingenious genetic systems acting in eukaryote cells.

2 Assay methods for oncornaviruses

Before the development of cell culture methods for the propagation and assay of tumour viruses, they were titrated by inoculating animals to determine the lethal dose or the dose that induced tumours in 50 per cent of the hosts. This method still has to be used as the bioassay for the mouse mammary tumour virus because no reliable system for infecting cells in culture has yet been found for this type of virus, which is very dependent for replication on the state of differentiation and hormonal regulation of the host mammary gland tissue. However, the leukaemia and sarcoma viruses can readily be grown in monolayer cultures of embryonic fibroblasts or of established cell lines. With cell culture techniques the number of infectious units of virus can be titrated rapidly and accurately, and the virus can be cloned and mutants selected; while essential for virological analysis, cell culture methods are, of course, further removed from studies of viral disease.

Unlike the majority of animal viruses, RNA tumour viruses do not typically kill the host cells when they replicate, for progeny virus particles bud off from the cell membrane without causing lysis. Therefore, apart from a few exceptions, it has not been possible to develop bioassays based on cytopathic plaque titrations, as used for lytic animal viruses and bacterio-phages. However, the infected cells may become 'transformed' into a malignant state and simultaneously produce virus. Transformation of the host cell can be employed as a criterion of infection; it is used in the bioassay of sarcoma viruses by enumerating focal areas of transformation in monolayer cultures. In fact, the first focal transformation assay was devised in 1938 for Rous sarcoma virus when Keogh found that he could obtain small, discrete tumours on the vascular chorioallantoic membrane of embryonated chick eggs, by adding appropriate dilutions of virus to the membrane through a hole made in the shell. The tumours developed within seven days, and the number of tumours per membrane was directly proportional to the inoculum dose. This *linear dose–response* indicated that each tumour was initiated by a single infectious viral particle, which was not only important conceptually, but meant that any sarcoma virus stock could be titrated in terms of transforming or *focus-forming units*. Subsequently it was observed that Rous sarcoma virus would transform embryonic cells in culture; this was developed into a focal assay system by Temin and Rubin in 1958 and remains the standard bioassay of sarcoma viruses today. Fibroblastic cultures are infected with serial dilutions of virus, and nutrient culture medium containing molten agar is then poured over the cell monolayer and allowed to set. The agar prevents the secondary spread of progeny virus and of loose, transformed cells. After 7–10 days' incubation, the foci — as the colonies of transformed cells are called — can be counted by eye or with a low-power light microscope. The appearance of foci of transformed cells viewed by scanning electron microscopy is shown in Fig. 2.1.

(a)   (b)

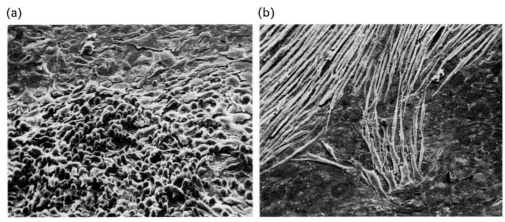

Fig. 2.1 Scanning electron micrographs of foci of cells transformed by Rous sarcoma virus (× 180). (a) Edge of a focus of piled up, loose, rounded, transformed cells; compared with flattened monolayer of normal fibroblasts at top of picture. (b) Edge of a focus of cells transformed by a fusiform variant of Rous sarcoma virus; note that fusiform transformed cells do not pile up on each other, but migrate over the normal fibroblast monolayer. (Photographs by A. Boyde and R. Weiss.)

Similar focus assays to that of Rous sarcoma virus have been devised for mammalian sarcoma viruses. It is more difficult, however, to titrate leukaemia viruses and non-pathogenic oncornaviruses. Although most strains of leukaemia virus replicate well in fibroblastic cells, the cells do not usually become morphologically changed, so that the infection cannot be measured by the appearance of the culture. Focal transformation of haemopoietic cells in culture by infection with viruses that cause acute erythroblastic or myeloblastic leukaemia can be seen, but this is difficult to achieve with the lymphocytic leukaemia (or lymphoma) viruses, which represent the common field strains in most host species. Some strains of murine leukaemia virus will induce cell fusion in the rat cell line XC to form multinucleated syncytia, and this observation has been developed into a quantitative, *syncytial plaque* assay. There are also some mammalian cell lines infected with defective murine sarcoma virus that only assume a transformed morphology when superinfected with a leukaemia helper virus, so that these cells can be used for *focus assays* of the leukaemia virus. Another method is to infect normal fibroblasts with leukaemia virus and to challenge subsequently with a closely-related sarcoma virus; the presence of excess leukaemia virus can block the entry of sarcoma virus into the host cell, thus enabling the titration of leukaemia virus as *interfering units.*

Where these plaque, focus, or focus inhibition techniques are inappropriate, virus production can be measured by *biochemical* and *immunological* techniques. The presence of virus released into the nutrient culture medium of infected cells can be detected by assaying the specific viral enzyme, reverse transcriptase, by using radioactive precursors to RNA or proteins to reveal the synthesis of virus particles, or by using immunological methods to detect viral antigens in the infected cells or in the culture medium. If separate culture plates are infected with serial dilutions of virus, and are then maintained or passaged for several days or weeks to allow the virus to replicate to high titre, these methods can be used as an end-point titration of the original virus stock.

Viral components inside the cells in productive or nonproductive infections are detected by two major methods (see Fig. 2.2).

A. *Viral nucleic acids* are assayed by specific *molecular hybridisation* to nucleic acid sequences prepared from the virus particles. The viral RNA can be used directly, after rendering it radioactive. This can be useful for detecting the presence of complementary DNA provirus sequences in the host cell. It cannot be used to monitor viral mRNA production because the virion RNA has the same 'sense' or strandedness as the mRNA and will therefore not hybridise. However, complementary DNA (cDNA) can be prepared *in vitro* from the virion RNA by setting off the virion's own reverse transcriptase activity. This cDNA can then be used to detect either DNA or RNA homologous to viral RNA.

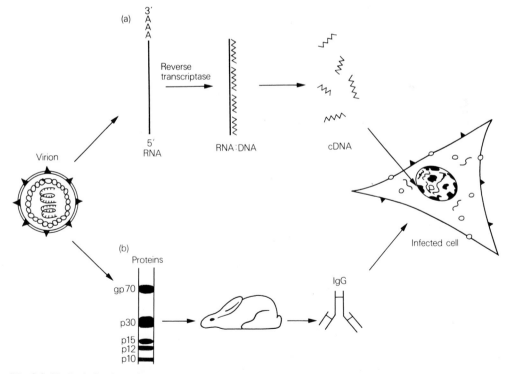

Fig. 2.2 Methods for detecting presence and activity of virus in cells.
(a) Preparation of complementary DNA to probe for viral DNA or RNA in the infected cell. (b) Preparation of antibodies to viral antigens to monitor viral protein synthesis.

B. The presence of *viral proteins* can be monitoried by preparing *specific antisera* to the proteins and using these as immunological probes. Oncornaviruses are composed of six or possibly seven distinct polypeptides which, after disruption of the virus particles with detergents, can be separated by polyacrylamide gel electrophoresis or by chromatography on guanidine hydrochloride columns. The separated proteins can then be used to immunise rabbits or other hosts to obtain specific antisera. The amount and location of viral proteins can be determined using these antisera in various immunological methods, of which immunofluorescence and radioimmunoassays are perhaps the most widely practised.

3 Structure and replication

3.1 Composition and function of the virion

An indication of the architecture of an oncornavirus particle is shown in Fig. 2.2, though the details vary from one virus to another and the precise structure has yet to be elucidated. The virus particle, or virion, is enclosed by a lipoprotein envelope derived from the plasma membrane of the host cell. Inside the particle is a nucleoprotein core surrounded by a core shell. When examined by electron microscopy of thin sections (Fig. 2.3), the core and core shell appear as an electron-dense nucleoid; in immature (freshly budded) particles this may appear as a ring structure, but it soon becomes uniformly dense. Morphologically, the oncornavirus particles are classified either as C-type viruses possessing a central nucleoid, which are typical of leukaemia viruses, or as B-type viruses possessing an eccentric nucleoid, which are typical of the mammary tumour virus. Within infected cells, particularly in mammary tumour cells and myeloma cells, A-type particles are sometimes apparent, which resemble cores without the outer envelope but may lack RNA.

Spikes or knobs protrude from the virion envelope, and they have been identified with the virus-coded glycoproteins. The larger glycoprotein has a molecular weight of about 85 000 for avian leukaemia-sarcoma viruses, 70 000 for murine leukaemia virus (MuLV), and 52 000 for

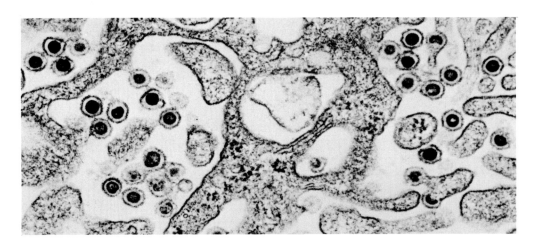

Fig. 2.3 Transmission electron micrograph of section through cultured cells producing C-type virus particles (× 48 000). Mature particles have dense nucleoids, freshly budded particles have ring-shaped nucleoids, and budding particles have crescent-shaped nucleoids. (Reproduced from Teich *et al.* (1975), *Nature*, **256**, 551–5.)

murine mammary tumour virus. In addition to the glycoproteins, a protein with a molecular weight of about 15 000 is also associated with the envelope, at least of murine C-type viruses.

The glycoproteins promote adsorption to the host cell membrane as the initial step in infection, and recognise specific receptor sites at the cell surface which are necessary for the penetration of the virus into the cell. The glycoproteins are important, therefore, in determining the host range of the virus, which cannot infect a cell that lacks the appropriate receptor sites. How the virion becomes uncoated is still not clear; the outer envelope may remain at the plasma cell membrane releasing the core into the cytoplasm, or the entire particle may be engulfed by the cell.

The inner core of the virion is surrounded by a core shell composed of the major internal protein, which varies between 27 000 to 31 000 molecular weight, according to the virus strain. The structural proteins of the virions are named by their approximate molecular weight. Thus the major internal protein of murine leukaemia virus is designated *p30* and the major external glycoprotein is *gp70*. Within the core shell are the smaller proteins, the reverse transcriptase and the RNA.

The viral proteins are recognised as *antigens*. Upon infection of adult hosts the glycoproteins in the virus envelope elicit neutralising antibodies which are usually type-specific, i.e. the antibodies recognise only that particular virus strain and strains with closely-related envelope glycoproteins. Early studies also showed that antibodies may also be generated that react with the internal viral proteins. These usually exhibit a broad cross-reaction with all viruses of the group – say avian leukaemia viruses, or murine leukaemia viruses – and the antigens were therefore termed group-specific. Now it is realised that each protein may have multiple antigenic sites, some of which are type-specific, others group-specific, and yet others shared between different groups of virus (interspecies antigens), e.g. common antigenic sites on the p30 proteins of murine, feline and simian C-type viruses. Table 2.2 lists the major structural proteins of murine C-type viruses and their antigenic properties. This has been elucidated by the application of highly sensitive and specific immunological techniques which are also useful in probing the expression of viral antigens in infected cells.

The *genetic material* of oncornavirus particles is single-stranded linear RNA, and the particles carry two related genome subunits, each of about $2 \cdot 7 \times 10^6$ daltons. In addition to the genomic RNA, there are smaller RNA species that are derived from the host cell. In particular, a species of transfer RNA is associated with the genomic RNA and acts as a primer for the reverse transcriptase. The primer for chicken leukaemia viruses is tRNA^{try} and for mouse leukaemia virus is tRNA^{pro}. The two genome strands of RNA and their associated primers can be extracted from the particles as an RNA complex with a sedimentation value of 60 to 70S, and this complex is commonly known as 70S RNA. Recent electron-microscopic studies indicate that the two genome subunits are linked near their 5' termini. On heating, the complex irreversibly dissociates into the 4S tRNA species and two 35S genome subunits which have poly(A) at the 3' termini.

Evaluations of genome complexity by hybridisation and by oligonucleotide estimation of T_1 digests indicate that the two genomic subunits are identical or closely related. In other words, the virus particle appears to be *diploid*. All the viral genes must therefore be carried on each subunit, which, having a molecular weight of nearly 3×10^6, could code for polypeptides totalling approximately 300 000 daltons. This coding capacity accommodates the known structural genes and a possible transformation gene, or 'oncogene', without leaving substantial spare sequences. The viral genome therefore appears to be economically organised, although the reason for diploidy is not clearly understood. The gene order on the linear RNA chromosome has recently been mapped for Rous sarcoma virus and will be discussed in section 5.

Table 2.2   Murine leukaemia virus proteins

Precursor polypeptide	Protein	Molecular weight	Location	Major antigenic determinants		
				Type	*Group*	*Interspecies*
pr65	p10	10 000	core	–	+	–
	p12 (phosphoprotein)	12 000	RNA (core)	+	–	–
	p15	15 000	core	+	–	–
	p30	30 000	core shell	–	+	+
pr90	p15E	17 000	envelope	+	?	?
	gp70 (glycoprotein)	69 000–71 000	envelope	+	+	+
	reverse transcriptase	70 000	core	+	+	–

3.2 The provirus hypothesis

It took Temin six years to convince molecular biologists that his provirus hypothesis was correct. There was considerable resistance to the idea that DNA might be synthesised on an RNA template, for it represents a reversal of the usual flow of biological information. Moreover, there was no precedent for this mode of replication among RNA bacteriophages and there was suspicion that an animal virus (or virologist) might be more innovative than phages. Reverse transcription does not really contradict the central dogma of molecular biology, which states that nucleic acid sequences embody the code for amino acid sequences in proteins, but not vice versa. Knowing the precise base-pairing involved in the transcription of RNA from a DNA template, as well as in the replication of double-stranded DNA itself, it is not difficult to envisage the evolution of a polymerase that will recognise RNA as a template for the synthesis of DNA. Perhaps it was felt that if such enzymes were commonplace, the conservation and organisation of the cellular genome would be difficult in the presence of new pieces of DNA copied from various RNA transcripts.

Temin was intrigued by the persistence of Rous sarcoma virus (RSV) and its transforming effect on cells through numerous mitotic cell divisions with no apparent loss or dilution, even in non-permissive cells that did not produce progeny virus. This suggested some intimate association of the virus with the host genome, as was known at that time for temperate phages. Subsequently, Bader and Temin independently found that inhibitors of DNA synthesis interfered with infection of cells by RSV, demonstrating a requirement for DNA synthesis. Once infection was established, however, DNA inhibitors did not suppress further virus synthesis, but inhibitors of RNA transcription such as actinomycin C did so, indicating a continuous requirement for DNA-dependent RNA synthesis. These experiments did not prove the veracity of the DNA provirus hypothesis, for it was difficult to distinguish a dependence on host DNA functions from a direct effect of DNA inhibitors on the viral genome itself; this was shown later by analysing the effect of drugs on the infection of non-proliferating cell cultures, where there is negligible chromosomal DNA synthesis. Attempts were also made during the 1960s to reveal the provirus by molecular hybridisation, but the techniques were not yet sensitive enough, and the unknown presence of host-inherited viral genomes gave positive results in the control of 'uninfected' cells which confused the issue. Later the discovery that viral antigens or functions were present as host Mendelian traits in uninfected cells (see section 4) lent strong support to the provirus hypothesis, at least for those who were aware of these studies. But it was the dramatic discovery of *reverse transcriptase* (more properly called RNA-directed DNA polymerase) in the virions of murine leukaemia virus by Baltimore and of RSV by Mizutani and Temin that changed nearly everyone's mind. Subsequently, the observation that mutants of RSV with temperature-sensitive reverse transcriptase activity were also transiently temperature-sensitive for an early function in the host cell, showed that reverse transcription was an obligatory process in viral replication. Formal proof of the formation of a complete DNA provirus was provided by Hill's and Hillova's demonstration that DNA purified from RSV-transformed cells when introduced into normal chick cells caused cell transformation and synthesis of progeny RSV, a phenomenon known as *transfection*.

Following the discovery of reverse transcriptase, it has become fashionable to seek similar transcriptional processes for the reorganisation of the genome in normal eukaryotic cells; for example, in shunting variable and constant regions of immunoglobin genes during lymphocyte differentiation. Reverse transcriptase, however, has not been detected in normal cells except in association with retroviruses, although we should remember that most vertebrate species

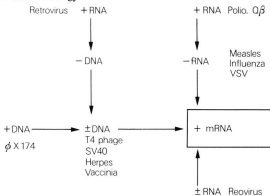

Fig. 2.4 Diversity of viral genomes and their strategies for making mRNA. (Adapted from D. Baltimore (1971), *Bacteriol. Rev*, **35**, 235–41.)

studied carry inherited viral genomes, including a gene coding for the enzyme, and that these latent viruses endogenous in the host may express functions useful to that host.

If one stops to consider it, all RNA viruses are exceptions to conventional cellular RNA synthesis because the RNA *replicates*. It is true that only the retroviruses replicate by forming a DNA provirus, but the formation of complementary RNA on a viral RNA template is equally strange and also requires a transcriptase not present in the host cell. Baltimore has usefully classified viruses into different groups according to the relationship of the viral genome in the particle to the messenger RNA that is ultimately necessary for the synthesis of viral proteins (see Fig. 2.4). Viruses with double-stranded (ds) DNA genomes, such as T-even bacteriophages or the DNA tumour viruses discussed in the previous chapter, can utilise cellular polymerases for replication and transcription, as can single-stranded (ss) DNA viruses, such as ϕX 174, once the complementary DNA strand is transcribed early in infection. RNA viruses, however, require special polymerase functions. These may involve the synthesis of a small polypeptide which modifies a host polymerase to enable it to use RNA as a template, as is the case with Qβ and related RNA phages; or they may involve the provision of a complete, virus-coded, RNA replicase, as is the case with many animal viruses.

Animal RNA viruses have evolved in many different forms (see Fig. 2.4). There are viruses with dsRNA, such as reoviruses (where the genome is segmented, each segment representing a single gene), and viruses with ssRNA such as polio (non-segmented) and influenza (segmented). The ssRNA viruses are called positive-strand viruses when the virion RNA has the same sense or strandedness as the mRNA (e.g. polio virus) and are called negative-strand viruses when the virion RNA is complementary to the mRNA (e.g. myxo-, paramyxo- and rhabdo-viruses). Now the positive-strand virion RNA can itself serve as mRNA once the virus has infected the cell and has become uncoated. Thus, early in infection the polio virion RNA becomes associated with ribosomes and is used as mRNA to translate viral proteins. One of these proteins is an RNA polymerase that promotes the synthesis of complementary, negative RNA strands, and then copies these back into positive strands which function as more mRNA and are also packaged as genomes in progeny virus. On the other hand, the RNA of negative strand and of dsRNA viruses cannot be utilised as mRNA, yet mRNA synthesis is, of course, a prerequisite of viral protein synthesis. Hence the virus needs to synthesise mRNA from a negative-strand RNA template before it can synthesise viral proteins; yet a viral protein, an RNA-directed RNA polymerase, is needed for mRNA synthesis. This paradox was solved when it was realised that negative-strand viruses incorporate the RNA-directed RNA polymerase into the virion as a component of the ribonucleoprotein core. Upon infection, then, the polymerase is already associated with the virion RNA and can begin to synthesise mRNA. Since the polymerase is assembled into the virus particles, its presence and activity can be assayed in enzyme reactions utilising purified virions.

The genomes of RNA tumour viruses are positive-strand RNA. Indeed, they bear poly(A) at the 3′ termini like cellular mRNAs and can be translated *in vitro*. There is also recent evidence that the virion RNA, when purified, is also translated on introduction into cells in culture, but there is no evidence of early translation in natural infection. With the discovery of virion polymerases in negative-strand RNA viruses it seemed reasonable to look for reverse transcriptase inside the virion, and once sought there, it was readily detected. Treatment of purified virus particles with mild detergent, and addition of radioactive precursors of DNA and a divalent cation such as Mg^{2+} or Mn^{2+}, sets off the reverse transcriptase activity and allows the detection of radioactive DNA product.

3.3 Reverse transcription and integration

Figure 2.5 outlines the life-cycle of an oncornavirus. Following adsorption and penetration via a specific cell surface receptor (1), and uncoating of the virus (2), it is thought that the provirus is synthesised in the cytoplasm (3), migrates to the nucleus and becomes integrated into a host chromosome (4). Then transcription (5), protein synthesis (6) and maturation of the virus at the cell surface (7) occurs.

Details of early events of infection are unclear, though much is now known about reverse transcription *in vitro* (Fig. 2.6). The reverse transcriptase is primer-dependent, i.e. a sequence of nucleic acid complementary to the RNA template is needed to initiate the polymerisation. The natural primer is a cellular tRNA, a part of which has complementary bases to a sequence on the virion RNA, and DNA synthesis proceeds in the 5′ to 3′ direction as a covalent extension of the primer, to form an RNA–DNA hybrid molecule. When the complementary DNA strand has been synthesised, the enzyme exhibits two further functions: it specifically hydrolyses the RNA of the hybrid (RNase H activity) and it polymerises a second strand of DNA to form a dsDNA molecule. In this last reaction the enzyme is using DNA as a template, behaving as a DNA-directed DNA polymerase (Fig. 2.6(b)). This reaction is inhibited by actinomycin D so

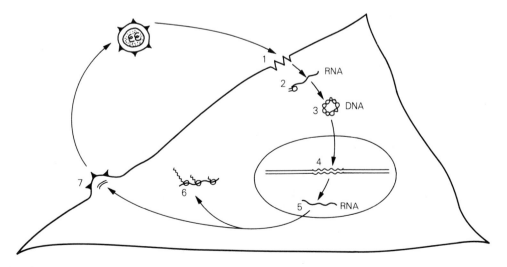

Fig. 2.5 Replicative cycle of an RNA tumour virus.

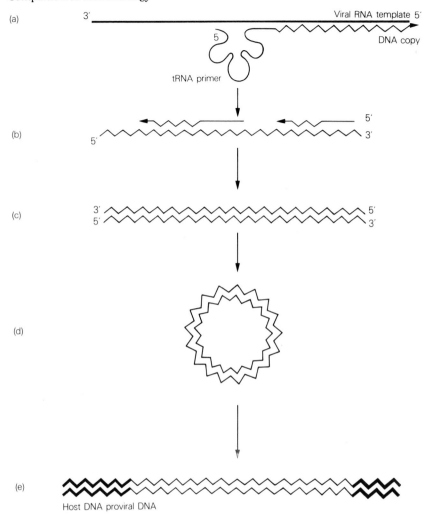

Fig. 2.6 Synthesis and integration of DNA provirus. (a) Synthesis of first DNA strand. (b) Degradation of RNA and synthesis of second DNA strand using the first strand as template. (c) Linear ds DNA provirus. (d) Circular provirus. (e) Integration of provirus.

that by allowing the enzyme to react in the presence of the drug ssDNA can be prepared pure. Synthetic template-primer molecules are useful in monitoring reverse transcriptase activity, of which poly(C)-oligo(dG) is the most specific. Until recently it was possible to obtain only short strands of DNA (100 to 1 000 nucleotides) with the reverse transcriptase *in vitro*, but with changes in conditions such as ionic strength it is now possible to synthesise full length provirus, which is copied so faithfully and completely that it is infectious.

The dsDNA provirus has a molecular weight of approximately 6×10^6 and therefore appears to represent one subunit only. Whether both RNA subunits are necessary to make the haploid provirus is not clear, but it is possible that reverse transcription is initiated on one subunit and proceeds onto the other. This is an attractive idea because it could then transcribe into DNA on the second RNA strand the sequences that are represented by the primer molecule on the first, which, in the oversimplified scheme depicted in Fig. 2.6, would not be synthesised in the provirus. Furthermore, although reverse transcription proceeds in the $5'$ to $3'$ direction for the nascent strand, the primer appears to be situated near the 5' end of the template molecule, so that only a small proportion of that molecule would be copied.

There is evidence that *in vivo* the provirus assumes a closed circular form (Fig. 2.6(d)), resembling the papovavirus genome. This form would aid integration into a linear host chromosome because a single recombinational event only would be required. There are also claims that the provirus may integrate as an RNA–DNA hybrid, in which case the second strand of DNA may only form during replication of the host chromosome. If this is normally the case, however, it does not explain the RNase H activity of the viral enzyme and the detection of viral dsDNA in the cytoplasm.

Virtually nothing is known about the integration sites in the host DNA. This problem is now being actively studied in several laboratories. The use of restriction enzymes (see Ch. 1) to cut out segments of DNA containing integrated provirus should tell us something about host sequences adjacent to the viral DNA. We should soon know whether integration sites occur randomly or at specific sites in both viral and host genomes; whether different viruses integrate at different host sites; and whether the site of integration affects viral gene expression. Those are important problems that can now be tackled experimentally. Probably retroviruses are the only RNA viruses able to cause cancer because the integrated DNA provirus allows a stable interaction with the host cell.

3.4 Transcription and maturation

The integrated provirus is adopted by the host cell as an extra set of genes. It may remain latent, in which case the cell is unaffected by the virus, or the genes may be expressed. Viral gene transcription depends on host RNA polymerase. Careful studies of infection of synchronised cell populations indicate that transcription is not initiated until the host cell passes through both the S-phase (DNA synthesis) of the cell division cycle and mitosis into the next G1 phase. Once initiated, however, viral RNA transcription continues at all phases of subsequent cell division cycles.

Unilateral transcription of the provirus provides both viral RNA for progeny virions and mRNA for translation into viral proteins. Both kinds of RNA become polyadenylated and it is not clear whether they represent a single kind of RNA. Probably the mRNA is synthesised (or subsequently cleaved) as smaller molecules than the 35S genome RNA segments. It is possible that there are three different mRNA species representing genes for the virion structural proteins. One gene, known as *gag* (for *g*roup-specific *a*ntigen) is translated as a large precursor polypeptide, which is proteolytically cleaved in progressive steps into the non-glycosylated internal virion proteins p10, p12, p15 and p30 (for murine leukaemia virus). The reverse transcriptase may be translated from another mRNA species (it is synthesised in much smaller amounts) and a third may determine the envelope glycoproteins. Thus oncornavirus transcription appears to be intermediate between viruses like polio, in which the entire genome is translated into a giant polypeptide and is then cleaved into functional proteins, and viruses like vesicular stomatitis, in which separate mRNA species are transcribed from the genome for each viral protein.

Little is known about the assembly and maturation of oncornavirus particles. With C-type viruses nothing is apparent until the particle begins to bud from the plasma membrane, when the nucleoid condenses as an electron-dense crescent just beneath the membrane (see Fig. 2.3). With B-type viruses, the nucleoid condenses as a complete core under the cell membrane and becomes enveloped. Maturation of virus particles is not lytic; infected cells may chronically produce virus particles without any diminution of cell growth or metabolism.

4 Transmission of oncornaviruses

4.1 Horizontal and vertical infection

In the previous section we discussed the replication of oncornaviruses as infectious particles which enter cells through specific receptor sites, and form DNA proviruses that integrate into the host genome. The infectious virus may be transmitted from one host to neighbouring hosts by contagion, but another mode of transmission is from parent to offspring. Gross, who first demonstrated that the 'familial' leukaemia of mice was in fact a viral disease, coined the terms *horizontal* and *vertical* transmission to denote the two routes of infection.

Some viruses depend on vertical transmission for their perpetuation; these viruses are generally not very harmful to their hosts, for it is apparent that neither host lineage nor virus would survive if they were. Other viruses which are typically contagious may incidentally be transmitted vertically, as in the congenital infection of the human foetus with rubella virus (german measles). Yet other viruses may be horizontally transmitted in one host and vertically transmitted in an alternate host. The arthropod-borne viruses, such as the tick-borne encephalitis virus and the mosquito-borne yellow fever virus, are highly pathogenic in some mammalian hosts but are usually harmless to the arthropod. The viruses are transmitted horizontally from one mammal to another via the blood-sucking tick or mosquito. The mammalian host is generally called the *reservoir* species and the arthropod, the *vector*. Many of these viruses are transmitted passively in the gut or salivary fluid of the vector to the next mammalian host, but some actively replicate in the vector's tissues. These viruses may be transmitted through the vector's eggs and various larval stages of development until the adult bites a mammalian host again. In such a case, the vector species also constitutes a reservoir of the virus. A similar situation is found with rickettsial infections, such as Rocky Mountain spotted fever (related to typhus). As Darlington pointed out, 'the terror of its *infection* in man has closed our eyes to the fact of its *inheritance* in the tick'.

RNA tumour viruses are transmitted both horizontally and vertically. Furthermore, we must distinguish between two modes of vertical transmission, congenital infection and genetic transmission, which are quite different at the molecular level and consequently are subject to different biological controls.

Congenital infection is the type of transmission that we have already discussed, which occurs when infectious virus particles released by the mother infect the offspring. The virus may infect the ovum or zygote directly, or it may be transmitted via the placenta or milk. The first example described for oncornaviruses was Bittner's observations in 1936 that the mouse mammary cancer agent was transmitted to the offspring in the milk. He had noticed that the high incidence of tumours characteristic of C3H mice, was inherited maternally but not paternally. When he took baby C3H mice immediately after birth and nursed them on foster mothers of a low cancer strain, the mice did not subsequently develop mammary tumours; indeed, the offspring's offspring were also free of the agent. Thus the mice could be 'cured' of an apparently heritable disease by avoiding the mother's milk for one generation, for this broke the chain of infection. Conversely, if baby mice of a low incidence strain were foster-nursed on C3H mice, the females subsequently developed mammary tumours and their offspring in turn acquired this trait through the milk.

Genetic transmission, on the other hand, involves the passage of the viral genome from one host generation to the next as a DNA provirus, integrated as part of the genetic complement of the gametes. Bentvelzen found that the high incidence of mammary tumours in GR strain mice, in contrast to C3H mice, was inherited through either the father or the mother, and that nursing baby GR mice on low-incidence foster mothers did not affect their propensity to develop mammary cancer. The high cancer incidence was inherited as a simple Mendelian trait, with dominant expression; yet the lactating females also secreted mammary tumour virus in their milk which was oncogenic in baby mice of low incidence strains. Apparently, the viral genome is inherited in GR mice as a set of host genes but becomes spontaneously activated in lactating mammary gland tissue to produce infectious virus.

Whereas congenital infection usually occurs through the mother, genetic transmission is both maternal and paternal. Furthermore, genetic transmission bypasses such restrictions to infectivity as a lack of cell-surface receptors, or defective reverse transcription because the viral genome is not transmitted as a virus particle. It resembles an integrated prophage (hence the term provirus) passed in the germ line of the multicellular, eukaryotic host. Bentvelzen called the inherited viral genome the *germinal provirus* in order to distinguish it from *somatic proviruses* acquired by infection, and the term *virogenes*, coined by Huebner and Todaro, denotes the same. Inherited viral genomes are also generally known as *endogenous* viruses, in apposition to *exogenous* viruses that infect the host, as extrinsic agents.

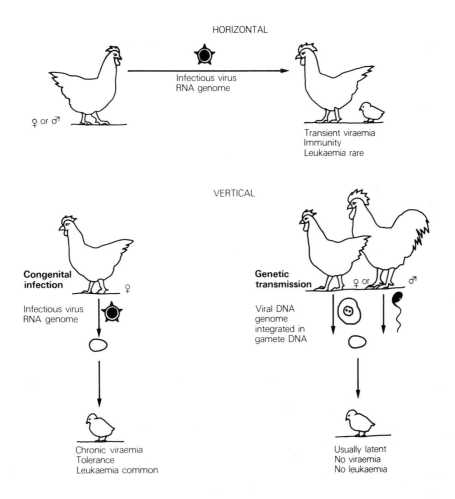

HORIZONTAL

Infectious virus
RNA genome

♀ or ♂

Transient viraemia
Immunity
Leukaemia rare

VERTICAL

Congenital infection

Infectious virus
RNA genome

♀

Chronic viraemia
Tolerance
Leukaemia common

Genetic transmission

♀ or   ♂

Viral DNA
genome
integrated in
gamete DNA

Usually latent
No viraemia
No leukaemia

Fig. 2.7 Transmission of avian leukaemia viruses.

The C-type viruses are transmitted by all three modes, horizontal, congenital and genetic. Figure 2.7 illustrates the transmission of avian leukaemia viruses in the domestic fowl (leukaemia viruses are usually termed leukosis viruses by veterinarians). Chickens can become horizontally infected with virus as adults or juveniles, in which case the virus is not generally harmful, and leukaemia is only a rare consequence. Following a transient viraemia (a general infection in the circulation), the host mounts an immune response, chiefly by producing type-specific antibodies to the viral envelope glycoproteins which neutralise the infectivity of the virus. However, if a hen is laying eggs during the phase of transient viraemia, infectious virus may be congenitally passed into the egg. One of the host tissues that most actively secretes virus is the ovalbumin-secreting epithelium of the oviduct, and substantial titres of infectious virus can be found in the egg albumin. This virus infects the chick embryo early in development and replicates in most of the embryonic tissues. It is remarkable that development proceeds with no apparent anomalies and a healthy chick hatches. These chicks, having been exposed to viral antigens as embryos, typically do not recognise the virus as distinct from host, and thus remain immunologically tolerant and chronically viraemic. Later on in life, usually when they are adult, they are likely to develop lymphocytic leukaemia (a B-cell leukaemia of cells derived from the bursa of Fabricius). If a congenitally infected hen lives long enough to lay eggs herself, those eggs in turn will be congenitally infected and vertical transmission of the virus may be continued for many generations in this way.

It is important to note that the horizontal and congenital modes of virus transmission are not mutually exclusive. Horizontal infection of a laying hen can initiate a vertical line of transmission, and conversely, congenitally infected birds are an important source of horizontal infection since they continually shed virus. Thus when leukaemia virus is endemic in a flock of domestic fowl there is a balance between (a) congenitally infected, tolerant virus-shedders; (b) horizontally infected, immune fowl, and (c) non-infected susceptible chicks.

In addition to the infectiously transmitted leukaemia viruses, all chickens apparently carry an endogenous genome representing a closely related virus, though there is no evidence that this virus induces leukaemia. In mice, the naturally occurring C-type viruses which induce leukaemia are all transmitted genetically and horizontal or congenital infection is uncommon. Leukaemia in cats and cattle, on the other hand, is typically transmitted horizontally; infection of adult hosts, although it frequently invokes immunity, alternatively may induce leukaemia, or sometimes a disease resembling aplastic anaemia. Thus the epidemiology of oncornaviruses in different hosts varies widely. Where the disease is transmitted by horizontal infection, local epidemics of leukaemia occur; where the disease results only from congenital infection, the time—space clustering of the disease is not so apparent; where the disease is induced by an endogenous virus, it is familial, and the fact that a virus is involved may not be apparent at all.

4.2 Detection and expression of endogenous viruses

Table 2.3 lists the lines of evidence that betrayed the presence of endogenous viral genomes in the normal host genome. When these began to emerge in the late 1960s, the provirus hypothesis was not yet generally accepted and many of the observations at first seemed confusing. This is exemplified by three lines of investigation in different laboratories that indicated the presence of an endogenous virus in chickens, the results of which were so unexpected that they were initially misinterpreted. First, in attempts to test the provirus hypothesis by identifying viral DNA in cultures infected with Rous sarcoma virus, DNA that hybridised to viral RNA was also found in the uninfected control cultures, which cast doubt on the specificity of the viral RNA used for hybridisation. Second, normal chicken embryos, apparently free of any congenitally

Table 2.3   Evidence for the presence of endogenous
RNA tumour viruses in normal cells

DNA homologous to viral RNA
Viral antigens expressed as Mendelian traits
Complementation of defective exogenous viruses
Spontaneous virus production
Induction of virus by mutagens and mitogens
Specific immunity to reinfection
Rescue of endogenous viral genes by recombination
    with exogenous viruses

infecting leukaemia virus, possessed an antigen that cross-reacted in complement fixation tests with antibodies prepared against the internal proteins of the virus, which similarly cast doubt on the specificity of the immunological assay. Third, a strain of Rous sarcoma virus that had previously been regarded as defective in envelope antigens and was therefore uninfectious unless complemented by a helper virus (see section 5), actually synthesised infectious virus possessing a new host range; it was mistakenly concluded that previous theory, rather than the virus itself, had been defective. In fact, it was soon realised that these three observations — hybridisation of viral RNA to uninfected chick cell DNA, presence of viral antigens in uninfected chick embryos, and the ability of uninfected chick cells to complement a defective oncornavirus — all reflected the presence and expression of endogenous viral genes in normal host cells.

The clearest evidence for genetically transmitted avian oncornavirus came from studies of two highly inbred lines of chickens. One line was positive for the internal viral antigen p27 and was also positive for the helper factor that complemented defective Rous sarcoma virus, which was later shown to be viral glycoprotein coded by the endogenous genome. Another inbred line was negative for both these viral gene products. Cross-breeding experiments showed that both the viral proteins were expressed as a single Mendelian dominant trait. It looked, therefore, as if the positive inbred chicken line carried a heritable provirus which the negative line lacked, but this proved not to be the case. Careful molecular hybridisation studies showed that both kinds of chicken possess DNA in equal amounts (one or two provirus copies per haploid host genome), whereas viral RNA is only detectable in the positive line. Thus the Mendelian locus that distinguished the two inbred chicken lines does not necessarily represent the site of the provirus itself, but acts as a regulatory gene for transcription of viral structural genes. It is interesting to note that viral RNA and protein synthesis is a dominant trait, which indicates that the regulatory gene does not act as a repressor.

Chickens of most commercial breeds express viral antigens without producing complete virus particles, but two viral proteins, p15 and reverse transcriptase, cannot be detected in the chicken cells. One strain of chicken, though, spontaneously produces infectious C-type particles which bear the same antigenic specificities as the viral proteins present in the latent infections of other host strains. These virus-producing chickens are chronically viraemic (i.e. infectious virus is continuously detectable in their blood) but the chickens do not show a high incidence of leukaemia; presumably this virus has little or no oncogenic potential. A single Mendelian locus controls virus production, and this locus does not appear to be linked to the locus which determines partial virus expression. Whether these genetic loci represent different proviruses or different regulatory genes acting on the same provirus is not yet known.

All chickens and mice appear to carry complete DNA proviruses which usually remain latent. However, cells of these and several other species can be induced to release complete, infectious virus, by treatment with chemical mutagens and carcinogens or with ionising radiation. The activation of endogenous viral genomes has been studied most thoroughly in mice, in which several distinguishable proviruses may be induced. Halogenated pyrimidines, such as 5-iodo-deoxyuridine, appear to be the most efficient inducing agents, although a variety of mutagens also work, as they do for induction of temperature bacteriophages. This led to the idea, called the *viral oncogene hypothesis* by Huebner and Todaro that many chemical and physical carcino-gens may act by triggering-off latent, inherited proviruses. It was already known in the late 1950s that X-rays induced a lymphoma in C57BL mice which was henceforth virally trans-missible, but its significance was not fully realised at that time owing to our ignorance of the nature and molecular biology of the oncornaviruses. Where chemical and physical carcinogens caused cancer without inducing viruses, Huebner and Todaro suggested that the provirus may be partially activated, so that its oncogenic potential was expressed. While this view of tumour induction might be pertinent to certain cancers, particularly leukaemia, it seems improbable that activation of 'viral oncogenes' is the universal mechanism of carcinogenesis. However, the oncogene concept is a useful one and will be discussed in more detail in section 6.

The expression of latent oncornaviruses is also influenced by the differentiated state of the cell and the physiological state of the host animal. The mouse mammary tumour virus is only produced in substantial quantities by the lactating mammary gland and in some mouse strains, mammary tumours induced by the virus only appear if the female mouse has experienced many pregnancies, or has been artificially stimulated with oestrogen and prolactin. In male mice, the viral antigens of mammary tumour virus are synthesised in the epididymis. Glucocorticosteroid hormones also enhance production of both B-type and C-type viruses. Studies on the effect of dexamethasone, a synthetic glucocorticosteroid hormone, on mammary tumour cells in culture indicate that the hormone exerts an early and direct effect by elevating the rate of RNA transcription from the provirus. C-type viruses in mice are activated in mouse lymphoblasts not only by mutagens but also by inhibitors of protein synthesis, and by stimulation of cell proliferation induced by plant lectins or by immune reactions.

It is not understood why these viruses are so readily activated, nor whether they serve a useful function to the host. One view is that viral proteins, particularly the envelope glyco-proteins, which are expressed at the cell surface when the virus is partially expressed, may act as cell recognition markers during differentiation, and it is true that glycoproteins representing different proviruses are found in mice in different tissues or at various stages of development. Furthermore, these viral antigens may be closely related to cellular antigens but are apparently not linked genetically to them. Whether the viruses have evolved by picking up and modifying genes for cell-surface proteins or whether the host has adopted and adapted viral envelope antigens is not clear, for the evolution of viruses that are inherited as host genetic elements is obviously complex. It has also been suggested that the induction of viral antigen expression in tumours would enhance any immunological surveillance by the host against that tumour. Certainly the expression of p30 viral antigen in the surface of mouse lymphoma cells is an important factor in cellular immunity directed against that tumour. So it is conceivable that an inherited provirus might remain latent in normal cells but may be activated if a cell becomes malignant, and its antigens might lead eventually to the elimination of that cell or its progeny. In this way the inherited latent virus would be advantageous to the host, and one could say that cancer induces viruses.

4.3 Host range

Oncornaviruses exhibit a restricted host range of species, and strains within species, which are susceptible to infection. This has been studied in some detail with avian oncornaviruses for which the major determining factor is the interaction of the envelope glycoprotein with the cell-surface receptor. The avian oncornaviruses can be classified into several distinct host range subgroups that correlate with a serological classification of their envelopes. Leukaemia and sarcoma viruses that are subject to cross-neutralisation by antisera share a common host range, because that particular kind of envelope glycoprotein must utilise a particular cell receptor, which may be present in one chicken strain and absent from another. The receptors are genetically determined, cell-surface components which have not yet been identified biochemically. Dominant alleles code for the receptors and genes for different receptors are located on different chromosomes. Thus in cross-breeding experiments, it is found that susceptibility to one virus subgroup segregates independently of susceptibility to another virus subgroup.

Outbred populations of chickens will frequently be polymorphic at several receptor gene loci, yielding a complex pattern of susceptibility and resistance. This will affect the transmission of infectious virus particles whether by horizontal or congenital routes, and some effort has been devoted to breeding flocks that are homozygous for several recessive receptor alleles, and hence are resistant to infection and to leukaemia. The inherited provirus will, of course, be passed on from one generation to the next regardless of the susceptibility of the host at the receptor level. Indeed, most chickens are resistant to reinfection with their own endogenous virus, either because they genetically lack the appropriate receptor, or because, when receptors are present, they are competitively blocked by endogenous, viral glycoprotein synthesis. The result is that activated endogenous virus neither spreads horizontally from host to host bird, nor from cell to cell within one host. However, the endogenous chicken virus can infect cells of related birds, such as pheasants, quails and turkeys, that have the appropriate receptors which are not blocked by endogenous glycoproteins. This has been shown in experimental host range studies but it is not known whether infection spreads naturally from one avian species to another. It does appear to be evolutionarily significant that chickens are usually resistant to the infectious spread of their own inherited virus.

In mice, C-type viruses can also be classified into host range subgroups. While all murine C-type viruses appear to be naturally transmitted genetically, some when activated will reinfect mouse cells and others will only infect cells of other species. The viruses that will reinfect cells of their own species are termed *ecotropic*, signifying that they replicate in their home species, while the viruses that only infect foreign hosts are termed *xenotropic*. Recently, C-type viruses have been isolated from wild mice which have a wide host range in foreign species and among mice; these are termed *amphotropic*. As with the avian viruses, the three host ranges of the murine viruses are chiefly determined by the presence and absence of appropriate specific cell-surface receptors. Both ecotropic and xenotropic viruses are naturally transmitted in the DNA of the gametes of the mice. When activated by physiological events or mutagens, they may infect, at least under experimental conditions, new host cells, either of mice if they are ecotropic viruses, or of foreign species such as humans and rabbits if they are xenotropic. There is no evidence that mice represent a natural reservoir of infection for man, but this has not been closely studied. A spontaneously active xenotropic virus in mice of the New Zealand Black strain is associated with an autoimmune disease resembling systemic lupus erythematosis in man. Although xenotropic viruses have not yet been shown to induce cancer, they have been isolated from other species such as cats and baboons, and the endogenous chicken virus discussed earlier is in essence xenotropic.

Ecotropic mouse viruses can be further classified on the basis of host range into *N-tropic* and *B-tropic* viruses, according to whether they replicate in cells of the NIH—Swiss strain or the BALB/c strain. There are also laboratory selected virus strains, termed *NB-tropic*, which replicate equally well in either strain. Any mouse can be classified as N-type or B-type according to their susceptibility to infection by N-tropic or B-tropic viruses, respectively. The determining host factor in N and B host range is a single gene, *Fv-1*, with two alleles, *n* and *b*, denoting the respective susceptibility. This gene is not related to a cell-surface receptor, for N-tropic and B-tropic viruses adsorb to and penetrate susceptible and resistant cells equally well. Some later event in the infectious cycle is blocked, possibly integration of the provirus.

4.4 Inheritance of exogenously acquired viral genomes

The origin of oncornaviruses is not known. Like most viruses that carry little genetic information, we may assume that retroviruses originally developed from host genetic material that evolved into autonomously replicating systems. Temin has postulated that they are continuously evolving and that induction of endogenous virus is really the last stage of that evolution. However, I feel that these viruses are so readily activated into fully infectious form with precisely coded functional proteins, such as envelope antigens and reverse transcriptase, that the endogenous viral genomes have been guilty of infection during their past history. How these genomes first entered the germ line of the host is not clear, particularly for xenotropic viruses, but it is quite likely that they may have been acquired when the host was not resistant to infection, and that the host species has acquired resistance to reinfection as a consequence of harbouring the virus.

Using a laboratory strain of ecotropic murine leukaemia virus, Jaenisch has shown that experimental infection can result in genetic transmission. He infected early mouse embryos by inoculating virus into blastulae which he implanted into foster mothers. Several of the embryos developed normally and did not become viraemic, though the provirus could be detected in several tissues. Most of the mice developed leukaemia but at least one mouse was bred before dying of the disease. It was apparent that the offspring have inherited the viral genome, which is distinguishable from pre-existing endogenous viruses by several markers.

The inheritance of exogenously acquired oncornavirus genomes has also occurred on occasions in nature. Benveniste and Todaro have investigated the extent of nucleic acid homology of viral genomes present in diverse host species. They found that the infectiously transmitted virus of gibbons is related to an endogenous virus of an Asian species of mouse and has apparently been transmitted from the mouse, just as it is possible that man could become infected with the xenotropic virus of the house mouse. Similarly an endogenous virus of pigs is related to the Asian mouse virus and the exogenous leukaemia virus of cats is at least partially related to an endogenous virus of rats.

The most striking example of transfer between unrelated host species is the endogenous C-type virus of the domestic cat. This is a readily inducible, xenotropic virus which is not related antigenically or by genome homology to the infectiously transmitted feline leukaemia virus group. It is, however, closely related to an endogenous virus of baboons. The baboon virus is representative of viral genome sequences present in the normal DNA of all African monkeys; the cat virus is endogenous only in North African and European cats and is lacking in the DNA of Asian species of the same genus. It appears that a common ancestor of the domestic cat and other Mediterranean species must have acquired the monkey virus after the Asian cats of the genus *Felis* had diverged, and that the virus then became inherited through the germ line. Thus a viral genome carried as an integrated genetic element in one host species has been transmitted into the genome of another, unrelated host. This is akin to transduction by temperate bacteriophage. It is not known how common a phenomenon it is, nor whether it has provided a significant means of genetic exchange between vertebrate species.

5 Genetics of oncornaviruses

Much can be learned about oncornaviruses and their effects on host cells by employing genetic techniques. Mutants have been isolated and existing variants exploited to analyse the replication of the viruses and their capacity to induce malignant transformation. Defective viruses have been characterised and used both for genetic mapping and for studying virion assembly. The discovery of recombination between RNA tumour viruses has extended the range of experimental genetic manipulation and has posed questions about their evolution. Most genetic studies have focused on avian oncornaviruses, and on Rous sarcoma virus in particular, because (*a*) RSV has for long been an experimental tool, (*b*) genetically determined phenotypic markers soon became apparent, and (*c*) non-defective strains of RSV exist.

5.1 Mutants and markers

Temin showed in 1960 that the transformed phenotype of the RSV-infected cells was influenced by a genetically determined function of the virus. While most foci of transformed fibroblasts in culture consisted of piled-up cells of a rounded morphology, occasional foci were composed of elongated cells which he called fusiform (see Fig. 2.1). Virus harvested from fusiform-cell foci bred true for the fusiform morphology, and virus harvested from rounded-cell foci bred true for the round morphology, with rare mutation to the fusiform type. In other words, the virus determined the morphology of the cell, which was the first evidence that a specific viral gene affected malignant transformation.

Ten years later, Martin isolated mutants of RSV that were temperature-sensitive (ts) for cell transformation. When infected chick fibroblast cultures were incubated at 35°C they appeared to be typically transformed, but when they were incubated at 41°C (physiologically normal for chickens) the cells kept a normal morphology and physiology (see Fig. 2.10). If cultures transformed at 35°C were shifted up to 41°C they reverted to normal, while cultures shifted from 41°C to 35°C assumed a transformed phenotype within a few hours. Furthermore, virus production occurred at both the permissive and the non-permissive temperatures for transformation, indicating that this type of mutant affects cell transformation without influencing virus replication. Thus these ts mutants are temperature-sensitive for *maintaining* the host cell in a transformed state. By providing conditional malignancy in the Petri dish they have proved most valuable for analysing physiological and biochemical changes involved in transformation.

Non-conditional mutants of RSV that are unable to transform fibroblasts while retaining their capacity to replicate have also been isolated. These mutants all appear to be genetic deletions, and Vogt has called them transformation-defective (td) mutants. Double infection of cells with ts and td mutants show that there is no complementation between these types of mutants, indicating that they probably belong to a single gene. Since viral replication is not affected by this gene, it appears to determine a luxury function as far as the virus is concerned, though it is essential for malignant transformation. We may therefore call it an *oncogene*. No protein has yet been associated with this gene and its possible functions will be discussed in section 6.

Many mutations which affect replication functions have also been identified in recent years. The ts reverse transcriptase mutants, isolated by Wyke, Mason and Linial, have already been mentioned as providing evidence that reverse transcription is indeed necessary for viral replication. Other mutations with ts functions have not been unequivocally identified with particular proteins but several exist that affect the assembly and maturation of virus particles. There are also deletion mutants or variants that lack one or more structural proteins, and these will be discussed next.

5.2 Defective viruses and helper viruses

The non-defective strains of RSV are rather unusual in being able to carry enough genetic information in a single virus particle both to replicate and to transform fibroblasts. Most sarcoma viruses, including some strains of RSV, and the murine, feline and simian sarcoma viruses lack one or more genes coding for components of the virion. This is also true of many of the leukaemia viruses that cause acute malignant disease. Apparently these viruses have acquired oncogenes at the expense of losing viral genes.

The best known example of defectiveness is the Bryan strain of RSV. In the early 1960s both Hanafusa and Temin independently observed that foci of transformed cells that were propagated from high-dilution infections did not release infectious progeny. At first it was thought that these transformed cells released no virus particles at all and hence they were called non-producer cells, but later it was found that non-infectious virions are synthesised and released from the cell surface. The defective virions apparently lack envelope glycoproteins. If the transformed cells are superinfected with a leukaemia virus, infectious RSV is rescued, and now bears the envelope glycoproteins coded by the leukaemia virus. The RSV is distinguishable from the leukaemia virus because it can transform fibroblasts, and the leukaemia virus is known as a *helper virus*. This kind of *genetic complementation* in which a helper virus provides components utilised in the assembly of the sarcoma virus is known as *phenotypic mixing*. The 'wolves in sheep's clothing', as we may regard the sarcoma viruses bearing envelopes of the leukaemia virus, are known as *pseudotypes*. Figure 2.8 illustrates phenotypic mixing of defective RSV with a helper virus. Infection depends on an RSV pseudotype in the first place, but if the cell is not co-infected with a helper virus present in the RSV stock, the cell (which becomes transformed) produces defective virions only. If these cells are then superinfected with helper virus, infectious RSV pseudotypes are synthesised which bear the envelope specificity and hence the host range (see section 4.3) of the helper virus.

Other defective strains of transforming viruses are similarly complemented by phenotypic mixing. For instance, Hanafusa has studied a variant of Bryan strain RSV that is not only defective for envelope antigens but also for reverse transcription. This virus can also be rescued by helper viruses and then represents a pseudotype in respect to both envelope glycoproteins and reverse transcriptase. Many of the mammalian sarcoma virus strains are even more defective, and are unable to synthesise even non-infectious particles. Nevertheless, virus can

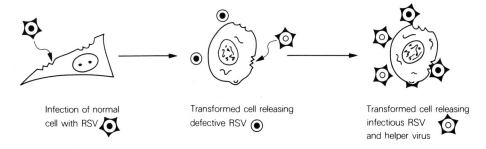

Infection of normal cell with RSV

Transformed cell releasing defective RSV

Transformed cell releasing infectious RSV and helper virus

Fig. 2.8 Defective Rous sarcoma virus is complemented by leukaemia virus which provides envelope glycoproteins to form sarcoma virus pseudotypes.

usually be rescued from transformed, genuine non-producer cells by helper viruses. Recently it has been found that phenotypic mixing of envelope antigens can occur between quite unrelated enveloped viruses, such as oncornaviruses and the rhabdovirus, vesicular stomatitis virus. Apparently the unrelated glycoprotein is assembled onto the budding virion and becomes a functional component of the envelope. Phenotypic mixing between two non-defective viruses frequently yields virions bearing a mosaic of the envelope antigens.

From the foregoing discussion, it can be seen that stocks of defective sarcoma viruses must also contain helper viruses if they are to remain infectious. The titre of the helper virus is usually in excess of the sarcoma virus. The helper viruses themselves may be oncogenic; those of defective RSV cause lymphoid leukosis. Defective leukaemia viruses also represent a mixture of transforming virus and helper virus. Thus the murine Friend virus, which causes acute erythro-leukaemia, and the avian myeloblastic leukaemia virus (AMV), which also causes acute disease, both depend on helper viruses for replication. The helper viruses by themselves cause lymphoid leukaemia which normally takes much longer to develop. In fact the standard strain of AMV contains two distinct helper viruses, belonging to different host range subgroups, one of which causes lymphoid leukaemia and kidney tumours, the other of which causes lymphoid leukaemia and osteopetrosis. From these stocks, variants can be selected which primarily cause one kind of disease or another. Field isolates of oncornaviruses also frequently comprise mixed stocks carrying defectives and helpers with different host ranges, different antigenic properties, and disease spectra of their own. Thus the natural population genetics of oncornaviruses is complex, and a particular strain, known by the disease with which it is associated, may change its properties during propagation.

5.3 Mapping the genome

Most non-conditional defective mutations of oncornaviruses are deletion mutants. This was first realised when Duesberg and Vogt showed that the RNA genome subunits of a transformation-defective (td) mutant of RSV were about 15 per cent smaller than the RNA of the non-defective RSV from which the mutant was derived. The td mutants really resemble leukaemia viruses, both in genome size and in pathogenicity. Replication-defective (rd) viruses, such as the Bryan strain RSV, also tend to have smaller genomes.

These deletion mutants have been exploited by Weissmann's and Duesberg's research groups to map the genome by means of oligonucleotide markers. When the purified 35S RNA genome subunits are digested to completion with bacterial T_1 endonuclease, which recognises guanine residues, most of the RNA will be digested to very small fragments containing less than 10 nucleotides. Occasional tracts exist in the genome where there is no G in the sequence for 10–20 nucleotides and these oligonucleotides will be preserved in the digests. The chances that any of these larger oligonucleotides have exactly the same nucleotide composition as another in the same genome is extremely small, so that they can be used as markers of specific regions of the genome. They are recognised in two-dimensional electro-phoresis–chromatography of the T_1 digests as unique fingerprints. When the fingerprint pattern of a non-defective virus is compared with that of a deletion mutant, certain fingerprints may be missing from the deletion mutant. In this way oligonucleotides can be identified with regions of the genome essential for certain viral functions or proteins; in other words, the oligonucleotides can be assigned to particular viral genes. The gene order can then be allocated by using a physical fractionation technique. Because the oncornavirus genome has a poly-A tract at the 3' end of the molecule, the RNA will hybridise to poly-U. If the genomes are sheared into many fragments, and then passed through a poly-U–sepharose column, only those fragments bearing poly-A will bind to the column. These may be eluted from the column and separated into different size classes. The smallest size fragments contain only the sequences that are immediately adjacent to the poly-A tract at the 3' end, larger fragments will contain more sequences and the largest represent most of the genome. Now the different size classes of poly-A-containing fragments may be digested with T_1 enzyme and fingerprints prepared. The oligonucleotides that are situated nearest the poly-A tract should be present in all the fingerprint patterns, while those that are located near the 5' end of the genome will be present in digests of only the largest fragments. In this way, the location of the oligonucleotides may be assigned in a linear 3' to 5' order. Combining this knowledge with the allocation of the oligonucleotides to specific genes, a reliable genetic map can be constructed.

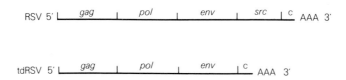

Fig. 2.9 Genetic maps of Rous sarcoma virus and a transformation-defective mutant.

Figure 2.9 shows the gene order of Rous sarcoma virus, and of a transformation-defective (td) deletion mutant. Near the 3' end of the genome is a constant (c) region of unknown function which is present in all viable deletion mutants. Then comes the *src* gene, standing for *sarc*oma function, because this region is necessary for transformation of fibroblasts in culture and for induction of sarcomas *in vivo*. The td mutant replicates perfectly well without *src*, and induces lymphoid leukosis *in vivo*; it has the genome size and biological properties of a leukaemia virus. Adjacent to the *src* gene is *env*, which codes for the *env*elope glycoprotein; this is the gene that is deleted from the Bryan strain RSV discussed earlier. The td mutants can serve as helper viruses for rd (replication-defective) viruses; hence the two kinds of deletion mutants are complementing. The gene *pol* (*pol*ymerase) codes for the reverse transcriptase and the gene *gag* codes for the precursor polypeptide that becomes cleaved into the internal virion proteins. It has only recently been shown that *pol* is 3' to *gag*, because the further away from the poly-A tail the gene is situated, the less accurate is the oligonucleotide method of mapping.

Now that complete, high-fidelity DNA proviruses can be made *in vitro*, it will soon be possible to make a more detailed structural analysis of the oncornavirus genome using restriction endonucleases, as described for papovaviruses in Chapter 1.

5.4 Genetic recombination

Vogt showed in 1971 that related strains of oncornavirus undergo high-frequency genetic recombination following mixed infection. We have already seen that mixed infection of sarcoma and leukaemia viruses yields phenotypically mixed progeny which have assembled each others components, and that defective viruses depend on this kind of complementation for their continued propagation. Vogt studied the progeny of mixed infections between non-defective avian sarcoma viruses of host range A and leukaemia viruses of host range B. Most of the progeny were phenotypically mixed, which can mask the genetic identity of the virus, but by cloning the viruses their genetic properties may be revealed. The sarcoma virus was cloned by propagating virus from single foci of transformation after infecting fibroblast cultures with high dilutions of the mixed virus stock. About 20 per cent of the RSV clones exhibited host range B in place of A, and this property was stably inherited in these clones. This meant that among the progeny of the mixed infection there were many non-defective sarcoma virions which inherited the *src* gene from the sarcoma virus and the *env* gene (which determines host range, as discussed in section 4.3) from the leukaemia virus parent.

At first it seemed that this was an example of reassortment of genetic markers on different genome segments, as occurs with influenza viruses. It was not clear at that time whether RSV carried 2 to 3 genome segments, each bearing different genes (like different chromosomes in eukaryote cells) or whether the virions carried 2 to 3 copies of the same segment (i.e. it was diploid or triploid for a single chromosome). If the recombinant sarcoma virus had one segment inherited from a sarcoma parent and one from the leukaemia parent, segments of the reassorted virus should be distinguishable by size, the segment derived from the leukaemia virus being smaller than the sarcoma virus segment. Duesberg and Vogt found only one size of segment in the recombinant virus, being that characteristic of non-defective sarcoma viruses. Estimations of genome complexity, based on molecular hybridisation and on oligonucleotide frequency, indicated a unique sequence of about 10 000 bases, which is roughly equivalent to a single subunit (3×10^6 daltons). From these data, it was concluded that the genome of RSV contained two copies of one kind of segment. This signified that the genome segments of the recombinant virus, bearing the transformation marker of one parent and the host range of the other, represented a genuine covalent recombinant RNA molecule. Later this was shown for several recombinant clones derived from parents possessing distinguishable oligonucleotides. Recombinant viruses not only possessed phenotypic markers derived from both parents but also carried oligonucleotides characteristic of each parent. Indeed, mapping oligonucleotides in recombinants according to distance from the $3'$ poly-A terminus revealed which regions of the genome were derived from each parent, and showed that many recombinant viruses had more than one crossover site.

Thus mixed infections of related oncornaviruses yield progeny with genuinely recombinant genomes. The mechanism of recombination is not well understood and its high frequency seems puzzling. Weiss, Mason and Vogt have suggested a two-stage process in the formation of recombinants, based on the discovery of RSV particles that behave as unstable heterozygotes. We studied recombination between RSV and the inherited viral genome of chicken cells. It was conceivable that recombinants could arise either if the provirus of the exogenous RSV integrated into the host chromosome adjacent to or within the site of the endogenous viral genome (for integration is a recombinational event) or if progeny virus picked up RNA genome segments from both parents. The latter appeared to be the case because recombinants only occurred after propagation of RSV in cells that were transcribing endogenous viral RNA (see section 4.2). Therefore we think that, first, heterozygous virus particles, which bear one genome segment from each parent, are formed. During the next replicative cycle a true recombinant is generated when the DNA provirus is formed.

Recombination allows rapid exchange of genetic information between related viruses. The host range may be changed by recombination independently of an oncogene or a gene coding for internal virion proteins, which complicates the population genetics of oncornaviruses and also their transmission. The separate lines of transmission depicted in Fig. 2.7 are an oversimplification, for as we have seen, an infecting virus may recombine with endogenous viral genes. This raises the possibility that oncornaviruses may also pick up cellular genetic information. Indeed, viral oncogenes appear to have been derived in this way.

6 Oncogenesis

6.1 Acute and slow oncogenesis

How do oncornaviruses transform cells into the malignant state? It is useful to distinguish those viruses that carry oncogenes such as the *src* gene of RSV from those that show no evidence of specific genes involved in transformation. As far as we know, the majority of naturally-occurring oncornaviruses belong to the second category. The mammary tumour virus of mice, and the lymphoid leukaemia viruses of cats, mice and chickens apparently do not induce transformation as soon as they infect the appropriate target cell. Rather, they are present in the target tissues for many weeks or months before malignancy develops. When a tumour does appear, it is usually clonal, i.e. all the tumour cells are derived by mitosis from a single transformed cell, even though the virus is actively replicating throughout the target tissue, and often in other tissues too. We may refer to this type of virus as a slowly-oncogenic virus. We have little or no idea how the virus promotes the transformation of these occasional cells. It is possible that these viruses have no specific oncogenic effect, acting merely as mutagens, increasing the likelihood that a malignant clone will arise. Such a non-specific effect, does not, however, explain the high degree of target cell specificity of the eventual malignancy – for instance, the thymic lymphoma of mice in contrast to the bursal lymphoma or leukaemia of chickens – nor does it explain why some actively replicating oncornaviruses are evidently not oncogenic at all, and why others are associated more with autoimmune disease than with malignancy.

In contrast to the slowly-oncogenic viruses, the acutely-oncogenic viruses appear to carry specific oncogenes, genes which are necessary for initiating and maintaining the transformed state of the cell. With the exception of the non-defective strains of Rous sarcoma virus, the acutely transforming viruses are defective in replication, depending on helper viruses for their propagation, as described in section 5.2. Apparently these viruses have acquired oncogenes (*onc*) at the expense of losing viral genes. Acutely-transforming viruses typically transform the host cell into a neoplastic state after one cell-division cycle following infection. When helper virus is also present, as is usually the case, the transforming virus will replicate, infect neighbouring cells and in turn transform them. Thus the tumours that develop are usually not clonal, but grow as much by recruitment and transformation of new host cells as by mitosis of those that have already been transformed. This has been studied in some detail for the growth of Rous sarcomas in chickens and for the development of erythroblastic leukaemia in mice infected with Friend leukaemia virus. Focal colonies of erythroblastic cells can be distinguished in the spleens of mice within a few days of Friend virus inoculation, but these incipiently leukaemic colonies are not clonal in cellular origin.

6.2 Properties of transformed cells

One of the major problems of cancer research is that we still do not have a clear idea of the characteristic properties of cancer cells or what the malignant state really means. Of course, experimental oncogenesis with viruses may help to elucidate these properties, but it also raises doubts as to whether what we think to be an important feature of malignancy might not be crucial to understanding cancer at all. This is particularly worrisome in studying the transformation of cells in culture by acutely-oncogenic viruses, which surely represents an unusual situation. The problem is whether the degree of analysis gained by this special situation justifies its distance from 'real life' disease. I think that it does.

What do we mean by cancer? Basically cancer is a derangement of cellular growth and organisation. Because tumour cells proliferate when normal tissue would be quiescent, and even more because malignant cells migrate to new sites and set up secondary colonies, cancer may be regarded as a developmental disease. Until we understand more about normal growth control, cell proliferation and migration, and intercellular recognition, we shall not gain much insight into cancer. Yet cell culture systems and acutely-transforming viruses provide us with tools that allow us to mimic the more complex *in vivo* situation in a simplified system. Embryonic fibroblasts transformed in culture by sarcoma viruses show many of the features of cancer cells. Whereas normal fibroblastic cells assume an ordered arrangement in the Petri dish, transformed cells are more haphazardly arranged (see Fig. 2.1); they proliferate to higher densities, will grow in less aerobic conditions, and will grow in suspension when normal fibroblasts depend on anchorage and spreading on a solid substratum.

The plasma membrane of transformed cells has many changed properties. It is more fluid, so that membrane glycoproteins easily float laterally; this is reflected in the sensitivity of cancer cells to agglutination by plant lectins because the membrane glycoproteins aggregate into patches. There is an increased rate of transport across the cell membrane of ions, nucleotides, sugars, etc., changes in cyclic nucleotide metabolism, and increased activity of membrane projections, such as ruffles and microvilli. None of these features is unique to cancer cells; some are characteristic of proliferation in general, or of cells in a particular stage of the cell division cycle. In many cancers new cell-surface components are evident. If they are antigenic, the tumour may be susceptible to immunological attack, but this does not appear to be an important mechanism in the natural control of potential tumours. Many new components actually represent proteins expressed at an earlier stage of development or differentiation (foetal antigens) and cancer may be regarded as arrested or retrogressive development.

One of the most interesting features of cancer cells currently under study is the loss or absence of cell-surface components present in normal cells. Chief among these is a glycoprotein of about 250 000 daltons which is a major component of the surface of fibroblasts and other solid tissue cells. It is called LETS protein, standing for *l*arge, *e*xternal, *t*ransformation-*s*ensitive. LETS protein may act as an intercellular glue and it may also have a transmembrane role in the organisation of cytoplasmic proteins such as actin fibrils. Transformed cells probably synthesise LETS protein, but it is not conserved at the cell surface, apparently being shed into the plasma or the culture fluid. This loss of LETS protein may be a result of another activity of malignant

(a)   (b)

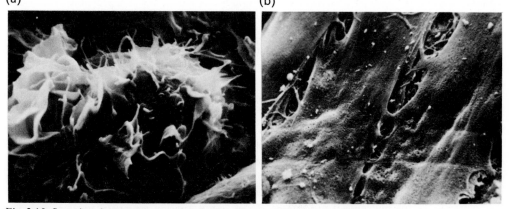

Fig. 2.10 Scanning electron micrographs of cells infected with a mutant of Rous sarcoma virus that is temperature-sensitive for cell transformation (X 3 500). (a) Transformed cell incubated at 35°C. (b) Revertant cells 24 h after shift to 41°C. (Photographs by A. Boyde and R. Weiss.)

cells, the enhanced production of a cell-surface protease. This protease cleaves plasminogen to active plasmin, thereby setting off a cascade of proteolytic reactions. By activating proteolysis, cancer cells may increase their invasiveness of surrounding normal tissues.

The synthesis of plasminogen activator and the loss of LETS protein, again, are not unique to malignant cells, but are characteristic of normally invasive cells, such as macrophages, and of non-adhesive cell surfaces, the lumen border of some epithelia, for example. It is difficult to find any one property that is exclusive to malignant cells. Cancer cells do, however, usually show an abnormal karyotype; chromosomal anomalies and rearrangements occur and may reflect functional genetic changes. Oddly enough, there is little or no evidence that this happens in cells infected with acutely-transforming oncornaviruses, though of course new genetic material, the provirus, is added to the cell.

6.3. Is there a virus-coded oncoprotein?

Nearly all the physiological changes of tumour cells discussed in the previous section feature in the transformation of fibroblasts by sarcoma viruses. And all of them apparently depend on the function of the *src* gene. Figure 2.10 shows scanning electron micrographs of fibroblasts infected with a transformation temperature-sensitive mutant of RSV incubated (a) at the permissive temperature for transformation and (b) at the non-permissive temperature. The rounded cell in a, with its many membrane ruffles, would in the living state exhibit the biochemical and physiological characteristics of the cancer cell, while the flattened cells in b, though infected with RSV, look and behave like normal fibroblasts.

How can one viral gene affect all these characteristics of malignant cells? Obviously nearly all of them must be secondary effects, so we must try to identify the primary effect. It can be argued that the *src* gene does not actually code for a polypeptide, but specifies a site of integration at which the virus deranges the cellular gene expression, or is transcribed into RNA species that interfere with cellular processes. But these are unattractive models as, in almost all other systems in prokaryotes and eukaryotes where temperature-sensitive mutants have been isolated and examined, they code for proteins which cannot function at the non-permissive temperature. This is likely to be true of the *src* ts mutants too. Therefore, it is most important to attempt to identify the *src* gene product, the oncoprotein that transforms the cell.

Many laboratories are currently seeking the oncoprotein. As the *src* gene is not required for viral replication, the oncoprotein will not be found in the virion, and it is probably synthesised only in small amounts in the transformed cell. The most likely sites of interaction of a viral-transforming protein in the host cell are the chromatin, where it may cause changes in gene expression, or the cell membrane, where it may interfere with cell recognition mechanisms. In cells infected with DNA tumour viruses, a new nuclear antigen is evident, which binds to DNA, for instance the T antigen of papovaviruses (see Ch. 1). This antigen also represents an essential early functional protein for viral replication. But no nuclear antigen has been found in association with oncornavirus transformation. We might, therefore, look at the cell surface, where an antigen of avian oncornaviruses has been identified in specific association with the transformed state. This is known as the tumour-specific surface antigen, or TSSA for short. It appears to be virally coded, because it is expressed on the cell surface of different species, e.g. chickens and rats. It is a glycoprotein of about 100 000 daltons, which is larger than the coding capacity of the *src* gene of RSV (about 40 000 daltons), though the exact contribution of carbohydrate to the molecule is not known. TSSA is not a virion structural protein, and it is only expressed on cells transformed by avian oncornaviruses. However, it does not appear to be exclusive to sarcoma viruses. Myeloid leukaemia cells transformed by AMV also express TSSA.

While AMV may have its own *onc* gene, it does not carry homologous sequences to *src*.

Without a suitable immunological marker, it will be difficult to identify oncoproteins. One approach is to translate viral RNA *in vitro* and to compare the products of RSV and transformation-defective RSV. This has not succeeded so far; with the location of *src* near the 3′ end of the genome, translation will be much less efficient than for *gag*, located near the 5′ end, where translation is initiated, as internal initiation sites may not be recognised. Another approach is to seek normal cellular functions that may be mimicked by *src*. For instance, Todaro, De Larco and Cohen have suggested that the *src* function of murine sarcoma virus (MSV) is analogous to epidermal growth factor (EGF), a specific mitogen that stimulates proliferation of epithelial and fibroblastic cells, as does MSV. This synthesis of small amounts of EGF in cells that would normally not be proliferating would trigger off the cell-division cycle. Normal cells have specific receptors for EGF on the cell surface which can be estimated by measuring the adsorption of radioactively labelled EGF. Whereas normal mouse or rat fibroblasts, and cells transformed by SV40 or chemical carcinogens, express EGF receptors, MSV-transformed cells do not bind EGF, suggesting that their receptors might be blocked by endogenously synthesised EGF from the MSV provirus. EGF is a polypeptide of 6 000 daltons and could easily be coded by a part of the MSV *src*. This would not explain the overall behaviour of transformed cells, but one important property, unregulated cell division, would be accounted for.

Different acutely-transforming viruses probably carry different *onc* genes; even the *src* sequences of different sarcoma viruses are not closely related. According to Todaro's model, different transforming viruses may mimic different specific growth factors, which would help to explain the tissue-specificity of some transforming viruses. If oncoproteins are functionally related to normal cellular growth or migratory factors, we might also expect to see sequence homology. While this cannot yet be studied at the protein level, there is, indeed, some evidence that viral oncogenes are related to host cell sequences.

6.4 Origin of oncogenes

Recently, specific hybridisation probes have been prepared for the *src* sequences of Rous sarcoma virus. Stehelin *et al.* (1976) prepared complementary DNA (cDNA) from RSV RNA by reverse transcription and hybridised the cDNA with excess RNA prepared from a transformation-defective deletion mutant of the RSV (see Fig. 2.9). After exhaustive hybridisation to tdRSV RNA, the only cDNA sequences remaining unhybridised were those specifically transcribed from the *src* gene. These $cDNA_{src}$ sequences could then be used as a specific probe for homologous *src* sequences in normal and transformed cells. Stehelin and his colleagues found that normal chicken DNA contained *src* sequences. They were also present in all other species of birds tested. Examination of the melting temperatures of hybrids formed between viral $cDNA_{src}$ and normal cellular DNA indicated mismatching of sequence homology, with increasing divergence according to taxonomic distance from chickens. Nevertheless the cellular *src* sequences are relatively well conserved in evolution. The host *src* sequences do not appear to be genetically linked with endogenous viral genomes. Their function is unknown, and it will be interesting to see whether they are related to specific growth factors. The $cDNA_{src}$ can also be used to monitor *src* RNA expression in cells. Low-levels of *src* RNA have been found in all proliferating tissues, with no obvious correlation with particular differentiating tissues or with spontaneous and chemically-induced malignancy.

Similar studies utilising cDNA probes are being conducted with MSV, but they are more difficult to interpret because one cannot generate the isogenic td mutants necessary for precise cDNA$_{src}$ preparation from a virus that is already replication-defective. However, MSV cDNA can be exhaustively hybridised to its helper virus, and the remaining sequences studied. Two strains of MSV originated after passage of mouse leukaemia viruses in rats, and these strains contain nucleotide sequences homologous to normal rat cells. It seems likely that avian and mammalian sarcoma viruses have evolved when leukaemia viruses picked up host genetic sequences and incorporated them into the viral genome. Thus viral oncogenes may be derived from diverse host elements which would help to explain the variety of acutely-transforming viruses. These viruses promise to yield exciting results that may change our concepts of malignancy and normal growth control.

7 Human RNA tumour viruses

Despite intensive research, the presence of oncornaviruses as natural infections of humans remains equivocal and elusive. Human tumours do not produce virus particles and the incidence of cancer in time and space does not resemble an infectious disease, so there would seem to be little reason to implicate viruses in human cancer. But we should bear in mind the various modes of transmission and examples of non-productive infection associated with animal oncornaviruses. Moreover, there are numerous reports of circumstantial evidence suggesting the presence of oncornaviruses in human tissues:

a Occasional particles resembling oncornaviruses have been observed in human malignant tissue.

b Molecular hybridisation studies indicate the presence of cytoplasmic RNA in human leukaemic tissues that shows some homology to either or both baboon and woolly monkey C-type viruses.

c Radioimmunoassays of oncornavirus antigens indicate the presence of related antigens, at least of interspecies sites, in human leukaemia.

d Perhaps the most telling indirect evidence is of reverse transcriptase activity detected in human leukaemia cells which is inhibited by antibody specific to the woolly monkey and gibbon group of C-type viruses.

Four independent laboratories have recently claimed the isolation and propagation of C-type viruses from long-term cultures of human tissues and there are also two reports of B-type viruses associated with human mammary cancer. All the C-type virus isolates are so closely related to the woolly monkey virus, that it has been suggested that they might represent laboratory contaminations of the latter. It seems improbable that this should have happened in four different laboratories; alternatively it is possible that the woolly monkey virus itself is really a human virus because the only woolly monkey from which a virus has been isolated was a human family pet. The gibbon C-type viruses are also related to the human/woolly monkey viruses. There are several isolates from widely separated gibbon colonies. In each case, however, some gibbons in these colonies have been inoculated with human tissues (malarial blood, diseased brain, etc.) so it is uncertain whether the oncornaviruses are natural to gibbons or acquired from humans.

The two putative human C-type virus isolates that have been studied most extensively — those reported by Gallagher and Gallo, and by Panem and Kirsten — are remarkable in comprising two distinct virus populations, one resembling the woolly monkey virus mentioned already, the other being closely related to the endogenous virus of baboons. Perhaps infectious virus is only produced in circumstances of double infection as with complementation of defective viruses. There is no doubt that the baboon virus is a genuine virus of that host species and one wonders why a closely related virus should also be associated with human tissue. But this is the virus that long ago became transmitted to cats (see section 4.4), and it might also have been transferred to humans. Molecular hybridisation of DNA from many normal human tissues does not reveal sequences that are closely homologous with woolly monkey or baboon viruses, indicating that if those viruses are indeed associated with humans they are not genetically transmitted. The possibility of congenital transmission remains, however, especially as virus particles have been reported in human placentae, and the isolate studied by Panem and Kirsten came from cultures of foetal cells.

While there is no doubt that the 'human' oncornavirus isolates are typical C-type viruses (Fig. 2.3 represents Gallo's isolate), many virologists remain sceptical about their origin. Even if they do prove to be natural infections of man, as I am inclined to believe, their relation to disease is obscure. Leukaemias and lymphosarcomas are under suspicion of having a viral aetiology, if only because that is the case in most animal species, but the disease most implicated with C-type viruses is systemic lupus erythematosus, a disease involving autoimmune haemolytic anaemia and glomerular nephritis. An analogous disease of the NZB strain mice is strongly associated with a murine xenotropic virus. Serological studies of humans are as controversial as the reports of virus isolations. Some groups claim that most adult people have antibodies that react with woolly monkey and baboon viruses, whereas other groups have difficulty in confirming these observations. It is most likely in my opinion, that C-type viruses are present in most people as non-clinical infections. It is too early to say whether there is any association of antibody titres with disease.

While we may look forward to rapid developments in the elucidation of human oncornavirus infections, it is not surprising that a long-lived species like ourselves, should, if we harbour tumour viruses, keep them latent for long periods. This is certainly the case with DNA virus infections, such as various herpes viruses and papovaviruses, which may remain latent for decades and only reappear during immunological stress. We need not be surprised if human oncornaviruses are equally or more elusive. Even if oncornaviruses eventually prove to be of little consequence to human cancer, studying them as model systems for oncogenesis and cellular genetics will continue to be rewarding.

References

General references

TOOZE, J. (Ed.) (1973) *The Molecular Biology of Tumour Viruses.* Cold Spring Harbor Laboratory.
    A second edition will be published in 1978. This is the standard text book on tumour viruses.
BECKER, F. F. (Ed.) (1975) *Cancer: A comprehensive treatise, Vol. 2, Viral Carcinogenesis.*
    This volume includes review articles by J. M. Bishop and H. E. Varmus on the molecular biology of RNA tumour viruses, H. Hanafusa on avian oncornaviruses, and M. M. Leiber and G. J. Todaro on mammalian oncornaviruses.
VOGT, P. K. (1977) 'Genetics of RNA tumour viruses', in *Comprehensive Virology, Vol. 9,* (Eds. H. Fraenkel-Conrat and R. R. Wagner.)
    This is a long, detailed, yet clear review of oncornavirus genetics.
BALTIMORE, D. (1976) 'Viruses, polymerases, and cancer', *Science,* **192**, 632–6.
    Baltimore's Nobel Prize lecture provides an excellent perspective of the subject.

TEMIN, H. M. (1976) 'The DNA provirus hypothesis. The establishment and implications of RNA-directed DNA synthesis', *Science,* **192,** 1075–80.
    Temin's Nobel Prize lecture describes the evolution and implications of his hypothesis.
COLD SPRING HARBOR LABORATORY (1975) *Symposia on Quantitative Biology, Vol. 39, Tumor Viruses.*
    Contains many useful original and review papers on tumour viruses.

Selected references mentioned in text

BALTIMORE, D. (1971) 'Expression of animal virus genomes', *Bacteriol. Rev.,* **35,** 235–41.
BENVENISTE, R. E. and TODARO, G. J. (1974) 'Evolution of type-C viral genes; inheritance of exogenously acquired viral genes', *Nature,* **252,** 456–9.
HILL, M. and HILLOVA, J. (1972) 'Virus recovery in chicken cells tested with Rous sarcoma cell DNA', *Nature New Biol.,* **237,** 35–9.
JAENISCH, R. (1976) 'Germ line integration and Mendelian transmission of the exogenous Moloney leukemia virus', *Proc. Natl. Acad. Sci. USA,* **73,** 1260–4.
MARTIN, G. S. (1970) 'Rous sarcoma virus: a function required for the maintenance of the transformed state', *Nature,* **227,** 1021–3.
STEHELIN, D., VARMUS, H. E., BISHOP, J. M. and VOGT, P. K. (1976) 'DNA related to the transforming gene(s) of avian sarcoma viruses is present in normal avian DNA', *Nature,* **260,** 170–3.
TEMIN, H. M. and RUBIN, H. (1958) 'Characteristics of an assay for Rous sarcoma virus and Rous sarcoma cells in tissue culture, *Virology,* **6,** 669–88.
TODARO, G. J. and HUEBNER, R. J. (1972) 'The viral oncogene hypothesis: new evidence', *Proc. Natl. Acad. Sci. USA,* **69,** 1009–15.
TODARO, G. J., DE LARCO, J. E. and COHEN, S. (1976) 'Transformation by murine and feline sarcoma viruses specifically blocks binding of epidermal growth factor to cells', *Nature,* **264,** 26–31.
WEISS, R. A., MASON, W. S. and VOGT, P. K (1973) 'Genetic recombinants and heterozygotes derived from endogenous and exogenous avian RNA tumour viruses', *Virology,* **52,** 535–52.

3

Control of DNA synthesis and cell division in bacteria

R. James
School of Biological Sciences, University of East Anglia

1 A system for studying differentiation

One of the most exciting unresolved problems of modern biology is the search to explain cellular differentiation. How is it that embryonic cells are programmed to differentiate into muscle cells, liver cells, brain cells, etc. and thus make up the characteristic multicellular organs of animals? So far, study of this problem has been largely constrained to the generation of hypotheses. As yet we have no hard biochemical evidence for a single compound or group of compounds being key regulators of eukaryotic differentiation. In order to resolve this intractable problem considerable attention has been focused upon the study of differentiation in apparently 'simpler' systems.

The ordered morphological and biochemical events which make up the bacterial cell-division cycle constitute a 'simple' system in which to investigate differentiation at the cellular level as a vehicle for generating ideas applicable to the problem of differentiation at the multicellular level. This hypothesis is largely based upon the principle of universality, that regulatory mechanisms which are successful in ordering the periodic events of the bacterial cell cycle will also be found in the apparently more complex process of differentiation in eukaryotic systems. There are also a number of other good reasons for using bacterial cell division as a model system, including (a) the rapid growth rates of bacteria in defined medium; (b) the relative ease of synchronisation of bacterial cell division cycles; (c) a well-characterised genetic system and the availability of mutants in particular cell cycle events. In addition the spectacular advances of prokaryotic molecular biology in elucidating the fine detail of individual metabolic pathways have generated considerable momentum towards an understanding of how these pathways are integrated with the growth and physiology of whole cells. It is as if, having completed several scattered sections of a jigsaw puzzle, we are now looking for the pieces which connect them together.

Analysis of a cell-division cycle must seek to establish knowledge of: (1) the easily recognisable landmarks of the cycle, both morphological, e.g. cell division, and biochemical, e.g. DNA replication; (2) the relationship between the landmarks of the cycle, e.g. whether they are strictly sequential in nature or can occur independently; and (3) the genes which control the separate landmarks of the cycle. When we can describe the specific role of individual gene products essential for any specific cell cycle landmark then we can begin to study the factors which determine the timing of synthesis of those gene products – the 'biological clock' which regulates cell division. It is anticipated that the principal mechanisms of such a biological clock will have wide application in biology, even in differentiation at the multicellular level.

In this chapter I will seek to describe our progress in piecing together the regulatory mechanisms of the bacterial cell-division cycle, concentrating to a large extent on two of the major landmarks of the cycle, namely, the initiation of DNA replication and cell division. Since, partly for historical reasons, much of the work in the area of bacterial cell division has used the Gram-negative bacterium *Escherichia coli,* I will limit this chapter to a consideration of this organism with reference to work with other species where appropriate.

2 The *E. coli* cell-division cycle

The morphological landmarks of the *E. coli* cell-division cycle are relatively few. Basically a newly divided cell doubles its mass during the cycle before dividing. Cell division occurs by the formation of a central septum which is distinguishable by light microscopy in its late stages of formation and by electron microscopy at a significantly earlier stage of formation (Fig. 3.1). An unusual feature of the morphology of the *E. coli* cell cycle is that the doubling in mass is achieved solely by extending the length of the long cylindrical axis of the cell. Although *E. coli* cells growing at different growth rates have a different cell width, at any specific growth rate there is no change in cell width during the cell cycle. This imposes obvious constraints upon theories of how *E. coli* cells accommodate an increase in mass by surface extension of the envelope, the structure which maintains shape and integrity in spite of the high internal osmotic pressure of bacterial cells.

2.1 Synchronisation techniques

Since much of the experimental work on cell division in *E. coli* is facilitated by using synchronous cultures with cells being at a similar cell cycle stage, I will digress briefly to consider methods of obtaining cell synchrony.

To be useful a synchronisation technique must be able to generate a sufficient number of cells for biochemical analysis at a similar stage of the cell-division cycle, without introducing artifacts due to metabolic distortion. Early methods relied upon 'induction' synchrony in which the temporary exposure of a population of exponentially growing cells to an inhibitor which blocks cell division at a specific stage in the cell cycle results in the piling up of cells at the block point. Upon release of the division block, cells resume progress towards division in a synchronised manner. It is apparent that to achieve good synchrony the exposure time to the inhibitor must be at least equal to the interdivision time; it is this that leads to considerable metabolic distortion of cells synchronised by an induction method. Cells close to the block point when the inhibitor is added are not comparable to cells which have just transversed the block point.

The more recently introduced methods of 'selection' synchrony rely upon a physical separation of the smallest cells, i.e. the newly divided cells, in an exponentially growing culture. This is obviously preferable for avoiding artifacts due to metabolic distortion. The two specific methods of selection synchrony in common use are membrane elution and density gradient separation (Mitchison 1971).

Membrane elution is based upon the adsorption of large numbers of exponentially growing *E. coli* cells to the surface of a large diameter membrane filter so that all 'binding sites' on the filter are occupied by cells. After inversion of the filter, fresh, warm growth medium is slowly

(a)

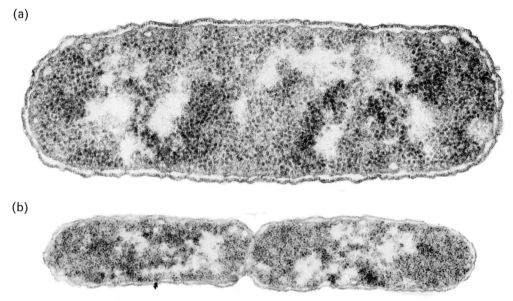

(b)

Fig. 3.1 Electron micrographs of (a) newly divided *E. coli* cell (× 85 000), (b) dividing *E. coli* cell (× 80 000).
Two membranes are clearly visible, together with the clear areas representing the genome.

run through. Once all the surplus unbound cells are eluted, a regular stream of small, newly divided cells are released in the eluant. These represent the daughter cells which are not attached to the filter (Fig. 3.2). On adsorption to the filter, cells close to division rapidly release the unattached daughter cell into the eluate, whereas cells which had divided before adsorption to the filter release a daughter cell into the eluate after one generation period. With a sufficiently large filter a reasonable number of synchronous cells may be obtained by this technique.

Density gradient separation of the smallest cells in an exponentially growing culture is achieved by centrifuging a concentrated cell suspension into a linear sucrose gradient. Theoretically, cells at any age can be obtained by suitable fractionation of the resulting broad band of cells. However in practice, good synchrony is only achieved by removing the top 5 per cent of the band, representing the smallest cells. Improving the synchrony obtainable by this technique, by removing a smaller percentage of the band of cells, results in a considerable sacrifice in the numbers of cells available for subsequent biochemical analysis.

2.2 Biochemical landmarks of the *E. coli* cell cycle

Using cells synchronised by the membrane elution technique a general model has been proposed comprising (*a*) the biochemical landmarks of the *E. coli* cell cycle, and (*b*) the relationships between the biochemical landmarks (Cooper and Helmstetter 1968). The main features of the model are the existence of three sequential periods, each biochemically distinct, which are essential requirements of normal cell division (Fig. 3.3). The first, called the I period, consists of events which are necessary for the capacity to initiate a round of DNA replication. The second, or C period, is the time required for the entire replication of the *E. coli* genome, and the third, or D period, is the time between termination of DNA replication and the subsequent cell division.

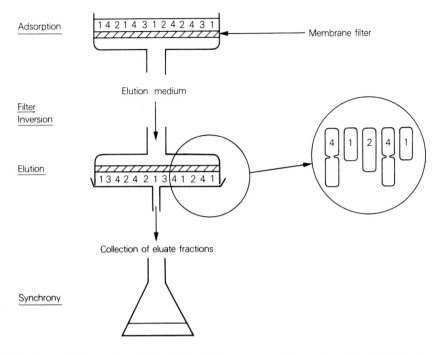

Fig. 3.2 Membrane elution synchronisation of *E. coli*. An exponentially growing culture (individual cells at different cell cycle stages are represented by numbers 1 to 4) is adsorbed onto the surface of a membrane filter. After inversion, fresh warm growth medium is used to elute the newly divided daughter cells (numbered 4). A magnified view of part of the membrane filter demonstrates the rationale for selective elution of newly divided cells.

The main experimental evidence to support this model comes from indirect measurements of the rates of DNA synthesis during the cell cycle of synchronous cultures of *E. coli*. In a typical experiment an exponentially growing culture was pulse-labelled with radioactive thymidine to label DNA in cells in which DNA synthesis was in progress. After synchronisation the amounts of label were monitored in the newly divided cells eluted from the membrane filter. The first cells eluted are the result of division of a 'mother' cell which was very close to division during the period of pulse-labelling, whereas cells eluted at later times are the products of division of

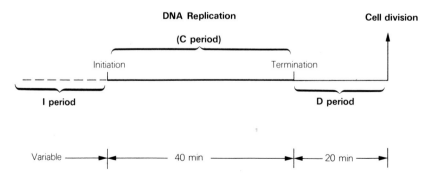

Fig. 3.3 General model for the *E. coli* cell cycle. For convenience the three sequential periods of the cell cycle are represented as a linear sequence of events.

'mother' cells which were further from division at the time of labelling. If DNA synthesis is continuous and taking place at the same rate throughout the cell cycle then the specific activity of thymidine (counts/cell) should be constant in the eluted cells for at least one generation period.

In fact, a two-fold jump in the specific activity of thymidine was observed at a definite stage of the cell cycle, thus implying that the rate of synthesis of DNA doubles at this point. Assuming that the rate of movement of a replication fork is constant as it traverses the genome, this result is consistent with a doubling in the number of replication forks (initiation sites) at a specific cell-cycle time. By varying the growth media of the *E. coli* cultures used in the experiment, it was demonstrated that the cell-cycle time at which the doubling in rate of DNA synthesis occurred was dependent on growth rate. Detailed analysis of the relationship between the time of the initiation of replication and physical cell division demonstrated that, for cultures growing with a doubling time between 20 and 70 min., the C period was constant at 40 min. and the D period was constant at 20 min.

The apparent paradox that duplication of DNA, a vital cell component, requires a minimum time at least twice as long as the time for *E. coli* cell doubling in very rich media was resolved when multiple replication forks were observed, i.e. a new round of replication may start before an existing round terminates. In a cell growing with a doubling time of 60 min. a round of DNA replication commences at the beginning of the cell cycle and is completed at 40 min., followed by the 20-min. D period before septation occurs. With a doubling time of 40 min. DNA synthesis is continuous because rounds of replication start in the middle of the previous cell-division cycle; at division, each daughter cell thus contains a half-replicated genome (Fig. 3.4).

The logical conclusion of a fixed sequential C and D period is that after DNA replication is initiated the cell will divide 60 min. later. It could thus be argued that the timing of initiation of DNA replication is a key, cell-division regulatory event. Thus, we wish to answer the

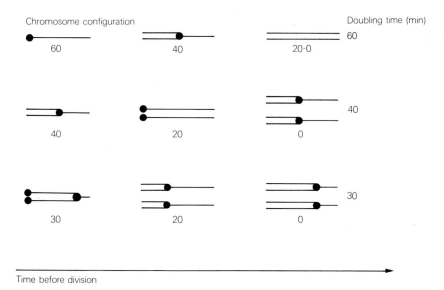

Fig. 3.4 The chromosome configuration of *E. coli* cells, growing with a doubling time of 30, 40 or 60 min., represented at various times before division. The single lines represent the *E. coli* circular genome and the closed circles replication forks. With a doubling time of 40 min. DNA replication is completed 20 min. before division and replication initiates at that time for the next cell division cycle

question: What is the nature of the 'biological clock' which determines the frequency of initiation of DNA replication?

Summary

The *E. coli* cell-division cycle consists of three distinct sequential periods, two of which, the C and D periods, are of constant duration, irrespective of growth rate, for cultures growing with a generation time between 20 and 70 min. Thus a critical variable in cell division is the timing of the beginning of the C period of the *E. coli* cell cycle.

3 Regulation of DNA initiation

Initiation of replication of *E. coli* DNA starts at a precise point, the origin, located at 67 min. on the circular map of the genome. DNA replication then proceeds bidirectionally until the replication forks meet at the chromosome terminus 40 min. later.

Measurements of average cell mass and DNA content of cultures of *Salmonella typhimurium* growing exponentially in a variety of culture media at different growth rates gave results similar to those shown in Fig. 3.5. The exponential relationship between mass/cell or DNA/cell and growth rate implies that fairly simple rules correlate these variables and allows for the testing of any model formulated to explain the biological clock for DNA initiation.

Once it was apparent that the variable rate of DNA synthesis at different growth rates is determined by changes in frequency of initiation, rather than by a change in the velocity of replication forks, several hypotheses were proposed based upon the premise that the frequency

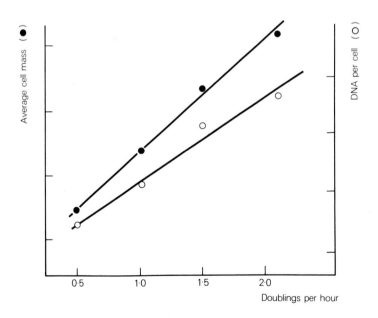

Fig. 3.5 The relationship between average cell mass, DNA content per cell and growth rate of exponentially growing cultures of *Salmonella typhimurium*. (Redrawn from data of Schaechter, M., Maaløe, O. and Kjeldgaard, N. O. (1958), *J. Gen. Microbiol.*, **19**, 592–606.)

of initiation is related to frequency of doubling of cell mass. Perhaps the most elegant version of this hypothesis comes from Donachie (1968).

Using data derived from work with *S. typhimurium* on the relationship between cell mass at division and growth rate, together with the rule that $C + D = 60$ min., Donachie was able to calculate the mass at initiation of DNA replication for cultures of *E. coli* growing with a doubling time between 20 and 60 min. (Fig. 3.6). The perhaps surprising result is that the mass of initiation M_i is a constant for doubling times between 30 and 60 min., and twice this value for doubling times between 20 and 30 min.

Since it can readily be derived from the Cooper–Helmstetter model of the *E. coli* cell cycle that the number of initiation sites N_i is two for doubling times between 30 and 60 min. and 4 for doubling times between 20 and 30 min., there is thus a constancy between M_i and N_i for any growth rate between 20 and 60 min. which could constitute a biological clock.

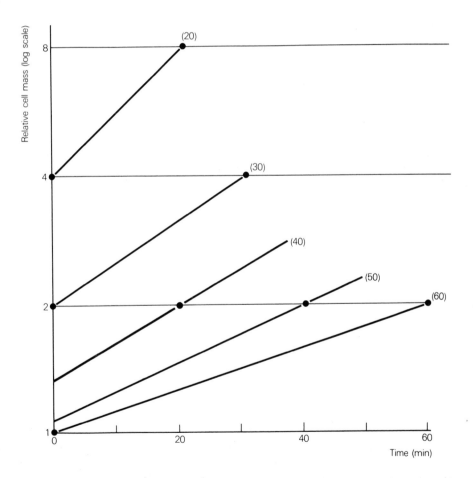

Fig. 3.6 The increase in mass of individual *E. coli* growing at different growth rates (doubling times of 20–60 min.). The initial mass at time 0 (represents time of last division) is proportional to the experimentally recorded average mass of a population of cells growing at the same growth rate. The next division occurs after the initial mass has doubled. Since periods C and D equal 60 min. it is possible to calculate the time of initiation of DNA replication relative to cell division, shown by closed circles. The initiation masses at different growth rates are the same, or multiples of this constant value. (Redrawn from Donachie (1968), *Nature*, **219**, 1077–9.)

3.1 Initiator proteins

The simplest explanation of the constancy between cell mass at initiation and the number of origins is that an initiator protein has to be built up during the I period, at a rate proportional to the growth rate, to reach a critical concentration at the time of initiation. There is some support for this hypothesis in experiments in which protein synthesis is inhibited, either by chloramphenicol, or by starvation of an auxotroph of *E. coli* for an essential amino acid. DNA replication in progress proceeds to termination when protein synthesis is inhibited, but no new rounds of replication are initiated. This presumably reflects (*a*) the stability of the protein(s) (enzymes) required for *continuing* DNA replication which are already present in the cell before inhibition of protein synthesis, and (*b*) the fact that protein(s) required for *initiation* of DNA replication are not stable and/or must be synthesised for each new round of replication.

Inhibition of protein synthesis for a sufficient time, as described above, results in alignment of the chromosomes of *E. coli*. Thus, upon resumption of protein synthesis, the majority of cells initiate rounds of DNA replication over a short period. Re-addition of chloramphenicol at various times after the resumption of protein synthesis has indicated that there may be more than one initiator protein. A low concentration of chloramphenicol, although still capable of inhibiting most protein synthesis, stopped DNA initiation only if added at least 30 min. before initiation would have occurred (Lark and Lark 1964). At later times the addition of a low concentration of chloramphenicol was capable of blocking initiation up to 15 min. before it would have occurred. It is suggested that this implies a role for two initiation proteins whose synthesis is sensitive to different concentrations of chloramphenicol and occurs at different times in the I period. Similar results were found in identical experiments performed with synchronous cultures *E. coli* obtained by membrane elution. They cannot therefore have been the result of metabolic distortion induced during the period of amino-acid starvation necessary to align the chromosomes.

The identity of the initiator protein(s) is at present unknown but an approach to this problem has become apparent with the isolation and characterisation of temperature-sensitive DNA initiation mutants, or *dnaA* mutants of *E. coli*. At the permissive temperature (30°C) such mutants grow normally. However, upon transfer to the non-permissive temperature (42°C) they exhibit characteristics identical to those observed during inhibition of protein synthesis in an exponential culture of *E. coli*, i.e. DNA replication in progress terminates but no new rounds of replication are initiated. It is tempting to speculate that the temperature-sensitive phenotype of *dnaA* mutants is the result of thermolability of the *dnaA* gene product, a protein which presumably has a key role in initiation of DNA replication. Thus the biochemical characterisation of the *dnaA* gene product may add much to our understanding of initiator proteins.

3.2 Role of the cell envelope in regulating DNA initiation

The *E. coli* cell envelope consists of (*a*) the cytoplasmic or inner membrane, (*b*) the rigid peptidoglycan or murein layer and (*c*) an outer membrane. The inner membrane appears to be a typical 'fluid bilayer' and contains many enzymes of the electron transportant chain, together with permeases and enzymes responsible for the synthesis of the components of the exterior layer. The murein layer consists of covalently-linked repeating units of *N*-acetyl glucosamine — *N*—acetyl muramic acid — pentapeptide (Fig. 3.7) assembled in a layer one molecule thick which covers the whole cell (murein sacculus). The outer membrane of *E. coli* appears to be a much more rigid structure than the inner membrane and also contains many fewer protein species. One of the major proteins of the outer membrane (in terms of numbers of molecules per cell) is

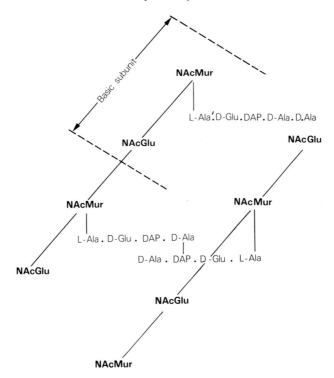

Fig. 3.7 Murein configuration in *E. coli*. The basic murein subunit of *N*-acetyl glucosamine (NAcGlu) linked to *N*-acetyl muramic acid (NAcMur) via a glycosidic bond, with a pentapeptide side chain (L-alanine–D-glutamic acid–diaminopimelic acid–D-alanine–D-alanine), is inserted into the growing murein sacculus. Covalent cross-linking of two pentapeptide chains (via D-alanine and diaminopimelic acid) occurs by an enzyme-catalysed cleavage of the terminal D-alanine and formation of a peptide bond.

the lipoprotein (Braun 1975). Approximately 1 in 3 lipoprotein molecules of the outer membrane are covalently attached to murein by displacement of alanine from the diamino-pimelic acid molecule. The precise role of this 'bound lipoprotein' is unknown, although it has been suggested as constituting a means of coordinating the synthesis and growth of the murein and outer membrane.

The replicon hypothesis (Jacob, Brenner and Cuzin 1963) suggested that DNA–membrane (presumably cytoplasmic membrane) attachment could have a key role in the regulation of DNA initiation and in symmetrical segregation of replicated DNA into the two daughter cells (Fig. 3.8). Although there is some electron microscopy evidence, especially from Gram-positive bacteria, which is at least consistent with this hypothesis (Ryter 1968), the strongest support comes from biochemical analysis of DNA–envelope complexes.

Gentle lysis of *E. coli* cells with a mixture of lysozyme and detergent followed by sucrose gradient centrifugation results in the isolation of a rather broad band containing DNA, protein and lipid. This rapidly sedimenting complex (RSC) may be seen by electron microscopy to consist of DNA associated with a piece of cell envelope. Treatment of RSC with phospholipase leads to the loss of DNA from the complex, thus suggesting some specificity of DNA envelope attachment involving cell envelope phospholipid.

Experiments involving pulse-labelling to identify small sections of the genome have demonstrated that the origin of DNA replication in *E. coli* is associated with the cell envelope in RSCs. A culture of *E. coli leu* was synchronised with respect to initiation of DNA replication by

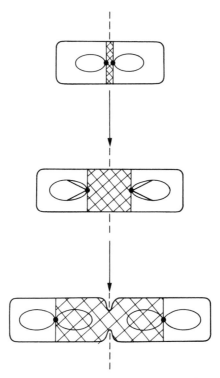

Fig. 3.8 The replicon hypothesis of DNA membrane attachment in *E. coli*. At an early cell cycle stage the proposed DNA membrane attachment site duplicates and thus signals the events which lead to the initiation of DNA replication. Cell membrane growth between the duplicated attachment sites (shown by cross hatching) leads to segregation of the replicating genomes before subsequent cell division. (Redrawn from Jacob *et al.* (1963), *Cold Spring Harbor Symp. Quant. Biol.*, **28**, 329–48.)

leucine starvation for a period sufficient to allow all rounds of replication in progress to terminate. Upon addition of leucine, DNA replication rapidly initiates, and therefore a short pulse-label with ^{3}H-thymidine effectively labels the origin of DNA replication. Isolation of RSCs from part of a culture labelled in this way revealed the presence of a considerable amount of ^{3}H activity. As the pulse is 'chased out' of the remaining culture by the addition of excess unlabelled thymidine and subsequent growth, isolation of RSCs still revealed a large amount of ^{3}H bound to the cell envelope. Suitable controls were performed to show that non-origin regions of the genome do not persist in RSCs during a chase. The interpretation of these results, that the genome origin of replication is firmly fixed to the cell envelope, is obviously consistent with the replicon hypothesis; however we require much more specific information about the nature of the binding of DNA to the cell envelope before we can be certain that the *in vitro* results obtained with RSCs correspond with the *in vivo* situation in *E. coli*.

Using a technique for the isolation of RSCs which did not require the use of detergents, Heidrich and Olsen (1975) have analysed the protein composition of the cell envelope of RSCs. One specific protein of 80 000–90 000 daltons was found to disappear from the RSC upon incubation with trypsin, a treatment which also released DNA from the complex. If one assumes that the two events are related, then the trypsin-sensitive RSC protein could be a DNA-binding protein.

The *in vivo* demonstration of DNA-binding properties for any specific protein is difficult to achieve; however, an indirect method uses treatments which lead to cross-linking of protein molecules in close proximity to DNA. Ultraviolet irradiation of *E. coli* cells which had incorporated bromodeoxyuridine into their DNA results in such cross-linking. Subsequent analysis by electrophoresis has revealed two proteins cross-linked to DNA, one of which has a molecular weight of 80 000 daltons and may be identical to that observed by Heidrich and Olsen. Much further work is required to determine the pattern of synthesis and incorporation of this envelope protein during the cell cycle and its precise relationship with the initiation of DNA replication at the origin of the genome.

Summary

The timing of the initiation of DNA replication in *E. coli* appears to correlate with a doubling in cell mass. Since it is difficult to see how a bacterial cell can readily titrate its mass, the simplest explanation involves the build-up of an initiator protein (at a rate proportional to cell mass) until it reaches a critical level to allow initiation to proceed. Some very preliminary evidence suggests a role for a cell-envelope protein in DNA attachment at the origin of replication.

4 Regulation of cell septation

Termination of DNA replication is an essential prerequisite for subsequent cell septation in *E. coli*. The addition of an inhibitor of DNA replication, e.g. nalidixic acid, to an exponentially growing culture of *E. coli* allows a limited amount of cell division for approximately 20 min. A similar experiment with synchronous cultures demonstrates that nalidixic acid will only inhibit cell division if added some 20 min. before cell division would have occurred (Fig. 3.9), a time which coincides with the end of the C period of the cell cycle. Thus cells which had terminated DNA replication upon addition of nalidixic acid went on to divide. In rapidly growing cultures in which DNA replication is continuous, i.e. one round starts before the previous round terminates, the same experimental observations were made. Thus septation is specifically coupled to the round of DNA replication which initiated 60 min. previously, and is unaffected by inhibition of subsequent initiated rounds of replication. This has obvious implications in constructing an hypothesis to explain the link between termination of DNA replication and subsequent septation. It has been generally assumed that a signal is generated at the termination of DNA replication which initiates a series of events during the D period of the division cycle of *E. coli* which result in septation.

4.1 Termination of DNA replication and septation

An appealing mechanism of generating a signal for septation at the termination of replication is based on the idea that duplication of a gene(s) located at or near the genome terminus allows the production of sufficient of its gene product(s) to catalyse the initiation of septation. This mechanism implies a 'gene dosage' effect whereby the synthesis of 'termination protein(s)' is critically dependent upon the number of copies of its relevant gene(s). In general, the synthesis of most of the enzyme systems of bacteria so far studied seems to be relatively independent of gene dosage. Their rate of synthesis appears to be repressed by control systems which affect the rate of transcription of messenger RNA.

   An alternative proposal is that some physical event associated with termination of DNA replication is the signal for subsequent septation. Recently, considerable excitement was generated by the hypothesis that at termination the *E. coli* genome is released from the cell

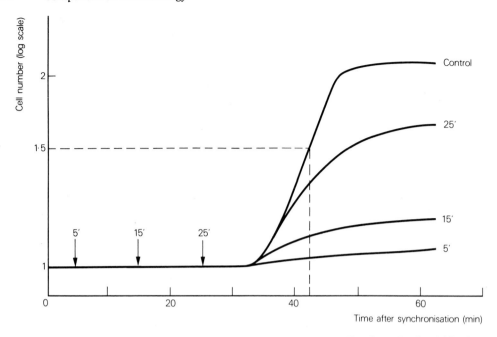

Fig. 3.9 The effect of nalidixic acid on division of a synchronous culture of *E. coli*. At 5, 15 and 25 min. after synchronisation, nalidixic acid was added to part of the culture and subsequent cell division followed by Coulter counting. Division of the untreated control part of the culture occurred at 43 min. after synchronisation (determined from the mid-point of the synchronous increase in cell number). The addition of nalidixic acid, more than 20 min. before division would have occurred (i.e. at 5 min. or 15 min. after synchronisation) blocks subsequent division.

envelope. Experimentally it was demonstrated that the majority of *E. coli* DNA was isolated as an RSC on a sucrose gradient when using a higher temperature of lysis than that described above (see section 3.2). However, after inhibition of protein synthesis, *E. coli* DNA was only obtained in an envelope-free form, i.e. a form which sediments much more slowly. Since it was known that inhibition of protein synthesis prevents initiation of new rounds of DNA replication but does not affect rounds of replication in progress, it was suggested that at termination the genome is released from the cell envelope attachment site.

It has since become apparent that these results are an artifact of the experimental conditions used. After inhibition of protein synthesis, *E. coli* DNA actually sediments faster through a sucrose density gradient, i.e. it does not release from the cell envelope at termination of replication.

It should be borne in mind that there is no compelling reason why the regulation of septation imposed by DNA replication should be 'positive', i.e. a gene product which actively initiates the events of the D period leading to septation. It is entirely feasible to generate a model of 'negative' control in which septation is not directly dependent upon DNA replication, but in which inhibition of DNA replication generates a signal which blocks septation. Such a system could constitute a defence mechanism of *E. coli* cells to prevent septation before complete genome replication, a potentially lethal event for the cell.

At present we cannot distinguish between positive and negative control; however, the latter system would necessitate a second biological clock to control the frequency of septation.

A separate clock for regulating septation has been proposed on the basis of experiments which separated the C period of the *E. coli* cell cycle from the parallel period of protein synthesis known also to be necessary for subsequent septation (Jones and Donachie 1973). Under these conditions septation started less than 5 min. after termination of DNA replication, instead of the normal D period of 20 min. Thus it was concluded that many of the events of the D period are not dependent upon termination of DNA replication.

These authors therefore propose that septation is largely controlled by a pathway of events (responsible for the synthesis of 'division proteins') which runs in parallel with genome replication. By assuming that the pathway for synthesis of division proteins is initiated at the same time as the pathway of DNA replication the problem of how the two pathways are integrated to achieve balanced growth is avoided. At present there is little experimental evidence to support this model but it is appealing in that it has strong similarities to one recently proposed for the yeast cell cycle (Hartwell 1974).

A significant factor which is delaying progress in elucidation of the mechanism which coordinates termination of a round of DNA replication and subsequent septation is the failure to isolate cell-division mutants of *E. coli* which are blocked in intermediate stages of septation. Although there are undoubtedly numerous biochemical events occurring during the D period, and presumably some of the many cell-division mutants of *E. coli* so far described are specifically blocked in one of these events, the final morphology of nearly all such mutants is filament formation with no evidence of partly completed septa. Without mutants defective in *septum completion* rather than *septum initiation* it is difficult to dissect the precise relationships between termination of replication and the biochemical events of the D period of the cell cycle.

4.2 The cell envelope and septation

The events of the D period of the *E. coli* cell cycle result in a change in the pattern of synthesis of the cell envelope. Instead of the integrated extension of all three layers of the cell envelope during cell elongation, septation is associated with invagination of the inner membrane and murein layers of the cell envelope (James *et al.* 1975). It is only very late in septation that the outer membrane becomes invaginated and leads to physical separation of two daughter cells.

Since the *E. coli* outer membrane is covalently cross-linked to the rigid murein layer by large numbers of lipoprotein molecules it is assumed that the observed physical separation of the outer membrane and other cell envelope layers during the early stages of septation may reflect a localised change in this covalent lipoprotein coupling. At present the (presumably) enzyme systems for the covalent attachment of lipoprotein and murein are poorly understood. Although it is far from certain that the localised separation of outer membrane from murein is an important event in septation, it is of interest because it provides a system in which spatial regulation of enzyme activity can be studied. It should be possible to determine whether the lipoprotein—murein attachment system is (*a*) distributed equally throughout the cell envelope, thus implying a localised modification at the septum site, or (*b*) localised at the presumptive septum site with a cell-cycle-dependent switching mechanism to regulate its activity. We cannot yet distinguish these two possibilities for generating a localised cell envelope structure, but the techniques are rapidly becoming available with which to explore this problem, e.g. a specific antibody against lipoprotein molecules.

Because of the possibility that lipoprotein has an important role in septation, attempts have been made to isolate lipoprotein-deficient mutants of *Escherichia coli.* The procedure used for the isolation of such mutants relies upon the earlier observation that the lipoprotein lacks five common amino acids, including tryptophane, histidine and proline. Thus starvation of *E. coli trp, his, pro* for these amino acids results in continued synthesis of lipoprotein, even though synthesis of the majority of cellular proteins is inhibited. If, during the amino-acid starvation, a radioactively labelled amino acid (which is found in lipoprotein, e.g. arginine) is present in the culture medium, then cells which are synthesising lipoprotein will incorporate large amounts of radioactivity. On prolonged storage, cells which had synthesised lipoprotein will tend to die because of the lethal effects of the incorporation of highly radioactive arginine. Cells which had not synthesised lipoprotein (or any other proteins) during the period of amino-acid starvation would survive storage. Thus lipoprotein mutants in the original culture are enriched.

Lipoprotein-deficient mutants isolated by a suicide selection technique (Torti and Park 1976) were found to be inhibited in septation, as predicted. However, careful analysis of the composition of the cell envelope proteins of these mutants and of the parent strain suggests that the lipoprotein may not be the only envelope protein missing in the mutants. It is conceivable that the mutation which exists could prevent the orderly integration of a group of proteins, any one or more of which could be essential for septation.

Further progress in our understanding of the role of lipoprotein in cell division will await the isolation of a specific deletion mutant in the structural gene coding for the transcription of lipoprotein mRNA. An analysis of the morphology of such a mutant under non-permissive growth conditions will provide much information on the possible cellular role of lipoprotein.

4.3 Murein synthesis and septation

Since the murein layer is the principal shape-maintaining structure in *E. coli*, any consideration of the role of the cell envelope in septation must consider the mechanisms by which the rigid murein layer is signalled to invaginate at the precisely located septum site on receipt of a signal generated by termination of DNA replication. Obvious questions to be answered are:

(*a*) Is the murein of the septum different from that of the cylindrical side walls?

(*b*) What enzymes are involved in synthesising murein for cell elongation and for septation?

(*c*) If there are different enzyme systems for murein elongation as a rod and for the formation of hemispherical murein caps during septation, how are they switched on and off during the cell cycle?

Answers to some of these questions have now started to become apparent because of the use of β-lactam antibiotics as specific probes of septation.

Penicillins, together with cephalosporins, are members of the β-lactam group of antibiotics. Typical β-lactams, such as Penicillin G, inhibit septation of *E. coli* at low concentrations, thus resulting in the formation of long multinucleate filaments, and cause cell lysis at high concentrations. The two morphological effects of Penicillin G have been suggested to reflect the existence of at least two modes of murein synthesis: (1) septum specific, and (2) basic murein synthesis for cell expansion. It is assumed that the enzyme systems responsible for the two modes differ in their sensitivity to Penicillin G. At low concentrations of Penicillin G enzyme system (1) might be inhibited, thus preventing septation but not expansion, whereas at higher concentrations enzyme system (2) might also be inhibited, thus inducing lysis as a result of the weakened murein layer of the envelope.

Recently a third morphological effect on *E. coli* has been observed with FL1060 (Mecillinam®), an amidino derivative of Penicillin G. Over a very wide concentration range FL1060 induces the formation of spherical cells, without specifically inhibiting cell division or inducing lysis. We have suggested that the timing and morphological result of the action of FL1060 on *E. coli* is consistent with inhibition of murein synthesis necessary for initiating a cell-cycle pathway necessary for cell envelope expansion as a rod (James *et al.* 1975).

Support for the hypothesis that synthesis of septum murein is distinct from murein for cell elongation as a rod comes from several sources. The synthesis and distribution of murein can conveniently be followed by analysing the incorporation of a radioactively labelled unique murein component, diaminopimelic acid (DAP), using high-resolution autoradiography (Schwarz *et al.* 1975). For easier analysis of incorporation patterns the complete murein layer of the envelope (murein sacculus) was isolated, still retaining its typical rod shape, after a short pulse-label with ^{3}H-DAP. Areas containing labelled murein then showed up as dark silver grains against the lighter background of the murein sacculus, thus allowing measurements of the number of grains per cell and of their distribution in the complete sacculus.

A computer analysis of the distribution of grains in sacculi isolated immediately after pulse-labelling an exponential culture of *E. coli* with ^{3}H-DAP revealed distinct zones of incorporation, e.g. in sacculi from cells undergoing septation there was a heavily labelled central zone, accompanied by two lateral zones which were generally broader and less intensely labelled. In sacculi isolated from cells of a temperature-sensitive, cell-division mutant of *E. coli* during incubation at the non-permissive temperature (e.g. when filamenting), the central growth zone was missing but not the two lateral zones. After shifting the culture back to the permissive temperature, cell division resumed coincidently with the reappearance of the central zone of incorporation. The authors therefore suggest that the central zone of incorporation of diamino-pimelic acid represents septum-specific murein synthesis, whilst the two lateral zones reflect murein synthesis associated with cell elongation as a rod.

4.4 Penicillin-binding proteins and septation

Although β-lactam antibiotics inhibit, *in vitro*, several enzymes thought to have a role in murein synthesis, i.e. carboxypeptidase I, transpeptidase, and endopeptidase (Fig. 3.10), there is no conclusive evidence to support a specific role *in vivo* for any of these enzymes in septation or elongation. An alternative approach to dissecting the β-lactam-sensitive enzyme systems for murein synthesis *in vivo* has been developed, based upon their binding of ^{14}C-Penicillin G (Spratt 1975).

The addition of radioactively labelled Penicillin G to whole cells or to purified envelopes of *E. coli* allows the subsequent separation by acrylamide electrophoresis of six protein bands which co-electrophorese with labelled Penicillin G, e.g. six penicillin-binding proteins. These are assumed to reflect the existence of six separate penicillin-sensitive enzymes, all bound to the inner membrane, inhibition of one or more of which gives rise to a characteristic morphological response.

By binding unlabelled FL1060 to whole cells or isolated envelopes of *E. coli* before addition of labelled Penicillin G, Spratt was able to show that FL1060 specifically bound to only one of the six penicillin binding proteins (binding protein 2). Analysis of the pattern of binding of ^{14}C-Penicillin G to an FL1060-resistant mutant of *E. coli* revealed that binding protein 2 was missing in the mutant. The FL1060-resistant mutant grew as spherical cells in the absence of FL1060, thus suggesting a role for the FL1060-binding protein 2 in cell shape.

The observation that cephalexin, a cephalosporin member of the β-lactam group of antibiotics, inhibited septation of *E. coli* but did not induce immediate lysis (or rounding-up), whatever the concentration used, suggested that cephalexin was a more specific probe of murein synthesis associated with septation than Penicillin G. Pre-binding of cephalexin to *E. coli* envelopes before the addition of labelled Penicillin G indicated that penicillin-binding protein 3 appeared to bind cephalexin specifically. A cephalexin-resistant mutant of *E. coli* appeared to lack this same binding protein, thus implying a role for penicillin-binding protein 3 in septum-specific murein synthesis.

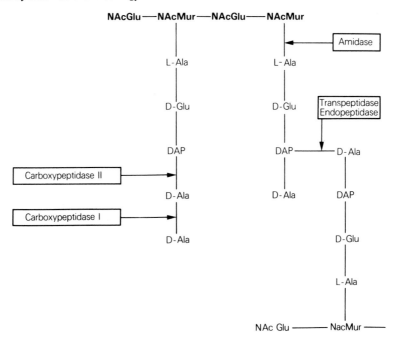

Fig. 3.10 Some enzymes with roles in the biosynthesis or degradation of murein. Carboxypeptidase I and II cleave the terminal or sub-terminal alanine. Transpeptidase catalyses the formation of covalent cross links between diaminopimelic acid and D-alanine. Endopeptidase cleaves these covalent cross-links, whereas amidase cleaves the entire peptide side-chain. Although the individual *in vitro* enzymatic activities are known, the precise *in vivo* role of each enzyme, or the interactions between them, are largely unknown.

Thus Spratt has proposed that the morphological effect of any β-lactam antibiotic is determined by its affinity for the penicillin-binding proteins of the inner membrane of *E. coli*. In the future it is clear that the isolation, purification and identification of the specific enzymatic activity of the principal penicillin-binding proteins will facilitate our understanding of the different types of murein synthesis associated with septation or cell elongation.

4.5 Murein-synthesising enzymes and the cell cycle

A different approach to the problem of the mechanism of regulation of septum-specific murein synthesis is to study, during the cell cycle, the activity of enzymes known to have a role in murein biosynthesis in an attempt to identify an enzyme which exhibits a cyclic variation in activity corresponding with the time of septation. Using cultures of *E. coli*, synchronised by density gradients, Beck and Park (1976) determined the activity of three murein-synthesising enzymes – carboxypeptidase I, carboxypeptidase II and amidase (see Fig. 3.10). The enzyme assays used required making the cells permeable immediately before the assay (by toluene treatment) so that relatively large synthetic substrates could penetrate the cell and be acted upon by the enzymes present. The activities of carboxypeptidase I and amidase were constant throughout the division cycle but detectable carboxypeptidase II activity varied, being highest at the time of septation by a factor of three. Further support for a correlation between carboxypeptidase II activity and septation came from studies using a temperature-sensitive, cell-division mutant of *E. coli*, BUG 6. After a shift from 32°C to 42°C the specific activity of carboxypeptidase II in BUG 6 cells was only one-third of that in control cells maintained at the permissive temperature 32°C. Some 15 min. after return to 32°C, septation of the BUG 6 cell rapidly resumed and was accompanied by a tenfold increase in the specific activity of carboxypeptidase II.

Although premature, it is possible to speculate that carboxypeptidase II may have an important role in triggering septum-specific murein synthesis, or some other modification of murein associated with septation. It is unlikely that the penicillin-binding protein 3 of Spratt is related to carboxypeptidase II because the former is firmly bound to the inner membrane whereas the latter appears to be a periplasmic enzyme. It is therefore apparent that much detailed study is still required before we can unravel the precise role of the multitude of enzymes acting on murein and their relationship to events of the cell cycle.

Summary

Our knowledge of the cell septation in *E. coli* is limited to a tentative dissection of some of the enzymes of murein biosynthesis into those likely to be involved in septum-specific murein synthesis or in murein synthesis involved with shape maintenance and cell elongation. Until we define the precise roles of murein biosynthetic enzymes in the generation of a septum we cannot anticipate much progress in determining the mechanisms of the spatial and temporal regulation of septation.

5 Reflection

It is hoped that this chapter has illustrated the problems of trying to understand the apparently trivial problem of the integration of morphological and biochemical events of the bacterial cell-division cycle. Although simple to work with experimentally, bacteria still provide a wealth of intellectual problems.

Those of us who work with prokaryotic systems can say with some conviction that until we understand how they 'tick' we cannot hope to understand the further complexity of eukaryotic systems with their multiple chromosomes. Our colleagues, who have already taken the step into the void of eukaryotic differentiation, justly reply that since we have had such limited success with our simple, single-chromosome systems they may as well start to ask trivial questions of their tissue culture systems, etc. It will be interesting to look back in ten years' time to see which cause has advanced the furthest.

References

BECK, B. D. and PARK, J. T. (1976) 'Activity of three murein hydrolases during the cell division cycle of *Escherichia coli* K12 as measured in toluene-treated cells', *J. Bacteriol.,* **126**, 1250–60.

BRAUN, V. (1975) 'Covalent lipoprotein from the outer membrane of *Escherichia coli*', *Biochim. Biophys. Acta,* **415**, 335–77.
   A comprehensive review of this *E. coli* major outer membrane protein.

COOPER, S. and HELMSTETTER, C. E. (1968) 'Chromosome replication and the division cycle of *Escherichia coli* B/r', *J. Molec. Biol.,* **31**, 519–40.

DONACHIE, W. D. (1968) 'Relationship between cell size and the time of initiation of DNA replication', *Nature,* **219**, 1077–9.

HARTWELL, L. H. (1974) '*Saccharomyces cerevisiae* cell cycle', *Bacteriol. Rev.,* **38**, 164–98.

HEIDRICH, H-G. and OLSEN, W. L. (1975) 'Deoxyribonucleic acid—envelope complexes from *Escherichia coli*', *J. Cell Biol.*, **67**, 444–60.

JACOB, F., BRENNER, S. and CUZIN, F. (1963) 'On the regulation of DNA replication in bacteria', *Cold Spring Harb. Symp. Quant. Biol.*, **28**, 239–48.

JAMES, R., HAGA, J. Y. and PARDEE, A. B. (1975) 'Inhibition of an early event in the cell division cycle of *Escherichia coli* by FL1060, an amidinopenicillinanic acid', *J. Bacteriol.*, **122**, 1283–92.
Electron micrographs of septation.

JONES, N. C. and DONACHIE, W. D. (1973) 'Chromosome replication, transcription and control of cell division in *Escherichia coli*', *Nature*, **243**, 100–3.
Describes experiments which lead to the authors' proposals for modification of the Cooper—Helmstetter model of the *E. coli* cell cycle.

LARK, C. and LARK, K. G. (1964) 'Evidence for two distinct aspects of the mechanism regulating chromosome replication in *Escherichia coli*', *J. Molec. Biol.*, **10**, 120–36.

MITCHISON, J. M. (1971) *The Biology of the Cell Cycle*, Cambridge University Press, London.
A text covering synchronisation techniques and a comparative description of prokaryotic systems.

RYTER, A. (1968) 'Association of the nucleus and membrane of bacteria: a morphological study', *Bacteriol. Rev.*, **32**, 39–54.
Excellent electron micrographs of both Gram-positive and Gram-negative bacteria.

SCHWARZ, U., RYTER, A., RAMBACH, A., HELLIO, R. and HIROTA, Y. (1975) 'Process of cellular division in *Escherichia coli*: differentiation of growth zones in the sacculus', *J. Molec. Biol.*, **98**, 749–59.

SPRATT, B. G. (1975) 'Distinct penicillin binding proteins involved in the division, elongation and shape of *Escherichia coli* K12', *Proc. Natl. Acad. Sci. USA*, **72**, 2999–3003.

TORTI, S. V. and PARK, J. T. (1976) 'Lipoprotein of Gram-negative bacteria is essential for growth and division', *Nature*, **263**, 323–6.
The conclusion of this paper is speculative but it illustrates the use of mutants to investigate the cell cycle.

4

The biology of plasmids

P. A. Williams

Department of Biochemistry and Soil Science, University College of North Wales, Bangor

1 Introduction

Plasmids are circular DNA molecules which exist in the bacterial cell independently of the chromosome; they are replicated and inherited by both daughter cells on cell division. In addition to carrying genes which determine a wide range of biological functions, many also specify the ability to transfer themselves from the host cell (donor) to a recipient by *conjugation,* which involves direct cell-to-cell contact. It is this feature which distinguishes conjugation from the other mechanisms by which genes can be transferred between bacteria, *transduction* and *transformation*, in which bacteriophages and naked DNA respectively are intermediates.

The sex factor F (denoting *f*ertility) in *E. coli* K-12 was the first extrachromosomal genetic element recognised, as a result of its ability to carry chromosomal genes with it into a small proportion of recipient cells during conjugation; this property is shared by many other bacterial plasmids since discovered. The sex factor F has one particular property which distinguishes it from many other plasmids, namely, the ability to integrate covalently into several possible positions in the *E. coli* chromosome, where it acts as a set of chromosomal genes during replication and cell division. This dual existence, either as an autonomous unit or integrated into the chromosome, has in the past led to it being considered as a rather distinct entity from a plasmid, and it has consequently been referred to as an episome. It is now more usual to think of F as a plasmid which has an additional phenotypic character, the ability to integrate into the chromosome: this recognises that the similarities between F and other plasmids are greater than between F and bacteriophage λ, which has also been termed an episome because of its ability to exist either autonomously or integrated.

F not only has an important position in the history of microbiology but also has been a powerful tool in bacterial genetics. Much of this importance is a direct result of its integrated state. First, its property of promoting an ordered conjugational transfer of the entire chromosome from Hfr strains (*h*igh *f*requency of *r*ecombination), in which it is in the integrated state, is one method of mapping the *E. coli* chromosome. Secondly, the F′ plasmids, which result from faulty excision of F from the integrated to the autonomous state, carrying with it some covalently linked chromosomal genes, have enabled these genes to be analysed by complementation (e.g. the use of F′lac in elucidating the lac operon), because they can be used to construct strains which are partial diploids. These properties and uses of F are covered in basic textbooks on microbiology and microbial genetics. The intention of this chapter is to describe some of the other characters which are determined by plasmid genes and to outline the importance and properties of these plasmids. F cannot be ignored because it is the most studied of the extrachromosomal elements and because many other plasmids share features in common with it. Here it will be considered where relevant, but only in the context of its autonomous state.

2 Characterisation of a function as plasmid-coded

The initial clue that a particular phenotypic character may be determined by plasmid genes is often that there is an unusually high spontaneous loss of the character, or that it is spread to or shared by other organisms from the same environment. Subsequent demonstration that the character is specified by a plasmid requires careful verification by the following criteria.

2.1 Curing

No naturally-occurring plasmids are essential to the viability or growth of the host cell under normal culture conditions: in other words, all the essential functions of a cell (energy metabolism, biosynthesis, cell division, etc.) are coded for by chromosomal genes. Plasmid-coded functions are additional to these and give the host cell a survival or growth advantage under particular environmental conditions. For example, an antibiotic-resistance plasmid (R plasmid) will enable a cell to grow in an otherwise lethal concentration of one or more antibiotics; a degradative plasmid will enable a cell to grow where there is a supply of a particular growth substrate.

It follows that in a non-selective medium (i.e. one in which the plasmid does not confer any advantage) a cell which loses the plasmid will not be at a disadvantage and will proliferate at least at the same rate as plasmid-containing cells. Although plasmids are stably inherited and replicate in synchrony with cell division, there is usually some small spontaneous loss of plasmid, presumably as a result of a defect in plasmid replication in a small number of cells. It is this spontaneous loss of a particular phenotypic character in a minority of cells in a culture which has very often given the first clue to the plasmid-specified nature of the character.

Such spontaneous loss may be very infrequent. However, experimental conditions can sometimes be used which considerably enhance the natural frequency of plasmid loss. The most common practice is to grow a culture in the presence of a chemical which produces a high frequency of plasmid 'curing'. Compounds which have been used as curing agents include the acridine dyes, ethidium bromide, mitomycin C and rifampicin, but the effectiveness of any individual compound varies greatly from plasmid to plasmid. Alternatively, changes in the physical conditions of growth, such as elevated temperatures or thymine starvation, can also result in plasmid curing. By one or other of these methods the population, after only a few generations, may consist almost entirely of cured cells.

Care must be taken with curing agents since many may also act as mutagens. Cells which appear to be cured must be screened to ensure they are not merely mutants in the appropriate function. First, they should not be altered from the parent strain in any chromosomally-determined functions (e.g. growth rate on non-selective media, prototrophy, colony morphology) and, secondly, they must not revert to the parental phenotype (which is assumed to be plasmid-specified) at a detectable frequency.

A high frequency of curing with no reversion or other character suggestive of mutation is good presumptive evidence for plasmid involvement.

2.2 Conjugational transfer

Many of the larger plasmids with molecular weights in excess of 20×10^6 carry genes which enable the host cell to transfer the plasmid by conjugation to a recipient cell; these are termed *conjugative* plasmids. Plasmids with molecular weights smaller than 20×10^6 do not seem to be sufficiently large to carry the genes necessary to code for conjugal transfer, and some larger plasmids are also unable to promote their own transfer and are consequently called *non-conjugative*. Conjugational transfer of a phenotypic character is strong evidence in favour of plasmid involvement, although, because some plasmids are non-conjugative, the absence of such transfer does not rule out the possibility of a plasmid location of the genes.

In initial experiments a cured derivative of the wild type is usually used as recipient and is conjugated with a derivative of the wild type as donor. Conjugation may be carried out by mixing donor and recipient cultures on the surface of a solid medium, but for a quantitative assessment the two cultures are usually mixed in a non-selective liquid medium for a specified period, after which the culture is serially diluted and plated on to a selective medium. Because it is only the number of *transconjugant* cells (recipient cells which have acquired the plasmid) which is required, donor and recipient cells (or donor cells alone) are genetically marked so that neither grows on the final selection medium (Table 4.1). Results are usually expressed quantitatively in terms of the *number of transconjugants produced per donor cell,* and this frequency varies from system to system from the lower limits of detection (about 10^{-8}) up to and even exceeding 1.

To confirm definitively that the transfer is by conjugation it may be necessary to demonstrate (1) that no cells with the putative plasmid phenotype arise when supernatants from donor cultures are mixed with recipients (to exclude transduction) and (2) that the apparent frequency of transfer is not altered in the presence of DNAase (to exclude transformation).

Table 4.1   Examples of experimental protocol for conjugational plasmid transfer

*Donor phenotype**	*Recipient phenotype**	*Transconjugant phenotype**	*Medium in selection plates†*
(a) pro⁻R⁺ × R⁻ ⟶		R⁺	Glucose-minimal + antibiotic (resistance to which is coded by R)
(b) CAM⁺ × strʳCAM⁻ ⟶		strʳCAM⁺	Camphor-minimal + streptomycin

* Phenotype designations: pro⁻, requirement for proline; strʳ, chromosomally coded streptomycin resistance; R⁺,R⁻, with and without an R plasmid; CAM⁺, CAM⁻, with and without a plasmid specifying the ability to grow on camphor.

† The selection medium is chosen such that transconjugants will grow, but not donor or recipient Two procedures are illustrated which can achieve this: (a) an auxotrophic donor and an unmarked recipient or (b) a wild type donor and recipient with chromosomally-determined drug resistance.

Table 4.2   Mobilisation of non-conjugative plasmid

Strains required

(a) Donor containing a 'transfer factor', pD
(b) Intermediate containing a non-conjugative plasmid, pI, with a selectable phenotype
    (e.g. a drug-resistance marker)
(c) A genetically marked recipient strain

Procedure

(a) Incubate donor and intermediate cultures together for 6 h
    Donor (pD^+) × Intermediate (pI^+) $\longrightarrow$ Intermediate (pD^+pI^+)
(b) Add culture of recipient strain and incubate for further 18 h
    Intermediate (pD^+pI^+) × Recipient $\longrightarrow$ Recipient (pD^+pI^+)
(c) Plate on to medium which selects for cells of recipient strain which
    have acquired the phenotype specified by pI

Uses

(a) As a means of transferring a plasmid or putative plasmid which is unable to mediate its own
    transfer
(b) In order to determine whether the donor strain carries a 'transfer factor' able to moblise a
    known conjugative plasmid

Some plasmids are non-conjugative and cannot effect their own transfer. Where this is the case two alternative methods can be used. One is to use a transducing phage as vector, as in the case of the staphylococcal plasmids (section 3.9). Alternatively, conjugative plasmids can act as 'transfer factors' and effect the mobilisation of non-conjugative plasmids (section 3.1), and the procedure shown in Table 4.2 can be used. In both cases care must be taken in interpreting the results because both phages and conjugative plasmids can mobilise chromosomal genes as well as plasmids. However the stable survival of a transferred chromosomal gene depends on its recombination with the recipient chromosome, whereas plasmids are autonomous and do not need to recombine. Recombination in a recipient only occurs if it has a functional recombination system (rec), the biochemical nature of which is not fully understood. Derivatives with mutations in this function (rec⁻) are unable to incorporate homologous DNA by recombination: in *E. coli*, where a more detailed genetic analysis has been carried out, several genes are known to be involved in the rec system but mutants in one of these, the *recA* gene, are the most effective and are normally used. If transfer by transduction or mobilisation occurs with equal efficiency to both rec⁻ mutants and wild type strains, then a plasmid must be involved.

2.3 Physical characterisation of plasmid DNA

Plasmids are present in a cell as closed circles of double-stranded DNA which, in the resting state, have a right-handed superhelical coil structure as if the two ends of a linear double helix were twisted several times before being covalently linked: the number of superhelical turns is

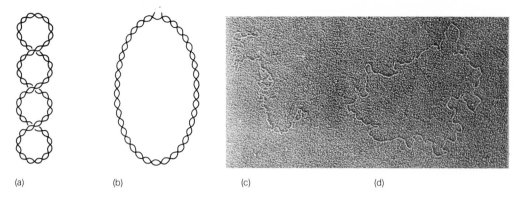

(a)   (b)   (c)   (d)

Fig. 4.1 Molecular forms of plasmid DNA. (a) and (c) Covalently closed circles (ccc) with right-handed superhelical twists. (b) and (d) Open nicked circles with one strand broken releasing the superhelical twisting. (Electron micrograph from Clowes (1972), *Bacteriol. Rev.*, 36, 361–405.)

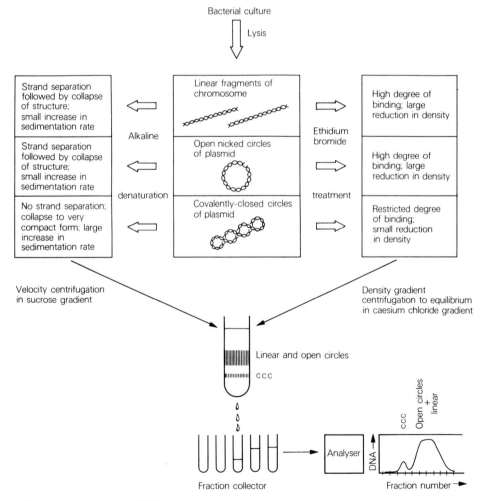

Fig. 4.2 Purification of plasmid DNA.

about one for every 400–600 base pairs ($1\cdot7$–$2\cdot5 \times 10^5$ molecular weight). This twisted conformation is known as covalently closed circular (CCC) DNA.

Gentle lysis of a culture containing a plasmid releases the plasmid DNA either as CCC DNA or in a form in which one of the chains is cleaved or 'nicked', thus releasing the superhelical twists to give an open circular form (Fig. 4.1). The chromosomal DNA is much larger than the plasmid and, in handling, becomes sheared to produce fragments of linear duplex.

The CCC form of the plasmid DNA endows it with two properties which facilitate its separation from the fragments (Fig. 4.2).

Ethidium bromide binds to DNA by intercalating between the base pairs: this causes a slight unwinding of the double helix and a reduction in its buoyant density. With CCC DNA there is a very definite physical constraint on the extent to which uncoiling can occur, which limits both the binding of ethidium bromide and the consequent reduction in density. There is no such constraint with either open nicked circles or with linear DNA; in the presence of ethidium bromide their density is reduced much more, to such an extent that the CCC DNA can be separated from them by centrifugation to equilibrium on a caesium chloride density gradient.

Another property of CCC DNA can also be exploited to enable it to be separated from linear DNA. At pH $11\cdot5$–12, the double-stranded DNA in the linear and open nicked circular forms is denatured and dissociates, and at about pH $12\cdot5$ the resultant single strands collapse to a random coil. Strand separation cannot occur with CCC DNA, and at high pH its collapsed structure is far more compact, and its rate of sedimentation is from 3–4 times greater than the single-stranded forms. Velocity centrifugation on an alkaline sucrose gradient thus enables the faster moving band of CCC DNA to be isolated.

If it can be demonstrated by either of these two methods that the wild type strain contains CCC DNA then it can be assumed to harbour a plasmid. This does not definitively prove that the phenotypic trait under investigation is coded on the plasmid, but additional evidence in support will be obtained (1) if cells of the cured derivative do not contain any plasmid DNA and (2) if the phenotypic character is regained when the purified DNA is reintroduced into the cured strain by transformation.

Isolation by ethidium bromide–caesium chloride or by alkaline sucrose gradient centrifugations also permits characterisation of the plasmid in terms of base composition and molecular weight. The latter may be measured by centrifugal analysis but more often an electron micrograph is made by spreading the preparation onto a carbon grid coated with the basic protein cytochrome C, which facilitates spreading and adsorption of the DNA molecule. Nicks are produced in the plasmid DNA during the preparation and subsequent handling, so that most of the adsorbed molecules are in the open circular form. The grids are shadowed and the contour lengths of open circles measured: it is assumed that each μm corresponds to a molecular weight of about $2\cdot1 \times 10^6$.

3 Classes of plasmid

3.1 Transfer and sex factors

In addition to being able to determine their own transfer between bacteria, conjugative plasmids can also cause the transfer of other genes from donor to recipient during conjugation. Many can transfer a non-conjugative plasmid present with the conjugative plasmid in the donor cell, in which case the latter is said to have *transfer factor* activity and to be responsible for *mobilising* the former. The mechanism of the process is not known, but it does not necessarily involve a passive co-transfer of the non-conjugative element since instances are known where a proportion of the recipient cells receive the non-conjugative plasmid but not the plasmid responsible for its mobilisation. One experimental procedure which has been used to demonstrate transfer factor activity is shown in Table 4.2; any of the recipient cells which acquire the non-conjugative plasmid pI by the end of the experiment can only have done so as a result of its mobilisation by pD. Using this test as a means of detecting the presence of transfer factors, it has been shown in a number of experiments that a very large proportion (often more than half) of natural isolates carry plasmids with transfer factor activity.

Some conjugative plasmids are also capable of causing the transfer of chromosomal genes during conjugation. This can be demonstrated by mating a prototrophic donor containing the plasmid with an auxotrophic recipient. If chromosomal genes are transferred, then a fraction of the recipient cells into which the plasmid has been transferred also become prototrophic: in these cells the chromosomal fragment transferred will have recombined with the recipient chromosome. Plasmids capable of this are often referred to as *sex factors*. Again the mechanism is unknown, but in all instances in which transfer of chromosome occurs, the plasmid is also transferred.

It was, of course, the ability to mobilise chromosomal genes which led to the original discovery of F in *E. coli* K-12. It is known that F can integrate into the chromosome and, from the resultant Hfr strains, transfer of chromosomal genes can occur with very high frequency. Some of the chromosomal mobilisation by a F^+ strain can be attributed to a very small number of Hfr cells within the population, but undoubtedly the greater part is due to a totally different mechanism which involves F in its autonomous state only. In this process there is no permanent or semipermanent association between F and the chromosome. Certainly many other conjugative plasmids which do not appear to integrate are capable of sex factor activity.

Because this ability to mobilise unlinked genes is shared generally by conjugative plasmids of all classes, transfer and sex factors should not really be considered as a separate class of plasmids. Rather they should be looked at as conjugative plasmids which either do not possess any phenotypic character other than the ability to mobilise unlinked genes or else for which no other character has yet been found.

3.2 R plasmids

If a pure culture of an antibiotic-sensitive strain is grown in gradually increasing concentrations of the antibiotic or plated on a medium containing a lethal concentration of it, spontaneous mutants can be selected which have an increased resistance. The mutations in such strains are chromosomal and usually have the target site of action of the antibiotic modified. For example, the streptomycin-resistant mutants often used as genetically-marked derivatives of a strain (see Table 4.1) are produced in this way and have modified ribosome structures.

Most antibiotic-resistant bacteria isolated from the wild are not chromosomal mutants but carry the resistance genes on plasmids (R plasmids) and the mechanism of resistance is usually quite distinct from that of a chromosomal mutant resistant to the same antibiotic. These plasmids cause considerable medical and veterinary problems because (1) many of them are conjugative and can be transferred, particularly between different genera of enteric bacteria, and (2) many of them specify resistance to more than one drug, and in some cases to as many as six or seven unrelated drugs.

3.2.1 Ecology and epidemiology

R plasmids were first discovered in Japan in the late 1950s in strains of *Shigella* during successive epidemics of dysentery. The drug resistance of clinical isolates of this organism increased from the introduction of antibiotic therapy in 1946 until by 1964, 50 per cent were simultaneously resistant to all four of the major drugs used at that time against dysentery, namely, the sulphonamides, streptomycin, tetracycline and chloramphenicol. One of the first indications of the extrachromosomal nature of the resistance was the observation that *E. coli* in patients often carried a similar spectrum of resistance to the infecting pathogen and that their resistance could be transmitted to drug-sensitive *Shigella* spp. by conjugation both in a patient and under laboratory conditions. Publication of the work stimulated considerable investigation and within a few years the widespread distribution of R plasmids was demonstrated.

Two features of the Japanese dysentery epidemics illustrate the serious medical implications of plasmid-mediated drug resistance. First, conjugational transfer can result in a drug-sensitive pathogen becoming resistant by contact with a drug-resistant non-pathogenic member of the normal gut flora. Secondly, pathogens readily acquire new R plasmid resistances during the course of epidemics, as found in many other instances. A carefully documented example which demonstrates both features was a serious outbreak of bovine *Salmonella typhimurium* infection in Britain in the early 1960s.

The causative organism was a particularly virulent strain, phage type 29. During the epidemic, there was a considerable increase in the incidence of resistance in all *S. typhimurium* strains examined. In 1961–62 only about 3 per cent were drug-resistant, but by 1964–65, this proportion had risen to 61 per cent. At the same time the spectrum of resistance in type 29 strains increased from involving only sulphonamides and streptomycin to include ampicillin, chloramphenicol, kanamycin, neomycin and tetracycline. It seems likely that one particular herd became infected with the virulent strain and served as a focus for its spread, which was undoubtedly facilitated as the organism acquired multiple resistance. The source of the new resistances was probably by conjugation with resistant *E. coli* or other normal flora which had been selected by the widespread use at that time of antibiotics as a generalised prophylactic treatment against the onset of salmonella enteritis. Not only was this outbreak of agricultural and economic importance, causing up to 50 per cent mortality in infected cattle, but it had human health implications as well. Cattle have been shown to be a major source of human salmonellosis by way of meat and milk. During 1965, at the same time as the bovine epidemic, the commonest *S. typhimurium* isolated from human infections was also type 29 and showed the same spectrum of resistance as bovine isolates. Because of this resistance, many patients failed to respond to therapy and six deaths resulted.

This example demonstrates that intensive and unrestricted use of antibiotics selects for resistant populations of normal flora which can transfer their R plasmids to any infecting pathogens. Antibiotics appear to act in three ways to aid this transfer: (1) by providing an environment in which only drug-resistant bacteria can grow; (2) by destroying most of the resident gut bacteria, which are more suited to the conditions of the gut and normally compete effectively against invading organisms; (3) as a result of disturbance of the normal microbial population, by producing conditions of pH and oxygenation which greatly improve the frequency of conjugational transfer of plasmids.

The prevalence of R plasmids under drug selection is apparent in several surveys comparing patients in and out of hospitals. For example, up to 20 per cent of the *E. coli* present in sewage from Bristol hospitals were shown to carry resistance to one or more antibiotics, whereas in the normal city sewage the figure was only about 2 per cent. One result of such intensive selection in hospitals has been an increase in the incidence of hospital-acquired infections by R plasmid-carrying strains of Gram-negative organisms which normally have little or no pathogenicity but can become opportunistic pathogens in patients in a weakened condition: bacteria such as *E. coli, Pseudomonas aeruginosa, Serratia marcescens, Klebsiella* spp. and *Proteus* spp. have all

Table 4.3　Some typical R plasmids

Plasmid	Molecular weight ($\times 10^6$)	Resistance determinants*	Incompatibility group†	fi character§	Conjugation	Copy number‖	Source
R1	62	ApCmKmSuSm	IncFII	+	+	1 to 3	*Salmonella typhimurium*, Britain
R6	65	CmKmNmSuSmTc	IncFII	+	+	1 to 3	*Salmonella typhimurium*, Germany
R100¶	58	CmSuSmTc	IncFII	+	+	1 to 3	*Shigella flexneri*, Japan
R64	79	SmTc	IncI	−	+	1 to 3	*Salmonella typhimurium*, Britain
RP4	36	ApKmTc	IncP	−	+	1 to 3	*Pseudomonas aeruginosa*, Britain
N3	33	SuSmTc	IncN		+	1 to 3	*Shigella flexneri*, Japan

* Ap, ampicillin, Cm, chloramphenicol; Km, kanamycin; Nm, neomycin; Su, sulphonamides; Sm, streptomycin; Tc, tetracycline.
† See section 4.1 for discussion of incompatibility.
§ See section 4.1 for fi character.
‖ See section 4.3 for copy number.
¶ Also called 222 and NR-1.

acquired clinical significance with their R plasmids, in particular in surgical wards and burns units.

After removal from antibiotic selection, the resistant organisms in the gut appear to be at some disadvantage, since their number drops quite rapidly from virtually 100 per cent to a much lower figure. However, it needs very small numbers of residual organisms to re-establish a resistant population on resumption of antibiotic treatment and, undoubtedly, there is a considerable gene pool of R plasmids in the normal community not under any medical treatment, as shown by the results from Bristol quoted above.

3.2.2 The molecular nature of R plasmids

The size of isolated and characterised R plasmids varies considerably but the majority of conjugative plasmids have molecular weights from 30 to 70 × 10⁶ (Table 4.3).

The plasmid DNA in bacterial strains with multiple resistance has been found in two forms which have been called *co-integrates* and *aggregates*.

In plasmid co-integrates, all the resistance genes, together with the genes necessary for transfer, are covalently linked in a single circular molecule. Experimentally, these are distinguished by all the resistance determinants being co-transferred on conjugation and simultaneously lost on curing; only one species of plasmid DNA is in the host cell. An example of a co-integrate is R100 (Cm, Sm, Su, Tc) isolated from *Shigella flexneri* in the Japanese dysentery outbreak which led to the discovery of R plasmids.

In plasmid aggregates, the various genes reside for at least part of the time on separate molecules. During conjugation, or on curing, the determinants segregate independently and a number of plasmids of different size can be isolated from the host strain. The multiple resistance described by Anderson and Lewis in one isolate of *Salmonella typhimurium* type 29 falls into this category (Fig. 4.3). This strain with phenotype (ΔSAT)⁺ was resistant to sulphonamides and streptomycin (S), ampicillin (A) and tetracycline (T) and was able to transfer them (Δ). On conjugation with *E. coli* K-12 as recipient, three different transconjugant

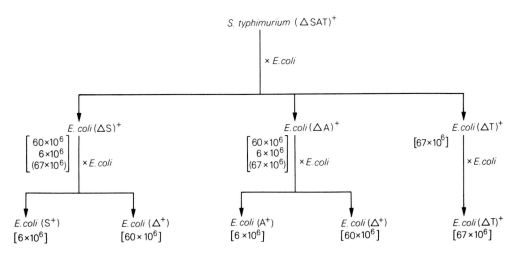

Fig. 4.3 Segregation of resistance markers from *Salmonella typhimurium* plasmid aggregate. Phenotype designation: S, sulphonamide–streptomycin resistance; A, ampicillin resistance; T, tetracycline resistance; Δ, ability to transfer or act as transfer factor.

The figures in square brackets represent the molecular weights of plasmid DNA species present; in the (Δ A)⁺ and the (Δ S)⁺ strains, the 67 × 10⁶ molecular weight plasmid is a minor component, <5 per cent of total.

phenotypes were obtained, each with a single but transmissible resistance $(\Delta A)^+$, $(\Delta S)^+$ and $(\Delta T)^+$. Using $(\Delta A)^+$ and $(\Delta S)^+$ as donors in further crosses, the drug resistance determinant and the transfer genes segregated, so that a small number of the transconjugants were only transfer positive Δ^+ or carried non-conjugative drug resistance, either S^+ or A^+. In contrast, the $(\Delta T)^+$ strain was stable and no segregation of the transfer function and resistance marker was found.

Examination of the plasmid DNA in the various strains showed at least four molecules present: a plasmid in the Δ^+ cells (M.W. 60×10^6), two non-conjugative plasmids both of molecular weight 6×10^6, specifying ampicillin- and sulphonamide–streptomycin-resistance respectively, and a conjugative plasmid in the $(\Delta T)^+$ cells (M.W. 67×10^6). This latter plasmid could be either a distinct species or a stable co-integrate of the 60×10^6 molecular weight plasmid with a tetracycline determinant T. In the $(\Delta A)^+$ and $(\Delta S)^+$ strains the conjugative and non-conjugative plasmids may have been in equilibrium between the co-integrated and the separate states, since a very few molecules (less that 5 per cent) had a molecular weight of 67×10^6, equal to the sum of the two component plasmids.

The co-integrate and aggregate states may only be extremes of a continuous spectrum of association. Several instances are recorded where a plasmid appears as a stable co-integrate in one organism, but on transfer to a different genus (particularly to *Proteus*) dissociates into two or more distinct molecules. This behaviour may be a result of the way new drug resistance genes are acquired during the infectious spread of a plasmid. One possible mechanism for this assumes that there is a pre-existing gene pool of conjugative and non-conjugative plasmids. If a conjugative plasmid is transferred into a cell containing a non-conjugative R plasmid, the presence of antibiotics would select for co-transfer of the R plasmid with the conjugative plasmid: either the two plasmids might remain separate as an aggregate, or alternatively they might undergo recombination to form a covalent co-integrate. In the latter case it is not difficult to imagine that the stability of the recombinant plasmid might be affected by the cellular environment and subsequently dissociate. Co-integrates can also result from transposition of resistance markers (section 4.3).

3.2.3 Biochemical mechanisms of plasmid-coded resistance

R plasmids coding for resistance to virtually all the clinically used antibacterial drugs have been described. They include sulphonamides, chloramphenicol, tetracycline, trimethoprim, the β-lactams (cephalosporins and penicillins) and the aminoglycoside group (streptomycin, neomycin, kanamycin, etc.). The biochemical means by which the resistance is achieved differs from one drug to another but for any particular drug the mechanism appears almost universal, irrespective of the organism in which the plasmid is isolated. Four distinct mechanisms have been found (Table 4.4).

Resistance to chloramphenicol and the β-lactam antibiotics are the result of plasmid-coded enzymes which chemically modify the antibiotic to a non-toxic form. Chloramphenicol is converted by acetyl-CoA dependent-chloramphenicol acetyltransferase, mainly to the 3-O-acetyl derivative and the penicillins and cephalosporins are hydrolysed by β-lactamases. The substrate specificities of the β-lactamases vary between plasmids, but fall into two main groups, the cephalosporinases which have little activity on penicillins, and the penicillinases which have a much broader specificity, hydrolysing all the β-lactams.

Superficially, plasmid-specified resistance to the aminoglycoside antibiotics also appears to be by detoxification, since modifying enzymes are present in R^+ strains: these enzymes are all transferases which either acetylate, phosphorylate or adenylate the drug at hydroxyl or amino groups (Fig. 4.4). A number of such enzymes have been described and each is non-specific to some extent and confers resistance to a number of related drugs. However, unlike chloram-

Table 4.4   Biochemical mechanisms of plasmid-coded drug resistance

Mechanism	*Examples*
Detoxification	Chloramphenicol, β-lactams
Permeability barrier	Aminoglycosides, tetracycline
Modification of target site	Erythromycin, lincomycin
Replacement target site	Trimethoprim, sulphonamides

phenicol and the β-lactams, no chemical modification of the drug takes place in the culture medium and some of the modified forms of the drug produced by the enzymes retain antibacterial activity. It would appear that their role is not one of detoxification. The enzymes are located in the periplasmic space between the inner and outer membranes, and it has been suggested that the antibiotic is modified once it has permeated the outer membrane and that the modified form blocks the system which would transport the drug through the inner membrane to the site of action at the ribosome: on this hypothesis, the enzymes are part of a permeability barrier.

R plasmid-mediated tetracycline resistance also appears to involve a permeability barrier: a specific and inducible protein located in the cell envelope is synthesised, but its exact role is uncertain.

Linked resistance to erythromycin and lincomycin is the only known instance in which the target site in the cell is modified. Plasmids in *Staphylococcus aureus* and *Streptococcus faecalis* conferring this resistance code for an enzyme which methylates a specific guanine on the 23S ribosomal RNA: this modification prevents binding of the drug to the ribosome, its site of action.

The fourth mechanism is found when the antibacterial drug acts by inhibiting an essential bacterial enzyme: the R plasmids code for a second enzyme which has identical activity but is not sensitive to inhibition by the antibiotic, which takes over the metabolic role of the

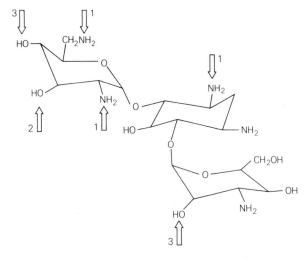

Fig. 4.4 R plasmid mediated enzyme modifications of the aminoglycoside antibiotic, Kanamycin B. This antibiotic can undergo modification by a number of different R plasmid specified enzymes, at the sites indicated by arrows: 1, acetylation; 2, phosphorylation; 3, adenylation.

chromosomal enzyme. The enzyme involved for trimethoprim is dihydrofolate reductase and for sulphonamides it is dihydropteroate synthetase.

3.3 Plasmids conferring pathogenicity to mammals

Although *Shigella, Salmonella* and *Vibrio cholerae* are of great importance in enteric disease, they are not the only causative organisms. Many diarrhoeal infections are caused by entero-pathogenic strains of what are normally considered to be harmless members of the gut flora, especially *E. coli*. Infections by *E. coli* are particularly dangerous during the first months of life, and are still an important cause of death in human babies in undeveloped countries, although rare in countries with comprehensive medical services. In modern intensive agriculture they can be devastating amongst young cattle and pigs: in Europe it has been estimated that between three and five per cent of all newborn pigs (approximately 5 million per year) die of diarrhoeal infection, mainly due to *E. coli*, within the first six weeks of life.

Although not all strains isolated from infections can be shown to carry plasmids, many do. Three distinct types of plasmid are found in stains isolated from human, bovine and porcine infections which are not found in non-pathogenic strains, and therefore appear to be implicated:

1 Ent plasmids code for synthesis of the enterotoxins described in Chapter 6. All the plasmids which have been investigated are responsible for production of either the ST toxins (heat *s*table *t*oxins) alone or of both ST and LT toxins (*l*abile *t*oxins) simultaneously: no naturally-occurring plasmids coding for LT toxin alone have been found (see Ch. 6 for a description of ST and LT toxins). In the Ent(ST + LT) plasmids, the syntheses of the two toxins appear to be coded on a single plasmid molecule and no segregation of them occurs on conjugation.

   The Ent(ST + LT) plasmids from both animal and human sources appear very similar with molecular weights from $55-60 \times 10^6$ and with considerable DNA homology. The Ent(ST) plasmids on the other hand are a more heterogeneous group with molecular weights from $20-80 \times 10^6$; they share little homology with the Ent(ST + LT) plasmids.

2 Hly plasmids specify the production of α-haemolysins which cause lysis of erythrocytes. Neither the molecular nature nor the mode of action of the haemolysins is known, nor has much work been carried out on the plasmids.

3 K plasmids are responsible for the production of antigens on the cell surface of *E. coli*, which can be detected by reaction with specific antisera. The antigens are species specific, the K88 antigen being found in infections of pigs and the K99 in infections of cattle. They are large filamentous proteins found on the surface of cells as a covering of fine pili varying from $0.1-1.5$ μm long and $70-100$ Å in diameter.

Investigation of the Ent and K plasmids is experimentally difficult; because of their phenotypes it is not possible to devise simple media capable of determining whether a clone harbours a plasmid or not, as is possible, for example, with the R plasmids where replica plating on media with or without antibiotic is used. For the K plasmids each clone must be tested by an agglutination test against specific antisera. The only screening procedures for Ent plasmids involve bioassays. For ST production a culture of each clone is fed to a suckling mouse, which is sacrificed 4 h later and examined to see whether there is massive accumulation of liquid in the intestine, which is indicative of the presence of an ST toxin. Until recently the synthesis of LT could only be tested by a similar *in vivo* test on a ligated section of the intestine of an adult rabbit, but a simpler procedure is now possible which depends on morphological changes in cells of a tissue culture of Chinese hamster ovary brought about by the LT toxin as a result of its action on adenyl cyclase. These procedures make the screening of colonies after a conjugation experiment more difficult than is usual in microbial genetics experiments.

The role played by Ent and K88 plasmids in causing porcine diarrhoea was elucidated by Smith and Lingwood, who constructed a number of derivatives of the same *E. coli* strain with different plasmid components. Cultures of these were then tested for pathogenicity by feeding to piglets. The only strain effective in producing diarrhoea was Ent⁺ K88⁺; the K88⁺ strain was only marginally effective and, perhaps surprisingly, the Ent⁺ strain was ineffective. On examining the accumulation of organisms in the alimentary tract they found that strains containing K88 were able to reside and multiply on the small intestine wall, normally a fairly sterile tissue, whereas strains without K88 were rapidly washed through. Presumably the filamentous pili specified by the K88 plasmid enable the bacteria to populate the intestinal wall. The K88⁺ strain produced mild symptoms due to the unusually high bacterial concentration in the small intestine, but the Ent⁺ K88⁺ strain had a major effect because it was able to remain there and simultaneously produce the enterotoxin. The presence of an Hly plasmid made no difference to the pathogenicity of a strain, so these experiments do not elucidate its role nor explain its widespread occurrence in enteropathogenic *E. coli*.

The plasmids in *E. coli* responsible for diarrhoeal disease are the only well-characterised example of plasmid participation in animal disease, but may prove to be the first of several such examples. There are reports of possible correlations between plasmid content and pathogenicity in a number of other genera, for example *Clostridia, Staphylococci* and *Vibrio parahaemolyticus*.

3.4 Col plasmids

Many bacterial strains produce compounds, the bacteriocins, which are lethal to other bacteria of the same or related genus. The ability of certain *E. coli* strains to synthesise colicins, active against other enteric bacteria, has been documented by the work of Fredericq and his collaborators; such strains are termed colicinogenic. The synthesis of colicins is invariably plasmid mediated. There are a number of different classes of colicins, each of which is distinguished by a letter and the nomenclature of the plasmid reflects the colicin type. Thus Col E1-K30 and Col V-K94 specify production of colicins of type E1 and V, respectively.

Col plasmids fall into two main groups. The conjugative plasmids have molecular weights from 60 to 100×10^6 and are carried in small numbers in the cell, from 1 to 3 copies; examples are Col I and Col V-K94. The second group are small (M.W. $\sim 5 \times 10^6$), non-conjugative and are present in about 10–30 copies per cell; an example is Col El-K30. Colicinogeny is the subject of Chapter 5.

3.5 Degradative plasmids

The saprophytic microbial population of the soil and natural waters is essential to all living organisms. The amount of carbon on the planet which is available to the biosphere is only a small fraction of the total carbon and it is essential that it is continuously recycled and does not become permanently fixed in any one form. Phototrophic organisms — the plants and algae — fix carbon from the reservoir of CO_2 present in the atmosphere and oceans into organic compounds. Some of this enters food chains via herbivores and carnivores, but eventually much finishes up as 'dead' organic matter as a result of excretion or death. Only microbes have the metabolic ability to degrade and utilise this organic matter for biosynthesis and energy, thus ensuring that it is recycled and not eventually fixed in this form.

The capacity of the natural microbial population for degradation is vast. If a few grams of a

Table 4.5   Degradative plasmids in *Pseudomonas*

Plasmid	Molecular weight (× 10^{-6})	Primary growth substrate		Plasmid coded pathway
CAM	160	camphor	(structure)	camphor ⟶ acetate + isobutyrate
OCT	N.D.*	octane	$CH_3(CH_2)_6CH_3$	octane ⟶ octanoate
SAL	51	salicylate	(structure)	salicylate
NAH	49	naphthalene	(structure)	naphthalene ⟶ salicylate ⟶ catechol ⟶ acetaldehyde + pyruvate
TOL†	60 to 170	toluene	(structure)	toluene ⟶ benzoate ⟶ catechol
		m-xylene	(structure)	
		p-xylene	(structure)	

* Not determined.

† Molecular weight range for a number of different isofunctional plasmids.

healthy soil are incubated in a simple medium containing a single natural product, a population of microorganisms which can effect the complete degradation of the compound is usually selected in a few days. Often a number of strains are present each of which is capable of using the compound as a sole carbon and energy source. Fortunately for modern industrial society this degradative ability extends to many synthetic compounds which are released into the environment, although certain compounds, in particular heavily chlorinated molecules such as the insecticides aldrin and dieldrin, are almost totally resistant to microbial attack.

These degradations are carried out by all genera of heterotrophic soil and water microbes, both fungal and bacterial, but perhaps the most prominent are the *Pseudomonas* group, in particular *P. putida, P. multivorans, P. acidovorans* and *P. fluorescens.* Strains of pseudomonads have been isolated which are individually capable of growing on about 100 different organic compounds as sole energy and carbon sources. Strains such as these must have a considerable proportion of their chromosome devoted to degradative enzymes compared, for example, with *E. coli* which has a chromosome of similar size but a very limited range of possible growth substrates. In addition, however, to an obviously wide range of chromosomally-coded degradative enzymes, the pseudomonads also appear to owe some of their versatility to the existence of plasmids which code for both structural and regulatory genes for the catabolic pathways of a number of the more complex growth substrates (Table 4.5). All the pathways known to be determined by these degradative plasmids involve fairly complete degradations of the growth substrate to a compound or compounds close to central metabolism: this ensures that transfer of the plasmid into a different strain will enable the recipient to grow on the primary growth substrate, which can be converted by the plasmid-coded enzymes to compounds which the recipient is able to metabolise further by means of its own chromosomal enzymes. In the case of CAM, NAH, SAL and TOL at least nine enzymes, together with the associated regulatory proteins and genes (since all the enzymes are inducible), are determined by the plasmid. Interestingly, three of the plasmids, namely, NAH, SAL and TOL, specify a common metabolic sequence, the conversion of catechol, a common intermediate, to acetaldehyde and pyruvate.

SAL, CAM, NAH and some of the TOL plasmids are conjugative co-integrates and can be transferred to a range of other *Pseudomonas* spp. with varying frequencies. In its natural host OCT is a plasmid aggregate and on curing or on conjugation segregates into three components — a plasmid with sex factor activity, a non-conjugative plasmid determining the octane-degrading enzymes and a conjugative plasmid conferring resistance to mercury. CAM and some of the TOL plasmids are noteworthy because of their size; they are amongst the largest naturally-occurring plasmids reported.

Only for TOL has any assessment of the environmental significance of plasmid-specified catabolism been made. A number of independent strains of *P. putida* which possess the same pathway, all carry the pathway genes on plasmids. These plasmids differ in their molecular properties, their ability to transfer and in the biochemistry of the pathway. This suggests that, at least within the saprophytic pseudomonads, plasmids which specify catabolic pathways are more common than might be indicated by the limited number of examples so far reported.

Most degradative pathways in bacteria are inducible. This property ensures that bacteria do not synthesise enzymes unless the substrate is present, thereby preventing unnecessary biosynthetic activity. Degradative plasmids extend this a stage further since the genes coding for the catabolic enzymes need be maintained only in a very few members of the population, thus reducing the biosynthetic burden of the population as a whole. The possibility of plasmid transfer ensures that, when there is a source of the particular substrate, the genes are available to a wider range of strains.

In addition to these plasmids coding for complete catabolic pathways, a few examples have been described where some fermentation reactions are plasmid-specified in enteric bacteria — e.g. the ability to ferment lactose and other carbohydrates in strains of *Salmonella, Proteus* and *Yersinia enterolytica.* The existence of such plasmids can cause confusion in bacteriology laboratories since the ability or inability to ferment lactose is routinely used as a diagnostic character in classifying clinical isolates. The presence of a lac plasmid in any of these genera which do not normally carry out this reaction could lead to an incorrect classification with serious medical consequences. Superficially, at least, these plasmids resemble the F′ plasmids of *E. coli* which result from faulty excision of F from an Hfr strain, taking with them some of the adjacent chromosome. These plasmids may well have arisen from a similar event in a lactose-fermenting bacterium, followed by transfer by conjugation into a different organism.

3.6 Plasmids conferring resistance to mercury

Plasmids which enable the host organism to withstand otherwise toxic concentrations of heavy metal ions have been found in the enteric bacteria, *Pseudomonas* and *Staphylococcus aureus.* In *S. aureus,* plasmids which confer resistance to a wide selection of toxic anions and cations are common, but plasmid-coded resistances to only nickel, cobalt, mercury and arsenate have been described in *E. coli* and *Salmonella*, and to only mercury in *Pseudomonas.* Most experimental work has been carried out on mercury resistance, probably because of the demonstration that the relatively harmless inorganic mercury compounds in the wastes from a number of industrial processes can be converted to the neurotoxic organomercurial methylmercury compounds by the action of anaerobic bacteria in sediments.

Mercury resistance appears to be a fairly common feature of both *Pseudomonas* and enteric bacterial plasmids which have been isolated without any selection for the character. From 800 R plasmids in the Hammersmith Hospital Collection, which were originally isolated from a range of different antibiotic-resistant enteric bacteria, 200 conferred mercury resistance, including for example R100 (Table 4.3). In *Pseudomonas,* resistance has been found on sex factors, R plasmids and as a component of the plasmid aggregate which included the OCT degradative plasmid. It is difficult to see a selective advantage in the character which would account for its common occurrence, especially in enteric organisms which seldom encounter mercury.

Some plasmids confer resistance only to toxicity by Hg^{2+} ions, whereas others protect against both Hg^{2+} ions and organomercurials such as methyl or phenyl mercuric salts. The biochemical mechanism of resistance is the same in both instances. The mercury is reduced by an inducible enzyme system to elemental mercury which can evaporate from solution: growth of the bacterium only starts once this detoxification has reduced the concentration of the mercury compound below a critical level:

$$Hg^{2+} \rightarrow Hg \uparrow \qquad R{-}Hg^+ \rightarrow RH + Hg \uparrow$$

The reductase system requires NAD(P)H, a thiol compound and probably two or more protein components.

3.7 Tumour-inducing plasmid in *Agrobacterium tumifaciens*

Crown gall is a disease which can attack a wide range of dicotyledonous plants. It is bacterial in origin, caused by infection of wounded tissue by *Agrobacterium tumifaciens,* which stimulates a tumorous growth of the plant tissue around the infected wound.

The virulence of *A. tumifaciens* has been correlated with the presence of large plasmids, which in different strains have molecular weights from $112-156 \times 10^6$. The loss of virulence is always accompanied by loss of the plasmid, and naturally-occurring avirulent strains do not possess such plasmids. The transfer of the plasmid to avirulent strains always confers virulence: this has been carried out *in vitro* both by transformation with the purified DNA and by mobilisation with the R plasmid RP4, and *in vivo* by inoculating a wound with a virulent donor at the same time as an avirulent recipient, and isolating a virulent form of the recipient at a later stage from the infected plant.

An interesting feature of this system is that there appear to be two types of tumour which can be distinguished biochemically by a high concentration of an unusual amino acid, either octopine or nopaline, which is produced in the tumour tissue. Which particular type of tumour is produced depends on the infecting bacterial strain, some causing octopine-rich tumours and some nopaline-rich tumours. There is also a correlation between the type of tumour and a degradative function which is plasmid-specified. Those plasmids found in strains which cause octopine-rich tumours, also code for the ability to utilise octopine as sole source of carbon and nitrogen, and similarly for nopaline. An explanation for this has not been found.

Once a tumour has been induced in plant cells it is possible to maintain a tissue culture of the tumour cells in the absence of bacteria. A very intriguing possibility is that part or all of the plasmid becomes permanently incorporated into the plant cells, either within the chromosome or else as an autonomous piece of DNA, and can then be stably inherited on cell division.

3.8 Cryptic plasmids

Many instances have been reported of circular DNA of low molecular weight being present in bacteria of many genera. For the most part the molecular weights are less than 10×10^6 and they have no sex or transfer factor activity or other discernible phenotype or biological role, although the failure to detect one does not exclude the possibility of there being one. These plasmids are termed cryptic for obvious reasons.

3.9 Staphylococcal plasmids

In many respects the plasmids found in *S. aureus* are analogous to those in Gram-negative bacteria, and similar functions are found: antibiotic resistance, resistance to heavy metal ions and production of enterotoxins and haemolysins. They differ in one major respect in that all are non-conjugative and all are specific to *Staphylococcus*. They seem to be a distinct line which has evolved in parallel to the plasmids in Gram-negative bacteria.

Their non-conjugative nature imposes experimental limitations, but transfer between strains can be achieved *in vitro* by transduction; this also appears to be the mechanism by which plasmid transfer occurs *in vivo*.

The penicillinase plasmids are the largest in size, with molecular weights of $18-21 \times 10^6$. Like the plasmids of Gram-negative bacteria, the penicillin resistance is conferred by the detoxifying action of a β-lactamase, but the enzyme differs in being extracellular and being induced only in the presence of penicillin. Penicillinase plasmids all code for a variety of other resistances, in particular to ions such as cadmium and arsenate; resistance to erythromycin may also be found. Not all penicillinase plasmids can coexist in a cell and they can be divided into at least three incompatibility groups by this criterion (see section 4.1).

Resistance to other antibiotics can be plasmid-specified but the plasmids involved differ from the penicillinase group in both incompatibility grouping and size. Resistance to kanamycin and neomycin is found on plasmids with molecular weights around 15×10^6 and resistance to tetracycline, streptomycin and chloramphenicol is found singly on plasmids of about 3×10^6 molecular weight. Only the mechanism for chloramphenicol resistance is understood. As in Gram-negative bacteria, it is by the action of a chloramphenicol acetyltransferase.

4 Some important features of plasmid biology

4.1 Relationship between plasmids

The sex factor F is extremely efficient at promoting its own transfer: in a few hours the number of recipient cells converted to F^+ can exceed the number of donor cells. Examination of F^+ cells in the electron micrograph shows the presence on the cell surface of organelles, the sex pili, which are not present on F^- cells (Fig. 4.5): these pili play an important part in the mechanics of conjugation, since rapid mixing of an F^+ culture shears the pili from the cell and removes the donor ability, which is only regained when new pili have been synthesised.

Most other plasmids are much less efficient at conjugational transfer — at least in short mating experiments — and this is reflected in the lower number of piliated cells in donor cultures. The reason for this is an efficient system for repression of pilus synthesis present on most plasmids, but not on F. It is possible to make mutants in which pilus synthesis is derepressed and the transfer efficiency much increased (drd mutants).

Two common types of sex pili are found. Some plasmids specify production of pili very similar in morphology to the F pilus, which are termed F-like. These plasmids have the additional property that, if they can be transferred to a cell containing F, they repress synthesis of the F pilus and drastically reduce the conjugational transfer frequency of F. Plasmids with this phenotype are termed fi^+ (*fertility inhibition*), and include members of all classes found in enteric bacteria, R plasmids, Col plasmids, Hly, Ent and K88. Other plasmids specify morphologically different pili, which are similar to those found on cells containing Col I plasmids (e.g. Col Ib−P9), and are called I-like (Fig. 4.5). They differ in being shorter than F-like pili, about $2\,\mu$m as against $20\,\mu$m, and seldom have the central axial hole often seen in F-like pili. They also differ in their antigenicity and in the adsorption of pilus-specific (or male-specific) phages: several phages are specific for F-like pili (e.g. f_1, f_2, fd) and one, If_1, is specific for I-like pili (Fig. 4.5). Most plasmids which produce I-like pili have no effect on the fertility of F if present with F in the same cell and are termed fi^-.

It was at first thought that the fi phenotype could be used as a primary method of grouping plasmids into only two classes, the fi^+ and fi^- plasmids. As more plasmids were isolated and characterised it became apparent that this was an oversimplification.

A more precise method of classification resulted from observations that there are barriers which prevent the transfer of certain plasmids into a cell containing a different plasmid. This superinfection immunity is the result of two distinct phenomena. The first is a *surface exclusion* by a resident plasmid, which does not affect the formation of a mating pair but does inhibit the conjugational transfer of DNA into the cell. Secondly, there is *incompatibility* between plasmids: even if an incompatible plasmid is introduced into a cell already containing a plasmid by sufficiently strong selective pressure, the two plasmids interact by inhibiting replication of each other. The net result is that upon removal from any selective pressure which is ensuring that both plasmids coexist the plasmids rapidly segregate during a few generations to give a mixed population of progeny containing either one or the other plasmid.

(a)

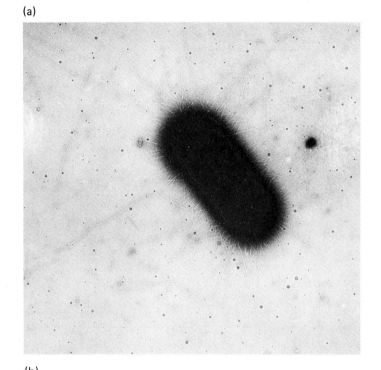

(b)

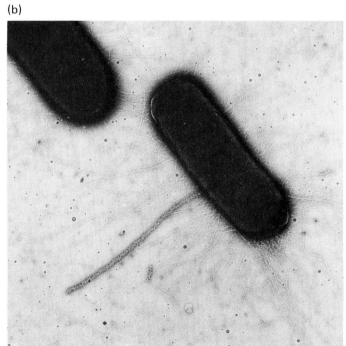

Fig. 4.5 F-pili on *E. coli* cells. (a) *E. coli* cell containing an F′ plasmid, showing large numbers of common pili and one longer and thicker F-pilus. (b) Cell of same *E. coli* strain infected with the pilus- (or male-specific) RNA phage R17, which binds only to the sex pilus. (Electron micrographs kindly supplied by Dr. S. To.)

Study of this property of incompatibility was extended, in particular by Datta, and is now the major method of classifying plasmids. An *incompatibility group* is defined by the following experimental criterion. Two plasmids, p1 and p2, are assigned to the same group if a strain, usually *E. coli* K-12, into which both plasmids have been transferred (p1$^+$ p2$^+$) segregates progeny which are either p1$^+$ or p2$^+$ but no stable (p1$^+$ p2$^+$) lines. If p1 and a third plasmid p3 are similarly shown to be incompatible, then p3 is assigned to the same group and p2 and p3 will also prove incompatible. The various groups are designated Inc groups with a letter characteristic of the group. By this criterion, the fi$^+$ plasmids which specify F-like pili are not a single class, but are made up of five incompatibility groups, designated IncFI-IncFV. The fi$^-$ plasmids fall into about 14 groups, of which those producing I-like pili constitute about 5. One group which deserves some special mention are the IncP plasmids (e.g. RP4, Table 4.3) the original members of which were R plasmids isolated from *P. aeruginosa* but representatives of which have now been isolated in a wide range of genera. This range of parental strains is paralleled by a very wide range of genera into which IncP plasmids can be transferred, and they may prove to be a major vector for transfer of genetic information across wide generic boundaries in the wild.

The various incompatibility groups reflect wide differences in genetic structure. The degree of DNA homology between plasmids can be assessed experimentally by two methods, *hybridisation* and *heteroduplex analysis*. In hybridisation the fraction of single-stranded DNA from one source which will bind to single-stranded DNA from a second source (i.e. is complementary) is measured. This gives the overall gross homology without specifying its distribution. One procedure is to bind an excess of the purified DNA of one plasmid onto a nitrocellulose membrane: prior to this the DNA is sheared to small fragments by sonication and denatured to dissociate the double strands. The membranes are then incubated with a small amount of sheared, single-stranded DNA of the second plasmid which has been labelled radioactively by growing the host strain on a radioactive medium. The fraction of the total radioactivity which becomes bound to the membrane is a measure of the homology between the plasmids since the only DNA which will bind is that which has base sequences complementary to the membrane-fixed fragments.

Heteroduplex analysis is far more specific and can detect short regions of homology of only 100–200 base pairs, and locate them on the plasmid. The purified DNA from two plasmids, most of which will be present as open nicked circles, is mixed, denatured to separate the strands and re-annealed. Many of the original complementary strands come together to form homo-duplexes, but where there is homology between the plasmids, some heteroduplexes between opposite strands of the two plasmids form (Fig. 4.6). The renatured mixture is then prepared for electron microscopy and examined. Homoduplexes are seen as circles of thick, smooth curves of double-stranded DNA. Heteroduplexes are seen as circles with alternate sections of double-stranded DNA and single-stranded DNA, which appears thinner and less smoothly curved (Fig. 4.7): the double-stranded sections correspond to homology and the single-stranded sections to non-homology. Heteroduplex analysis is also a powerful tool for mapping plasmid genes: examination of heteroduplexes formed between deletion mutants of the same plasmid can lead to a very precise relative location of the genes because of the characteristic loops which result from deletions in one of the duplex strands (Fig. 4.8).

Hybridisation has shown that plasmids of the same incompatibility group share considerable DNA homology whereas much less is shared with plasmids of other groups. Where heteroduplex analysis has been carried out, the homology is more precisely defined. F itself (IncFI) shares about 44 per cent homology with many fi$^+$ plasmids such as R1, R6, R100 (all Inc FII, and therefore compatible with F) and with Col V-K94 (fi$^-$, Inc FI) and the common area of

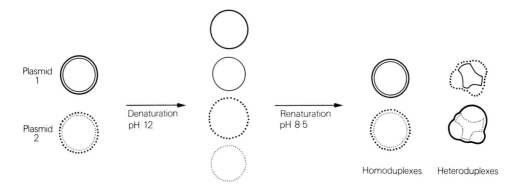

Fig. 4.6 Diagrammatic representation of formation of plasmid heteroduplexes. Strand separation of plasmid duplexes only takes place if one or both strands are nicked; due to shearing forces involved in handling, this is almost invariably true for most of the molecules.

homology is in one long section of the F molecule. By combining this with genetic mapping it can be shown that this section corresponds only to the genes responsible for conjugational transfer on each plasmid, and not to any of the other functions of either F (e.g. chromosomal integration) or of the other plasmid (e.g. drug resistance or colicin production). These results imply that the IncF plasmids have evolved as a result of a plasmid with genes capable of coding for its own conjugational transfer (i.e. a sex factor) which, during transfer between strains, is able to acquire other genes which give it additional phenotypic characteristics and thus increase the selective advantage it bestows on the host cells.

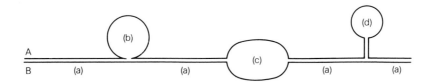

Fig. 4.7 Structural features seen in heteroduplex molecules. Double-stranded DNA is seen as thicker, less kinked structures than single-stranded DNA. (a) Areas of complete homology between A and B. (b) An I–D loop, resulting from an insertion in A or a deletion in B. (c) Non-homology between A and B. (d) a transposition with terminal inverted repeat sequences, present in A but not in B (see section 4.4).

Heteroduplex comparisons amongst R plasmids from different locations can show surprisingly extensive homology. R1, R6 and R100 were originally isolated in England, Germany and Japan, respectively, and are all IncFII plasmids (Table 4.3). R6 appears to be identical to R100 except for three additional sections which may be partly attributed to its additional resistance to kanamycin and neomycin. Eighty-five per cent of the sequences present in R1 are also found on R6. It must be concluded that all three are descendants of the same plasmid, and that the remarkable similarity between R6 and R100 is a result of carriage of one or both of these between Japan and Germany prior to their isolation. If this is so, it demonstrates the immense possibilities for dissemination afforded by an R plasmid (which can be infectious between bacteria) present in bacteria (which are infectious between humans) and magnified by the ease and rapidity of international transport.

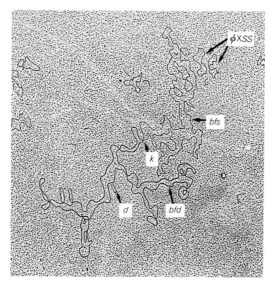

Fig. 4.8 Electron micrograph of a heteroduplex molecule formed from two F′ plasmids, which are homologous over most of their length (bfd). One of the plasmids however has two additional sequences of DNA not present in the other, which are seen as single-stranded I–D loops, a small one (d) and a large one (bfs) which comes out of the duplex DNA (bfd) at (k). Also present are circles of the single-stranded DNA from phage φX174 (φXSS), included as an internal molecular weight standard. (From Sharp *et al.* (1972), *J. of Molec. Biol.*, 71, 471–97.)

4.2 Conjugation

The process of conjugation can be divided into two phases. First is the *formation of a mating pair,* the feature which distinguishes conjugation from transduction and transformation. The sex pilus specified by the plasmid in the donor is essential in this phase, since removal of the pili by shearing removes the ability to form mating pairs. The second phase of conjugation is *DNA transfer and metabolism.* One specific strand of the plasmid duplex is cleaved at a unique site (the origin) and passed from the 5′ end into the recipient cell. Both the strand remaining in the donor and the strand introduced into the recipient are replicated to give intact duplex plasmids in both cells (Fig. 4.9).

The role of the pilus in the DNA transfer is uncertain. Because of the axial hole apparent in the F-like pili it has been suggested that the pilus acts as a tube down which the DNA is passed. However, I-like pili do not appear to have an axial hole, and no DNA within the pilus has ever been detected. An alternative hypothesis is that the pilus brings the cells into direct contact and that the DNA is transferred across the walls at the contact point.

It is only in IncF plasmids that the genetics of the transfer system have been investigated in any detail; the genetic organisation in other incompatibility groups is likely to be considerably different although there may be analogies.

The genes in F responsible for conjugation are located on a continuous section of the plasmid and are transcribed as a single operon, the tra (*tra*nsfer) operon (Fig. 4.10) in which there are 12 structural genes. The first eight of these, *traA* to *traH*, are responsible for pilus synthesis. Mutations in any of these genes result in cells without pili, which are unable to transfer. The *traS* gene determines the surface exclusion property (not incompatibility) which reduces the ability of a like plasmid to be transferred into the cell. Both *traD* and *traI* are

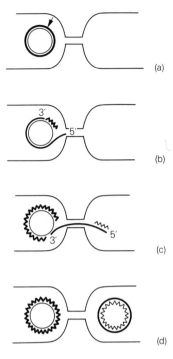

Fig. 4.9 Conjugation between donor and recipient cells. (a) Formation of mating pair. Nick introduced at origin of transfer in one specific strand, denoted by arrow. (b) and (c): The nicked strand transferred from 5′ end into recipient possibly through the pilus. Replication of single strands in donor and recipient cells (new DNA denoted by jagged line). (d) Recircularisation of both plasmids.

involved in the DNA metabolism in the donor during conjugation and *traG* is bifunctional and mutations in it are defective in both pilus formation and DNA transfer.

The regulation of the *tra* operon is complex and shows some interesting features (Fig. 4.11). The gene product of *traJ* is a positive regulator of the entire operon which is only transcribed in its presence. However, there are three additional genes which determine whether or not the *traJ* gene product is synthesised: *traO, finO* and *finP* (for *f*ertility *in*hibition). When the gene products of *finO* and *finP* are both present, they combine at the *traO* gene to block transcription of *traJ* and hence production of its gene product: thus *traO* acts like an operator gene for *traJ* and the *finO* and *finP* gene products as *cooperative* negative regulators.

The net result is that normally, when both *finO* and *finP* gene products are synthesised, the *traJ* gene product is not synthesised, the tra operon is not transcribed and pilus synthesis is repressed. In mutants with either a defective *finO* or *finP* gene, transcription of *traJ* is not blocked, pilus synthesis is constitutive and the cells show the derepressed (drd) phenotype.

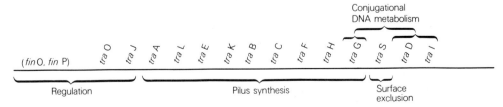

Fig. 4.10 The tra operon of F-like plasmids.

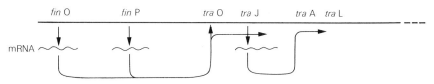

Fig. 4.11 Regulation of tra operon in F-like plasmids; the *tra* J gene product acts as a positive regulator for the tra operon. The *fin* O and *fin* P gene products act together at *tra* O to block transcription of *tra* J, so that normally the *tra* J gene product is not synthesised and the tra operon is repressed.

This regulatory mechanism accounts for three properties of F and other IncF plasmids.

(1) F itself lacks a functional *finO* gene and hence shows derepressed pilus synthesis and transfers at high frequency.

(2) The fi$^+$ character of IncF plasmids which are compatible with F is due to a lack of specificity of their *finO* gene products (Fig. 4.12). Each *finP* gene product is specific for its own plasmid, but it can combine with the *finO* gene product of another IncF plasmid to block its *traO* gene. Thus when another IncF plasmid is present in the same cell as F, its own *finO* gene product combines with the *finP* gene product of F to repress the normally derepressed transfer system of F.

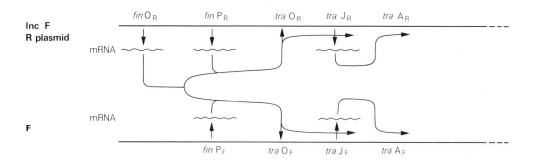

Fig. 4.12 Fertility inhibition of F through interaction of the regulatory systems of an Inc F plasmid and F itself. F has no fuctional *fin* O gene and consequently expression of the tra operon is normally depressed. The *fin* O gene products of other compatible Inc F plasmids are non-specific: when one of these is present in a cell with F, its *fin* O product can interact with the *fin* P product of F to repress expression of the F tra operon.

(3) The efficiency of transfer of fi$^+$ plasmids in the short term is poor because of the repression of the tra operon. However, if donor and recipient cultures are left together for longer periods, say overnight, an unexpectedly high extent of transfer is obtained. This is because pilus synthesis is not repressed in a newly infected cell, which can further transmit its plasmid at high frequency. The explanation is that in a newly infected cell, there is no preferential expression of either *finO*, *finP* or *traJ* genes. Consequently there is some transient synthesis of the *traJ* gene product and this only slows down and ceases as the *finO* and *finP* gene products build up and block transcription of *traJ*. Because the *traJ* gene product is a positive regulator, as long as it is present in the cell the *tra* operon is expressed and high frequency transfer occurs. The concentration of *traJ* gene product in a newly infected cell only drops as cell division dilutes it out, and after a period of about 4 h the progeny are once again in the normal repressed low transfer state.

This feature is advantageous since it ensures that in a stable population the ability to transfer is repressed with a resultant saving in biosynthesis. When a recipient population becomes available and once some small amount of transfer takes place, the depression of newly infected cells ensures an efficient infectious spread of the plasmid.

There is no genetic relationship between the transfer system of Inc F plasmids and those of other Inc groups, and transfer deficient (tra⁻) mutants of one group are not complemented by plasmids of another. There are likely to be analogies, however, since, for example, IncI plasmids also change from being repressed to derepressed in newly infected cells.

4.3 Replication

Replication of plasmid DNA takes place at two different points of time. *Transfer replication* occurs during conjugation, when the single strand remaining in the donor and the complementary strand transferred to the recipient are replicated to form a duplex plasmid DNA in both cells. *Vegetative replication* occurs during the normal cell cycle ensuring that both daughter cells inherit the plasmid on cell division. The two processes are similar in that both rely to some extent on the host cell replication mechanisms, though the detailed biochemistry of neither is known. They also differ in a number of important respects. They start at different positions on the plasmid; the transfer origin which determines the start of replication during transfer is distinct from the replication origin where vegetative replication begins. They also appear to involve different plasmid genes, which in both processes probably determine the control and initiation of replication.

Control of vegetative replication is dependent on plasmid size. Large plasmids with molecular weights from about 30×10^6 appear to exist in only a small number of copies, from one to three, in the host cell: these are said to be under *stringent* replication control. Small plasmids with molecular weights below about 30×10^6 are often found in higher numbers (> 10 per cell) and are said to be under *relaxed* control. The copy number of these plasmids also varies, depending on the state of the cells when harvested, and cells in stationary phases are found to harbour a greater number of plasmid copies than when growing exponentially. The relationship between plasmid size and copy number suggests that relaxed control is a result of a less sophisticated or even a total lack of the replication control genes which in stringently controlled plasmids keeps vegetative replication in phase with chromosome replication.

The replicon hypothesis originally proposed by Jacob, Brenner and Cuzin in 1963 may explain some of the features of vegetative replication. They proposed that replicons (DNA molecules capable of replication, i.e. chromosomes or plasmids) are attached to unique sites on the inner membrane of the bacterial cell. This site of attachment is involved in the initiation of replication and during cell division; duplication of the attachment site results in duplication of the attached replicon, one of which is inherited by each daughter cell.

There is evidence that the plasmid DNA is membrane-attached at least during some phase of the cell cycle since, under some conditions of cell lysis, the plasmid DNA is found in a very rapidly sedimenting species attached to membrane fragments. After lysis of bacterial cells containing Col E1 with non-ionic detergents in the 'cleared lysate' procedure, Helinski and his colleagues have found that the plasmid DNA is present as CCC molecules to which are attached a complex of three proteins with molecular weights 60×10^3, 17×10^3 and 11×10^3. This DNA–protein complex, called the *relaxation complex*, is unstable and under conditions which affect protein structure, such as treatment with the ionic detergent sodium dodecyl sulphate, dissociates to give protein-free, open nicked circles. The attachment of the protein complex and the nick in one of the circles are at the same position, which is close to or at the replication origin, suggesting that the proteins may be involved both in attachment to the membrane site and in the initiation of replication.

Replication of the small plasmids under relaxed control has not been fully explained in terms of the membrane attachment hypothesis. From experiments using labelled DNA it appears that replication is a random process in which any one of the plasmid copies is as likely to replicate as any other, rather than a specific process with one plasmid acting as a master copy from which all others are replicated. This could be explained if there are a number of membrane sites for these plasmids. Alternatively, if there is only one such site, the attached plasmid must replicate and then dissociate from the site, allowing any other plasmid in the pool to bind there and subsequently replicate.

The membrane attachment theory offers a possible explanation for incompatibility, if it is assumed that plasmids from the same Inc group compete for the same single membrane site: segregation of the two plasmids into different daughter cells would therefore occur on cell division. On this hypothesis plasmids from different Inc groups have different membrane attachment sites and therefore coexist without competition.

4.4 Transposition of plasmid genes

In 1974 Hedges and Jacob showed that the genes on plasmid RP4 which code for the β-lactamase which confers ampicillin resistant (Ap) can be transposed on to other compatible plasmids present in the same bacterial host. The acquisition of the Ap determinant by the other plasmid is accompanied by an increase in its molecular weight of about 3×10^6, showing that the transposition involves acquisition of new DNA. The plasmid which has gained the new resistance can pass it on further to any other compatible plasmid:

$$RP4(Ap, Km, Tc) + p1 \longrightarrow RP4(Ap, Km, Tc) + p1(Ap)$$
$$p1(Ap) + p2 \longrightarrow p1(Ap) + p2(Ap)$$

The fragment transferred has been termed a *transposon* and the particular element on RP4 is called transposon A (TnA).

TnA has terminal sequences of DNA of about 140 base pairs which are repeated at either end, inverted and on the complementary strand (Fig. 4.13(a)). As a result the transposon can be seen on electron micrographs of either single strands of the DNA or heteroduplexes between plasmids with and without the element, as single-stranded loops (that include the Ap determinant) on a double-stranded neck (the inverted sequences from the same strand) (Fig. 4.13(b)).

As well as being transposable in the laboratory, TnA has obviously undergone considerable transposition in the wild and has been found in a number of plasmids of several incompatibility groups which have been isolated from bacteria of a range of genera in different geographical locations. Other transposons have subsequently been found, and are usually characterised by terminal inverted repeat sequences. One example is the tetracycline (Tc) resistance determinant common to both R6 and R100 (section 4.1) which is bounded by inverted repeats of about 1 400 base pairs. The existence and widespread distribution of these transposons offers one explanation for the way plasmids are found to acquire new resistance genes during the course of epidemics where heavy selection pressure is being applied through the use of antibiotics.

The mechanism of transposition is not known. The inverted repeats must play some role, but normal reciprocal recombination does not appear to be involved, since the rec system required for normal recombination need not be functional: transposition in *E. coli* occurs just as readily in *recA* mutants as in wild type hosts. There is no evidence that the transposon can exist

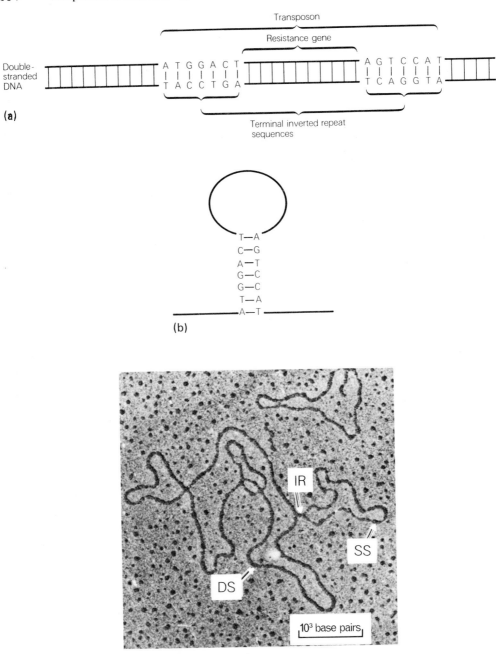

Fig. 4.13 (a) Hypothetical structure of a transposon (n.b. The inverted repeats found on transposons correspond to more than 100 base pairs, not the 7 shown diagrammatically here). (b) Hairpin loop of transposon as seen on electron microscopy due to hydrogen bonding by terminal inverted repeat sequences on same strand (cf. Fig. 4.8). (c) Electron micrograph of heteroduplex between two plasmids which differ only by the insertion of a transposon into one of them. The double-stranded segment (DS) represents the region of homology between the two plasmids: the inserted transposon is seen as a single-stranded loop (SS) with a short inverted repeat (IR) at its base. (Electron micrograph kindly provided by Dr S. N. Cohen.)

autonomously although attempts have been made to identify such a unit. Some specificity in the point of insertion appears to be necessary since TnA has been found to integrate into the plasmid RSF 1010 in at least 12 sites, suggesting that a specific but not uncommon base sequence is required. The insertion of a transposon has no overall effect on the genome into which it becomes integrated unless it is inserted into a gene or operon when it is mutagenic.

Inverted repeat sequences are not only found in plasmids: the sequence bounding the Tc resistance gene in R6 and R100 is homologous with *IS3* (insertion sequence) found in the *E. coli* chromosome. It seems likely that transposons evolved by combination between a gene and an inverted repeat sequence. Their incorporation into plasmids ensures transfer through a wide range of bacterial species: the low specificity required at the integration site can lead to their incorporation into heterologous DNA, unlike normal recombination where a considerable degree of homology is required. Such exchange can help to maintain the genetic flexibility of the microbial population as a whole and be an important factor in microbial evolution.

Plasmids may be important in this process but need not be the only vectors since transposons can integrate into phage or chromosomal DNA. Inverted repeat sequences are found in both prokaryotic and eukaryotic chromosomes and their occurrence on plasmids may be only part of a wider role in genetic exchange across species boundaries.

4.5 Plasmids in genetic engineering

The fact that bacterial genera have retained their own unique characteristics shows that under normal environmental conditions genetic exchange across generic boundaries does not take place to any great extent although, as has been shown, under conditions of intense selective pressure, as in modern usage of antibiotic therapy, such exchange can take place and plasmids and transposons are important vectors. It is now possible, however, to introduce genes from diverse sources into plasmids *in vitro*. Such genetic engineering can be of considerable scientific and technological importance. It can enable scientists to study genes at the molecular level in an organism such as *E. coli* with a well-characterised genetic system; these genes can be taken from other organisms, in particular from eukaryotes, which have a genetic organisation too complex for present experimental techniques. Alternatively, it can be used to incorporate genes into a bacterium which will either make it produce proteins or enzymes of scientific or technological interest, or give it some new and useful character, such as the ability to fix nitrogen (e.g. the nif plasmid in Ch. 14, although in this particular case, the plasmid was constructed *in vivo*).

On the other hand, there are very distinct dangers in such genetic manipulations. It could result in the incorporation of genes, for example from an animal virus, which might make any host bacterium potentially lethal to man, animals or the environment in general. Linkage to a plasmid with its additional properties of transmissibility and the ability to acquire antibiotic resistance would only increase the potential dangers. Concern over such possibilities has led to a very strict code of practice regarding these experiments being introduced by the National Institutes of Health in the USA and equivalent bodies in other parts of the world in order to eliminate the chances of any potentially hazardous plasmid or bacterium being released into the environment.

The experimental methods for introducing new genes into plasmids require two classes of enzymes, the *restriction endonucleases* which hydrolytically cleave DNA at specific sites, and the *DNA ligases* which will link together hydrolysed fragments.

The bacterial restriction endonucleases are responsible *in vivo* for hydrolysing foreign DNA which is introduced into a bacterial cell. The specific sites on the DNA which are recognised and cleaved are short sequences of four to six base pairs, which are palindromic in that the same

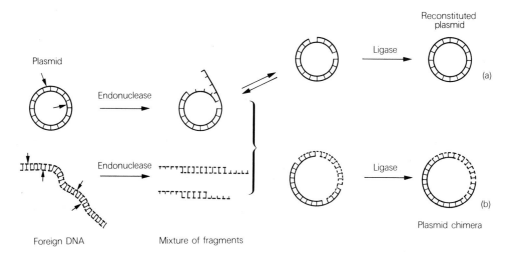

Fig. 4.14 Use of endonuclease and ligase enzymes to (a) cleave and reconstitute a plasmid with a single endonuclease target site and (b) incorporate genes from foreign DNA into the plasmid.

sequence is present on the complementary strand but in the opposite direction. The interesting property of many of the enzymes is that they cleave the DNA at staggered positions on the two strands (see Ch. 1). This produces identical single-stranded lengths of DNA on the two ends which, because of the symmetry of the cleavage sequence, are 'sticky' and will readily hydrogen-bond together or with any other end which has been produced by the same restriction enzyme. If a plasmid is carefully chosen such that it has only one cleavage site for a particular restriction enzyme, the hydrolysed plasmid with its two sticky ends can be recircularised by the action of a DNA ligase (Fig. 4.14). Alternatively, if the cleaved plasmid is mixed with fragments of another DNA produced by the same endonuclease (and therefore with the same complementary sticky ends) then some of the foreign fragments will hydrogen-bond into the plasmid, and on incubation with a DNA ligase will become covalently linked there. The plasmid chimera (from the mythological beast, the Chimera, which was part lion, part goat and part serpent) containing the DNA from two sources can be reintroduced into a bacterial host by transformation.

It is likely that the foreign DNA used to make the fragments for incorporating into the plasmid will be large, particularly if prepared from eukaryotic cells, and will therefore contain a large number of restriction enzyme cleavage sites. This will result in a very heterogeneous mixture of fragments all with sticky ends. As there are as yet no general ways of purifying the fragment containing the gene of particular interest from the mixture, an equally complex mixture of different plasmid chimeras will be produced: selection must be applied to the resultant transformed bacteria to pick out those organisms which have been transformed by a plasmid containing the gene of interest.

Using these techniques, a number of laboratories have made plasmid chimeras using vectors such as pSC 101, a small R plasmid conferring tetracycline resistance, and Col E1, both of which have single cleavage sites for the Eco R1 enzyme. DNA from such varied sources as other R plasmids, staphylococcal plasmids, and genes coding for ribosomal RNA from the toad *Xenopus laevis*, has been successfully ligated into these plasmids without any effect on the other properties of the plasmid.

Plasmids are not the only vectors which can be used for cloning segments of DNA and considerable work has been devoted to developing a comparable system using the phage λ. Each system has particular advantages but both are likely to contribute to our understanding of genetic organisation in higher organisms if used with responsibility under carefully controlled experimental conditions.

5 Concluding remarks

Plasmid biology is a rapidly expanding area of study at the present time. Recent discoveries that plasmid genes are involved in a wide range of bacterial functions have brought together scientists from disciplines which might otherwise have little in common. The aim of this chapter has been to point out the relevance of plasmids not only to genetics and molecular biology but also to medicine, soil microbiology, plant pathology, microbial evolution and even, through the use of genetic engineering, to the study of eukaryotic genetics. It seems likely to the author that in the near future an even wider range of different functional plasmid types will be discovered and that plasmids will prove to be, not an exceptional or rare phenomenon, but one which is common or even general amongst natural populations of microorganisms.

6 Bibliography

This is a rapidly expanding area of research and any list of references is going to date quickly. Readers should therefore keep an eye open in relevant review journals for articles written since this list was compiled (January 1977).

General textbooks

FALKOW, S. (1975) *Infectious multiple drug resistance.* Pion, London.
MEYNELL, G. G. (1972) *Bacterial plasmids.* Macmillan, London.

Books covering conferences, with a wide variety of articles by different authors

SCHLESINGER, D. (Ed.) (1974) *Microbiology, 1974,* American Society for Microbiology, Washington. Also later volumes in this series.

Review articles

ANDERSON, E. S. (1968) 'The ecology of transferable drug resistance in the Enterobacteria', *Ann. Rev. Microbiol.,* **22**, 131–80.
CHAKRABARTY, A. M. (1976) 'Plasmids in Pseudomonas', *Ann. Rev. Genetics,* **10**, 7–30.
CLOWES, R. C. (1972) 'Molecular structure of bacterial plasmids', *Bacteriol. Rev.,* **36**, 361–405.
COHEN, S. N. (1975) 'The manipulation of genes', *Scient. Am.,* July, 24–32.
COHEN, S. N. (1976) 'Transposable genetic elements and plasmid evolution', *Nature,* **263**, 731–38.
HELINSKI, D. R. (1973) 'Plasmid-determined resistance to antibiotics: molecular properties of R factors', *Ann. Rev. Microbiol.,* **27**, 437–70.

MEYNELL, E., MEYNELL, G. G. and DATTA, N. (1968) 'Phylogenetic relationships of drug-resistance factors and other transmissible plasmids', *Bacteriol. Rev., 32*, 55–83.

NOVICK, R. P. (1969) 'Extrachromosomal inheritance in bacteria', *Bacteriol. Rev., 33*, 210–63.

NOVICK, R. P., CLOWES, R. C., COHEN, S. N., CURTISS, R., DATTA, N. and FALKOW, S. (1976) 'Uniform nomenclature for bacterial plasmids: a proposal', *Bacteriol. Rev., 40*, 168–9.

RICHMOND, M. H. (1973) 'Resistance factors and their ecological importance to bacteria and man', *Progr. Nucl. Acid. Res. Molec. Biol., 13*, 191–248.

SMITH, D. and DAVIES, J. (1978) 'Plasmid-determined resistance to antibiotics', *Crit. Rev. of Microbiol.* (in the press)

WHEELIS, M. L. (1975) 'The genetics of dissimilatory pathways in *Pseudomonas*', *Ann. Rev. Microbiol., 29*, 505–24.

WILLETTS, N. (1972) 'The genetics of transmissible plasmids', *Ann. Rev. Genetics, 6*, 257–68.

5
Bacteriocins

K. G. Hardy
Biological Laboratory, University of Kent at Canterbury

1 Introduction

Bacteriocins are protein antibiotics produced by bacteria. Several bacteriocins differ from most other antibiotics because they kill only bacteria which are closely related to strains which produce them. For example, the most extensively studied group of bacteriocins, the colicins, are antibiotic proteins produced by some strains of *Escherichia coli* and closely related enterobacteria such as *Shigella sonnei*. The colicins kill only strains of *E. coli* and its close relatives; they are ineffective against Gram-positive bacteria and even against most Gram-negative bacteria.

The genes for the various types of colicin proteins are on circles of DNA called *Col plasmids*. Many Col plasmids also determine the production of sex pili and are able to transfer themselves to other bacteria by conjugation (see Ch. 4 for a discussion of plasmids and their transfer).

Substances analogous to colicins are produced by many genera of bacteria, including several Gram-positive strains. For example, some strains of *Streptococcus* produce streptococcins; and staphylococcins are produced by certain strains of *Staphylococcus*. In each case the active component of the bacteriocin is protein. However, a wide variety of inhibitors are produced by Gram-positive bacteria so that it is difficult to decide which of them can usefully be called a bacteriocin. Recent research indicates that several bacteriocins produced by Gram-positive bacteria are specified by plasmids.

Killer strains are also found among eukaryotic microorganisms. There are killer strains of yeast which kill other strains of yeast (Wickner 1976). And *Paramecium* strains which have kappa particles (endosymbionts which appear to have evolved from Gram-negative bacteria) are able to kill other strains of Paramecia (Preer *et al.* 1974).

A large proportion of *E. coli* strains produce colicins. Usually between 10 and 70 per cent of *E. coli* strains isolated from human or animal faeces produce one or more types of colicin. Strains which produce colicins are said to be *colicinogenic*, or Col⁺. Colicin production can be simply demonstrated on nutrient agar plates as shown in Fig. 5.1. Strains which produce a particular type of colicin are not themselves sensitive to its lethal effects; they are *immune* to the colicin they produce.

This chapter is inevitably concerned mainly with colicins because much more is known about this group of bacteriocins than about other groups. Much still remains to be explained about colicinogeny, but as our understanding of colicins and Col plasmids increases they

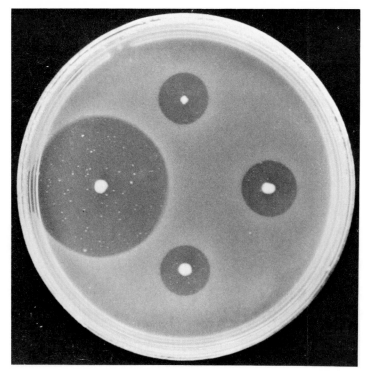

Fig. 5.1 Colicin production by strains of *E. coli*. Four colicinogenic strains of *E. coli* were inoculated onto an agar plate and incubated for 48 h. The colonies were killed with chloroform and the plate overlaid with a colicin-sensitive strain of *E. coli*. Colonies of colicin-resistant mutants can be seen in the largest inhibition zone.

become increasingly useful in studying many aspects of microbiology and molecular biology. In particular, as a result of investigations by D. R. Helinski and co-workers on the Col plasmid, ColE1–K30 (Hershfield *et al.* 1974), derivatives of this plasmid are among the most frequently used vectors in genetic engineering (see Ch. 4, section 4.5). Colicins may also be very useful in investigating the energised state of the cytoplasmic membrane (Luria 1975).

2 Classification and chemistry of colicins

Colicins (and the Col plasmids which code for them) are given a name which comprises the colicin type, such as A, B or I, and the name of the strain which was first shown to produce the colicin. Thus, colicin B-K98 is a type B colicin which was originally detected in *E. coli* strain K98. Even when the ColB-K98 plasmid is transferred to other strains of bacteria, it retains its original name.

Colicins were classified into about 20 different types by Fredericq about 25 years ago on the basis of their action on a set of colicin-resistant bacteria. Colicin-resistant mutants can readily be isolated; they form colonies in the inhibition zones produced by colicins, as shown in Fig. 5.1. Most colicin-resistant mutants have lost their colicin-receptors (see section 5.1). Mutants which are resistant to the colicins produced by different strains of bacteria can be isolated to provide a set of colicin-resistant mutants. Colicins of a particular type can then be defined as all those colicins which do not kill a particular colicin-resistant strain. In practice, classifying colicins in this way is not completely straightforward because certain colicin-resistant mutants are insensitive to more than one colicin type, and some strains of bacteria produce more than one type of colicin.

Subdivisions of the colicin types are based on colicin immunity. For example, colicins E2 and E3 are indistinguishable on the basis of their action on colicin-resistant mutants. However, bacteria which have a ColE2 plasmid are immune to other E2 colicins but not to E3 colicins. (The mechanism of immunity to colicin E3 is described in section 5.3). Similarly, bacteria which produce colicin Ib (i.e. those which have a ColIb plasmid) are immune to other Ib colicins but not to Ia colicins.

2.1 Chemistry of colicins

Colicins are proteins with molecular weights ranging from about 27 000 to 90 000 (Table 5.1). There are wide variations in the amino acid compositions of different colicins, but a few are similar, indicating that certain colicins have evolved from a common ancestor. For example, colicins E2-P9 and E3-CA38, and cloacin DF13 (determined by a plasmid found in *Enterobacter cloacae*) have similar amino acid compositions. The amino acid sequences of colicins E2-P9 and E3-CA38 are also similar, as are those of colicins Ia-CA53 and Ib-P9 (Herschman and Helinski 1967; Konisky 1973).

Table 5.1   Molecular weights of colicins

Colicin	Molecular weight
D-CA23	90 000
E1-K30	56 000
E2-P9	60 000
E3-CA38	60 000
Ia-CA53	80 000
Ib-P9	80 000
K-K235	45 000
M-K260	27 000

3 Colicin synthesis

Not all Col$^+$ cells produce detectable amounts of colicin. In a broth culture of ColE2$^+$ bacteria, for example, only about 1 per cent of the cells produce detectable amounts of colicin, even though there is an average of about ten ColE2 plasmids in every cell. It is, of course, possible that the remaining 99 per cent of cells produce so few colicin molecules that they cannot be detected by means of existing techniques.

A small proportion of ColE$^+$ or ColK$^+$ cells release sufficient colicin to form small inhibition zones (lacunae) in a lawn of sensitive bacteria on an agar plate. To demonstrate lacunae, a culture of Col$^+$ bacteria is treated with chloroform to kill the cells, mixed with sensitive bacteria and soft agar and then poured onto an agar plate. Small inhibition zones, which look like phage plaques, are seen in the lawn of sensitive bacteria after overnight incubation. The percentage of cells which produce colicin, as determined by the lacuna technique, in a broth culture of ColE2$^+$ bacteria varies from a minimum of about 0·01 per cent to a maximum of about 10 per cent, depending on the growth-phase of the culture. As the culture approaches 'stationary phase', the percentage of lacuna-forming cells increases. Not all types of Col$^+$ bacteria produce and release sufficient colicin to form lacunae. For example, the low titres of colicin V produced in broth cultures remain largely cell-bound even when the cells do not have colicin receptors (section 5.1) which might adsorb any colicin V that is released.

The apparent repression of colicin synthesis suggests an obvious parallel between colicinogenic bacteria and lysogenic bacteria which harbour prophages. In populations of lysogenic bacteria, only a small proportion of the cells are spontaneously derepressed and eventually lyse to release bacteriophage particles.

The parallel between colicinogeny and lysogeny extends to the phenomenon of *induction* by agents such as ultraviolet irradiation (UV). When bacteria containing the lambda prophage are irradiated with a suitable dose of UV, almost all the prophages are induced; the phage DNA replicates rapidly and many phage proteins are produced. The cells eventually lyse to release phage particles. Many inhibitors of DNA synthesis induce prophages. Similarly, irradiation of Col$^+$ bacteria with UV greatly increases the colicin titre of the culture, and almost all the cells become able to form lacunae. The antibiotic mitomycin C has the same effect. However, this increased colicin titre does not result from an increased rate of replication of the Col plasmid.

The mechanisms of prophage induction and of the induction of colicin synthesis are unknown. A fragment of the repressor protein can be detected shortly after lambda lysogens are induced, suggesting that induction leads in some way to destruction of the repressor which then allows the prophage to enter the lytic pathway of development. However, if this is indeed the mechanism, the reasons for destruction of the repressor after induction are unknown.

After induction of colicin synthesis the cell wall does not completely disintegrate as it does after induction of a lambda lysogen. However, the synthesis of proteins determined by chromosomal genes is inhibited in ColE$^+$ bacteria when they are treated with concentrations of mitomycin C which induce the synthesis of colicins, but which have no effect on protein synthesis in Col$^-$ cells. Induced cells release colicins E1 or E2 for several hours but are unable to form colonies. Since low doses of UV or mitomycin C, which are insufficient to kill Col$^-$ cells, induce the lethal synthesis of E colicins, ColE$^+$ bacteria are more sensitive to these inhibitors than are Col$^-$ bacteria. The few cells which spontaneously produce colicins E or K in cultures which have not been treated with an inducer, are also probably killed as a result. However the synthesis of certain types of colicin does not necessarily result in cell death. For example, the induced synthesis of colicin Ib does not kill ColIb$^+$ cells; indeed, cells containing a ColIb plasmid are more resistant to UV than Col$^-$ cells, despite the induction of colicin synthesis.

4 Col plasmids

The genes for the colicin proteins are on Col plasmid DNA molecules which range in size from about 0·2 per cent to about 3·5 per cent of the length of the *E. coli* chromosome. No strains having a chromosomal gene for colicin have been isolated from a natural environment, although such strains can be made in the laboratory.

There are two major groups of Col plasmids: Group I Col plasmids have molecular weights of about 5×10^6, enough DNA to code for about 10 average-sized proteins; Group II Col plasmids have much higher molecular weights, about 70×10^6, which is enough DNA to code for about 100 average-sized proteins (Table 5.2). Group I Col plasmids are maintained in greater numbers

Table 5.2   Molecular weights of Col plasmids

Group I		Group II	
Col plasmid	Molecular weight ($\times$ 10^6)	Col plasmid	Molecular weight ($\times$ 10^6)
A-CA31	4·6	B-K77	70
D-CA23	3·1	Ib-P9	61·5
E1-K30	4·2	V-K94	85
E1-16	5·0		
E2-P9	5·0		
E3-CA38	5·0		
K-K235	5·0		

per cell than Group II Col plasmids. Cells usually maintain an average of about ten copies of Group I Col plasmids, but only one or two copies of Group II Col plasmids.

The smaller Col plasmids (Group I) are non-conjugative; they do not determine the synthesis of sex pili and are unable to transfer themselves to other bacteria by conjugation. However, these plasmids are often transferred at high frequency if a conjugative plasmid is present in the same cell as a Group I Col plasmid.

The larger Col plasmids (Group II) are almost invariably conjugative plasmids (see Ch. 4). ColB and ColV plasmids determine the production of F-like sex pili, pili very similar to those determined by the fertility factor, or F plasmid, of *E. coli* K-12. Indeed, the sex pili determined by ColV-K94 are indistinguishable by immunological methods from those determined by the F plasmid. Cells containing ColV-K30 or ColV-K94 further resemble F^+ cells in that almost all the cells produce sex pili and can transfer plasmid DNA. However, synthesis of sex pili determined by several ColB plasmids is repressed in most ColB$^+$ cells. From genetic evidence, it appears that ColB plasmids determine the production of a repressor (analogous to that which controls transcription of the Lac operon in *E. coli*) which inhibits the synthesis of sex pili and other products required for conjugation in most ColB$^+$ cells. When ColB-K98 is present in the same cell as the F plasmid or ColV-K94, conjugative ability determined by these plasmids is also greatly reduced. Because of this effect on F^+ cells, the ColB plasmid is said to be responsible for *fertility inhibition* and is called a fin$^+$ (or fi$^+$) plasmid. Presumably, the F plasmid and ColV-K94 have retained an operator which can be controlled by the ColB repressor even though they do not themselves determine a repressor.

A segment comprising about 40 per cent of the F plasmid DNA has a sequence of nucleotides which is so similar to that of ColV-K94 that a single-stranded molecule of F DNA (i.e. one half of a double helix) forms a double helix with the complementary single-stranded DNA of ColV-K94 (Sharp *et al.* 1973). The hybrid molecule so formed is called a heteroduplex. (Further details of the formation and analysis of plasmid heteroduplexes are given in Ch. 4, section 4.5.) The heteroduplex technique has also been used to demonstrate that about 80 per cent of the base sequences of ColE2-P9 and ColE3-CA38 are very similar (Inselburg 1973).

The ColI plasmids determine a sex pilus which is shorter and a little thinner than the F-like pilus (Ch. 4, section 4.1). Synthesis of sex pili is repressed in most ColI$^+$ cells. Many of the plasmids which code for the synthesis of I-like sex pili and the synthesis of type I colicins also confer resistance to several antibiotics, so these Col plasmids could be called R plasmids.

Further details about the replication of Col plasmids, and their incompatibility and exclusion relationships, features which are common to many types of plasmid, are discussed in Chapter 4 and in other reviews on plasmids (Falkow 1975; Hardy 1975; Helinski 1976; Meynell 1972).

5 Effects of colicins on sensitive bacteria

A distinctive feature of colicins is that they have a narrow spectrum of action, killing only strains of *E. coli* or bacteria which are closely related to *E. coli*, such as *Shigella sonnei*. As we shall see, this narrow spectrum seems to be due to the need for specific protein receptors in the outer membranes of bacteria.

In outline, colicins kill cells as follows. First, colicins bind to protein receptors in the outer membranes of bacteria (Stage I). Colicins belonging to different types bind to different receptors. Following adsorption to receptors, colicins do not bring about any irreversible changes leading to cell death unless the cytoplasmic membrane is energised. If membrane energy is available, colicin-treated cells enter Stage II which leads irreversibly to cell death. Finally, not all colicins bring about the same biochemical changes in sensitive cells; different colicins can kill cells by affecting different biochemical targets. For example, Colicin E3 inactivates ribosomes, colicin E2 affects DNA, whereas colicins E1, K and I disrupt many energy-dependent reactions in the cytoplasmic membrane. Colicin E3 (or perhaps a part of it) probably penetrates the cells to inactivate ribosomes directly. The killing of cells by colicins follows single-hit kinetics, indicating that a single colicin molecule can kill a sensitive cell. But hundreds of colicin molecules are usually needed to kill a cell, perhaps because most of them adsorb to 'non-lethal' receptors which bind colicins but do not allow them to initiate a chain of events leading to cell death.

5.1 Colicin receptors

The first step in the killing of cells by colicin is the binding of colicin to receptors on the external surface of the cell. The receptors for colicins B, D, E, I, K and M are mainly, if not exclusively, composed of proteins in the outer membranes of sensitive bacteria. The outer membrane of *E. coli*, the outermost layer of the cell wall, comprises a bilayer of phospholipids as well as proteins and lipopolysaccharide.

Most of the proteins in the outer membrane which act as colicin receptors have other functions as well. In particular, several colicin receptors are involved in transporting various metabolites into the cell (see Ch. 12 for a discussion of bacterial transport). The outer membrane acts as a barrier to the penetration of molecules which have molecular weights of more than about 600, so *E. coli* appears to have evolved special mechanisms for transporting several types of large molecule into the cell. For example, the colicin E receptor, a glycoprotein with a molecular weight of 60 000, is also a receptor for vitamin B12. Vitamin B12 (M.W. 1 355) binds to this outer membrane protein before being transported into the cell in an energy-dependent reaction. (Vitamin B12 is not an essential nutrient for *E. coli*). The colicin K receptor is probably involved in the transport of nucleosides, even though these molecules have molecular weights of about 250 and therefore should be able to penetrate the outer membrane. Colicin K-resistant mutants, which lack protein receptors for colicin in their outer membranes, are deficient in the uptake of nucleosides.

Certain colicin receptors are also necessary for the uptake of iron by *E. coli*. Iron is usually taken up in the form of a complex between iron and a ligand which chelates iron in the environment of the bacterium. Enterochelin (a trimer of 2,3-dihydroxybenzoylserine) is an iron-chelating ligand which is excreted by *E. coli* when the intracellular iron concentration falls too low. The receptor for colicin B is an outer membrane protein involved in the accumulation of ferric enterochelin. Similarly, the colicin M receptor is an outer membrane protein (M.W. 85 000) which is a receptor for ferrichrome (M.W. 740), another ligand which binds iron. Because these colicin receptors are involved in the accumulation of iron, colicin-resistant mutants which have lost colicin receptors are deficient in iron uptake. Several mutants which are insensitive to colicin even though they have colicin receptors, namely, colicin-tolerant mutants (see section 5.4), are also deficient in iron uptake. This suggests that other components of the cell envelope, in addition to colicin receptors, are required both for the accumulation of iron and for the action of certain colicins. The positions of the mutations in these colicin-tolerant mutants and the locations of the genes which determine various colicin receptors have been mapped on the *E. coli* chromosome.

Several colicin receptors are also receptors for bacteriophages. For example, phage T1 adsorbs to the same outer membrane protein as colicin M; phage BF23 adsorbs to the colicin E receptor; and phage T6 binds to the colicin K receptor. The fact that certain colicins and phages adsorbed to the same receptor suggested that colicins and Col plasmids might have evolved from phages, but there are no antigens which are common to colicin K and to the proteins of phage T6.

Spheroplasts made from mutants which are resistant to colicin K because they lack receptors in the outer membrane, are sensitive to colicin K. This suggests that the function of colicin receptors is to bring the colicin into contact with other parts of the cell envelope, perhaps the cytoplasmic membrane, which it can reach when outer parts of the cell envelope are disrupted, as in spheroplasts.

After colicins have adsorbed to receptors in the outer membrane, a sequence of events leading to cell death is initiated only if the cytoplasmic membrane is energised (see Luria 1975). Colicin adsorbed for several hours to cells which have been treated with an uncoupler (an agent which dissipates the energised state) can be destroyed with trypsin. When the uncoupler is also removed the cells are then able to grow and form colonies, demonstrating that colicin cannot kill cells in the absence of membrane energy. The reason for this dependence on the energised state of the cytoplasmic membrane is unknown. Perhaps colicins must be transported or rearranged after adsorption to bring them into contact with the cytoplasmic membrane, or perhaps they interact with a protein which is in a suitable configuration only when the membrane is energised.

5.2 Biochemical effects of colicins

5.2.1 Colicins which affect energy-dependent processes in the cytoplasmic membrane

It seems likely that several colicins, including E1, K and I, come into direct contact with the cytoplasmic membrane and that they are bactericidal because they disrupt the energised state of membrane. The high-energy state of the membrane is, therefore, both essential for the action of these colicins and also appears to be their 'biochemical target'.

The energy for the active transport of several amino acids, some sugars and some ions can be provided by electron transport, via the energised state of the membrane, without the necessity for ATP or other phosphorylated compound as an intermediate (see Ch. 12). These active transport processes are inhibited in cells treated with colicins such as E1 or K. For example,

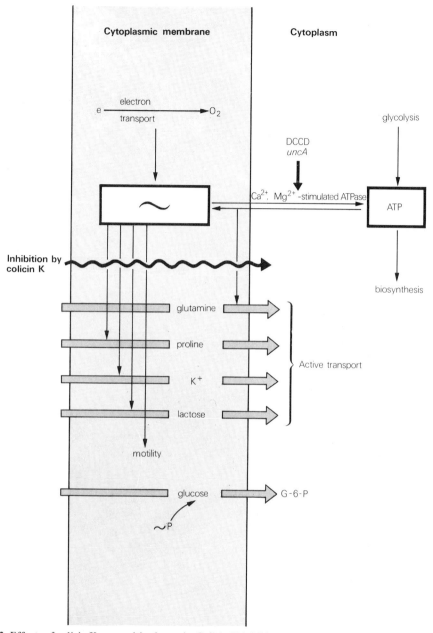

Fig. 5.2 Effects of colicin K on sensitive bacteria. Colicin K inhibits many energy-dependent processes which take place in the cytoplasmic membrane. Its sites of action are indicated by the black wavy line. The active transport of K^+, proline and lactose, all of which are dependent on the energised-stage ($\sim$) of the cytoplasmic membrane, is inhibited. (The active transport of glutamine is inhibited although transport of this amino acid is coupled directly to ATP.) Transport of glucose, mediated by the phosphoenolpyruvate–phosphotransferase system, is not inhibited by colicin K.

The effect of colicin K on energy-dependent reactions in the membrane leads to a decrease in the concentration of ATP, which is coupled to the energised state via the Ca^{2+}, Mg^{2+}-stimulated ATPase. When the activity of the ATPase is inhibited (either by the ATPase inhibitor, N,N' dicyclohexylcarbodiimide (DCCD), or by the use of an *uncA* mutant which lacks ATPase activity), the ATP concentration in colicin K-treated cells does not decrease. (After Luria (1975), *Scient. Am.*, **233**, 30–7.)

ions such as K^+ and Mg^{2+}, which have been accumulated by active transport, are immediately released from the cell. This, by itself, might be expected to be sufficient to kill the cell because K^+ and Mg^{2+} are co-factors for many enzymes. However, there are further consequences of the action of colicins on the energised state of the membrane (Plate *et al.* 1974; Luria 1975). The ATP concentration also decreases in cells treated with colicins E1 or K, and consequently the synthesis of DNA, RNA and protein stops. This fall in the concentration of ATP may occur because the cells seek to maintain high intracellular concentrations of K^+ for example, which has been accumulated by active transport through the expenditure of ATP. Evidence for this is provided by the effect of inhibiting the ATPase which couples intracellular ATP to the energised state of the membrane. For example, adding the ATPase inhibitor, *N,N'*-dicyclo-hexylcarbodiimide, or using an uncA mutant (which lacks ATPase activity) prevents a fall in ATP concentration in colicin K-treated cells (Fig 5.2). Protein synthesis continues in these cells, provided the strain used is not dependent on an external supply of an essential amino acid. The importance of the losses of K^+ and Mg^{2+} in the killing of cells by colicin K, is demonstrated by the finding that cells are not killed when they are treated with colicin K in the presence of high concentrations of K^+ and Mg^{2+} (Kopecky *et al.* 1975).

The precise mechanism whereby colicins such as E1 and K inhibit reactions which depend on the high-energy state of the cytoplasmic membrane is not clear. The energy for processes such as active transport and motility is believed to be provided by a 'protonmotive force', i.e. this is the 'energised state', which comprises gradients of pH and electrical charge across the membrane. Uncouplers such as 2,4-dinitrophenol collapse this high-energy state by increasing membrane permeability to H^+. But colicin E1 does not increase the permeability of the membrane to protons. However, the release of other ions, such as K^+, could lead to a similar disruption of membrane potential, or perhaps colicins inhibit mechanisms which couple trans-membrane gradients to processes such as active transport.

5.2.2 Effects of colicin E3-CA38 and cloacin DF13 on ribosomal RNA

Colicin E3-CA38 and cloacin DF13, which have similar amino acid compositions, do not kill cells by disrupting processes which take place in the cytoplasmic membrane. In cells treated with colicin E3 or cloacin DF13 protein synthesis stops initially while the synthesis of RNA and DNA continues. Protein synthesis stops because these bacteriocins inactivate bacterial ribosomes by cleaving the 16S ribosomal RNA molecules, components of the 30S ribosomal sub units, about 50 nucleotides from the 3' end (Nomura *et al.* 1974). It seems likely that these two bacteriocins, or perhaps fragments of them, penetrate the cells and act as ribonucleases, for they have the same effect on ribosomes when mixed with them *in vitro*. Alternatively, they may activate a ribonuclease attached to ribosomes. Both the 30S and 50S ribosomal subunits must be present for colicin E3 to have any effect *in vitro*, perhaps because a specific ribosomal conformation is essential for ribonuclease activity. A further indication that colicin E3 penetrates cells is that when it is covalently attached to sepharose beads it is unable to kill cells, whereas colicin E1, which probably acts on the cytoplasmic membrane (section 5.2.1), can kill cells even when it is attached to sepharose. Possibly only a part of the colicin E3 protein penetrates cells. A fragment of this colicin (with a molecular weight of 18 000), isolated after the digestion of the protein with trypsin, inactivates ribosomes more efficiently *in vitro* than the intact colicin molecule, although the fragment does not kill sensitive cells of *E. coli*.

Colicin E3 and cloacin DF13 also inactivate ribosomes from several Gram-positive bacteria, so that their specificity in killing only certain enterobacteria is apparently because other bacteria lack colicin receptors and perhaps also a mechanism for transporting colicin into the cell.

Another property of the cloacin DF13 protein was recently discovered by Veltkamp *et al.* (1976). Many of the Clo plasmid DNA molecules isolated from Clo$^+$ cells had single molecules of cloacin protein attached to them. Under certain conditions, the cloacin protein acted as a nuclease and introduced a break into the circular Clo plasmid DNA, such that it was converted into a linear molecule. The immunity protein determined by CloDF13 (see section 5.3) prevented this nuclease activity. It remains to be seen whether this effect of the cloacin on the plasmid which codes for it has a role in the maintenance of the plasmid in the cell, perhaps in plasmid replication.

5.2.3 Effects of colicin E2-P9 on DNA

Although colicins E2 and E3 adsorb to the same receptor in the outer membrane and have similar amino acid sequences, the primary effect of colicin E2 is on DNA and not on RNA. The DNA in colicin E2-treated cells appears to be degraded by both endonucleases and exonucleases. Schaller and Nomura (1976) have recently shown that purified colicin E2 is itself an endonuclease which cleaves DNA molecules *in vitro* by first introducing single-strand breaks, followed by double-strand breaks. These correspond to the first two stages of DNA degradation seen in colicin E2-treated cells, but they are followed *in vivo* by a more extensive breakdown of DNA. It appears, therefore, that colicin E2 penetrates cells to act in a catalytic fashion as a DNA endonuclease and that exonucleases in the cell then degrade the DNA further.

5.3 Colicin immunity

Colicins are produced by strains of *E. coli* and kill strains of *E. coli*, but of course bacteria which have a particular Col plasmid are not highly sensitive to the particular colicin which it determines. When a Col plasmid is transferred into colicin-sensitive bacteria, the bacteria become *immune* to the corresponding colicin even though the cells retain colicin receptors and can adsorb as much colicin as they could before they became colicinogenic.

Cells containing a ColE3 plasmid or a CloDF13 plasmid are immune because the plasmids determine the synthesis of a small protein (M.W. 10 000) which binds very tightly to the bacteriocin protein and inactivates it (although not irreversibly) (Nomura *et al.* 1974). When colicin E3, excess colicin E3 immunity protein and ribosomes are mixed together, the ribosomes are not inactivated since each colicin E3 molecule binds one molecule of immunity protein and can no longer act on ribosomes. Presumably, therefore, all ColE3$^+$ cells are immune to colicin E3 because they all produce immunity protein which binds to the colicin molecules at some point after they have adsorbed to receptors. If enough colicin E3 is added, even ColE3$^+$ cells are killed presumably because the immunity proteins become saturated with colicin, leaving some colicin E3 molecules free to inactivate the ribosomes. Probably, genes for immunity proteins are present on all Col plasmids, but only the immunity proteins determined by ColE3, ColE2 and CloDF13 have been purified so far.

Each colicin E3 molecule which is released from ColE3$^+$ cells has a single molecule of colicin immunity protein bound to it. The binding of the immunity protein does not prevent adsorption to receptors or the killing of sensitive cells by colicin. Colicin E3 from which the immunity protein has been removed is as effective against sensitive cells as is the colicin which has immunity protein bound to it.

5.4 Colicin-tolerant mutants

There are two sorts of mutant which are insensitive to colicin. Many colicin-insensitive mutants do not adsorb colicin because the colicin receptors in their outer membranes have been lost or altered. However, *colicin-tolerant* mutants still adsorb colicin onto receptors but are nevertheless insensitive to colicin.

Several tolerant mutants are insensitive to more than one type of colicin. For example, tolA mutants are tolerant to colicins A, E1, E2, E3 and K even though some of these colicins adsorb to different receptors and have different biochemical effects on sensitive cells. Despite these differences in their mechanisms of action, the isolation of mutants which are insensitive to all of them indicates that there is a step after adsorption to receptors which is common to all these colicins. Analysis of tolerant mutants may be very useful in determining the sequence of reactions following adsorption of colicins to receptors, since the sequence of events which leads to cell death may be blocked at different points in different tolerant mutants.

6 Ecology of colicins

The properties of colicins as revealed by experiments in the laboratory raise important questions about the significance of colicins in the alimentary tract, a more usual environment for *E. coli.* Do colicinogenic strains of *E. coli* kill colicin-sensitive strains in the alimentary tracts of animals or man? Are pathogenic strains more likely to become established in the gut if they are colicinogenic? Are pathogens less likely to become established if the resident flora produces colicins which kill the pathogen? Important though these questions may be, we are far from being able to provide definitive answers to them. Few experiments have been made and it is sometimes difficult to assess the importance of colicins in relation to the many other factors which affect the competitiveness of strains in the alimentary tract. Indeed, on the basis of published data there is little, if any, evidence that colicins kill any strains of *E. coli* in the alimentary tract. However, recent research (Smith and Huggins 1976) has shown that one type of Col plasmid, ColV, increases the pathogenicity of certain *E. coli* strains, although not because it enables the bacteria to kill sensitive strains of *E. coli* (see section 6.2).

In considering how important colicins are likely to be in the ecology of enterobacteria in the intestine, it is worth considering the more obvious differences between conditions in the alimentary tract and the laboratory conditions which are usually employed in studies on colicins. The large intestine of man is anaerobic, and in this environment *E. coli* strains are greatly outnumbered by obligate anaerobes. In laboratory experiments on colicins and Col plasmids, *E. coli* is usually grown under aerobic conditions, often in a nutrient medium in which it grows with a mean doubling time of about 30 min. during the exponential phase. The doubling time of *E. coli* growing in the alimentary tract is unknown, but it is probably much longer. The mean doubling time of a pathogenic member of the *Enterobacteriaceae, Salmonella typhimurium,* was about 10 h when it produced an infection in mice. The concentrations of trypsin and other proteases which are secreted into the small intestine are probably high enough to inactivate colicins in this part of the gut. But most of the enterobacteria are in the large intestine where the trypsin concentrations – in pigs for example – are not high enough to inactivate colicins completely (De Alwis and Thomlinson 1975). Growth on the surfaces of the gut wall may also provide an important difference between conditions in the gut and those used in laboratory experiments. Lastly, the strain of *E. coli* used in almost all laboratory experiments, *E. coli* K-12, differs from freshly-isolated strains in that it lacks part of the lipopolysaccharide of the cell envelope and is consequently more sensitive to colicins than other strains. In view of the many differences between conditions in laboratory experiments and conditions in the alimentary tract, some of the properties of colicins as revealed by laboratory experiments may not necessarily provide a good guide to the role of colicins in the ecology of bacteria in the gut.

6.1 Selective advantage of colicinogeny: units of selection

In order to account for the existence of colicins in ecological and evolutionary terms, we should like to be able to answer the question: what is the selective advantage of colicinogeny? However, such a question immediately raises another: on what does colicinogeny confer a selective advantage? Presumably, the selective advantage is ultimately conferred on the gene for colicin, or on this gene and its associated colicin-immunity gene. However, a selective advantage may also be conferred on other entities (units of selection), i.e. on the Col plasmid and on the colicinogenic clone of cells. There seems little doubt that colicinogeny confers a selective advantage on plasmids which have the genes for colicin and for colicin immunity. However, the extent to which different types of Col plasmid increase the chances of survival of a clone of bacterial cells is less clear. Any advantage which is conferred does not seem to have been sufficiently great to select strains which have chromosomal genes for colicins.

We are far from being able to provide a reasonably complete picture of the ecology of colicins. Indeed, as noted above, there is little evidence to support the assumption that colicins kill enterobacteria in the alimentary tract and thereby increase the competitiveness of an *E. coli* strain when it is in competition with closely related bacteria.

Few investigations have been made to determine whether Col$^+$ bacteria kill colicin-sensitive bacteria in the alimentary tract. Germ-free animals would seem to be useful in testing the hypothesis that colicinogeny confers a selective advantage on bacteria in the gut, in that it at least allows the experimenter to control several variables, even though the conditions in the intestine are unlike those in normal animals. In the few published experiments in which Col$^+$ and Col$^-$ bacteria were allowed to colonise the alimentary tracts of germ-free animals, the colicinogenic strains coexisted with the colicin-sensitive strains for several weeks. Furthermore, in experiments where the colicinogenic strain had a selective advantage over the Col$^-$ strain, it also became predominant in competition with a colicin-resistant mutant of the same Col$^-$ strain, so the advantage conferred by colicinogeny could not be ascribed to the killing of the Col$^-$ bacteria.

Other studies have been made to discover whether strains which survive longest in the alimentary tract of a person or animal — 'resident strains' — are more likely to be colicinogenic. Some investigations indicate that resident strains — those maintained for perhaps several months — are more likely to be colicinogenic than transient strains. However, not all investigators have reached the same conclusion. It may be difficult to draw conclusions from these experiments unless the types of colicin produced by the bacteria are taken into account. As noted in the next section, certain types of Col plasmid have a marked effect on the ability of *E. coli* strains to survive in the human alimentary tract, while others have no apparent effect.

6.2 Presence of ColV plasmids in invasive strains of *E. coli*

It is perhaps rather curious, in view of the emphasis placed on the bactericidal properties of colicins in earlier sections of this chapter, that in the one case where an important ecological role can be ascribed to a Col plasmid, this seems to have nothing to do with the killing of bacteria by colicins. Several ColV plasmids increase the pathogenicity of *E. coli* strains for young animals and, more specifically, increase the invasiveness of these bacteria (Smith and Huggins 1976).

Strains of *E. coli* are an important cause of infection and death especially among young animals and in human babies. Their pathogenicity is usually in part due to properties determined by chromosomal genes (such as the components of the cell wall) and partly due to proteins determined by plasmid genes, such as enterotoxin which is synthesised by many enteropathogenic strains (see Ch. 4, section 3.3; Ch. 6, section 6.2). In contrast to enteropathogenic strains, which colonise the small intestine, when invasive strains produce disease they are not confined to the alimentary tract but invade the body and are found in large numbers in the blood and various organs such as the kidney, lymph nodes and liver.

The first indication that ColV plasmids contributed to the pathogenicity of invasive strains was the high incidence of colicin V synthesis among invasive strains which had been isolated from cases of generalized infection in animals (Table 5.3). The incidence of colicin V production among invasive strains was much higher than among non-pathogenic strains. The importance of ColV plasmids in pathogenicity was confirmed by removing ('curing') the plasmid from invasive strains and showing that the plasmid-free strains were far less invasive. When the Col plasmids were reintroduced, invasiveness was greatly increased. Neither ColE nor ColI plasmids increased the invasiveness of *E. coli* strains.

Table 5.3   Incidence of colicin V production among invasive strains of *E. coli*.* (From Smith and Huggins 1976)

Source of bacteria	Number of strains examined	Number of ColV$^+$ strains
Cattle	31	25
Sheep	5	3
Chickens	44	36

* *E. coli* strains of serotype 078:K80, which were responsible for independent cases of generalised infection in animals, were examined for colicin V production.

The reason for the increased invasiveness of ColV$^+$ bacteria is unknown and it is not known whether colicin V itself, or perhaps immunity to colicin V, is responsible. But in the light of these recent studies on ColV plasmids, it is interesting that the first colicin to be investigated in detail (by A. Gratia in 1925) was called colicin V because it was produced by a strain of *E. coli* (*E. coli* V) which was virulent for animals.

A further property of ColV$^+$ bacteria is their ability to be maintained for longer periods in the human alimentary tract in comparison with ColV$^-$ bacteria (Smith and Huggins 1976). It is not known whether this is in some way related to the increased invasiveness of ColV$^+$ bacteria. The bactericidal properties of colicin V were not responsible for the enhanced survival of ColV$^+$ bacteria. The introduction of a ColE plasmid into a strain of *E. coli* did not increase its ability to survive in the human alimentary tract, even though a ColV plasmid greatly increased the ability of the same strain to survive in the gut. The selective advantage conferred by Col plasmids other than ColV is therefore not at all clear.

7 Other bacteriocins: the problem of defining a bacteriocin

Much more is known about colicins than about other groups of bacteriocins. However, there are substances analogous to colicins produced by many groups of bacteria, both Gram-negative and Gram-positive. But in considering bacteria other than members of the *Enterobacteriaceae* it becomes increasingly difficult to define what we mean by a bacteriocin because such a wide variety of inhibitors are produced by bacteria. Some of the antibiotic proteins produced by Gram-positive bacteria kill a wide range of Gram-positive species. Furthermore, less is known about the importance of plasmids in specifying bacteriocins in Gram-positive bacteria, although plasmids in strains of *Staphylococcus, Streptococcus* and *Clostridium* have recently been shown to determine the synthesis of proteins which kill other Gram-positive bacteria.

The variety of antibiotic substances produced by bacteria also leads to practical difficulties when estimating the frequency of bacteriocin production. Inhibitors produced by bacteria include hydrogen peroxide, ammonia, lactic acid, fatty acids, as well as protein structures which are usually called 'defective phages' (because they resemble bacteriophage tails), protein toxins, haemolysins and enzymes such as lysostaphin which cause cell lysis. Precautions can be taken to prevent some of these substances, such as hydrogen peroxide, producing inhibition zones in tests for bacteriocin production, but it is nevertheless often difficult to decide how many of the inhibitors detected by such tests can usefully be called bacteriocins.

7.1 Bacteriocins produced by Gram-negative bacteria

Several members of the *Enterobacteriaceae* have been shown to produce bacteriocins, including: *Escherichia, Shigella, Salmonella, Klebsiella, Enterobacter, Proteus, Serratia* and *Erwinia*. Most of the well-studied colicins are produced by strains of *E. coli* and *Shigella*. Few of the bacteriocins produced by other genera of the *Enterobacteriaceae* have been studied in detail.

Other Gram-negative bacteria which produce bacteriocins include: *Pseudomonas, Yersinia (Pasteurella), Neisseria* and *Vibrio*. Several of the bacteriocins produced by these bacteria resemble the colicins in being inducible by mitomycin C and UV irradiation. Bacteriocins produced by strains of *Pseudomonas aeruginosa* are called aeruginocins or pyocins (*P. aeruginosa* was formerly called *P. pyocyanea*). Several structures which resemble the contractile tails of T-even phages are called pyocins, although phage-like structures produced by *E. coli* strains are now usually called 'defective phages'. Pyocins resembling phage tails are called R type pyocins and in fact some of them have been shown by immunological methods to be closely related to certain *Pseudomonas* phages. The smaller pyocins, which are proteins resembling colicins, are called S type pyocins.

Strains of *Neisseria meningitidis* which produce bacteriocins have been reported and a few instances described where a strain of *N. meningitidis* which inhibited *N. gonorrhoeae* was apparently responsible for preventing an infection by the gonococcus in man. But it is not known how common such a protection mechanism may be.

7.2 Bacteriocins produced by Gram-positive bacteria

Gram-positive bacteria which have been shown to produce bacteriocins include strains of *Streptococcus, Staphylococcus, Clostridium* and *Bacillus* (Tagg *et al.* 1976). The functions of these inhibitors in the biology of the plasmids or bacteria which determine their synthesis are unknown. They are classified as bacteriocins because they are bactericidal proteins which kill many groups of Gram-positive bacteria but no Gram-negative strains.

7.2.1 Staphylococcus

About 5 per cent of *Staphylococcus aureus* strains produce proteins which kill strains of

Staphylococcus and other Gram-positive bacteria. Some strains of *S. albus* also produce bacteriocins.

Staphylococcins are more difficult to purify than bacteriocins produced by Gram-negative bacteria because their synthesis cannot be induced by agents such as mitomycin C, and they are often difficult to remove from the cells which produce them. Three staphylococcins have been purified and have molecular weights of about 9 000, 12 500 and 15 000, respectively but the molecules readily associate into larger complexes. All three staphylococcins were composed of protein, lipid and carbohydrate and were inactivated by proteases.

Staphylococcin production is often an unstable character; cells which have lost the ability to produce staphylococcin occur frequently in cultures of staphylococcin-producing strains, suggesting that many staphylococcins are plasmid-determined. A bacteriocin produced by a Group II strain of *S. aureus* is determined by a plasmid which has a molecular weight of 33×10^6. The same plasmid also determines a toxin which is responsible for skin lesions known as scalded-skin syndrome.

Although there is no doubt that different strains of *Staphylococcus* interfere with each other's growth on skin and that this is an important feature of the ecology of the skin, the contribution made by staphylococcins to this antagonism is unknown. In fact, a bacteriocin was not responsible for the best-documented example of an antagonistic strain of *S. aureus* (strain 502A) which prevented colonisation of the skin by other staphylococci.

7.2.2 Streptococcus

Streptococcins which have been isolated resemble the staphylococcins in not being inducible and in being composed of small subunits which associate together to form larger structures. The low molecular weight forms of streptococcins produced by two group A streptococci had molecular weights of about 8 000 and 29 000. Bacteriocins produced by two group B streptococci both had molecular weights of about 10 000. All the streptococcins were inactivated by proteases.

The ecological importance of streptococcins is unknown. Both intra-species antagonisms and inter-species antagonisms are important factors in determining the numbers of the various species which colonise the human throat, but the significance of streptococcins in these interactions is unknown.

8 Uses of bacteriocins

The available data on colicins suggests that it is unlikely that they could have widespread therapeutic uses. There is no evidence that they kill bacteria in the alimentary tract, and the ColV plasmid actually increases pathogenicity. Most bacteriocins are relatively large protein molecules and many elicit the production of antibodies which neutralise their activity. Also, bacteriocin-resistant mutants arise at high frequency.

The possibility of replacing a pathogenic strain with harmless bacteria received much attention in the early years of the century, before the nature of bacteriocins was understood. A preparation of colicinogenic bacteria, known as 'Mutaflor', was used to treat intestinal diseases. Attempts have been made recently in Australia and California to control the plant pathogen *Agrobacterium tumifaciens,* which is responsible for Crown Gall (see Ch. 4, section 3.7), by inoculating plants with a non-pathogenic strain of *Agrobacterium* which produces a bacteriocin.

Bacteriocins are used in epidemiological investigations to type strains of bacteria, especially strains of *Shigella sonnei* and *Pseudomonas aeruginosa*. For example, to determine whether the same strain of *S. sonnei* is responsible for a number of cases of dysentery, bacteria isolated from each case are analysed to determine whether they all produce the same colicin type(s). Colicin typing of *S. sonnei* can be used in epidemiological investigations because the production of various types of colicin is a sufficiently common and stable feature of these bacteria.

9 Summary

Bacteriocins are protein antibiotics produced by bacteria. The bacteriocins which have been studied in most detail are the colicins, proteins with molecular weights of between about 27 000 and 90 000. The genes for colicins are on conjugative and non-conjugative Col plasmids found in strains of *E. coli* and related enterobacteria. Colicins kill only strains of *E. coli* and its close relatives and are classified into about 20 types according to their effects on colicin-resistant mutants.

Colicin synthesis seems to be repressed in most Col$^+$ cells; only a small proportion of cells containing a Col plasmid produce detectable amounts of colicin. Synthesis of colicins can be induced by ultraviolet irradiation or with the antibiotic mitomycin C. The mechanism of the induced synthesis of colicins is unknown, but it closely resembles prophage induction in *E. coli*.

Colicins adsorb to receptors, largely protein molecules, in the outer membranes of sensitive bacteria. An energy-dependent step following adsorption to receptors can be distinguished. Several colicins (for example, E1 and K) kill cells by disrupting processes which depend on the energised state of the cytoplasmic membrane. Colicin E3 and cloacin DF13 kill cells by cleaving the 16S RNA molecule in bacterial ribosomes. These two bacteriocins have the same effect on ribosomes *in vitro,* indicating that they enter cells. The DNA of colicin E2-treated cells is degraded. Colicin E2 is an endonuclease. Cells containing a ColE3 (or CloDF13) plasmid are immune to colicin E3 (or cloacin DF13) apparently because they contain a plasmid-determined immunity protein (M.W. 10 000) which binds to the bacteriocin and prevents it inactivating ribosomes.

Many aspects of the ecology of colicins are unknown. It is not known whether colicins kill significant numbers of sensitive bacteria in the alimentary tract. A ColV plasmid increased the ability of an *E. coli* strain to survive in the human alimentary tract, but a ColE plasmid did not enable the same strain to survive longer in the gut. Many ColV plasmids increase the pathogenicity of invasive strains of *E. coli.*

Antibiotic proteins are produced by most genera of Gram-negative and Gram-positive bacteria. But without further data it is difficult to decide on criteria which can be used to define bacteriocins, particularly for Gram-positive bacteria, because such a wide variety of inhibitors are produced. Bacteriocins isolated from staphylococci and streptococci have molecular weights of between 8 000 and 30 000, although they are also found as larger structures formed by the association of the smaller subunits. Several are composed of protein lipid and carbohydrate. The synthesis of staphylococcins and streptococcins is not inducible by mitomycin C. Many staphylococcins are probably determined by plasmids. The significance of bacteriocins produced by Gram-positive bacteria in the ecology of these bacteria is unknown.

Bibliography

Books and reviews

FALKOW, S. (1975) *Infectious multiple drug resistance,* Pion, London.
FREDERICQ, P. (1957) 'Colicins', *Ann. Rev. Microbiol.,* **11**, 7–22.
HARDY, K. G. (1975) 'Colicinogeny and related phenomena', *Bacteriol. Rev.,* **39**, 464–515.
HELINSKI, D. R. (1976) 'Plasmid DNA replication', *Fed. Proc. Amer. Soc. Exp. Biol.,* **35**, 2026–30.
HOLLAND, I. B. (1975) 'Physiology of colicin action', *Adv. Microbiol. Physiol.,* **12**, 55–139.
KONISKY, J. (1973) 'Chemistry of colicins', *Chemistry and Functions of Colicins* (Ed. L. Hager), pp. 41–58. Academic Press, New York.
LURIA, S. E. (1975) 'Colicins and the energetics of cell membranes', *Scient. Am.,* **233**, 30–7.
MEYNELL, G. G. (1972) *Bacterial plasmids.* Macmillan, London.
NOMURA, M., SIDIKARO, J., JAKES, K. and ZINDER, N. (1974) 'Effects of colicin E3 on bacterial ribosomes', *Ribosomes* (Eds. M. Nomura, A. Tissieres and P. Lengyel), pp. 805–14. Cold Spring Harbor Laboratory, New York.
PREER, J. R., PREER, L. B. and JURAND, A. J. (1974) 'Kappa and other endosymbionts in *Paramecium aurelia',* *Bacteriol. Rev.,* **38**, 113–63.
REEVES, P. (1972) *The bacteriocins.* Chapman and Hall, London.
TAGG, J. R., DAJANI, A. S. and WANNAMAKER, L. W. (1976) 'Bacteriocins of gram-positive bacteria', *Bacteriol. Rev.,* **40**, 722–56.
WICKNER, R. B. (1976) 'Killer of *Saccharomyces cerevisiae*: a double-stranded ribonucleic acid plasmid', *Bacteriol. Rev.,* **40**, 757–73.

Papers

DE ALWIS, M. C. L. and THOMLINSON, J. R. (1975) 'Some factors influencing colicin activity between pathogenic and commensal *Escherichia coli* from the pig', *Res. Vet. Sci.,* **19**, 63–70.
HERSCHMAN, H. R. and HELINSKI, D. R. (1967) 'Purification and characterisation of colicin E$_2$ and colicin E$_3$', *J. Biol. Chem.,* **242**, 5360–8.
HERSHFIELD, V., BOYER, H. W., YANOFSKY, C., LOVETT, A. and HELINSKI, D. R. (1974) 'Plasmid ColE1 as a molecular vehicle for cloning and amplification of DNA', *Proc. Natl. Acad. Sci. USA,* **71**, 3455–9.
INSELBURG, J. (1973) 'Colicin factor DNA: a single non-homologous region in ColE2-E3 heteroduplex molecules', *Nature New Biol.,* **241**, 234–7.
KOPECKY, A. L., COPELAND, D. P. and LUSK, J. E. (1975) 'Viability of *Escherichia coli* treated with colicin K', *Proc. Natl. Acad. Sci. USA,* **72**, 4631–4.
PLATE, C. A., SUIT, J. L., JETTEN, A. M. and LURIA, S. E. (1974) 'Effects of colicin K on a mutant of *Escherichia coli* deficient in Ca^{2+}, Mg^{2+}-activated adenosine triphosphatase', *J. Biol. Chem.,* **249**, 6138–43.
SCHALLER, K. and NOMURA, M. (1976) 'Colicin E2 is a DNA endonuclease', *Proc. Natl. Acad. Sci. USA,* **73**, 3989–93.
SHARP, P. A., COHEN, S. N. and DAVIDSON, N. (1973) 'Electron microscope heteroduplex studies of sequence relations among plasmids of *Escherichia coli*. II. Structure of drug-resistance (R) factors and F factors', *J. Molec. Biol.,* **75**, 235–55.

SMITH, H. and HUGGINS, M. B. (1976) 'Further observations on the association of the colicin V plasmid of *Escherichia coli* with pathogenicity and with survival in the alimentary tract', *J. Gen. Microbiol.,* **92**, 335–50.

TYLER, J. and SHERRATT, D. J. (1975) 'Synthesis of E colicins in *Escherichia coli'*, *Molec. Gen. Genet.,* **140**, 349–53.

VELTKAMP, E., POLS, K., VAN Ee, J. H. and NIJKAMP, J. J. (1976) 'Replication of the bacteriocinogenic plasmid CloDF13: action of the plasmid protein cloacin DF13 on CloDF13 DNA', *J. Molec. Biol.,* **106**, 75–95.

6
Molecular basis of toxin action

J. P. Arbuthnott

Department of Microbiology, Moyne Institute, Trinity College, University of Dublin

1 Scope of chapter

This chapter will deal with the mode of action of toxins produced by bacterial pathogens of man and animals. It will not include material on fungal or algal toxins or toxins produced by plant or fish pathogens. As bacterial toxins are dealt with only superficially, if at all, in most textbooks of microbiology, the early sections of the chapter will provide a general introductory background covering the definition of toxins, their importance in pathogenicity, the diversity of their biological effects, assay methods and a few general comments on production and purification. This is followed by an account of what is known of the molecular basis of action of some of the more important groups of toxins.

It is important to remember that studies on bacterial toxins touch on many different disciplines including microbiology, biochemistry, physiology and pharmacology and it is difficult to present an overall account of the present state of knowledge in a short chapter. The mechanisms of action of at least four bacterial toxins (diphtheria toxin, cholera enterotoxin, staphylococcal β-toxin and *Clostridium welchii* α-toxin) can be clearly defined in biochemical terms. The mechanisms of several others have been the subject of detailed study but as yet the molecular events responsible for their toxic action are not fully understood.

2 Toxins and pathogenicity

The concept that pathogenic microorganisms exert their inimical effects on man and animals by elaborating diffusible toxic substances is almost as old as the germ theory of disease. In the early 1870s Klebs suggested that 'sepsins' were responsible for lesions produced by staphylococci. About ten years later Robert Koch concluded that cholera was a toxicosis on the basis that the main symptom of the disease, a massive outpouring of fluid from the intestine, was not accompanied by invasion by the causative organism or by damage to the gut wall or neighbouring tissues. Around the same time Loeffler, who established the diphtheria bacillus as the cause of diphtheria, drew attention to the fact that although characteristic lesions could be found widely distributed in the organs of the body, these lesions were sterile and the diphtheria bacillus could be isolated only from the 'false membrane' in the throat. He therefore suggested that a poison produced at the seat of inoculation must have circulated in the blood. This prediction was verified later by Roux and Yersin who demonstrated that sterile filtrates from cultures of the diphtheria bacillus contained a poison or toxin which mimicked the symptoms of the disease when injected into experimental animals.

The search for other bacterial toxins began and met with early success. Within a few years tetanus toxin (tetanospasmin) and botulinum toxin were discovered. Proof of the importance of toxins as disease-producing agents was provided by the demonstration that experimental animals inoculated with small doses of attenuated tetanus and diphtheria toxins were rendered immune to these agents. (For an introduction to the historical aspects see van Heyningen 1970.)

Not surprisingly these findings raised hopes that the harmful effects of infectious diseases could be conquered by immunisation against the appropriate toxins. In fact the success of this approach has been limited and we now know that in many instances bacterial pathogenicity is a complex process involving successful survival of the pathogen on the body surfaces of the host, followed by penetration of host tissues and multiplication *in vivo* at the expense of non-specific and specific host defence mechanisms (see Smith 1976). Damage of host tissues by toxins is an important part of the pathogenic mechanisms in many infectious diseases. However there are still relatively few examples in which we can pinpoint the extent to which individual toxins are responsible for the symptoms of the disease. Table 6.1 lists most of the diseases in which the role of toxins in pathogenicity is well established.

The success of preventative measures together with chemotherapy and improved living standards has done much to eliminate infectious disease as a leading cause of death in many parts of the world. Paradoxically this success was probably responsible also for a decline in interest in research in the fields of pathogenicity and toxins. Changing patterns of infectious disease and the emergence of antibiotic-resistant organisms has led to a reawakening of interest but there are many fundamental questions concerning the role of toxins that remain to be answered. Medical considerations apart, bacterial toxins have a fascination of their own. They exert potent and diverse biological effects and knowledge of their mode of action at the molecular level has proved of interest in many areas of biology.

3 Diversity and special features of toxins

3.1 Summary of toxic properties

Bacterial toxins can be defined as a group of bacterial products whose principal common feature is that they are harmful to various sensitive hosts when administered in relatively small doses. It is difficult to give a more precise definition as toxins vary widely in potency and in the nature of the lesion or harmful effect produced. Here it is interesting to quote the somewhat rueful comments of one of the pioneers in the field, the late Professor Oakley: 'The word "toxin" has had its meaning fixed in such a way that it has ceased to be of much practical value. . . . However the word is so fixed in the literature and is so romantic in its sound that I have little hope of ejecting it.'

Most of the toxins to be described in this chapter are proteins. However an important toxic component of Gram-negative bacterial cell envelopes, lipopolysaccharide, consists of lipid and polysaccharide and here it is the lipid which is responsible for toxicity (see section 10). This lipopolysaccharide is commonly referred to as endotoxin because it is a structural component of the bacterial cell. Indeed the division of toxins into two classes, exotoxins which are secreted and endotoxins which remain associated with the cell, has proved useful especially to medical

Table 6.1   Toxins known to play an important role in pathogenesis

Disease	Causative organism	Toxin	Main biological effect of toxin
Diphtheria	*Corynebacterium diphtheriae*	Diphtheria toxin	Lethal, necrotising
Tetanus	*Clostridium tetani*	Tetanospasmin	Lethal, neurotoxic
Botulism	*Clostridium botulinum*	Botulinum toxins	Lethal, neurotoxic
Enterotoxaemia in domestic animals	*Clostridium welchii* types B, C and D	β-toxin ε-toxin	Lethal, necrotising
Cholera	*Vibrio cholerae*	Cholera enterotoxin	
Acute diarrhoeal disease in pigs and calves	*Escherichia coli* (certain animal strains)	*E. coli* enterotoxins	Defects in water and electrolyte transport in gut epithelial cells
Infantile gastroenteritis	*Escherichia coli* (certain human strains)	*E. coli* enterotoxins	
Food poisoning	*Clostridium welchii* type A	Clostridial enterotoxin	
Food poisoning	*Staphylococcus aureus*	Staphylococcal enterotoxins	Emetic and diarrhoeagenic
Staphylococcal scalded-skin syndrome	*Staphylococcus aureus*	Epidermolytic toxin (exfoliatin)	Cell separation in epidermis
Scarlet fever	*Streptococcus pyogenes*	Erythrogenic toxin	Erythema in skin

microbiologists. More recently, for a variety of reasons, this distinction has become less clear-cut although the term endotoxin can still be appropriately used to describe the lipopolysaccharide of Gram-negative envelopes.

There are several important points to be considered when comparing the biological activities of bacterial toxins.

1 Certain bacterial species are capable of producing a number of different toxins some of which have similar biological effects (Table 6.2).

Table 6.2   Examples of bacteria which produce several toxins

Organism	Toxin	Biological effect(s) of toxin
Staphylococcus aureus	α-Toxin	Membrane-damaging (lethal, dermonecrotic, haemolytic)
	β-Toxin	Membrane-damaging (haemolytic); sphingomyelinase C
	γ-Toxin	Membrane-damaging (lethal, haemolytic)
	δ-Toxin	Membrane-damaging (? lethal, haemolytic)
	Leucocidin	Membrane-damaging (affects leucocytes of man and rabbit)
	Epidermolytic toxin	Causes epidermal splitting
Streptococcus pyogenes	Streptolysin O	Membrane-damaging (lethal, cardiotoxic, haemolytic)
	Streptolysin S	Membrane-damaging (lethal, haemolytic)
	Erythrogenic toxin	Causes erythema
Clostridium welchii type B	α-Toxin	Membrane-damaging (lethal, necrotising, haemolytic); phospholipase C
	β-Toxin	Lethal, necrotising
	ε-Toxin	Lethal, necrotising
	κ-Toxin	Breakdown of muscle tissue; collagenase
	μ-Toxin	Spreading factor; hyaluronidase

2 Potency depends on the purity and concentration of the toxin as well as the route of injection and the species, age and weight of the test animal.

3 A few bacterial toxins are highly specific in their action but most are toxic for several different tissues and cell types. For instance staphylococcal epidermolytic toxin acts only on the epidermis of susceptible animals whereas staphylococcal α-toxin is a more general cytotoxic agent, being lethal, dermonecraotic and haemolytic. The range of toxic activity presumably depends on the distribution of substrate or target site in the susceptible host.

4 It holds for most toxins that the active site resides in a single molecular species. However there are a few examples where toxic action involves synergism between different components. For instance staphylococcal leucocidin is a toxic complex consisting of two protein components that act together to kill polymorphonuclear leucocytes and macrophages. Individually they have no toxic action on these cells.

These factors make it difficult to compare the relative toxicities of different bacterial toxins. In Table 6.3 are listed the potencies of some purified toxins in different assay systems. The most toxic in terms of lethal action, namely botulinum toxin, is about a million times more potent than strychnine, and is the most potent known poison.

Table 6.3   Potency of some bacterial toxins

Toxin	Toxic dose (µg)	Assay
Botulinum type A toxin	2.5×10^{-5}	Lethal action in mice
Tetanospasmin	4.0×10^{-5}	Lethal action in mice
Cholera enterotoxin	3.0×10^{-2}	Fluid accumulation in rabbit intestinal loops
Diphtheria toxin	6.0×10^{-2}	Lethal action in guinea pigs
Epidermolytic toxin	0.3	Epidermal splitting in neonatal mice
Staphylococcal α-toxin	1.0	Lethal action in mice
Streptolysin O	1.0	Lethal action in mice
E. coli enterotoxin	5.0	Fluid accumulation in rabbit intestinal loops
Staphylococcal enterotoxin B	5.0	Emesis in monkeys
Lipolysaccharide endotoxin (*Salmonella typhimurium,* 0901)	160	Lethal action in mice

3.2 Assay of toxins

Basically three types of tests can be used to assay bacterial toxins.

1 Measurement of biological potency in terms of the amount of toxin which will produce a particular indicator effect *in vivo* or *in vitro* such as death, necrosis, increased capillary permeability, fluid accumulation in intestinal preparations or haemolysis.

2 Determination of the amount of toxin which when mixed with a known amount of standardised antitoxin will produce an effect in an appropriate indicator system.

3 Measurement of antigen/antibody interaction.

The first of these is used most frequently and units are expressed as the smallest amount of toxin which will produce an indicator effect in a certain proportion of the animals, tissue preparations or cells tested. It is desirable that the end-point should lie on the steepest part of the dose response curve and, if for example, death is taken as the indicator then the unit employed is usually the LD_{50} i.e. the amount of toxin which kills 50 per cent of the animals tested. Similarly, haemolytic toxins can be assayed conveniently by determining the smallest amount of toxin which will cause 50 per cent haemolysis of a standardised suspension of erythrocytes. Variation between different animals must be taken into account in any bioassay. Also as mentioned previously there are considerable variations in the susceptibility of different animal species to a given toxin. For example, some samples of botulinum toxin are 6 000 times more toxic for guinea pigs than for mice. Haemolytic toxins also exhibit specificity: the α-toxin of *Staphylococcus aureus* is 100 times more haemolytic for rabbit erythrocytes than for human erythrocytes.

3.3 Toxin production and purification

Toxin production may be affected by a number of factors including the strain of the organism, the culture medium and the conditions of culture. The amount of toxin released into the culture fluid under optimal conditions varies for different toxins but is in the range $1-500$ mgl^{-1} and usually does not exceed 100 mgl^{-1}.

Different strains of the same species often vary in their capacity to produce toxin. In some cases, notably *Corynebacterium diphtheriae* ability to produce toxin is determined by the presence of a lysogenic phage which carries the genetic information coding for toxin production. Lysogenic conversion has also been implicated in the production of streptococcal erythrogenic toxin, staphylococcal β-toxin and type C and D botulinum toxin. Genes controlling toxin production can also be located in plasmid DNA. This has been established for *Escherichia coli* enterotoxin and staphylococcal epidermolytic toxin.

The choice of growth media and cultural conditions for toxin production have been largely empirical. Ideally the medium would be chemically defined, containing only low molecular weight components. This would facilitate the purification of the toxin and would avoid the necessity of having to separate the toxin from macromolecular components of the culture medium. Unfortunately factors influencing the synthesis and release of bacterial toxins are poorly understood and chemically defined media which support the growth of toxinogenic bacteria often do not promote good toxin production. This has led to the development of media based on acidic or enzymatic protein hydrolysates. These favour toxin production and are free of foreign macromolecules. Other factors such as pH, and composition of the gas phase must also be controlled to ensure good production of toxin.

The separation of individual toxins from other extracellular products of growth and from medium constituents is often a challenging problem. As we have seen, some organisms produce a number of toxins and enzymes. In defining the properties and mode of action of a toxin it is essential to obtain the toxin in question in a high state of purity. Contamination with other toxins and enzymes can affect the toxin's behaviour in certain test systems. Progress in this field has generally paralleled advances in protein purification technology. Toxins can be purified on the basis of molecular weight and molecular charge in many column chromatographic and electrophoretic procedures. Affinity chromatography and immunoabsorbent chromatography have added to the range of techniques available.

The degree of homogeneity of purified toxin preparations can be assessed by a variety of analytical techniques such as electrophoresis and isoelectric focusing. When combined with immunological methods or *in vitro* biological indicator systems, these techniques allow the detection of traces of contaminating proteins.

3.4 Toxins as proteins

Several toxins are released into the culture media as inactive *protoxins* resembling the zymogens, trypsinogen and pepsinogen. Such protoxins are relatively inactive biologically but can be converted into active toxins by the action of proteolytic enzymes which may be present in the culture fluid or in the gut of animals ingesting toxins. Protoxins have been demonstrated for the neurotoxins of *Clostridium botulinum* and for *Cl. welchii* ε- and ι-toxins. Activation of the botulinum protoxins occurs as a result of cleavage of the toxin molecule into at least two polypeptides which can be separated when the disulphide bonds linking them are reduced.

Molecular weight studies have revealed that several bacterial toxins can exist as aggregates of subunits. For example crystalline type A botulinum toxin has a molecular weight of 900 000 and is a complex consisting of neurotoxin components (M.W. 120 000–150 000) and haemagglutinin (M.W. 500 000). More commonly, aggregated forms of bacterial toxins result from polymerisation of identical subunits. The polymeric form of staphylococcal α-toxin contains six subunits arranged in hexagonal assay and has a characteristic morphology which can be seen in negatively stained preparations with the electron microscope. The conditions governing aggregation and dissociation are poorly understood. Mild heating, say to 60°C, often induces aggregation, and spontaneous aggregation sometimes occurs during purification leading to loss of biological activity during purification. The presence of aggregates has also been responsible for many discrepancies in molecular weight determinations performed in different laboratories. Aggregated forms can be dissociated by manipulating the pH and ionic strength of the solution and by the addition of dissociating agents such as urea or guanidine HCl. Recent work on diphtheria toxin (see section 5) and cholera enterotoxin (see section 6) has shown that the molecular structure of bacterial toxins has an important bearing on the molecular mode of action of these toxins and it is likely that future studies will lead to an understanding of structure function relationships for other toxins.

The technique of isoelectric focusing which separates proteins according to their isoelectric points has been widely used in the purification and characterisation of bacterial toxins. Almost all toxins so far examined have been found to exhibit multiple forms having different isoelectric points. Staphylococcal enterotoxin C_2 for example has been shown to contain up to eight different components. In only a few instances has the relationship between different molecular forms been investigated. Possible causes of multiple forms revealed by isoelectric focusing include aggregation, modification of the covalent structure of the toxin (such as deamidation or cleavage of peptides by proteases) and binding of ligands.

An important feature of bacterial toxins is that under certain conditions they may lose their toxicity without losing their antigenicity. Such atoxic derivatives of toxins are known as toxoids. Toxoiding may happen spontaneously (e.g. choleragenoid is a naturally-occurring toxoid of cholera toxin) or following treatment with one of several toxoiding agents such as formaldehyde or glutaraldehyde. Toxoids are valuable immunising agents and in developed countries almost everyone is now routinely immunised in infancy with diphtheria toxoid and tetanus toxoid. Numerous toxoid vaccines are employed in the prevention of infectious diseases in domestic animals and with the recent discovery of several enterotoxins there is renewed interest in developing additional toxoid vaccines.

4 Approaches to the study of mechanism of action

The problem of investigating the mechanism of action of bacterial toxins has been tackled at various levels — in the whole animal, in tissue and cell preparations and in cell-free systems.

Experiments carried out *in vivo* in the whole animal yield useful information about the physiological and biochemical changes occurring in toxin-damaged tissues. The different sites of action of neurotoxins such as botulinum toxin and tetanus toxin have been identified in this way. Also much useful information on the impairment of intestinal transport by the entero-toxins of *Vibrio cholerae, Escherichia coli, Clostridium welchii* and other organisms has been obtained by introducing these toxins directly into the intestine. It is important to stress once more the need to use an appropriate experimental system. Staphylococcal epidermolytic toxin for instance causes epidermal splitting in neonatal mice but the effect is not seen in certain other animals such as the rat and is difficult to demonstrate in the adult mouse.

In order to obtain information on the interaction between toxins and their receptor sites or substrates and to study the molecular basis of toxin action it is necessary to devise experimental models consisting of animal tissues, tissue extracts, organ cultures, cell cultures, or suspensions of cells and in some cases subcellular fractions have been used. At this point it would be useful to consider briefly how some of these model systems have been used to study the mechanism of toxin action.

Nerve muscle preparations have been used in studies of botulinum toxin and tetanus toxin. Extracts of nervous and intestinal tissues have led to the identification of gangliosides as likely receptors for tetanus toxin and cholera enterotoxin. The action of staphylococcal epidermolytic toxin can be mimicked in epidermal tissue maintained in organ culture. With cell cultures it has been possible to use several criteria such as morphological changes, cell lysis, degranulation, cytopathic effects, release of intracellular markers and metabolic changes to investigate toxin action at the cellular level. Erythrocytes, platelets, polymorphonuclear leucocytes, macrophages and lymphocytes are used extensively as test systems for the group of toxins known as cytolytic toxins which have the characteristic property of altering permeability properties of cell membranes. Studies of the inhibition of protein synthesis by diphtheria toxin both in whole cells and cell-free preparations has led to major advances in the understanding of the molecular basis of action of this toxin. Isolated cell membranes, and artificial lipid membranes have been used to study toxin membrane interaction by electron microscopic and physical chemical methods. However further progress in this field will require collaboration between 'toxicologists' and biophysicists.

Having considered the general properties of bacterial toxins as a group, I will now summarise our knowledge of the molecular basis of action of some individual toxins and groups of toxins. There is no entirely satisfactory system of classifying these agents but where possible I have grouped together toxins which produce similar pathophysiological effects.

5 Diphtheria toxin

Before dealing with the molecular basis of action of diphtheria toxin it is necessary to give a brief account of lysogeny and toxinogenicity in *Corynebacterium diphtheriae* because important advances in our knowledge of structure and function relationships of the diphtheria toxin molecule have been made by studying proteins related to diphtheria toxin produced by cells infected with mutants of the temperate corynebacteriophage phage *β*.

In 1951 Freeman found that certain strains of *C. diphtheriae* after exposure to a particular phage preparation became capable of producing diphtheria toxin. The majority of cells surviving treatment with phage were toxinogenic and lysogenic. The later finding that colonies from a lysogenic strain no longer carrying phage *β* had lost toxinogenicity strengthened the view that lysogeny and toxinogencity were related. Detailed genetic and molecular biological studies have established that phage *β* carries the structural gene (*tox*) for diphtheria toxin (see Singer 1976). In 1971 Uchida, Gill and Pappenheimer opened a new area of investigation by showing that a phage mutant directed production of a non-toxic polypeptide which cross-reacted with diphtheria antitoxin; this protein was designated cross-reacting material (CRM). Nine such CRM proteins have been described so far and their properties will be mentioned later. It is interesting to note that diphtheria toxin has no role in *β*-phage development or prophage maintenance. Also prophage induction and lytic growth is not a prerequisite for toxin synthesis by lysogenic strains.

Purified diphtheria toxin is lethal to sensitive eukaryotic species (e.g. guinea pig, rabbit, man). It also kills primary and established cell cultures derived from a variety of animal tissues. Strauss and Hendee (1959) were the first to show that diphtheria toxin inhibits protein synthesis. Collier, Gill, Pappenheimer and others (see Murphy 1976) played a major role in unravelling the events involved in the inhibition of protein synthesis by diphtheria toxin in cell-free systems. It was found that nicotinamide adenine dinucleotide (NAD) was required for inhibition and that protein synthesis was inhibited due to failure to transfer amino acids from aminoacyl-tRNA into growing polypeptide chains. The proteins known as Elongation Factors 1 and 2 (EF-1 and EF-2) are necessary for the binding of aminoacyl-tRNA to the ribosome and translocation, respectively. In 1967 Collier found that EF-2 activity but not EF-1 activity was impaired by diphtheria toxin. A further piece of evidence provided an explanation for these observations. Diphtheria toxin was found to catalyse a reversible enzymatic reaction in which the adenosine diphosphate ribose (ADPR) moiety of NAD is transferred to EF-2, preventing the latter from further participation in protein synthesis. Diphtheria toxin therefore possesses NAD:EF-2 ADPR transferase activity.

$$\text{NAD}^+ + \text{EF-2} \quad \underset{\substack{\text{NAD:EF2 Transferase} \\ \text{(diphtheria toxin)}}}{\rightleftharpoons} \quad \text{ADPR} - \text{EF2} + \text{nicotinamide} + \text{H}^+.$$

So far we have considered only the mode of action of diphtheria toxin in cell-free extracts. An understanding of the action of the toxin on whole cells requires that we consider the molecular structure of the toxin molecule.

Diphtheria toxin is synthesised as a single polypeptide chain having a molecular weight of 62 500. It contains two disulphide bridges and the peptide loop contained within the first of these bridges is extremely sensitive to proteolytic cleavage by trypsin, giving rise to what is known as 'nicked' toxin (Fig. 6.1). This comprises two fragments A and B, linked by the remaining disulphide bridge. In the presence of suitable reducing agents fragments A (M.W. 24 000) and B (M.W. 38 000) can be separated. It is fragment A which has NAD:EF-2 transferase activity. However when tested in animals or tissue culture systems fragment A is non-toxic by itself and requires fragment B for expression of toxic activity. The function of fragment B is specific binding of diphtheria toxin to the plasma membrane of susceptible cells. It seems that fragment B remains attached to this site while fragment A enters the cell and inhibits protein synthesis by its action on EF-2.

Support for this interpretation of the role of fragments A and B comes from studies of the CRM proteins produced by *C. diphtheriae* cells lysogenised with mutants of phage having defects in either the A or B region of the molecule (Table 6.4).

Table 6.4   Summary of properties of some CRM proteins

Protein	M.W.	Fragment containing defect	Enzymatic activity	Binding activity
CRM_{30}	30 000	B	+	−
CRM_{45}	45 000	B	+	−
CRM_{176}	62 000	A	±	+
CRM_{197}	62 000	A	−	+
CRM_{228}	62 000	A and B	−	±

6 The enterotoxins

In the last 10–20 years the discovery and characterisation of several enterotoxins responsible for the symptoms of diarrhoeal diseases (enteropathies) caused by enteropathogenic bacteria has represented a major advance in the field of bacterial toxins.

The enterotoxins of *V. cholerae*, *E. coli*, *Cl. welchii* and *S. aureus* have been studied in considerable detail. However the molecular basis of action has been elucidated only for *V. cholerae* and *E. coli* enterotoxins.

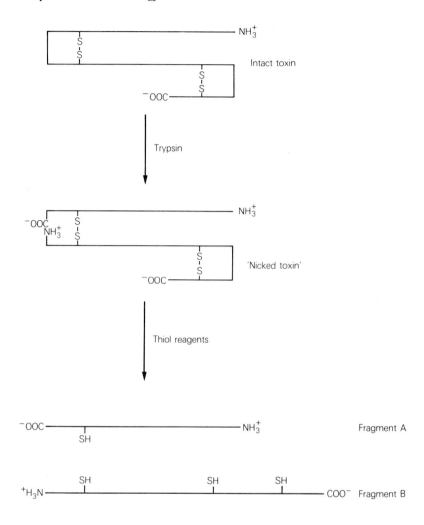

Fig. 6.1 Cleavage of diphtheria toxin into fragments A and B.

6.1 Cholera enterotoxin

Several reasons account for the delay in establishing the view expounded by Koch in the 1880s that cholera is a toxicosis. One of the most important was the lack of suitable experimental animal models. This problem was overcome by two groups of Indian investigators. In 1953 De and Chatterje showed that introduction of cholera vibrios into isolated segments of rabbit ileum (ileal loops) *in vivo* led to an outpouring of fluid which resulted in distention of the loop. Dutta and Habbu in 1955 used a different approach taking the development of fatal cholera-like diarrhoea in suckling rabbits that had been injected intra-intestinally, as an index of activity. It was soon shown that similar effects could be produced by cell-free material derived from *V. cholerae* organisms and by culture filtrates. Cholera enterotoxin was first purified in 1969 by Finkelstein and LoSpolluto and there followed a period of intensive research in a number of laboratories which has led to elucidation of the mode of action of cholera enterotoxin at the molecular level. This has proved important in increasing our understanding of the pathogenesis of cholera and other enteropathies. Also purified cholera enterotoxin is now used extensively as a probe for reactions mediated by adenosine-3'5' cyclic monophosphate (cyclic AMP).

Cholera enterotoxin (M.W. 84 000), like diphtheria toxin, consists of a biologically active region (A) and a second region (B) which binds to specific receptors on the surface of host cells (see Finkelstein 1976). In the presence of the dissociating agent sodium dodecyl sulphate the A region (M.W. 28 000) and the B region (M.W. 56 000) can be separated. Further study of the structure of cholera enterotoxin revealed that the A region consists of two disulphide-linked peptides designated A_1 and A_2 while the B region consists of six non-covalently linked peptide subunits of equal size. A naturally recurring toxoid known as choleragenoid contains only the B region of the molecule. The cholera toxin molecule may have a 'Y'-shaped structure (Fig. 6.2).

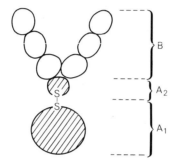

Fig. 6.2 Possible arrangement of subunits in cholera enterotoxin.

As a result of several independent studies the following important conclusions were made concerning the mode of action of cholera enterotoxin.

a Cholera toxin stimulated the activity of adenylate cyclase and cyclic AMP levels were elevated by cholera toxin.

b This could account for the hypersecretion of electrolytes and water from the intestine.

c These events were found to occur in human cholera.

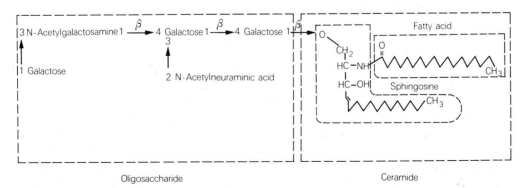

Fig. 6.3 The structure of G_{M1} ganglioside which is thought to be the receptor for cholera enterotoxin.

Further advances in our understanding of the action of cholera enterotoxin stemmed from the work of van Heyningen (1974) whose earlier studies on tetanospasmin (see section 7) led him to investigate the possibility that gangliosides (Fig. 6.3) serve as the receptors for cholera enterotoxin in host cell membranes. Gangliosides are acidic glycosphingolipids that consist of an oligosaccharide moiety (with one or more residues of *N*-acetylneuraminic acid (sialic acid) linked to ceramide which consists of the long-chain amino alcohol sphingosine connected by an amide linkage to a long-chain fatty acid. Over 20 different types of gangliosides have been identified and these differ in the number and relative positions of hexose and sialic acid residues. One particular ganglioside, namely the monosialosyl ganglioside G_{M1} was found to block the effects of cholera enterotoxin, and there is increasing evidence that the receptor for cholera toxin is a G_{M1}-like moiety. As mentioned above it is the B region of the toxin molecule which binds to this receptor. The mode of action of cholera toxin can be stated briefly as follows:

1 Toxin molecule consists of two regions, A and B.
2 Toxin binds to monosialosyl ganglioside G_{M1}-receptor in membrane through the B region.
3 A region (probably A_1) penetrates the plasma membrane and activates adenylate cyclase.
4 Increased levels of cyclic AMP lead to hypersecretion of electrolyte and water.

Many alternative model systems are now available for the assay of cholera enterotoxin and by virtue of its ability to simulate adenylate cyclase in a number of systems (see Table 6.5) the toxin is widely used to assess the role of cAMP in many cellular processes.

Table 6.5   Some properties of cholera enterotoxin

Fluid accumulation in rabbit ileal loop.
Increased permeability of capillaries in skin.
Enhanced lipolysis of epididymal fat cells.
Increased glycogenolysis in liver tissue and platelets.
Potent and long-lasting inhibition of delayed type hypersensitivity.
Reversal of mitogenic effect of lectins on lymphocytes.
Stimulation of steroidogenesis in adrenal cells.
Alteration of morphology of cultured adrenal and Chinese hamster ovary cells.
Melanin-stimulating hormone effect on melanoma cells.
Release of growth hormone from pituitary.

6.2 Other enterotoxins

The success achieved in establishing the pathogenic role and mechanism of action of cholera enterotoxin stimulated the search for production of enterotoxins by other enteropathogenic species, particularly *E. coli*. It has been known for many years that certain strains of *E. coli* cause diarrhoeal disease in animals (pigs and calves) and man and there is now convincing evidence that these enteropathogenic strains elaborate enterotoxins. Although considerable progress has been made in defining the properties and mode of action of these toxins the molecular basis of action is less well established than for cholera toxin. There are two reasons for this. Firstly, it has proved much more difficult to purify *E. coli* enterotoxins and secondly some *E. coli* strains produce two quite different enterotoxins. These are known as heat-stable enterotoxin, or ST (Smith and Halls 1967), which is a low molecular weight material and heat-labile enterotoxin, or LT (Gyles and Barnum 1969), which is a protein. In both cases the genes coding for toxin production are plasmid-borne (see Ch. 4). Most early work was done with animal strains but more recently it has been established that enteropathogenic strains of human origin produce LT. The question of production of ST by human strains at present is unresolved (see Finkelstein 1976).

LT enterotoxins from human and animal strains show a degree of antigenic similarity with cholera enterotoxin and resemble cholera toxin in biological activity, causing fluid accumulation in the rabbit ileal loop and activation of adenylate cyclase in several different systems. There is also evidence to suggest that the receptor for *E. coli* enterotoxin is the same as for cholera enterotoxin, namely ganglioside G_{M1}. There are however several differences between the two toxins and they cannot be regarded as identical (see Finkelstein 1976 and Taylor 1976).

Enterotoxins are also known to be produced by *S. aureus* and type A strains of *Cl. welchii*. In the case of *S. aureus* six well-characterised and serologically distinct enterotoxins have been described. Both clostridial and staphylococcal enterotoxins impair water and electrolyte transport in intestinal preparations but their mechanisms of action are not fully understood. It should be noted that staphylococcal enterotoxin B has been shown to possess neurotoxic properties in addition to its effect on intestinal transport. Experiments in monkeys have shown that emesis results from an action on sensory nerves in the abdominal viscera and the sensory emetic stimulus reaches the vomitting centre via the vagus and sympathetic nerves (see Taylor 1976). There is evidence from the production of enterotoxins by other organisms including *Shigella dysenteriae, Pseudomonas aeruginosa, Bacillus cereus,* and *Salmonella typhimurium*, but in most instances little is known about the role of these agents in disease and their mode of action.

7 The neurotoxins

Several bacterial toxins affect nervous tissue, but only tetanospasmin and the botulinum toxins are recognised as acting specifically on nerve elements.

7.1 Tetanospasmin

This toxin is responsible for the symptoms of tetanus (local or generalised muscular spasm). There is no simple *in vitro* assay for tetanospasmin; its biological activity can be demonstrated only in the whole animal or in nerve muscle preparations. This makes the task of investigating its mode of action at the molecular level more difficult.

The toxin can exist in two states, as a monomer (M.W. 67 000) which is toxic and as a dimer (M.W. *circa* 140 000) which is non-toxic. At the physiological level tetanospasmin is generally regarded as acting mainly on the central nervous system and it is well known that it produces uncoordinated muscular contractions of skeletal muscles by blocking inhibitory neurones which control the motor activities of the central nervous system (see van Heyningen 1971). However there is evidence that the toxin can also act peripherally at the level of neuromuscular transmission by interfering with the release of acetyl choline in a manner analogous to botulinum toxin and Diamond and Mellanby (1971) have shown that in the goldfish the peripheral action of tetanospasmin produces flaccid rather than spastic paralysis. Both *in vivo* and in nerve muscle preparations *in vitro* the toxin appeared to block neuromuscular transmission. In goldfish there was no effect on either inhibition or excitation of the central nervous system. Thus in this species only the peripheral effect of tetanospasmin is seen.

The molecular mechanism of the action of tetanospasmin is unknown but studies on the interaction of the toxin with extracts of nervous tissue have provided valuable clues. It has been known since the beginning of the twentieth century that tetanus toxin is fixed and inactivated by emulsions of brain tissue. The detailed studies of van Heyningen and his collaborators in the 1960s (see van Heyningen 1971) established that the substance in nerve tissue responsible for fixation of tetanospasmin is ganglioside. A comparison of the toxin-fixing properties of several gangliosides revealed that fixation of tetanospasmin depends on the number and the position of sialic acid residues in the ganglioside molecule (Fig. 6.4). Gangliosides G_{D1b} and G_{T1} were most active being capable of binding up to 20 times their weight of tetanospasmin. Gangliosides which contained only one sialic acid residue in the middle of the oligosaccharide chain had poor toxin-fixing activity. As a working hypothesis it is suggested that, as in the case of cholera enterotoxin, specific ganglioside receptors are responsible for the binding of tetanus toxin at its site of action on the cell surface of susceptible nerve endings. Subsequently events probably depend on the toxin molecule (or part of it) traversing the membrane.

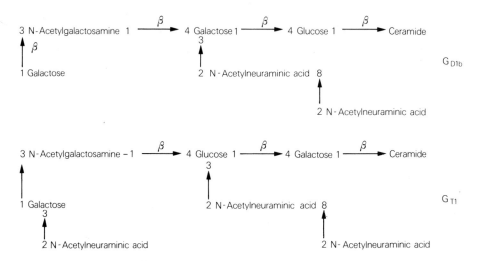

Fig. 6.4 Structure of gangliosides having a high affinity for tetanospasmin. (*Note:* N-acetyl neuraminic acid = sialic acid.)

7.2 Botulinum toxins

Botulism is caused by the ingestion of foodstuff containing preformed botulinum toxins. In humans the symptoms of the disease are difficulty in swallowing, difficulty in breathing, double vision and flaccid paralysis of the extremities.

Eight immunologically distinct neurotoxins (A, B, C_1, C_2, D, E, F and G) are produced by the *Cl. botulinum* group of organisms and cultures are typed on the basis of the serological type of toxin produced. Toxin types A, B, E and F have caused most cases of botulism in humans whereas types C, D and E predominate as causes of botulism in domestic and wild animal populations.

These toxins are the most potent poisons known; 1 μg of purified type A toxin contains about 250 000 minimal lethal doses for the mouse. The different serological types of toxin produce similar pharmacological effects.

Characterisation of the molecular properties of the botulinum toxins is complicated by the fact that purified toxins are complexes of neurotoxin (M.W. 150 000) and other components, notably haemagglutinin (see Boroff and Das Gupta 1971). These complexes can be dissociated at alkaline pH and at ionic strengths greater than 0·13. Also the toxins are released as protoxins which become maximally toxic only when activated by proteolytic enzymes. The toxins produced by proteolytic strains are activated by endogenous proteases present in the culture. Type E strains however are non-proteolytic and type E toxin must be activated by treatment with trypsin. Das Gupta and Sugiyama (1972) have suggested that proteolytic activation involves cleavage of at least one peptide bond in a peptide loop contained within a disulphide bridge. Subsequent reduction of the disulphide bridge results in the formation of two fragments (M.W. 50 000 and 102 000).

The site of action of botulinum toxin is the peripheral nervous system. The central nervous system does not seem to be affected. The physiological basis of the action of botulinum toxin is known. Flaccid paralysis is due to the action of the toxin on neuromuscular junction where the toxin binds specifically and inhibits the release of acetyl choline at motor end plates (see Boroff and Das Gupta 1971). The molecular mechanism by which this occurs is unknown. Neither synthesis nor degradation of acetyl choline is affected by the toxin. One suggestion is that botulinum toxin in some way antagonises serotonin-mediated transport of calcium ions across the cell membrane (Boroff and Das Gupta 1971). This in turn may affect acetyl choline release since the amount of acetyl choline released per nerve impulse is a function of the extracellular calcium ion concentration.

8 The cytolytic (membrane-damaging) toxins

We have seen that toxin—membrane interactions are central to the mode of action of all the toxins dealt with so far. In the case of diphtheria and cholera toxins the binding of toxin to membrane receptors triggers a series of intracellular events which lead to alteration of the metabolism of host cells. This general type of mechanism may hold also for tetanus and botulinum toxins. However a separate group of bacterial toxins exists whose cell-damaging activities depend on impairment of the permeability properties of host cell membranes. Here the molecular events involved in toxin damage occur in the plasma membrane and result from modification or rearrangement of membrane components (mainly lipids) required for structural integrity of the membrane.

Bernheimer (1970) introduced the term 'cytolytic' to describe this group of toxins. This term infers cell lysis and, although most members of this group are able to lyse erythrocytes, it is now clear that under certain conditions, permeability changes occur without consequent lysis. For this reason the term 'membrane-damaging' may be more appropriate.

A great deal is known about some membrane-damaging toxins and very little about others (see Avigad 1976, Freer and Arbuthnott 1976). It is clear that several different molecular mechanisms are involved in their mechanisms of action. Membrane disruption may be due to (*a*) enzymatic degradation of membrane phospholipids as in the case of β-toxin (sphingomyelinase C) of *S. aureus* and the α-toxin of *Cl. welchii* (phospholipase C); (*b*) a detergent-like solubilisation of membrane components by the δ-toxin of *S. aureus* and surfactin of *B. cereus*; or (*c*) the formation of 'functional pores' as a result of sterol-specific interaction by the so-called oxygen labile toxins (e.g. Streptolysin O and *Cl. welchii* θ-toxin). In this section I intend to summarise the molecular basis of the toxins for which the most significant event in toxin—membrane interaction is defined. I will then deal briefly with some toxins which have been thoroughly studied but whose mechanisms at the molecular level are as yet poorly understood. Advances in this field will require detailed physicochemical investigations of toxin—membrane interaction.

8.1 Staphylococcal β-toxin (sphingomyelinase C)

This toxin is haemolytic, the most sensitive erythrocytes being those of sheep, goat and cow. The work of Doery *et al.* (1963, 1965) and others established that staphylococcal β-toxin is an enzyme which hydrolyses sphingomyelin in the presence of Mg^{2+} ions to yield the ceramide, *N*-acylsphingosine, and phosphoryl choline

$$\text{Sphingomyelin} \xrightarrow{\text{Mg}^{2+}} \textit{N}\text{-acylsphingosine + phosphoryl choline}$$

The specificity for erythrocytes of different species seems to depend on the sphingomyelin content of the erythrocyte membrane. In sheep erythrocytes sphingomyelin accounts for 50 per cent of total phospholipid while in human erythrocytes which are approximately 100 times less sensitive to haemolysis by β-toxin, sphingomyelin represents only 26 per cent of total phospholipid.

It is now generally accepted that the inner and outer leaflets of the lipid bilayer of erythrocyte membranes are asymmetric and that most of the choline-containing phospholipids are located in the outer leaflet. Indeed β-toxin has been used as a tool in studies of the distribution of phospholipids in erythrocyte membranes (see Avigad 1976).

The phenomenon of 'hot—cold' lysis represents an interesting aspect of the action of β-toxin. When incubated with β-toxin at 37°C erythrocytes of sensitive species undergo little or no lysis even though up to 56 per cent of choline-containing phospholipids are hydrolysed under these conditions. Lysis only occurs when the treated cell suspension is chilled below 10°C; after chilling the haemolytic titre of potent purified preparations increases by up to a factor of 10^{6} to 10^{8}.

The changes in membrane structure which lead to hot-cold haemolysis by β-toxin are still not fully understood although it is generally accepted that the main biochemical event is hydrolysis of sphingomyelin in the outer leaflet of the lipid bilayer. The effects of substantial sphingomyelin hydrolysis on the morphology of the membrane have been investigated by several workers. However these studies have mainly involved the use of erythrocyte ghosts and it may not be possible to apply results obtained with ghosts directly to whole cells. With human and bovine erythrocyte ghosts, sphingomyelin hydrolysis leads to a process of invagination of membrane and accumulation in the interior of the membrane of solid droplets of material (see Freer and Arbuthnott 1976). This effect is more pronounced with bovine ghosts than with human ghosts, probably reflecting the higher content of sphingomyelin in the former. It has been postulated that a process of invagination with subsequent formation of vesicles would be expected to result from sphingomyelin hydrolysis if (1) sphingomyelin was preferentially located in the outer leaflet of the lipid bilayer and if (2) the hydrolysis product, ceramide, were to migrate laterally in the plane of the membrane rather than remain at the site of hydrolysis.

It seems likely that the mobility of ceramide within the plane of the membrane plays an important part in the haemolysis produced by β-toxin. But why is the altered membrane stable at 37°C and unstable when chilled?. One possibility is that the membrane lipids of the toxin-treated cell may undergo a phase transition when chilled and that this results in altered permeability properties. Another possible mechanism is that Mg^{2+} ions protect the membrane from lysis at 37°C owing to the formation of salt bridges between phosphate groups of residual unhydrolysed sphingomyelin. Such divalent ion stabilisation may be temperature-dependent. It is interesting to note that Smyth *et al.* (1975) have shown that chelators of Mg^{2+} ions induce lysis of β-toxin-treated erythrocytes at 37°C and that susceptibility to chelator lysis parallels susceptibility to hot—cold haemolysis.

8.2 *Cl. welchii* α-toxin

The α-toxin of *Cl. welchii* holds a special place in the field of bacterial toxins: it was the first toxin to be identified as an enzyme (MacFarlane and Knight 1941). Recent purification studies support the unitarian hypothesis that the biological activities of this toxin, namely, haemolysis, cytotoxicity, dermonecrosis, lethality and phospholipase C activity are properties of a single molecular species. The physicochemical properties of *Cl. welchii* α-toxin have also been studied in many laboratories. Although there is agreement on many properties such as isoelectric point and the existence of multiple forms there is as yet no close agreement on a value for the molecular weight.

As the major toxic product of type A strains of *Cl. welchii* which are commonly associated with gas-gangrene in man and animals, the α-toxin is considered by many to be responsible for damage to muscle tissue. It should be emphasised however that experimental evidence on this point is still conflicting.

As a phospholipase C the substrate specificity of *Cl. welchii* α-toxin is considerably broader than the β-toxin of *S. aureus* (see section 8.1). It hydrolyses phosphatidyl choline, sphingo-myelin and to a lesser extent phosphatidyl ethanolamine, phosphatidyl inositol and phosphatidyl serine. In experiments on the hydrolysis of phospholipids in erythrocyte ghosts the order of susceptibility to the toxin was phosphatidyl choline > sphingomyelin > phosphatidyl ethanolamine (Coleman *et al.* 1970). Not surprisingly this enzyme, like other bacterial phospholipases, has been used widely to probe the role of phospholipids in membrane structure and function (see Avigad 1976).

The hydrophobic product of phosphatidyl choline hydrolysis by α-toxin is diglyceride and this material collects in the form of phase-dense droplets ('black dots') in membranes exposed to the toxin. Like staphylococcal β-toxin, the enzyme requires divalent cations (in this case Ca^{2+} ions) for activity. It also exhibits the phenomenon of hot–cold haemolysis. Because α-toxin causes a greater degree of hot lysis than staphylococcal β-toxin the increase in titre observed after chilling is not as marked as with the latter toxin. This finding can be explained in terms of the distribution of substrate in the lipid bilayer. As has been stated earlier most of the choline-containing phospholipids in erythrocyte membranes are located in the outer leaflet; but some phosphatidyl choline is located in the inner leaflet and most of the phosphatidyl ethanolamine is located there. Substrates of *Cl. welchii* α-toxin therefore are present in both leaflets of the bilayer. For this reason the toxin would be expected to exert its effect on both sides of the membrane and this may explain why it is more haemolytic than staphylococcal β-toxin at 37°C.

8.3 Staphylococcal δ-toxin

Most of the properties of this toxin are in keeping with the suggestion that it affects membranes by virtue of its surface activity (see Bernheimer 1974, and Freer and Arbuthnott 1976). In particular it is soluble in chloroform–methanol, has a broad spectrum of activity, causing lysis of a wide range of cell types including bacterial protoplasts and spheroplasts, and it brings about gross dissolution of membranes in a manner resembling the detergent sodium deoxycholate. Its lytic activity differs from most other cytolytic toxins in that there is no lag period before membrane damage becomes apparent and the haemolytic reaction is independent of temperature.

The δ-toxin has proved difficult to characterise as it readily forms molecular aggregates, but when dissociated in the presence of 6 M guanidine HCl it has been shown to consist of subunits each having a molecular weight of 5 000. Amino acid analysis reveals a high content of hydrophobic amino acids and an absence of histidine, arginine, proline, tyrosine and cysteine residues.

We can postulate that a hydrophobic portion of the toxin molecule can penetrate the interior of the cell membrane, causing disturbances of normal protein–lipid interactions in this region which lead to altered permeability properties.

It is interesting to note that the activity of δ-toxin is inhibited by a range of phospholipids although no specificity was noted in the inactivation by phospholipids. More recently in a study of the effect of fatty acids on the haemolytic activity of δ-toxin, Kapral (1976) found that long-chain fatty acids (C_{21}–C_{23}) were inhibitory while shorter chain fatty acids (C_{13}–C_{19}) activated the toxin. Kapral suggests that fatty acid residues may be involved in the binding of δ-toxin to the erythrocyte membrane.

Finally I draw your attention to a new approach to the study of δ-toxin. In studies of the action of δ-toxin on perfused segments of rabbit and guinea pig small intestine, Kapral and his co-workers have shown that the toxin inhibits water absorption and increases the cyclic AMP content of guinea pig ileum. O'Brien and Kapral (1976) point out the similarity to known enterotoxins (see section 6) but interpret their results with caution: 'It does not necessarily follow that δ-toxin behaves in the same manner as cholera toxin. It is possible that δ-toxin inhibits water absorption by interfering with normal ion transport through a direct action on the cell membrane. The observed increase in cAMP levels may be coincident with such an effect and may not be the immediate cause of altered ion movement.'

8.4 The oxygen-labile toxins

The best-studied members of this group of 14 membrane-damaging toxins (Table 6.6) are streptolysin O, *Cl. welchii* θ-toxin and cereolysin. The features common to members of the group are (*a*) in solution they are inactivated in the presence of air; (*b*) the oxidised inactive toxins are reactivated by sulphydryl reducing agent; (*c*) despite the fact that they are produced by a range of different bacterial species they are immunologically related; (*d*) they are inhibited in the presence of low concentrations of certain sterols, including cholesterol. Detailed reviews of the properties and mode of action of this group of toxins have been given by Bernheimer (1976) and Freer and Arbuthnott (1976).

In considering the molecular basis of action we will explore further the interaction with cholesterol as this sterol is an important component of mammalian cell membranes. The inhibitory effect of cholesterol on the haemolytic activity of certain oxygen-labile toxins was established early on and in 1953 Howard *et al.* elucidated the chemical features which determine inhibitory activity. These are possession of a hydroxyl group in the β-configuration at carbon atom 3 (i.e. parallel to the plane of the sterol nucleus) and the presence of a hydrophobic side chain at position 17. Cholesterol, ergosterol, stigmasterol, cholestanol and corpostanol fulfil these criteria and act as inhibitors.

The importance of cholesterol in the mode of action of the oxygen-labile toxins is emphasised by the fact that all cell types known to be sensitive to streptolysin O contain cholesterol in the plasma membrane. Also artificial lipid membranes (liposomes) containing cholesterol inactivate oxygen-labile toxins whereas liposomes lacking cholesterol do not. Moreover a group of antibiotics known as polyenes including filipin, amphotericin B and nystatin (all of which exhibit sterol specificity) competitively inhibit the binding of oxygen-labile toxins to natural and artificial membranes. And several lines of evidence suggest that polyene antibiotics and oxygen-labile toxins produce similar functional lesions in biological membranes. These

Table 6.6   The oxygen-labile toxins

Producing organism	*Toxin*
Streptococci of groups A, B, C and G	Streptolysin O
Streptococcus pneumoniae	Pneumolysin
Clostridium tetani	Tetanolysin
Cl. welchii	θ-Toxin (perfringolysin O)
Cl. septicum	δ-Toxin (septicolysin O)
Cl. histolyticum	ϵ-Toxin (histolyticolysin O)
Cl. oedematiens Type A	δ-Toxin (oedematolysin O)
Cl. bifermentans	Bifermentolysin
Cl. botulinum Types C and D	Botulinolysin
Bacillus cereus	Cereolysin
Bacillus thuringiensis	Thuringiolysin O
Bacillus alvei	Alveolysin
Bacillus laterosporus	Laterosporolysin
Listeria monocytogenes	Listeriolysin

lesions or 'functional pores' are thought to result from redistribution of cholesterol molecules in relation to phospholipids in the membrane.

It seems likely that lysis proceeds in two stages:

1   Binding to cholesterol with the formation of toxin–cholesterol complexes in a reaction which is independent of temperature, pH and ionic strength.
2   Migration of toxin–cholesterol complexes in the horizontal plane of the membrane and the formation of 'functional pores' permeable to K^+ ions and (in erythrocytes) to haemoglobin. This second stage is inhibited by low pH, high ionic strength and divalent cations.

A feature of toxin–membrane interaction is the formation of ring-like structures having an outside diameter of 30–60 nm. The composition of these complexes and their relation to 'functional pores' is unknown at present.

8.5 Staphylococcal α-toxin

The α-toxin of *S. aureus* is haemolytic, dermonecrotic and lethal. In terms of pathogenicity it is probably the major tissue-damaging factor produced by this organism. The toxin has a molecular weight of 30 000 but can exist in various states of aggregation. A non-toxic polymeric form of the toxin (M.W. *circa* 170 000) consisting of six monomeric subunits has been studied in some detail. This component has a sedimentation coefficient of 12S and is designated α_{12S}-toxin. Negatively stained preparations of α_{12S}-toxin examined in the electron microscope have a characteristic ring structure (Arbuthnott, Freer, and Bernheimer 1967).

In attempts to define the molecular basis of action four different approaches have been followed.

1   Studies of the interaction of α-toxin with isolated cell membranes, liposomes and aqueous dispersions of single lipids.
2   Attempts to establish whether or not the toxin possesses enzymatic activity.
3   Studies of the release of intracellular markers from erythrocytes and cells in tissue culture.
4   Studies of binding of radiolabelled α-toxin to intact erythrocytes.

The subject has been reviewed recently by Freer and Arbuthnott (1976). At present, despite intensive investigation, not all the results obtained can be fitted readily within a single unified hypothesis to account for the toxin's mode of action at the molecular level. We do however have certain important clues.

There is a substantial body of evidence to indicate that staphylococcal α-toxin is surface-active and that its mode of action requires penetration and disorganisation of the hydrophobic region of the cell membrane. Under certain conditions of toxin concentration and ionic strength this results in polymerisation and formation of α_{12S}-toxin on the membrane. It is important to point out that there are important differences between the α-toxin and δ-toxin of *S. aureus* and that the former is more specific. The properties of these two toxins are summarised in Table 6.7.

Table 6.7   A comparison of staphylococcal α- and δ-toxins

Property	*α-Toxin*	*δ-Toxin*
Specific activity	2×10^4 haemolytic units mg^{-1}	200 haemolytic units mg^{-1}
Action on RBC	Lysies rabbit RBC > Sheep RBC > human RBC	Lysies RBC of different species to almost the same extent
Presence of a prelytic lag phase?	Yes	No
Lytic action for bacterial protoplasts?	No	Yes
Heat stability	Heat-labile	Heat-stable
Inactivation by normal serum?	No	Yes

The recent work of Cassidy and Harshman (1976) provides some valuable clues as to the mode of action of α-toxin. Working with radiolabelled toxin they carried out a detailed investigation of the kinetics of binding of α-toxin to erythrocytes and postulated the existence of a specific receptor (not yet identified) in the membranes of susceptible cells. Binding and membrane-damage however seem to be separate events and a working hypothesis to account for most of the available evidence would be that binding to a specific receptor is followed by penetration of the hydrophobic interior of the membrane which results in loss of normal permeability properties of the cell.

We have to view this 'receptor binding/penetration' hypothesis in the light of various claims that α-toxin is an enzyme. For instance Wiseman and his co-workers (see Wiseman 1975) have suggested that the toxin is a protease and is involved in a complex reaction which results in cell lysis due to hydrolysis of membrane proteins. In a recent study Dalen (1976) presents evidence to suggest that the formation of multiple forms of α-toxin can be explained in terms of proteolytic cleavage of the toxin molecule. I have already indicated that certain bacterial toxins are activated by cleavage of peptide bonds and it will be of interest to see whether future studies establish that a similar mechanism is involved in the action of staphylococcal α-toxin. At present the 'receptor binding/penetration' hypothesis and the 'enzymatic' hypothesis should not be regarded as mutually exclusive.

8.6 Other cytolytic toxins

Space does not allow me to deal with the remaining members of this group which includes staphylococcal leucocidin, staphylococcal γ-toxin, streptolysin S and the haemolysins produced by Gram-negative bacteria (e.g. *Vibrio parahaemolyticus, E. coli* and *Aeromonas hydrophila*). A brief account of these agents is given in the review of Freer and Arbuthnott (1976).

9 Staphylococcal epidermolytic toxin

Certain staphylococcal skin conditions of young children are characterised by the formation of blister-like lesions or 'bullae'. These lesions can be localised, as in staphylococcal bullous impetigo, or can be extensive, as in staphylococcal toxic epidermal necrolysis (TEN). In the latter condition the skin looks and feels as if it has been scalded and some workers refer to such superficial infections collectively as the staphylococcal scalded-skin syndrome. When examined histologically the affected areas show cell separation within the epidermis (epidermolysis) and there is little evidence of infiltration of leucocytes normally seen in staphylococcal lesions. By the late 1960s it was recognised that epidermolytic lesions developed as a result of infection with phage group II staphylococci and that the site of infection may be distant from the site of lesion formation, suggesting that intraepidermal splitting may result from the action of a diffusible staphylococcal toxin.

In 1970 Melish and Glasgow made a major advance when they found that phage group II staphylococci produced epidermal splitting when injected into neonatal mice. This provided an experimental model for the study of staphylococcal scalded-skin syndrome and within a short time it was shown (Arbuthnott *et al.* 1971, Kapral and Miller 1971) that the factor responsible was a previously undescribed staphylococcal toxin now known as epidermolytic toxin or exfoliatin. Epidermolytic toxin is a protein (M.W. 25 000) and to date two different serological types have been described. High levels of toxin production are associated with the presence of a plasmid and curing of the plasmid results in loss of toxin production. Also the strict association of toxin production and phage group no longer holds; some non-group II strains have been shown to produce either or both serological types of the toxin.

The toxin causes epidermal splitting in human skin and in the skin of many but not all species of experimental animals. Its effect is apparently restricted to the epidermis where it acts on a layer of cells known as the *stratum granulosum* which is made up of keratinocytes, cells destined to become the flattened dead cells of the skin surface. The molecular basis of action is not yet understood but electron-microscopic studies show that the toxin acts by disrupting the normal mechanisms of cell adhesion in the stratum granulosum. Cells in this layer are held together by 'intracellular cement' and by specialised organelles of adhesion known as desmosomes. At an early stage in cell separation, the small vesicles present in the intracellular space disappear and large spaces develop between opposing cell membranes. Cleavage of desmosomal junctions does not occur until later in the process. It is not yet possible to decide between several possible explanations of the toxin's mode of action. Does the toxin degrade a surface components involved in adhesion? Does the toxin activate an endogenous degradative enzyme? Does the toxin interact in some more subtle manner with the surface of susceptible cells through a direct or indirect effect on the cell membrane? These are challenging questions and the solution of the problem will probably involve the development of a suitable *in vitro* experimental system.

10 Lipopolysaccharide endotoxin

The importance of Gram-negative pathogens such as the salmonellae, shigellae, gonococci and meningococci is well known. Also several factors have led to the emergence of Gram-negative bacteria, such as *E. coli, Pseudomonas aeruginosa,* and *Klebsiella* spp., as the most important causes of hospital-associated bacteraemia. It has been estimated that such infections were responsible for approximately 18 000 deaths among 33 million hospital admissions in the USA in 1972 (Sanford 1975).

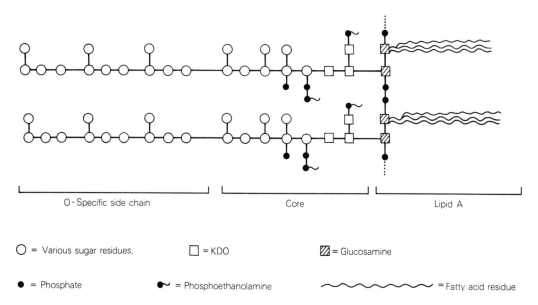

O - Specific side chain   Core   Lipid A

○ = Various sugar residues,   □ = KDO   ▨ = Glucosamine

● = Phosphate   ●～ = Phosphoethanolamine   〜〜〜 = Fatty acid residue

Fig. 6.5 Diagram of the structure of LPS. (From review of Reitschel *et al.* (1975), *Microbiology,* 307-13.)

In attempts to elucidate the mechanisms of pathogenicity of Gram-negative infections attention has focused on the role of lipopolysaccharide (LPS) endotoxin. This is a constituent of the outer membrane of the Gram-negative envelope and consists of heteropolysaccharide chains covalently linked to a lipid known as lipid A (Fig. 6.5). The polysaccharide component contains two regions, the O-specific side chain and the core. The O-specific side chain is represented by a polymer of oligosaccharide repeating units and carries the antigenic determinants responsible for species specificity. The core oligosaccharide is made up of a main chain of sugar residues which is substituted at various points by monosaccharides, phosphate, phosphoethanol- amine and pyrophosphoethanolamine. This core polysaccharide is covalently linked to lipid A which consists of a glucosamine phosphate backbone carrying long chain fatty acids in ester and amide linkages (Fig. 6.6).

There is now a considerable body of evidence which indicates that the toxicity of lipopoly- saccharide resides in the lipid A component (see Reitschel, Gelanos, and Luderitz 1975). The various biological activities of lipopolysaccharide endotoxin are summarised in Table 6.8.

These numerous biological activities of LPS endotoxin include many of the features observed in Gram-negative infections of man and experimental animals and it is tempting to conclude that the toxin is responsible for the symptoms of the shock which frequently accompanies Gram-negative sepsis and sometimes leads to death. Some workers hold the view that the extremely sensitive limulus test (a reaction based on the clotting of a lysate of amoebocytes from the horseshoe crab *Limulus polyphemus*) is useful in the diagnosis of sepsis and is predictive for shock and death. However this view is disputed by others and the longstanding controversy surrounding the role of LPS endotoxin in Gram-negative infections seems likely to continue for some time to come (see Shands 1975).

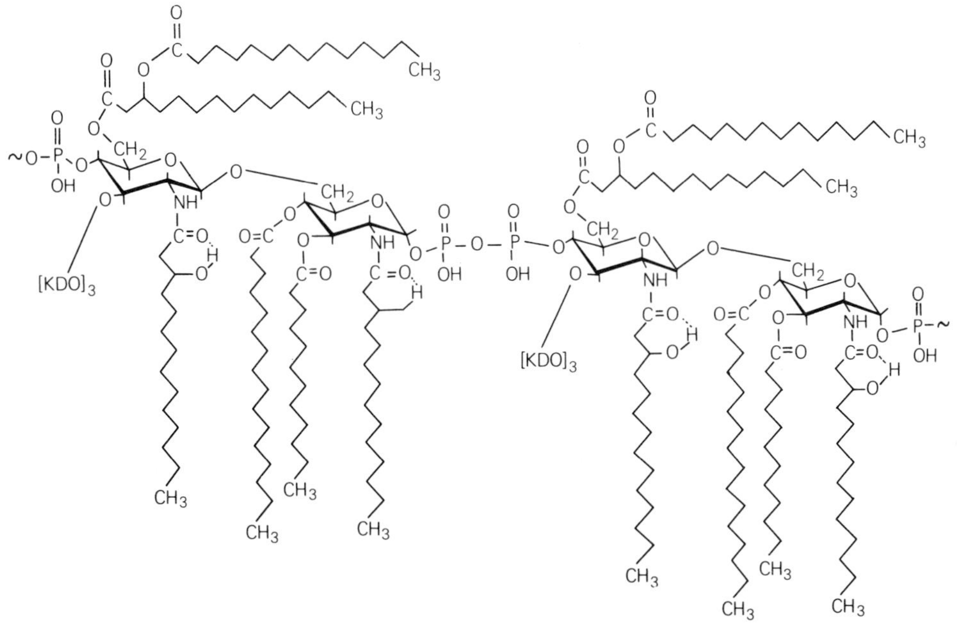

Fig. 6.6 The structure of lipid A component of *Salmonella* lipopolysaccharide. (From review of Reitschel *et al.* (1975), *Microbiology*, 307–13.)

Table 6.8 Biological activities of lipopolysaccharide endotoxin

Pyrogenicity (ability to induce a rise in body temperature)
Lethal action
Depression of blood pressure
Activation of complement
Intravascular coagulation
Leucopaenia and leucocytosis
Inhibition of glucose and glycogen synthesis in liver
Stimulation of B-lymphocytes
Macrophage inhibition
Interferon release
Induction of prostaglandin synthesis
Clotting of a lysate of amoebocytes from the horseshoe
    crab

At the molecular level little is known of the mode of action of LPS endotoxin. It seems unlikely that a single mode of action can account for its numerous pathophysiological effects, since some of these are mediated through the interaction with cells (possibly through binding to the cell membrane) while others result from its action on serum or plasma proteins. Although it is evident that the toxic component is lipid A, this lipid requires to be complexed with a hydrophilic carrier before its activity is expressed. The hydrophilic polysaccharide component of LPS endotoxin may be responsible for maintaining the molecular complex in the appropriate orientation.

Bibliography

Review articles on bacterial toxins

Articles appearing in *Mechanisms in Bacterial Toxinology* (Ed. A. W. Bernheimer). John Wiley, New York, 1976.

AVIGAD, G. 'Microbial phospholipases', pp. 99–168.
    A general review of the properties and modes of action of microbial phospholipases.
BERNHEIMER, A. W. 'Sulphydryl activated toxins', pp. 85–98.
    A review of the general properties of oxygen-labile toxins.
FINKELSTEIN, R. A. 'Progress in the study of cholera and related enterotoxins, pp. 53–84.
    A comprehensive review of the properties and mode of action of enterotoxins.
FREER, J. H. and ARBUTHNOTT, J. P. 'Biochemical and morphological alterations of membranes by bacterial toxins', pp. 169–94.
    A review dealing with the mode of action of membrane-damaging toxins.
MURPHY, J. R. 'Structure activity relationships of diphtheria toxin', pp. 32–51.
    A review of the molecular basis of diphtheria toxin action.
SINGER, R. A. 'Lysogeny and toxinogeny in *Corynebacterium diphtheriae*', pp. 1–30.
    A review of toxin synthesis and its relation to lysogeny.
TAYLOR, A. G. 'Toxins and the genesis of specific lesions: enterotoxin and exfoliatin', pp. 195–216.
    A review of the role of enterotoxins and staphylococcal epidermolytic toxin in pathogenesis.

Articles appearing in *Microbial Toxins* (Eds. S. J. Ajl, S. Kadis and T. C. Montie), vols. 1 and 2A. Academic Press, New York, 1970–71.

BERNHEIMER, A. W. 'Cytolytic toxins of bacteria', pp. 183–209.
    First comprehensive review of the cytolytic (membrane-damaging) toxins.
BOROFF, D. A. and DAS GUPTA, B. R. 'Botulinum toxin', pp. 1–68.
    A comprehensive review of the nature and mode of action of the botulinum toxins.
VAN HEYNINGEN, W. E. 'General characteristics', pp. 1–28.
    Historical introduction and summary of general properties of toxins.
VAN HEYNINGEN, W. E. 'Tetanus toxin', pp. 69–108.
    A comprehensive review of the nature and mode of action of tetanus toxin.

Articles appearing in *Microbiology 1975* (Ed. D. Schlessinger). American Society for Microbiology, Washington, 1975.

REITSCHEL, E. T., GALANOS, C. and LUDERITZ, O. 'Structure, endotoxicity and immunogenicity of the lipid A component of bacterial lipopolysaccharides', pp. 307–13.
    A summary of the chemical and biological properties of LPS endotoxin.
SANFORD, J. P. 'Pathogenic mechanisms in opportunistic Gram-negative bacillary infections: epidemiological and host factors', pp. 302–6.
    A brief but useful account of the clinical importance of Gram-negative infections.
SHANDS, J. W. 'Endotoxin as a pathogenic mediator of Gram-negative infection', pp. 330–5.
    An assessment of the role of endotoxin in pathogenicity.

Additional review articles

BERNHEIMER, A. W. (1974) 'Interactions between membranes and cytolytic bacterial toxins', *Biochim. Biophys. Acta,* **344**, 27–50.
Deals with the mode of action of certain cytolytic toxins including staphylococcal α-, β-, and δ-toxins and the oxygen-labile haemolysins.

SMITH, H. (1976) 'Survival of vegetative bacteria in animals', in *The Survival of Vegetative Microbes,* Symposium 26 (Eds. T. R. G. Gray and J. R. Postgate), pp. 299–326. Cambridge University Press for Soc. Gen. Microbiol., London.
General account of bacterial pathogenicity.

WISEMAN, G. (1975) 'The haemolysins of *Staphylococcus aureus*', *Bacteriol. Rev.,* **39**, 317–44.
A review of staphylococcal haemolysins which includes a summary of the protease hypothesis for α-toxin.

Original papers on bacterial toxins

ARBUTHNOTT, J. P., FREER, J. H. and BERNHEIMER, A. W. (1967) 'Physical states of staphylococcal α-toxin', *J. Bacteriol.,* **94**, 1170–7.
First description of the α_{12S} polymer of staphylococcal α-toxin.

ARBUTHNOTT, J. P., KENT, J., LYELL, A. and GEMMELL, C. G. (1971) 'Toxic epidermal necrolysis produced by an extracellular product of *Staphylococcus aureus*', *Br. J. Derm.,* **85**, 145–9.
Deals with discovery of epidermolytic toxin.

CASSIDY, P. and HARSHMAN, S. (1976) 'Studies on the binding of staphylococcal ^{125}I-labelled α-toxin to rabbit erythrocytes', *Biochem.,* **15**, 2348–55.
Describes evidence for the existence of an α-toxin receptor in sensitive erythrocytes and deals with the relation between binding, ion-leakage and osmotic lysis.

COLEMAN, R., FINEAN, J. B., KNUTTON, S., and LIMBRICK, A. R. (1970) 'A structural study of the modification of erythrocyte ghosts by phospholipase C', *Biochim. Biophys. Acta,* **219**, 81–92.
Deals with the effect of phospholipase C (*Cl. welchii* α-toxin) on the phospholipids of erythrocyte ghost membranes and the morphological changes which occur after treatment.

COLLIER, R. J. (1967) 'Effect of diphtheria toxin on protein synthesis: inactivation on one of the transfer factors', *J. Molec. Biol.,* **25**, 83–98.
Description of the inactivation of EF-2 by diphtheria toxin.

DALEN, A. B. (1976) 'Proteolytic degradation of staphylococcal α-toxin', *Acta Pathol. Microbiol. Scand.,* Sect. B, **84**, 309–14.
Proteolytic modification of the α-toxin molecule.

DAS GUPTA, B. R. and SUGIYAMA, H. (1972) 'A common subunit structure in *Cl. botulinum* type A, B and E toxins', *Biochem. Biophys. Res. Commun.,* **48**, 108–12.
Mechanism of activation of botulinum toxins.

DE, S. N. and CHATTERJE, D. N. (1953) 'An experimental study of the mechanism of action of *Vibrio cholerae* on the intestinal mucous membrane', *J. Path. Bacteriol.,* **66**, 559–62.
Description of the use of ligated intestinal loops in the study of cholera.

DIAMOND, J. and MELLANBY, J. (1971) 'The effect of tetanus toxin in the goldfish', *J. Physiol.,* **215**, 727–41.
The peripheral action of tetanus toxin in the goldfish.

DOERY, H. M. *et al.* (1963) 'A phospholipase in staphylococcal toxin which hydrolyses sphingomyelin', *Nature,* **198**, 1091–2.

DOERY, H. M. *et al.* (1975) 'The properties of phospholipase enzymes in staphylococcal toxins', *J. Gen. Microbiol.,* **40**, 283–96.
These two papers describe the sphingomyelinase activity of staphylococcal β-toxin.

DUTTA, N. K. and HABBU, M. K. (1955) 'Experimental cholera in infant rabbits: a method for chemotherapeutic investigation', *Brit. J. Pharmacol.,* **10**, 153–8.
Induction of fatal cholera-like diarrhoea in suckling rabbits.

FINKELSTEIN, R. A. and LOSPOLLUTO, J. J. (1969) 'Pathogenesis of experimental cholera: preparation and isolation of choleragen and choleragenoid', *J. Exp. Med.,* **130**, 185–202.
Purification of cholera toxin.

FREEMAN, V. J. (1951) 'Studies of the virulence of bacteriophage-infected strains of *Corynebacterium diphtheriae*', *J. Bacteriol.,* **61**, 675–88.
Original report of lysogeny in relation to diphtheria toxin production.

GYLES, C. L. and BARNUM, D. A. (1969) 'A heat-labile enterotoxin pathogenic for pigs', *J. Infect. Dis.,* **120**, 419–26.
Discovery of *E. coli* heat-labile (LT) enterotoxin.

HOWARD, J. G. *et al.* (1953) 'The inhibitory effects of cholesterol and related sterols on haemolysis by streptolysin O', *Brit. J. Exp. Path.,* **34**, 174–80.
Early account of properties of sterols inhibitory for oxygen-labile toxins.

KAPRAL, F. A. and MILLER, M. M. (1971) 'Product of *Staphylococcus aureus* responsible for the scalded skin syndrome', *Infect. Immun.,* **4**, 541–5.
Describes the discovery of epidermolytic toxin.

KAPRAL, F. A. (1976) 'Effect of fatty acids on *Staphylococcus aureus* delta-toxin haemolytic activity', *Infect. Immun.,* **13**, 114–9.
Deals with effect of fatty acids of different chain length on staphylococcal δ-toxin.

MACFARLANE, M. G. and KNIGHT, B. C. J. G. (1941) 'The biochemistry of bacterial toxins: I. Lecithinase of *Clostridium welchii* toxins', *Biochem. J.,* **35**, 884.
Description of *Cl. welchii* α-toxin as a phospholipase.

MELISH, M. E. and GLASGOW, L. A. (1970) 'The staphylococcal scalded skin syndrome: development of an experimental model', *New Engl. J. Med.,* **282**, 1114–9.
Injection of Phage Group II Staphylococci into neonatal mice produces epidermal splitting.

O'BRIEN, A. D. and KAPRAL, F. A. (1976) 'Increased cyclic adenosine 3′, 5′-monophosphate content in guinea pig ileum after exposure to *Staphylococcus aureus* delta-toxin', *Infect. Immun.,* **13**, 152–62.
Staphylococcal δ-toxin increases cyclic AMP levels.

SMITH, H. W. and HALLS, S. (1967) 'Studies on *Escherichia coli* enterotoxin', *J. Path. Bacteriol.,* **93**, 531–44.
Discovery of heat-stable enterotoxin (now designated ST toxin).

SMYTH, C. J. *et al.* (1975) 'Phenomenon of hot-cold haemolysis: chelator-induced lysis of sphingomyelinase-treated erythrocytes', *Infect. Immun.,* **12**, 1104–11.
Possible mechanism of hot-cold haemolytic action of staphylococcal β-toxin.

STRAUSS, N. and HENDEE, E. D. (1959) 'The effect of diphtheria toxin on the metabolism of HeLa cells', *J. Exp. Med.,* **109**, 144–63.
Original report of action of diphtheria toxin on protein synthesis.

UCHIDA, T. *et al.* (1971) 'Mutation in the structural gene for diphtheria toxin carried by temperate phage B', *Nature,* **233**, 8–11.
Isolation of cross reaction material (CRM).

VAN HEYNINGEN, W. E. (1974) 'Gangliosides as membrane receptors for tetanus toxin cholera toxin and serotoin', *Nature,* **249**, 415.
An appraisal of the possible role of gangliosides as surface receptors.

7

Efficiency of microbial growth

D. E. F. Harrison
Shell Research Ltd., Woodstock Laboratory, Sittingbourne

1 Introduction

Growth and multiplication of microbial cells represents a large decrease in the state of disorder or 'entropy' of the constituent molecules. In order that the second law of thermodynamics should not be defied, there must be, elsewhere in the system, a concomitant increase in entropy at least equal to that lost in organising the molecules to form a cell structure. This redistribution of entropy is achieved in living systems by coupling each synthetic step to a reaction which produces higher entropy molecules, thereby producing low-entropy structures. Those reactions which produce more elaborate (low-entropy) structures are referred to as 'anabolic processes' and reactions whose main products are more simple (high entropy) molecules are termed 'catabolic processes'. The efficiency of the growth processes in microorganisms will thus depend on the efficient coupling of anabolic and catabolic reactions. However, while this is simply conceived, it is not so easy to quantify exactly what is meant by 'efficiency' of growth.

In this chapter we will explore the methods of expressing growth efficiency and of assessing it. The influence of environmental factors on growth efficiency will be examined and an attempt will be made to draw some conclusions about the regulatory mechanisms involved in energy supply for growth. Finally we shall ask what, if any, is the practical importance of efficient growth in microorganisms.

2 Expression and measurement of growth efficiency

2.1 Thermodynamic considerations

The standard equation for expressing the free energy change when a system shifts irreversibly from one steady state to another is:

$$\Delta F = \Delta H - T \Delta S \qquad (1)$$

where ΔF (or often ΔG) is the total *free energy change*, ΔH is the change in heat content or *enthalpy*, T is the absolute temperature and ΔS is the change in *entropy*. Inasmuch as the growing microbial cell can be said to approximate to a steady state, and the process of growth is certainly irreversible (the cells cannot spontaneously reabsorb the heat and carbon dioxide lost during growth to recreate exactly the original, pre-growth status), equation (1) can be applied to microbial growth.

Consider a heterotrophic microorganism growing on glucose as the sole source of carbon. The difference in the heats of combustion of the sugar and the cell material is quite small, being about 120 and 116 kcal/g-atom of carbon, respectively. Thus the *enthalpy* change for rearranging sugar molecules into cell material would be very low. Thus, if the anabolic processes of growth on glucose could be isolated from the catabolic processes (which they cannot), the free energy change due to anabolism alone would approximate to the entropy change $T\Delta S$, the enthalpy change being very small. In order to calculate this entropy change it would be necessary to ascertain the precise state of each atom in the cell, clearly an impossibility.

Entropy changes in a system cannot be measured directly but enthalpy changes can, by means of a calorimeter which gives the heat output of the system at constant temperature and pressure. However, as catabolic processes and anabolic processes cannot, in reality, be divorced, only the enthalpy change in the *whole system* can be measured. The enthalpy change due to catabolic processes E_g only can be calculated from:

$$E_g = \Delta H - \Delta H_g \qquad (2)$$

where ΔH is the enthalpy change of the whole system as measured by calorimetry and ΔH_g is the enthalpy change due to anabolism, i.e. the conversion of the constituent monomers into cell structure. This latter quantity can be calculated approximately from the heats of combustion of the substrate and cells so that the enthalpy change due to catabolism can be calculated:

$$\Delta H = (C_s - C_m) \qquad (3)$$

where C_s is the heat of combustion of the substrates incorporated into cell material and C_m is the heat of combustion of cell material.

For a bacterium growing aerobically on glucose a typical mass and thermal balance might be:

1 mole glucose $\rightarrow$ 90 g cell + 114 g CO_2 (-293 kcal).

Taking the difference in heats of combusion of glucose and cells to be $+15$ kcal/g-atom carbon, the total energy change due to catabolic processes would be -308 kcal.

As the only source of energy in a heterotrophic organism is from catabolic processes, the free energy change due to the outflow of entropy from the growing cell cannot exceed that generated by catabolism. In the example above this would be -308 kcal per mole of glucose catabolised. However, the 'coupling' of catabolic and anabolic processes may be far from efficient in terms of energy, so the actual free energy change may be much higher than the minimum required for growth.

The true free energy efficiency of the growth process could be said to be the ratio of the free energy change involved in constructing the microbial cell from its constituent monomers, to that actually expended in the complete growing system. As the former cannot be measured or calculated, free energy efficiency has no practical meaning.

However, expressing growth efficiency in terms of heat changes can provide useful comparisons, especially for complex anaerobic systems, where the catabolic processes are unknown. In this case the heat output, under conditions of constant temperature and pressure, may be the only available estimate of energy change during growth. Yield can then be expressed in terms of heat as:

$$Y_{kcal} = \frac{\text{Weight of cells produced}}{\text{Heat evolved}} \qquad (4)$$

This term provides a useful comparison of different systems but must not be confused with true free energy efficiency.

2.2 Yield coefficient

The concept of a yield constant assumes a fixed stoichiometry between substrate utilised and cells formed. This concept was first quantified by Monod. As we shall see, such a fixed stoichiometry is not universal, particularly where the substrate in question is *not* limiting for growth. However, for a growth-limiting substrate in cultures grown under carefully defined conditions, the amount of cell mass formed for a given consumption of substrate tends to be constant.

The simplest way of expressing the efficiency of conversion of substrate to cells is on the basis of weight of cells produced per unit weight of substrate consumed. This dimensionless constant is usually called the *yield coefficient, Y*. Note that the term 'yield' is often used to denote the total mass of cells produced in a culture, sometimes over a given period of time but often, for a batch culture, with no time element. This latter usage leads to confusion and it is best if the term yield is used only to refer to the efficiency of conversion of substrate. The terms 'cell concentration' or 'cell density' should be used for the weight of cells per unit volume of culture and 'productivity' for the weight of cells produced for a given volume over a given time interval. Yield coefficient usually refers to a carbon substrate but it can be used to refer to any essential nutrient such as an inorganic nitrogen source. Yield is normally quoted for the substrate which limits the growth of the microorganism.

There are several other ways of expressing the efficiency of growth in common use and it will help if these are defined in turn.

a Molar yield coefficient. This is the yield coefficient expressed as g cells per g-mole of substrate consumed. It is usually expressed by suffixing the term Y with a letter or word denoting the substrate, e.g. Y_g or $Y_{glucose}$. This is a convenient expression for most research purposes as it more directly relates the efficiency of different substrates. For instance yields of microorganisms grown for the purpose of manufacturing single-cell protein on methanol and methane give yield coefficients of 0.5 and 0.8 g g^{-1}, respectively. At first sight this seems to indicate more efficient growth on methane but, expressed as molar yield coefficients, the values become $Y_{methanol} = 16$ g mole^{-1} and $Y_{methane} = 12.8$ g mole^{-1}.

b Carbon conversion efficiency. Molar growth yield has some serious shortcomings for quick comparisons of growth efficiency. For instance, comparing the growth of a microorganism on glucose and ethanol, the relevant molar growth efficiencies might be:

$$Y_{glucose} = 90 \text{ g mole}^{-1}; Y_{ethanol} = 23 \text{ g mole}^{-1}$$

Without knowing the molecular weights of the substrates these figures are not directly comparable (in each case $Y = 0.5$). 'Carbon conversion efficiency' can provide a more direct comparison as it expresses yield in terms of the ratio of carbon from the substrate which is incorporated into the cell to the number of gram-atoms of carbon in the substrate consumed.

$$\text{Carbon conversion (C.C.)} = Y \frac{\% \text{ carbon in dry cell mass}}{\% \text{ carbon in substrate}} \times 100 \tag{5}$$

For example, if $Y_{glucose} = 90$ g mole^{-1} ($Y = 0.5$),

$$\text{C.C.} = 0.5 \times 45/40 \times 100 = 56\%$$

for $Y_{ethanol}$ = 23 g mole^{-1}, C.C. = 43%.

Carbon conversion thus provides a very useful direct comparison of efficiencies *but* can really only be used where the substrate in question is the sole source of carbon. It reveals little about the energetics of the growth.

c Yield per average electron $(Y_{ave\,e^-})$. This expression was devised by Payne and co-workers (1970) in an attempt to relate the energy available from catabolism of different substrates to growth efficiency. Payne argued that any organic substrate may be characterised by 'its constitutive number of average electrons' (ave e$^-$), i.e. the average number of electrons donated by the substrate during the complete oxidation to CO_2 and water. For example, glucose would possess 24 ave e$^-$ and ethanol 1 ave e$^-$. Thus a $Y_{glucose}$ of 90 g mole^{-1} is equivalent to a $Y_{ave\,e^-}$ of 3·74 g per ave e$^-$, and a $Y_{ethanol}$ of 23 to a $Y_{ave\,e^-}$ of 1·8 g per ave e$^-$.

This would seem to be a very useful criterion for growth efficiency if all electrons donated were of equal value to the cell in terms of energy. Unfortunately this is *not* the case. For instance, the oxidations of NADH to NAD$^+$ and succinate to fumarate both involve the donation of 2 electrons but, whereas it is possible to generate 3 molecules of ATP from the oxidation of NADH, only 2 may be formed by the oxidation of succinate. The redox potential of the substrate system and the degree of 'coupling' to ATP formation both may affect the amount of energy conserved for growth. As the value of 'available electrons' will vary according to the pathways of metabolism of the substrate, the $Y_{ave\,e^-}$ cannot be used to predict yields with any degree of certainty.

d Yield based on heat production. The Y_{kcal} has been defined above. The advantage of this expression is that it can be used to compare growth on complex, undefined substrates, especially for anaerobic systems. There is considerable difficulty in measuring Y_{kcal} although expensive flow-through calorimeters now exist which are very accurate. For simple systems with defined growth substrates, heat production can be directly related to substrate used or, in aerobic cultures, to oxygen uptake and these provide a more direct and convenient measure of growth efficiency.

e Yield in terms of oxygen consumed Y_0. For aerobic growth Y_0 (i.e. g cells produced per g-atom of oxygen consumed) provides a most convenient measure of growth efficiency as oxidative phosphorylation is the chief source of energy for growth. However, Y_0 is very difficult to measure for batch cultures although it can be conveniently estimated for steady-state systems provided by continuous culture.

2.3 Y_{ATP}

All the methods discussed above for expressing yield have their specific applications but in the 1960s the need was recognised for comparing growth efficiency in a way that could be related to the organism's metabolism. Bauchop and Elsden (1960) suggested that yield should be expressed in terms of adenosine triphosphate (ATP) requirement as a convenient quantification of energy requirement for growth.

ATP can be considered as the currency for energy in living cells. It is in no sense a reserve of energy but, in the great majority of the reactions of living cells in which energy is required, ATP provides that energy. Potential energy sources are converted, via catabolic processes, into ATP, in which form the energy can be used to drive anabolic processes.

Thus ATP provides a convenient quantification of energy requirement, the molecule of ATP being used in this case as a unit of energy. It should always be borne in mind that ATP is *not* the only high-energy intermediate for driving growth processes; other compounds can provide the free energy, e.g. reduced compounds such as NADH and NADPH, ADP or other phosphorylated nucleotides and, also, ion gradients. Nonetheless, even when ATP is not directly involved, the energy source for the reaction can be expressed in ATP equivalents as ATP involvement in metabolism is ubiquitous. For instance, NADH or NADPH can be regarded as representing 3 ATP molecules (assuming 3 sites of oxidative phosphorylation), reduced flavoprotein as 2 ATP molecules and a K^+ ion pumped against a gradient as ½ ATP molecule. By this means the energy requirement of almost any set of biochemical reactions can be expressed in terms of molecules of ATP (or $\sim$ P) and growth can be expressed as weight, in grams, of biomass produced per g-mole of ATP expended, Y_{ATP}.

Bauchop and Elsden (1960) grew a number of microorganisms anaerobically on complex media in which all essential amino acids were supplied. They supplied glucose as sole energy source and, assuming that the amino acids were used only as carbon source and glucose only as a source of energy, calculated yields from glucose and obtained values of about 21 g cells mole^{-1}.

Assuming that glucose was catabolised via the Embden—Myerhof pathway so that 2 molecules of ATP were made available for each molecule of glucose used, a Y_{ATP} value of 10·5 was calculated. Similar experiments were made with *Streptococcus faecalis* using arginine as energy source. Arginine is catabolised to ornithine yielding only one ATP per mole. The molar growth yield obtained from arginine was 10·5, also giving a Y_{ATP} value of 10·5. From this and other data Bauchop and Elsden proposed the constant Y_{ATP} hypothesis, i.e. that to produce a given dry weight of cell mass the ATP equivalents that must be expended are constant for a wide variety of microorganisms. Further, a dry weight of cells that can be produced per g-mole of ATP expended is 10·5 g.

This is potentially a most useful concept; if the hypothesis is correct and dependable it should be possible to calculate the ATP generated by a microorganism during catabolism of the energy source, from the yield coefficient:

$$N = \frac{Y_{substrate}}{Y_{ATP}} \tag{6}$$

where N is the number of moles of ATP generated per mole of substrate catabolised. By this means it should be possible, by determining yields, to distinguish which of the main pathways of glucose metabolism are operational. Thus, *S. faecalis* had a $Y_{glucose}$ of 20 so that $Y/Y_{ATP} \simeq 2{\cdot}0$, indicating the operation of the Embden-Myerhof pathway, but *Pseudomonas lindneri* had a $Y_{glucose}$ of 8·3 so that $Y/Y_{ATP} \simeq 0{\cdot}8$, indicating that not more than one ATP is obtained, a result which is to be expected if the Entner—Doudoroff pathway is prevalent.

Minor discrepancies from the constant Y_{ATP} rule have been found but for very many organisms growing anaerobically the concept seems more or less valid. Results for aerobic growth tend to be far less consistent. Recently, however, there have been reports by Stouthamer and co-workers (1976) of very wide variations in the Y_{ATP} value. They have reported Y_{ATP} values as high as 28 for *Klebsiella aerogenes* growing on a complex medium.

Assessing the general validity or otherwise of the constant Y_{ATP} concept is difficult because the methods of determining it lend themselves to serious errors, which have not always been appreciated by workers in the field. Since the assessment of accurate yield coefficients is essential for any appraisal of the energetics of growth we shall discuss errors that may be involved.

2.4 Sources of error in the estimation of energy efficiency of growth

Much of the conflicting evidence in the literature on growth efficiency may be attributed to errors in estimating growth yields or in interpreting data. There are many possible sources of error and it will assist in appraising the evidence available to summarise the possible pitfalls and to suggest remedies.

1 Inaccurate estimations of substrate, cell mass and products. There is insufficient space here to describe all the available analytical techniques and their shortcomings. Suffice it to say that the only way to ensure that yield estimations are accurate is to base them on a full mass or carbon balance of the culture.

2 Additional energy sources are available to the organism which are not accounted for. This situation is most likely to occur in complex media where, for instance, amino acids which are present may be catabolised coincidentally with the supplied energy source. It is not sufficient to *assume* that catabolite repression prevents the use of such compounds; such suppositions must be tested experimentally. Also, it is not necessarily safe to use, as a control, the complex medium without added energy source and to subtract this from the yield obtained when the energy source is present; there is always the possibility of co-metabolism of minor compounds in the medium which only supply energy for growth in the presence of readily catabolised substrate. Mistakes can arise even from the use of defined media. These latter often contain chelating agents such as citrate which may be catabolised in the presence of glucose, especially when the latter is limiting for growth.

Again, the obvious guard against these errors is to carry out complete mass or carbon balances.

3 Wrong assumptions are made about the fate of the energy substrate. This is an obvious pitfall. It cannot always be assumed, for instance, that under aerobic conditions, glucose is fully oxidised to carbon dioxide and water and that the only other product is cell mass. Even organisms with an active aerobic metabolism such as *K. aerogenes* will, under certain conditions, exude other organic molecules into the medium when oxygen is in excess. Harrison and Pirt (1967) found that large amounts of pyruvate, or acetate could accumulate in chemostat cultures growing aerobically on glucose. Once again the only remedy is to account for all the organic material presented to the culture.

4 Wrong assumptions are made about the type of catabolism. The estimation of growth efficiency as Y_{ATP} from growth yields depends on predicting the number of ATP equivalents made available from catabolic processes. Therefore, it is important to be certain of the extent of aerobic or anaerobic metabolism. Normally this presents no difficulty in carefully controlled cultures where a full carbon balance is made. However, in anaerobic cultures, the presence of alternative inorganic electron acceptors such as nitrate or sulphate can lead to very large errors. Aerobic metabolism may yield 19 times more ATP than anaerobic metabolism so that the presence of only small amounts of oxygen or other electron acceptors will lead to a significant error.

5 Variations in cell composition. Not all components of cell mass require the same number of ATP equivalents for synthesis. For instance, glycogen formation from glucose must require less than 0·1 ATP per g. Thus, a culture accumulating significant quantities of glycogen from glucose will have a high Y_{ATP} value. Obviously the conclusion to draw is that the Bauchop–Elsden constant Y_{ATP} hypothesis cannot be correct for all cells. However, if cells of a similar protein content are compared then it would seem reasonable to assume that the energy requirement for growth is similar. Therefore, not only should cell mass be estimated, but also the protein content. The protein content of microorganisms varies with growth rate, but for energy-limited cultures this variation is not great. It is when growth is limited by factors other than energy source that large errors are likely to occur, because under such conditions storage products such as glycogen, poly-β-hydroxybutyrate or polyphosphate are likely to accumulate.

The remedies are to estimate the level of protein and storage material in the cell, or to ensure that the energy source is the growth-limiting factor.

6 Not allowing for energy source which is used as carbon source and incorporated into the cell. Where only one source of carbon and energy is available a simple calculation is used to determine that portion of the substrate used for cell material and that used for energy source (see Pirt 1975).

7 There may be a significant amount of energy used for 'maintenance' or wasted through 'negative growth'. These possibilities are discussed in the next section.

Summarising, the ideal conditions required for estimation of growth efficiency are:

a Continuous culture should be used in order to obtain defined growth conditions.
b The medium should be fully defined.
c The energy substrate should be limiting for growth.
d A complete carbon balance and as complete a mass balance as possible should be made.
e Dissolved oxygen should be measured or the electron acceptor clearly defined.
f The cells should be analysed for carbon and protein content.
g Cell mass should be determined directly from dry weight estimations or organic carbon analysis.
h Experiments should be made at different growth-rates to allow for non-growth associated energy utilisation.

3 Maintenance energy and 'true yield'

3.1 Definition of maintenance requirement

The concept of maintenance coefficient, then termed 'endogenous requirement', was first proposed by Herbert (1958) to explain the deviation in chemostat cultures from the Monod theory. Figure 7.1 shows the shape of the organism concentration–dilution rate curve normally obtained in an energy-limited chemostat, compared with that predicted by the Monod theory. Plotting the specific substrate utilisation rate $q_{substrates}$ against dilution rate yields a straight line which cuts the $q_{substrate}$ axis at a positive value (Fig. 7.2). Thus, at zero dilution rate, i.e. when there is no growth, there is a residual substrate demand. This was called by Herbert the 'endogenous' substrate requirement. Pirt (1965) later introduced the term 'maintenance ration' which is preferable because 'endogenous' is also used to refer to substrates stored within a cell.

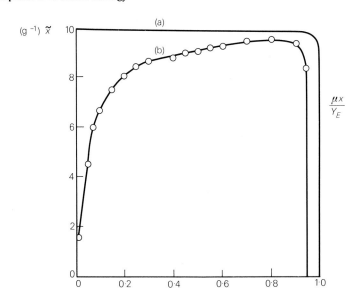

Fig. 7.1 The effect of maintenance requirement on response to dilution rate in a chemostat culture.
(a) Response of steady-state cell concentration $\tilde{x}$ in the absence of maintenance. (b) Response of steady-state cell concentration with a maintenance requirement of 0·10 g substrate per g cells per h.
    Other parameters used: $\mu_{max} = 1·0$ h^{-1}; substrate concentration in feed $(S_r) = 20$ g l^{-1}; $K_s = 0·02$ g l^{-1}; $Y_{EG} = 0·50$.

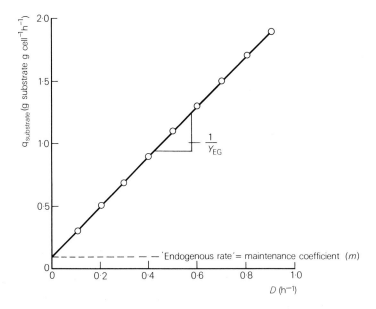

Fig. 7.2 Plot of specific substrate utilisation rate, $q_{substrate}$, against dilution rate. Values used to derive the plot are taken from Fig. 7.1, curve b.

Pirt devised a mathematical treatment for the estimation of the 'maintenance ration' which may be summarised as follows:

Total rate of substrate consumption	=	Rate of consumption for growth	+	Rate of consumption for maintenance

From this:

$$\frac{\mu x}{Y_E} = \frac{\mu x}{Y_{EG}} + mx \tag{7}$$

where μ = specific growth-rate (h^{-1}); x = organism concentration $(g\,l^{-1})$; Y_E = actual yield (corrected for substrate incorporated into cell material, $g\,g^{-1}$); Y_{EG} = the theoretical 'true' yield $(g\,g^{-1})$; and m = the maintenance energy requirement which is independent of growth rate $(g\,g^{-1}\,h^{-1})$.

Rearranging equation (7), we obtain:

$$\frac{1}{Y_E} = \frac{1}{Y_{EG}} + \frac{m}{\mu} \tag{8}$$

Thus a plot of the reciprocal of the measured yield against the reciprocal of dilution rate gives a straight line the slope of which is equal to the maintenance requirement and which cuts the yield axis at a point equal to the reciprocal of the 'true' yield.

Under steady-state conditions the residual substrate concentration in the chemostat is given by:

$$\bar{s} = \frac{K_s(D + mY_{EG})}{(\mu_m - mY_{EG} - D)} \tag{9}$$

where D is the dilution rate (h^{-1}).

Now this can be compared with the equation of Topiwala and Sinclair (1971):

$$\bar{s} = \frac{K_s(D + K)}{\mu_m - (D + K)} \tag{10}$$

This latter equation was based on the assumption that a constant rate of death or 'negative growth' K occurred such that the number of organisms dying was equal to $K \times t$. Comparing equations (9) and (10) it can be seen that $K = mY_{EG}$.

Thus, the net effect is the same whether it is assumed that the portion of substrate utilisation which is independent of growth-rate represents energy required to maintain cell integrity, or that it is a constant loss of cell mass through death and autolysis. In support of Topiwala's concept, earlier work had shown that the viability of cells in chemostat culture was as low as 70 per cent at dilution rates below 5 per cent of μ_{max}.

3.2 Nature of maintenance requirement

Typical values for the maintenance ration of microorganisms are shown in Table 7.1. For most cells grown under energy-limited conditions, the maintenance ration is low compared with the energy used at growth-rates of more than half the maximum rate. But in some cases, mostly where energy is not limiting for growth, a very large proportion of energy used seems to be required for 'maintenance'. Here it will be as well to make a clear distinction between the use of the terms 'maintenance substrate requirement' and 'endogenous metabolism'. The confusion

Table 7.1   Maintenance energy requirements m obtained for various organisms*

Organism	Limiting factor	Growth conditions	m (g substrate/ g cells/h)
Klebsiella aerogenes	Glucose	aerobic	0·042
	Glucose	anaerobic	0·50
	Tryptophan	anaerobic	3·69
Saccharomyces cerevisiae	Glucose	anaerobic	0·036
Azotobacter vinelandii	N_2	nitrogen-fixing, high O_2	1·5
	N_2	nitrogen-fixing, low O_2	0·15

* The references for these values are to be found in the reviews by Harrison (1976), Stouthamer (1976) and Pirt (1975).

arises because the term endogenous metabolism has always been used to denote that metabolism occurring *in the absence of growth*. This definition is not to be confused with that of maintenance requirement which is that utilisation of substrate which *is independent of growth-rate* (but occurs *during* growth).

The 'endogenous' metabolic rate may be higher than the 'maintenance' requirement, owing to mobilisation and utilisation of stored products in resting cells.

What then is the nature of the maintenance requirement? Thermodynamic considerations suggest that as long as a cell was not in an extremely hostile environment the energy required merely to maintain cell integrity would be very small indeed. Resting bacteria have been found to maintain viability for very long periods with negligible output of energy in the form of heat. Possibly maintenance energy is required to protect the integrity of cells only during growth, as active cells suffer some loss resulting from growth processes. Another explanation is that energy of maintenance represents a constant 'leakage' or 'wastage' of energy from active cells due to inherent inefficiencies, or, that it represents energy required to maintain certain concentration gradients across the cell membrane and that this is a function of cell mass and time but independent of growth-rate. A further possible explanation is that 'maintenance' represents a constant death rate of the culture as proposed by Topiwala and Sinclair. Which of these explanations is correct (and no doubt there are other possibilities) cannot be answered from the data presently available. Probably many or all of these mechanisms can contribute, at different times, to what is termed the 'maintenance' requirement.

The importance to our discussion here is that if maintenance is a real physiological parameter then in order to express the true efficiency of growth, the observed yield coefficient in a culture must be corrected for maintenance to obtain the 'true yield'.

3.3 'True yield'

From Table 7.1 it can be seen that the maintenance coefficient of cultures varies greatly. For cultures which are strictly energy source-limited, however, the maintenance metabolic rate is normally very low compared to the total rate at growth rates above half μ_{max}. In such cases the correction to the observed yield is insignificant. For example, *K. aerogenes* grown glucose-limited has a maintenance energy requirement of 0·22 mole glucose/g cell compared with a total $q_{glucose}$ at $\mu = 0.21$ h^{-1} of 12·8 mole glucose/g cell, so that the difference between $Y_{observed}$ and Y_{EG} would be only 1·7 per cent. On the other hand, when energy is not growth-limiting the maintenance ration can be very high indeed. Stouthamer and Bettenhaussen (see Stouthamer 1976) found a maintenance ration for a tryptophan-limited culture of *K. aerogenes* of 16 mmole glucose g^{-1}h^{-1} compared with a total $q_{glucose}$ of about 25 mmole g^{-1}h^{-1}. In this case the correction was very significant and the Y_{EG} was several times larger than $Y_{observed}$. This leads to an apparently paradoxical situation: although the actual yield observed in cultures which are not energy-limited is lower than that of energy-limited cultures, the Y_{EG}, after correcting for very high maintenance requirements, may be far greater.

This apparent paradox can be explained if the maintenance requirement is *not* always a constant. Consider Fig. 7.2: the specific substrate utilisation is made up of that used for growth $1/Y_{EG}$ and that not used for growth which is termed — misleadingly in this context — 'maintenance requirement' m. Therefore, we can write:

$$q_{substrate} = \frac{\mu}{Y_{EG}} + m \qquad (11)$$

Usually the plot of $q_{substrate}$ against D yields a straight line (Fig. 7.2), so it follows that if m is a constant Y varies, which is the assumption made in deriving equation (8). However, if it is assumed that Y is constant, then m must vary. On the face of it the former hypothesis would seem the more reasonable. However, the substrate uptake rate for a non-limiting substrate may not be regulated by its use as energy, although it is proportional to growth-rate; consequently more energy becomes available to the cell than is required for growth (growth being limited by another factor) and this excess energy is wasted. The rate of wastage (which appears as 'maintenance requirement') would decrease with growth-rate as the difference between uptake of the energy substrate and requirement for growth diminishes. This variable wastage of energy in the presence of excess energy source has been termed 'uncoupled' catabolism (Harrison 1976) or 'slip' catabolism by Tempest (Neijssel and Tempest 1976). It must be emphasised that this argument is invoked only where growth is *not* energy-limited and does not invalidate the arguments for constant maintenance, or constant death-rate, in energy-limited situations.

This issue is still open to argument and controversy: one school of thought states that the concept of calculated Y_{EG} has no meaning in reality as maintenance is not necessarily a constant, while another states that Y_{EG} represents more closely the growth efficiency of microorganisms than do observed yields. At present the reader must keep an open mind on this subject.

3.4 Y_{ATP}^{MAX}

Because there are so many reported exceptions to the constant Y_{ATP} hypothesis of Bauchop and Elsden, Stouthamer has proposed that the concept should be modified. He argues that there is a Y_{ATP}^{MAX} which represents the absolute maximum efficiency for growth but that this is rarely approached in reality because of maintenance requirement and inherent inefficiencies. Stouthamer calculated the Y_{ATP}^{MAX} from known metabolic pathways and the molecular composition of the microbial cell. He calculated the number of ATP molecules involved in producing all the main cellular constituents and from this has derived a value, termed Y_{ATP}^{MAX}, which was the maximum amount of cell material that could be produced per mole ATP. The values obtained for growth on glucose were about 30, far higher than that of 10·5 proposed by Bauchop and

Elsden. Stouthamer and his co-workers have reported values of Y_{ATP} approaching this high value, some being in excess of 25, but these were Y_{EG} values obtained by correcting for 'maintenance' rather than actual yields.

A serious criticism of this approach is that many essential energy-consuming processes connected with cell division are not accounted for because they do not involve the specific production of constituent molecules, e.g. membrane assembly and conformational changes in proteins. Also, it is assumed that all energy used for growth is actually mediated through ATP. This assumption neglects those processes which use, for instance, reducing power from NADH and NADPH which may be regarded as having an ATP equivalence.

Nonetheless, the painstaking approach of Stouthamer reveals some very interesting information about energy requirements. His calculations showed that very little energy is to be saved by supplying preformed amino acids and nucleic acid bases to cells. This is contrary to assumptions often made in the past. Also, he showed that growth on C_2 compounds such as acetate as the sole carbon source was certain to be far less efficient that growth on glucose.

This debate on the maintenance energy requirement and 'true yield' is central to the assessment of growth efficiency. The issue has not yet been resolved, but in the author's view the conflicting evidence can be reconciled if the 'constant maintenance requirement' concept is only invoked for energy-limited systems and where it is low compared with the total energy consumption for growth. Where maintenance requirement is high and in non-energy-limited systems, care must be taken to use a corrected value for yield. And of course, in the last analysis, the only yields of practical importance are those actually obtained rather than any hypothetical maximum values.

4 Factors affecting growth efficiency

4.1 Organism-dependence of yield

Table 7.2 shows the yield coefficients reported for a number of microorganisms grown on glucose under anaerobic conditions. The most frequent values fall between 20 and 25. Low yield values might reflect operation of the Entner–Doudoroff pathway for glucose catabolism

Table 7.2   Yield coefficients of
anaerobically grown microorganisms

Organism	$Y_{glucose}$ *(g cells/g-mole glucose)*
*Klebsiella aerogenes**	23·4 (after 1 day)
	14·4 (after 3 days)
*Streptococcus faecalis**	20·0
Lactobacillus plantarum	20·0
L. casei	42·0
*Saccharomyces cerevisiae**	20·0
S. rosei	23
Clostridium thermoaceticum	50

* Bauchop and Elsden (1960); other references are to be found in reviews by Harrison (1976), Stouthamer (1976) and Forrest and Walker (1971).

which generates only one ATP molecule against the two molecules generated by the glycolytic pathway. High values, if they are truly comparable and not a result of co-metabolism of other organic compounds in the medium, may be due to the extra ATP formed by anaerobic electron transport.

Table 7.3 shows yield coefficients for a number of microorganisms grown under methanol-limited conditions; each value should therefore be comparable. Clearly, there is a significant variation in yield and some of these differences can be attributed to cell composition. For instance, the yeast *Hansenula anomala,* although giving a yield of 0·54 on a w/w basis, comprised 40 per cent protein compared with 75 per cent in *Pseudomonas extorquens,* so that on a protein basis the yeast yield was lower. A lower yield from methanol is to be expected for yeasts, since it appears that the methanol is converted to formaldehyde by a direct flavoprotein oxidase which is not linked to ATP generation, while in bacteria the reaction is catalysed by a pteridine methanol dehydrogenase linked to at least one site of oxidative phosphorylation.

Table 7.3 Some typical growth yields obtained for microorganisms growing on methanol

Organism	*Optimised yield (g cells/g methanol)*
Protaminobacter ruber	0·25
Pseudomonas extorquens	0·40
'EN', obligate CH_3OH utiliser*	0·52
Pichia pinus	0·30
Hansenula anomala	0·54

* This organism was grown in a symbiotic mixed culture.

The difference in yield between the bacterium 'EN' and *Pseudomonas extorquens* is not surprising as it is known that the pathways of anabolism in these two species are different. *Pseudomonas extorquens* incorporates one-carbon units in the form of formaldehyde by means of the 'serine' pathway, while the organism EN uses the hexose phosphate pathway. The serine pathway is certainly more expensive in terms of ATP equivalents than the hexose phosphate pathway and this is reflected in the yields.

Thus, variation in yield coefficients of microbial species may reflect differences in catabolic or anabolic pathways, cell composition or so-called 'maintenance' ration. Whether the basic efficiency of growth, as reflected by Y_{ATP}, also varies is still an open question.

4.2 Effect of substrate-type on yield

4.2.1 Energy and carbon source

Obviously the energy source for growth is going to have a profound effect on growth yield depending on how much energy can be conserved during catabolism of the substrate and how much energy is required for anabolism. We have already seen that the number of available electrons in a substrate is only a rough guide to the amount of energy available to the organism for growth. The potential to produce ATP equivalents is a better indication of energy availability for growth. Table 7.4 shows the potential ATP production per molecule for a few typical substrates. There will also be differences in the energy requirement of anabolic processes depending on the main carbon source. If the latter is a molecule related to those found in cell

Table 7.4   Comparison of the ATP available from the catabolism of various substrates

Substrate	Conditions	ATP (mole ATP/mole substrate)
Glucose	Aerobic (P/O = 3)	38
Glucose	Anaerobic (products 1:1, formate + acetate)	3*
Glucose	Anaerobic (products ethanol + CO_2)	2*
Glucose	Anaerobic (products ethanol + CO_2)	1†
Pyruvate	Aerobic (P/O = 3)	30
Pyruvate	Anaerobic (products formate + acetate)	1
Pyruvate	Anaerobic (products ethanol + CO_2)	0
Palmitate (C_{16})	Aerobic	130

* Via the Embden–Myerhof pathway.
† Via the Entner–Doudoroff pathway.

structures, such as glucose, then the energy required for incorporation will be less than for chemically simpler molecules which require considerable elaboration before they can be incorporated into cell material. An extreme case is, of course, growth on CO_2.

4.2.2 Nitrogen source

It is common experience to find that the growth-rate and yield of microorganisms is greater on complex media than on simple media and that much of this benefit is derived from supplying essential amino acids. However, as we have seen from calculations of energy (as ATP) required to synthesise cell material, Stouthamer concluded that there would be negligible difference in the theoretical energy requirement during growth on sugars, whether amino acids are supplied or not. The improvement experienced may be due in part to the use of amino acids as an energy source through co-metabolism, although this seems unlikely to be the complete explanation. Possibly when no amino acids are supplied and all have to be synthesised, the demands made on certain pathways common to catabolism and anabolism lead to regulatory problems which in turn cause loss of efficiency. This would be relieved by the supply of preformed amino acids.

We can confidently predict an effect of inorganic sources of nitrogen on yield. If nitrate is supplied as the sole source of nitrogen for growth, it must be reduced to the level of ammonia before it is incorporated into cell material. The enzymes responsible are nitrate reductases and the overall reaction may be represented:

$$NO_3^+ + 10 H^+ \rightarrow NH_4^+ + 3 H_2O$$

If a phosphorylation efficiency, in terms of the P:(2 H^+) ratio, of 3 is assumed, then an equivalent of 15 extra ATP molecules is required for each gram-atom of nitrogen incorporated, compared with the incorporation of nitrogen from ammonia. For cells with a protein content of about 75 per cent of dry weight, the expected reduction in yield using nitrate as nitrogen source rather than ammonia, would be of the order of 10 per cent.

When ammonia is supplied as the nitrogen source, there are two known major pathways for its incorporation. For most microorganisms growing in the presence of excess ammonia, incorporation is via glutamate dehydrogenase:

$$
\begin{array}{ll}
\text{COOH} & \text{COOH} \\
| & | \\
\text{CO} & \text{HCNH}_2 \\
| & | \\
\text{CH}_2 + \text{NADH} + \text{H}^+ + \text{NH}_3 \rightarrow \text{CH}_2 + \text{NAD}^+ + \text{H}_2\text{O} \\
| & | \\
\text{CH}_2 & \text{CH}_2 \\
| & | \\
\text{COOH} & \text{COOH}
\end{array}
$$

However, in many bacteria, another mechanism is induced when growth is made under ammonia-limited conditions. This is the glutamine synthetase (GS)–glutamate synthase (GOGAT) system which may be represented:

$$
\begin{array}{l}
\text{NH}_3 + \text{ATP} \searrow \quad \nearrow \text{glutamate} \nwarrow \qquad \nearrow \text{glutamate} \\
\qquad\qquad) \;\; \text{GS} \;\; (\qquad\qquad) \;\; \text{GOGAT} \;\; (\\
\text{ADP} + \text{Pi} \nearrow \quad \searrow \text{glutamine} \nearrow \qquad \searrow \text{2-oxoglutarate}
\end{array}
$$

The overall balance then is:

$$\text{NH}_3 + \text{2-oxoglutarate} + \text{ATP} + \text{NADPH} \longrightarrow \text{glutamate} + \text{ADP} + \text{Pi} + \text{NADP}^+ + \text{H}_2\text{O}$$

Comparing this system with the glutamate dehydrogenase mechanism it can be seen that an extra ATP is required for ammonia incorporation. However, the K_m for the ammonia of glutamate synthetase is less than 1 mM while that of glutamate dehydrogenase is 3 to 4 mM. Thus, it would seem that when grown under limited-ammonia conditions the organism sacrifices ATP in order to take up ammonia from lower concentrations which would no doubt improve its competitiveness. The effect which this extra ATP would have on measured yield, however, is very small. For an organism of 75 per cent protein by dry weight and a Y_{ATP} of 10 there would be about a 2 per cent reduction in yield, well within the margin of error for most measurements.

The ability to fix atmospheric N_2 in the absence of any other source of nitrogen would clearly give a great selection advantage to a microbial species. Therefore, it is not surprising that some organisms have evolved a system for N_2 fixation although it is energetically expensive. For instance, in *K. aerogenes* a reduction in yield of 60 per cent on switching from growth on fixed nitrogen sources to N_2 fixation has been reported. *Azotobacter vinelandii* organisms, at their most efficient, appear to be more economical and it has been calculated that 5 molecules of ATP are required for each N atom fixed. On the other hand, *Azotobacter* species grow very inefficiently when oxygen is present in excess. Under these circumstances efficiency is sacrificed to protect the N_2-fixing enzyme, nitrogenase, against oxygen toxicity (Hill *et al.* 1972).

4.3 Environmental effects

4.3.1 pH

There have been few systematic studies on the effect of pH on growth efficiency. Generally, as far as growth *rate* is concerned, there is a restricted range for optimal pH and the μ_{max} decreases on either side of this point. The steepness of the curve varies greatly in different organisms. From the few data available it would seem that growth yield varies less with pH than growth rate. In experiments with glucose-limited cultures of *Escherichia coli*, no effect on yield was obtained on varying pH between 6·0 and 8·2 and only when the pH was lowered to 5·4 was there a significant decrease in growth yield (Harrison and Loveless 1971). Of course, in some cases the metabolism of a given substrate may be very much dependent on pH and this would be reflected in the growth yield.

4.3.2 Temperature

The growth-rate generally increases with temperature in accordance with the Arrhenius relationship up to an optimum temperature above which there is a decline in maximum growth rate. The effect on growth efficiency appears to be variable but is generally less pronounced. In some cases a decreased yield at temperatures above the optimum have been reported, with constant growth efficiency below this, but for other organisms low temperatures may diminish growth efficiency. In a systematic study of the effect of temperature on a chemostat culture of *K. aerogenes* Topiwala and Sinclair (1971) found that the fall in yield coefficient with increasing temperature could be wholly accounted for in terms of an increased specific rate of death, which would be equivalent to 'maintenance requirement'.

4.3.3 Dissolved oxygen

Since respiration is by far the largest source of energy in aerobic organisms, dissolved oxygen can be expected to have a great influence on growth efficiency. In fact, dissolved oxygen is undoubtedly one of the most important influences of the environment. No organism is completely indifferent to oxygen; it is essential for the survival of many species, but is very toxic to others and for all species it exerts an influence over metabolism. There is insufficient space here to discuss all the complicated effects oxygen may have on growth efficiency and the reader is referred to one of the reviews of this topic for more details (Harrison 1972). Instead a short résumé of the major points will be made.

For most aerobic and facultatively anaerobic microorganisms there is a range of dissolved oxygen tensions, usually between 10 and 150 mm Hg, over which dissolved oxygen has little or no effect on cell metabolism (air saturation is equivalent to about 156 mm Hg). A notable exception to this generality is provided by *Azotobacter* which, when growing and fixing N_2, demonstrates an increasing respiration rate and decreasing yield with dissolved oxygen over this range.

For most organisms there is a so-called 'critical dissolved oxygen tension' at about 1–10 mm Hg below which changes in dissolved oxygen greatly affect respiration rate and metabolism. For non-growing cells the respiration rate falls with decreased dissolved oxygen tensions below the critical value. This is often also the case for growing cells but not always. For *K. aerogenes,* for instance, the respiration rate *increases* at dissolved oxygen tensions just below the critical value and then falls off again as the oxygen tension is lowered further. This increased respiration rate at low oxygen tensions is accompanied by a fall in the yield coefficient from glucose and oxygen of 20–30 per cent.

As the oxygen tension is reduced in cultures of facultative anaerobes, fermentative pathways of catabolism eventually will function. These latter yield less ATP per molecule of organic substrate than respiration and so the yield expressed in terms of the organic substrate will fall. However, if the yield is expressed in terms of oxygen this may be expected to increase as fermentative pathways play a larger part in ATP generation.

High oxygen concentrations are toxic to all microorganisms, although the sensitive range varies considerably. Many organisms will grow at dissolved oxygen tensions of over 700 mm Hg, i.e. equivalent to saturation with a 100 per cent oxygen atmosphere. Some, however, are quite sensitive; a methanol-utilising bacterium, for example, demonstrated both decreased yield and growth-rate at oxygen tensions approaching air saturation.

4.3.4 Other environmental factors

Growth efficiency can be affected by many other factors such as the composition of mineral ions in the culture medium. Freshwater microorganisms need to accumulate potassium ions against a gradient and exclude sodium ions. If these organisms are grown in the presence of low potassium and high sodium concentrations significant amounts of energy will be expended in order to pump these ions against the concentration gradient. Marine organisms, on the other hand, will be adversely affected by low sodium concentrations and at least one non-halophilic bacterium, *Pseudomonas stutzeri,* has been reported to require a certain critical sodium concentration for maximum yield efficiency (Kodama and Taniguchi 1976).

The limitation of growth by any of the essential inorganic ions will influence metabolism and may affect yield, at least indirectly. Iron limitation, for instance, affects yield directly in aerobic yeast by causing the loss of site one of oxidative phosphorylation.

An interesting symbiotic mixture of bacteria when grown in a chemostat with methanol as sole carbon source not only has a faster growth-rate, but also a higher yield coefficient than a pure culture of the only methanol-utiliser present. The four heterotrophic symbionts were present in very small numbers ($<$ 10 per cent of the population) but had the effect of increasing the yield coefficient by more than 50 per cent. Whether the symbionts had their affect by removing inhibitory products of methanol metabolism or by supplying growth promoters has not yet been established (see Ch. 8).

5 The regulation of growth efficiency

5.1 Requirement for regulation

Microorganisms, when growing in a medium containing only one organic chemical which must function as both carbon and energy source, are faced with a dilemma. They must apportion the substrate between catabolic and anabolic processes, and the extent to which they balance the energy and carbon requirements will determine the growth efficiency or yield. The question is whether microorganisms do attempt to regulate energy metabolism in order to maximise growth efficiency, or, whether the whole system is just allowed to float. It would seem reasonable to suppose that the efficient conversion of substrate into cell biomass would confer some selective advantage on an organism and, therefore, we can expect to find regulatory mechanisms functioning to optimise growth efficiency. However, the mode of existence of most microorganisms, especially prokaryotic organisms, is such that, unlike the cells of higher organisms which are maintained in controlled environments, they are exposed to the vagaries of a changing environment. As survival is even more important than efficiency, at times micro-organisms must doubtless sacrifice growth efficiency for adaptability.

Regulation in microbial cells, as in all well-controlled systems, operates at several levels. The fastest and finest control is mediated by the inhibition and stimulation of key enzymes by means of allosteric mechanisms. The next level of control involves the induction and repression of enzyme synthesis. This is a coarser control and responds more slowly than allosteric regulation, but is capable of producing a greater change in rates. Yet a third level of control, it could be argued, is provided by strain selection, as microbial cultures usually show great genetic variability. However, for the purposes of our discussion we will restrict ourselves to the first two levels.

5.2 Regulation by induction and repression

An example of induction of new enzymes to adapt growth efficiency to environmental changes, is the response obtained when anaerobically growing cultures of facultative organisms are aerated. Facultative bacteria and yeast grown in the complete absence of oxygen, lack respiratory enzymes and have a restricted tricarboxylic acid (TCA) cycle. On aeration, cyto-chromes and TCA cycle enzymes are rapidly induced. In fact, in some organisms such as *E. coli* and *K. aerogenes* it requires only minute levels of oxygen to induce respiration fully. Regulation of enzyme synthesis is not necessarily an all-or-nothing effect; the cytochrome levels of cells can be modified by changes in the dissolved oxygen tension or growth-rate. Obviously, the induction of respiratory metabolism when oxygen becomes available enables the organism to grow more efficiently as more ATP can be generated.

An interesting example of regulatory control of growth efficiency is the repression of site 1 oxidative phosphorylation in *Saccharomyces cerevisiae* in the presence of excess glucose. This organism shows extreme repression of aerobic metabolism in the presence of high concentra-tions of glucose but it has been shown that, before respiration is entirely repressed, oxidative phosphorylation becomes less efficient, by 20 to 30 per cent because site 1 of oxidative phosphorylation is lost. What advantage this could possibly confer on the cell is open to conjecture but clearly under circumstances in which site 1 oxidative phosphorylation is lost, energy is not in short supply. On the other hand, under conditions where energy generation probably needs to be efficient, i.e. when glucose supply is limiting and oxygen is in excess, then site 1 oxidative phosphorylation is regained and growth efficiency maximised.

5.3 Feedback regulation

This type of control is perhaps best illustrated by the Pasteur effect in *Saccharomyces*. This effect is defined as the immediate stimulation of the glycolytic pathway when the cells are changed from aerobic to anaerobic conditions. Such a stimulation would enable the cells to maintain a high production rate of ATP, even though forced to use the less efficient fermenta-tion pathways for energy regeneration. Investigations by Pye and Chance (1966) showed that the control is achieved by allosteric regulation of the key enzyme, phosphofructokinase, by adenine nucleotides. Thus the trigger for stimulating this enzyme is a fall in energy level in the form of ATP.

It would seem logical that, as most of the energy utilised by the cell is used in the form of ATP, changes in ATP level should be a sensitive indicator of the state of energy metabolism. In fact, the actual amount of ATP in the actively growing cell is very small compared with the rate at which it is produced and used, i.e. 'turned over'. For example, in *E. coli* doubling its cell mass in about 0·5 h, the ATP 'pool' would turn over completely in a fraction of a second. Under these circumstances ATP is in no sense an energy reserve and the concentration of ATP *must* be regulated very accurately, for should it fall to zero, the cell could not carry out the substrate phosphorylation reactions essential to catabolism and would die. Therefore, it is not surprising to find that many key enzymes of catabolism and anabolism are subject to allosteric control by ATP, ADP or AMP, or any combination of these.

Atkinson (1971) proposed that the regulation was such that when the 'energy charge' was high, anabolic processes were favoured and when it was low, catabolic processes were favoured. Energy charge is defined as:

$$\frac{[ATP] + 0·5\ [ADP]}{[ATP] + [ADP] + [AMP]}$$

It is necessary to define energy charge in terms of all three adenine phosphates as the enzyme adenylate kinase renders them interchangeable:

$$2\ ADP \rightleftharpoons ATP + AMP$$

The concept of regulation by energy charge is not only logical but seems to be broadly substantiated by available data. Atkinson predicted that energy charge would be high in cells supplied with excess energy source and low in energy-limited cells. There are exceptions to this: *K. aerogenes* grown with glucose in excess has a lower energy charge than when grown glucose-limited. This does not invalidate the concept but it must be modified to take into account the fact that growing cells may radically alter their whole metabolic systems when adapting to environmental changes so that the 'set-point' for the control by energy charge may vary. A fuller discussion of regulation of energy metabolism has recently been published (Harrison 1976).

5.4 Mechanisms for changing growth efficiency

A fall in growth efficiency in terms of energy must indicate a fall in the efficiency of either energy generation as ATP equivalents or in the efficiency of utilisation of energy for growth. So far no one has shown with certainty which of these is responsible for the changes in growth efficiency observed, because the turnover rate of ATP cannot be determined.

A fall in the efficiency of energy utilisation for growth would imply either a fall in the Y_{ATP} value, or an increase in the 'maintenance requirement'. It is difficult to distinguish between these alternatives experimentally. Utilisation of ATP without useful work being carried out can be demonstrated by ATPase enzymes *in vitro*. Whether these enzymes are ever 'uncoupled' *in vivo* has not been ascertained. Certainly, there would seem to be no reason why such uncoupling of ATPase activity should not occur if it were beneficial to the cell. Similarly wasteful or 'futile' energy cycles can be conceived in which there is a net utilisation of ATP without net cell synthesis. There are many examples which can be constructed such as:

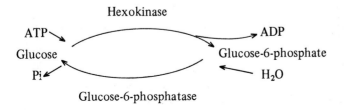

and

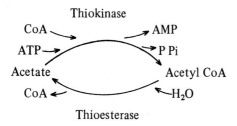

Thiokinase

Thioesterase

Where two or more enzymes are present which can create a futile cycle then they clearly have to be under rigid regulation by the cell to prevent a complete drain of energy. But they present a possibility for a drain of energy either due to imperfect control or deliberate 'uncoupling'. This has not yet been demonstrated in growing cells. A fall in the efficiency of ATP production would equally explain a decrease in growth efficiency. We have already stated that *Saccharomyces* can lose a site of oxidative phosphorylation when grown with glucose in excess or under oxygen limitation. Whether this occurs in other organisms is not known. Many chemical uncouplers of oxidative phosphorylation are known, such as 2,4-dinitrophenol, which effectively divorces respiration from ATP generation. There are no reports, however, of cells 'uncoupling' their own oxidative phosphorylation by production of such agents.

One hypothesis currently popular is that a partial 'uncoupling' of oxidative phosphorylation can be brought about in bacteria by the functioning of an alternative electron transport pathway deficient in one or more sites of energy conservation. Many bacteria possess several different cytochromes with the apparent properties of terminal oxidases. It would seem reasonable to assume that these represent the terminal members of alternative pathways for electrons to oxygen and that these are used to adapt to changes in environmental conditions. Ackrell and Jones (1971) have produced evidence of such a branched pathway in *Azotobacter* (see Fig. 7.3). They suggest that the very high rate of respiration, accompanied by a drastically decreased yield, demonstrated in these organisms when grown in the presence of excess energy source and oxygen and fixing atmospheric nitrogen, is caused by a switch of electron pathways from the one involving cytochromes a_1 and o to that involving cytochrome d. Because the latter is deficient in one site of oxidative phosphorylation, it is claimed that growth is effectively 'uncoupled'. However, although this is an attractive hypothesis, it cannot be the whole story, because loss of one site out of three would increase respiration and decrease yield by only 30 per cent, whereas the observed change was more than 5-fold. Probably, there is an increased ATP turnover also, perhaps due to the action of the nitrogenase system.

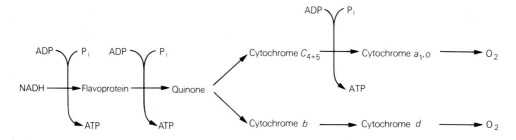

Fig. 7.3 The branched electron transport pathway for *Azotobacter* proposed by Ackrell and Jones.

The increased respiration rate and diminished yield observed in cultures of *K. aerogenes* and *E. coli* grown under oxygen limitation *were* of the order of 30 per cent and, therefore, could possibly be explained entirely by the loss of a site of oxidative phosphorylation. These organisms also possess three separate cytochromes which have been designated terminal oxidases. Thus this possibly represents a case where a branching of the electron transport system occurs and in which the less efficient system is favoured at low oxygen tensions. We can only speculate on the reason for this adaptation; perhaps the organism sacrifices efficiency at low oxygen concentrations for a system with a slightly greater affinity for oxygen. The diminished growth efficiency in *Azotobacter* at high oxygen tensions is easier to justify. The effect is to increase enormously the oxygen consumption rate which would reduce the oxygen concentration in the vicinity of the organism; as the nitrogenase enzyme system is highly sensitive to inhibition by oxygen, this provides a protection against oxygen toxicity.

6 The importance of growth efficiency

6.1 Ecological significance

At first consideration it would seem elementary that greater efficiency of growth must give an organism greater competitive advantage and that there would be selection for such a capacity. However, as we have seen, the growth efficiency of organisms does vary considerably and so efficiency would not always seem to be the most important aspect of microbial growth as far as natural selection is concerned.

The advantage to the organism of increasing growth efficiency in some circumstances must be balanced against advantages that might accrue by expending energy to increase the affinity for the substrate. As we have seen, some bacteria adopt the GOGAT pathway of ammonia utilisation under ammonia-limited conditions even though this costs more ATP. In this case the loss of growth efficiency is very small and presumably more than compensated by the ability to scavenge ammonia at lower concentrations.

Other circumstances may occur where it would benefit an organism to utilise a substrate rapidly even if this led to less efficient growth: for instance, if the energy source also happened to be toxic or inhibitory. Methanol-utilising bacteria have been found to grow far less efficiently when methanol is present in excess than when it is limiting for growth. Methanol is potentially toxic and its rapid oxidation is presumably desirable to the organism. Also in this category is the rapid and wasteful oxidation of substrate by *Azotobacter* which serves to protect the nitrogen-fixation mechanism from oxygen toxicity.

Although efficient growth confers an important selective advantage on an organism under most growth conditions, there are circumstances in which there are definite advantages to the organism in sacrificing growth efficiency. Perhaps, therefore, it is not surprising to find that some of the most adaptive and successful microorganisms have been found to alter their growth efficiency in order to adapt to the growth environment.

6.2 Significance of yield in industrial processes

6.2.1 Single-cell protein (SCP)

Recently much attention has been focused on the production of microbial biomass both for animal feed and as an addition to human diet. Several large commercial processes are now established for the manufacture of SCP from oil products, from methanol and from carbo-hydrate.

Obviously, where biomass is the desired product the yield coefficient is important in determining the substrate requirement of the process. However, even if the substrate is a waste substance the cost of which is negligible, yield is still economically the most important variable in the process. For an aerobic process growth efficiency is directly linked to oxygen consumption, thus yield will determine the oxygen supply rate required for the process. Productivity in an SCP plant will be limited by the gas-transfer capabilities of the fermenter; the fermenter will be designed for a maximum rate of oxygen transfer, so that as the oxygen requirement is a function of yield coefficient, the productivity of the plant will also be a function of growth efficiency. Thus, the size of the plant required is determined by the yield coefficient. We saw in section 2.1 that heat production is directly related to yield. Therefore, the amount of cooling required and cooling costs will depend on growth efficiency. Figure 7.4 demonstrates the control position played by yield coefficient as a determinant of process costs in the production of SCP. Therefore, research towards processes for the production of SCP must be heavily orientated towards maximising the growth efficiency of microorganisms.

6.2.2 Waste treatment

The aim of an efficient waste treatment process is the direct opposite of a single-cell protein process: it is desirable to oxidise as much as possible of the organic matter in a waste stream with minimum organism production (note that microorganisms growing on the waste form the 'sludge', the disposal of which is a very costly business). Thus, in this case growth conditions must be arranged to encourage inefficiency in growth and substrate utilisation. One way in

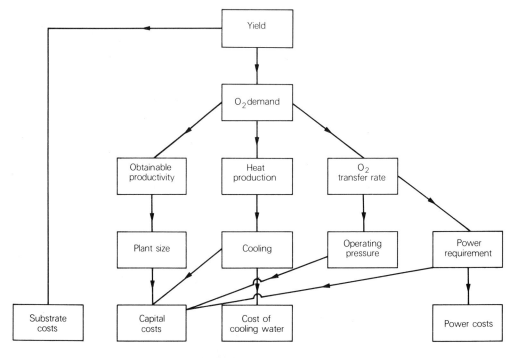

Fig. 7.4 Diagrammatic representation of the influence of growth efficiency (yield) on costs in a single-cell protein manufacturing process. Arrows indicate a direct influence.

which this could be achieved is by making the 'maintenance requirement' comparatively large. By growing organisms at a very low growth-rate the maintenance ration becomes very significant (see Fig. 7.1). An activated sludge plant for the aerobic treatment of waste-water functions similarly to a continuous culture with feedback of organisms. By recycling sludge into the aeration vessel, a very high throughput of waste water is obtained although the growth-rate of the organism, determined by the wastage rate of the sludge, may be very low. Sludge doubling times of 8 to 10 days are not uncommon.

In recent years an increasing effort has been put into improving traditional aeration-tank systems of waste-water treatment. These are aimed mostly at achieving higher rates of treatment and shorter contact times with the sludge. In many cases this leads to greater sludge growth-rate and more sludge production. Therefore, other methods must be sought to decrease the growth efficiency of sludge organisms.

6.2.3 Other processes and future prospects

For most industrial fermentation processes which are concerned with the production of secondary metabolites, e.g. an antibiotic, organic acid, or polysaccharide, it is important that as much of the substrate as possible is converted to the metabolite rather than to cell mass. Growth efficiency must, therefore, be low but it is equally important that the energy is used efficiently in the formation of the product and that the organic energy source is not oxidised in an 'uncoupled' manner to CO_2. A balance must be struck between regeneration of cells to provide the means of product formation and the almost complete conversion of substrate to product. The regulation of energy metabolism is clearly of no less importance in such systems and if this could be manipulated at will to divert the available energy either to cell production or metabolite formation then a full optimisation of the process could be obtained. In modern fermentation processes this is attempted by careful control of physical and chemical parameters, by manipulating the kinetics of growth, e.g. by deployment of fed batch systems (see Pirt 1975) or by genetic manipulation. Recent years have seen a great advance in our understanding of the genetic aspects of regulation of metabolism. So far advances in this area have been of an academic nature but already it has proved possible to upset deliberately the regulatory mechanisms of microorganisms so that virtually all of the organic substrate supplied is converted to one product. Looking forward, it would not be over-optimistic to predict that knowledge of energy metabolism regulation will enable organisms to be produced which can be manipulated to grow at maximum efficiency and subsequently induced to produce, as a high proportion of their cell mass, the necessary enzyme systems which can then be further manipulated to convert substrate into desired metabolites with similar maximum efficiency. Such systems represent the ultimate in self-generating industrial catalysts.

Bibliography

ACKRELL, B. A. C. and JONES, C. W. (1971). 'The respiratory system of *Azotobacter vinlandii* 2. Oxygen effects', *Eur. J. Biochem.,* **20**, 29.
    Evidence of a branched electron-transport pathway.
ATKINSON, D. E. (1971). 'Adenine nucleotides as stoichiometric coupling agents in metabolism and as regulatory modifiers. The adenylate energy charge', in *Metabolic Regulation* (Ed. H. J. Vogel), p. 1. Academic Press, New York.
    A review of energy charge.

BAUCHOP, T. and ELSDEN, S. R. (1960). 'The growth of microorganisms in relation to their energy supply', *J. Gen. Microbiol.,* **23**, 457.

The original paper on the constant Y_{ATP} theory.

FORREST, W. W. (1969). 'Energetic aspects of microbial growth', in *Microbial Growth* (Eds. Pauline M. Meadow and S. J. Pirt), 19th Symposium of the Society for General Microbiology, p. 65. Cambridge University Press, London.

A review with particular mention of thermodynamic considerations.

FORREST, W. W. and WALKER, D. J. (1971). 'The generation and utilization of energy during growth.' *Adv. Microb. Physiol,* **5**, 213.

HARRISON, D. E. F. (1972). 'Physiological effects of dissolved oxygen tension and redox potential on growing populations of microorganisms', in *Environmental Control of Cell Synthesis and Function* (Eds. A. C. R. Dean, S. J. Pirt and D. W. Tempest), p. 417. Society for Chemical Industry and Academic Press, London.

HARRISON, D. E. F. (1976). 'The regulation of respiration rate in growing bacteria', *Adv. Microb. Physiol.,* **14**, 243.

A detailed review of energetics and control of aerobic processes.

HARRISON, D. E. F. and PIRT, S. J. (1967). 'The influence of dissolved oxygen concentration on the respiration and glucose metabolism of *Klebsiella aerogenes* during growth', *J. Gen. Microbiol.,* **46**, 193.

An account of the phenomenon of decreased growth efficiency at low oxygen tensions.

HARRISON, D. E. F. and LOVELESS, J. E. (1971). 'The effect of growth conditions on respiratory activity and growth efficiency in facultative anaerobes grown in chemostat culture', *J. Gen. Microbiol.,* **68**, 35.

HERBERT, D. (1958). 'Some principles of continuous culture', in *Recent Progress in Microbiology VII Int. Congr. for Microbiology,* (Ed. G. Tunevall), p. 381. Almquist and Wiksell, Stockholm.

HILL, S., DROZD, J. W. and POSTGATE, J. R. (1972). 'Environmental effects on the growth of nitrogen-fixing bacteria', in *Environmental Control of Cell Synthesis and Function.* (Eds. A. C. R. Dean, S. J. Pirt and D. W. Tempest), p. 541. Society for Chemical Industry and Academic Press, London.

Contains a discussion of 'uncoupled' respiration as a protection against oxygen.

KODAMA, T. and TANIGUCHI, S. (1976). 'Sodium-dependent growth and respiration of a non-halophilic bacterium, *Pseudomonas stutzeri*', *J. Gen. Microbiol.,* **96**, 17.

NEIJSSEL, O. M. and TEMPEST, D. W. (1976). 'The regulation of carbohydrate metabolism in *Klebsiella aerogenes* NCTC 418 organisms growing in chemostat culture', *Arch. Microbiol.,* **106**, 251.

PAYNE, W. J. (1970). 'Energy yields and growth of heterotrophs', *Ann. Rev. Microbiol.,* **24**, 17.

Includes a treatment of $Y_{av.e^-}$.

PIRT, S. J. (1965). 'Maintenance energy of bacteria in growing cultures', *Proc. Roy. Soc. B.,* **163**, 224.

PIRT, S. J. (1975). *Principles of Microbe and Cell Cultivation.* Blackwell Scientific Publications, Oxford.

A comprehensive textbook on microbial growth.

PYE, E. K. and CHANCE, B. (1966). Sustained sinusoidal oscillations of reduced pyridine nucleotide in a cell-free extract of *Saccharomyces carlsbergensis*', *Proc. Natl. Acad. Sci. USA,* **55**, 888.

On control of glycolysis.

STOUTHAMER, A. H. (1970). 'Determination and significance of molar growth yields', in *Methods in Microbiology* (Eds. J. R. Norris and D. W. Ribbons), Vol. I, p. 269. Academic press, London.

A review.

STOUTHAMER, A. H. (1976). *Yield Studies in Micro-organisms*. Meadowfield Press Ltd., Durham.

A small review volume.

TOPIWALA, H. H. and SINCLAIR, C. G. (1971). 'Temperature relationships in continuous culture', *Biotechnol. Bioengng.*, **13**, 795.

On the effect of temperature on yield and maintenance requirement.

8
Interactions between microbial populations

J. H. Slater
Department of Environmental Sciences, University of Warwick, Coventry
and
A. T. Bull
Department of Applied Biology, University of Wales Institute of Science and Technology, Cardiff

'The complicated relationships between organisms which take place in nature have as their foundation definite elementary processes of the struggle for existence. Such an elementary process is that of one species devouring another or when there is competition between a small number of species in a limited microcosm.'

   G. F. Gause, 1934, in *The Struggle for Existence.*

1 Introduction

It is an axiom that populations of microbes, in common with all forms of life, rarely exist in natural environments in complete isolation from each other. However small an ecological niche is considered, the chances of finding more than one type of microbial population are high, and even greater if one considers the differences due to genetic variations within the same population. Gause was amongst the first to recognise that different organisms coexisted in nature and that it was improbable that individual populations could behave as independent entities. He argued that, as well as the constraints imposed on a particular population by the physical and chemical environment, the activity of neighbouring populations acted as additional external factors affecting the dynamics of growth and the physiological behaviour of that population. Gause was concerned principally with the effect of biological factors resulting in negative interactions between pairs of microorganisms − such as competition or prey−predator relationships − and with the mechanistic details of these relationships. The net result was a realisation that organisms had to struggle for their survival in the long term as a direct consequence of the activity of other populations and individuals.

   It is now quite clear, however, that beneficial relationships may also occur between pairs or groups of different species of microbe. Indeed it is quite likely that in nature mixed assemblages have developed to form distinct microbial communities structured on a network of potentially complex interactions. Microbial communities may be quite common and have the obvious advantage that as an integrated unit the component populations may adapt successfully to their habitat and respond efficiently to fluctuations and alterations in the environmental conditions.

   It is not the purpose of this chapter to catalogue in detail the many different interactions which have been observed. The interested reader can find specific information of this kind in the review articles cited in the general references. Instead we seek to outline and highlight some of the basic principles involved in interactions between microbes, particularly those processes which are thought to be important in nature and to illustrate these 'elementary processes' by a few selected examples.

2 Basic types of interaction

Many associations of two or more microbes have been described in mechanistic terms and are normally based on aspects of nutrition. All these associations, however, may be described by a few basic types of interaction which can be illustrated in the simplest interacting system, namely a two-membered mixed population containing organisms A and B. In fact there are only three possible responses a growing population − in this case let us say A − can make to the presence of the second population B.

1. The growth and metabolic activity of B may have a *beneficial* or *positive* effect on the growth of population A. For example, this could occur if population B excreted a compound which stimulated the rate of growth of population A, compared with its rate in the absence of population B. Alternatively the same positive effect might be achieved if population B utilised a compound initially present in the common habitat which was toxic to population A and hence restricted its development.
2. The presence of organism B could have a *detrimental* or *negative* influence on the growth of organism A. This situation could arise if organism B excreted a metabolite which was toxic to organism A.
3. It is possible that the growth of population B, the effect of its metabolic activities and its demands upon the resources of the common environment have no effect on the growth of population A. A *neutral* response of this kind would be shown by similar growth patterns for population A whether or not population B was present.

These fundamental interactions can occur in a reciprocal manner with organism B affecting the growth of organism A and *vice versa.* The various combinations between the two populations may be summarised in the simple matrix shown in Table 8.1. For two-membered mixtures, nine different responses could theoretically be established, illustrating a maximum of six fundamental types of interaction.

Table 8.1   Matrix of interactions of two microbial populations,
A and B

		The effect on the growth of organism A by the activity of organism B		
		+	0	−
The effect on the growth of organism B by the activity of organism A	+	+ +	+ 0	+ −
	0	0 +	0 0	0 −
	−	− +	− 0	− −

1 Neutralism, (0 0). In a strict sense the simple case of non-interaction between the component species ought to be excluded from a list of interactions and at present it is difficult to evaluate the significance of neutralism. On the one hand the growth of any population inevitably induces changes in the environment simply because essential minerals and other nutrients required for the growth of that population are sequestered from the habitat. Quite probably these changes will affect the growth of a second population. Alternatively in a limited number of cases it has been shown that neutralism does occur. Lewis (1967) deduced that the mixed growth of a *Lactobacillus* species and a *Streptococcus* species in a chemostat culture produced individual population sizes which were the same as those in separate monocultures under the same growth conditions. More recently M. Davies and his colleagues (personal communication) have isolated 50 bacterial strains from an activated sludge sample and shown that in over 400 pairwise tests only 59 (14·5 per cent) showed any kind of interaction. Thus for these organisms taken from a common habitat a very high proportion appeared to exhibit neutralism.

2 Mutualism, (+ +). This type of interaction occurs when both members of the mixture derive some advantage from each other's presence, in terms of increased growth rates or increased population sizes. Yeoh, Bungay and Krieg (1968) described a two-membered mixed culture of *Bacillus polymyxa* and *Proteus vulgaris* grown in a carbon-limited chemostat in a simple growth medium which could not sustain the growth of either population on its own. This indicated that each organism was in some way completely dependent on the other population to complement its minimum growth requirements. The mutualistic relationship depended on the *Proteus* species producing nicotinic acid which was an essential requirement for *B. polymyxa* growth, whilst the *Bacillus* reciprocated by excreting the vitamin biotin, thereby promoting the growth of *P. vulgaris* (Fig. 8.1). In theory a simple mutualistic interaction of this kind ought to

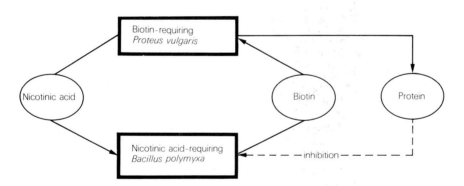

Fig. 8.1 Mutualism in a mixed culture of *Bacillus polymyxa* and *Proteus vulgaris*.

result in a steady-state culture when the two organisms are grown together in an open growth system (see Bull (1975) for a discussion of the merits of continuous-flow culture techniques). However, in this case regular oscillations were established and ascribed to the effect of a third interaction between the two populations. *Proteus vulgaris* produced a proteinaceous compound which inhibited the growth of *B. polymyxa* and caused a decrease in its population size. This in turn reduced the rate of biotin addition to the environment and, as its concentration declined, it could not maintain the original *Proteus* population size which also went into decline. At a later stage the concentration of the inhibiting protein was lowered sufficiently to cause a resurgence of the *B. polymyxa* population, completing the cyclical changes of the two

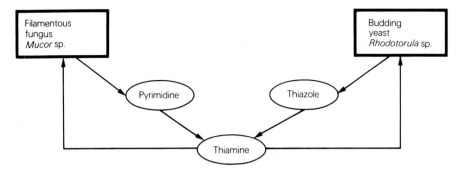

Fig. 8.2 Mutualism between a filamentous fungus and a yeast.

populations, which were then repeated. This example illustrates one of the major difficulties in population interaction studies, i.e. it is difficult to be certain that all the contributing interactions have been recognised or that the postulated simple single interactions have not been obscured by other interactions. These difficulties become particularly acute in attempting to construct quantitative models with mathematical formulations of the interactive forces between different populations.

In contrast, mutualistic relationships may be much more complex. One of the most interesting is the classic dual culture examined by Schopfer (1943) in which a filamentous fungus and a yeast cooperate as shown in Fig. 8.2 in the synthesis of the vitamin thiamine which is an essential growth requirement for both organisms.

Mutualistic relationships seem to occur most frequently on a growth factor–carbon source basis. However, a fascinating example of mutualism based on bioenergetics has been revealed in a naturally-occurring mixture of two bacteria, originally thought to be a monoculture of a photosynthetic bacterium called *Chloropseudomonas ethylica*. This 'organism' has now been shown to be a tight mutualistic relationship between a photosynthetic green sulphur bacterium, *Chlorobium limicola* and a species of sulphate-reducing bacterium, *Desulfovibrio* (Gray *et al.* 1973). It has not been elucidated precisely how many nutritional interactions have evolved between this pair of organisms but the most important interaction is known to be based on the need for an electron donor and an electron acceptor. For photosynthetic energy generation *Chlorobium* requires an accessory electron donor which for this organism is normally hydrogen sulphide, which becomes oxidised to sulphate during photosynthesis. The *Desulfovibrio* species utilises organic carbon compounds for its energy requirements and sulphate as the terminal electron acceptor, resulting in the formation of hydrogen sulphide. Thus in the growing association the sulphur compounds are repeatedly cycled between oxidised and reduced states. The advantage to both organisms lies in the fact that only small quantities of sulphur are required for growth since it is effectively used in a catalytic manner. This mechanism ensures that the sulphide concentration never rises to growth inhibitory levels, as it is extremely toxic to both bacteria.

3 Commensalism, (+ 0 or 0 +). This is the situation where one member of a community benefits from the presence of a second population which itself does not derive any advantage or disadvantage from the activity of the first organism. Commensalism is an extremely common interaction in nature and the process of organism succession can largely be thought of as a chain of commensal relationships; the growth of one population generating a particular set of conditions thereby enabling a second population to develop.

Megee *et al.* (1972) have studied an association of a yeast *Saccharomyces cerevisiae* and a bacterium *Lactobacillus casei* (see Fig. 8.3). The stability of the association depended on the production by the yeast of riboflavin needed by the bacterium for growth. This mixed culture was grown in an open growth system and the stability of the simple community was due to a second interaction, namely competition for the growth-limiting quantities of the carbon and energy source, glucose. The bacterium had a greater affinity for glucose than the yeast but if the *Lactobacillus* used too much of the carbon source it would cause a decrease in the size of the yeast population and hence a decrease in the rate of supply of riboflavin. Thus differences in glucose affinity of the two organisms ensured that the populations equilibrated to constant sizes under constant growth conditions. If the glucose affinities had been reversed the yeast would simply have been more competitive than the bacterium and would probably have caused the bacterium's removal from the growth vessel (see section 3.2). Again this comparatively simple example shows that single interactions rarely occur, particularly in communities of microorganisms isolated from the same natural habitat.

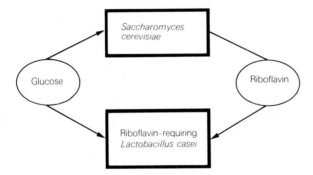

Fig. 8.3 Commensalism between the yeast *Saccharomyces cerevisiae* and the bacterium *Lactobacillus casei*.

Jensen (1957) described a simple community involving only one commensal interaction amongst some soil microorganisms growing on trichloroacetic acid as the major carbon source (see Fig. 8.4). An unidentified bacterium was able to metabolise trichloroacetic acid but could only grow in the presence of either of two species of *Streptomyces*. It was found that these actinomycetes excreted vitamin B_{12} which was an essential growth requirement for the bacterium. The two streptomycetes did not derive their carbon source for growth from any of the products of trichloroacetic acid metabolism and so had to be supplied with a second carbon source.

4 Amensalism, $(-0$ or $0 -)$. This is a little-studied form of interaction but one in which the growth of one population is restricted by the presence of a second which is unaffected by the metabolism of the inhibited population. Amensalism occurs when organisms produce anti-microbial compounds, such as antibiotics or colicins, or through non-specific effects, such as the elevation of the dissolved oxygen tension or changes in the hydrogen ion concentration. In a strict sense there ought to be no effect on the inhibiting population in amensal relationships; this would be manifest as unaltered growth characteristics whether or not the restricted population was present. Although there may be no direct effect on the inhibiting population there may well be an indirect advantage because, by limiting the assimilation of growth resources by the affected population, a greater proportion of these materials can be made available for growth of the inhibiting population.

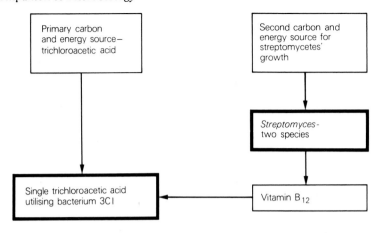

Fig. 8.4 Commensalism amongst soil microorganisms growing on trichloroacetic acid.

5 Prey–predator relationships, (+ − or − +). In this case one member of the mixed culture, the predator, gains directly at the expense of the second member of the culture, the prey, since the latter forms the complete nutritional requirements of the predator. In extreme cases the grazing of the predator may result in the elimination of the prey population but more usually a balanced situation develops between the two populations without necessarily establishing steady-state levels. Typically in open-growth systems continuous oscillations of the two populations can be established, with the increasing phase of the predator population lagging behind the increase in the prey population. This has been observed very clearly in the case of *Dictyostelium discoideum* amoebae feeding on *Escherichia coli* (Tsuchiya *et al.* 1972, Dent, Bazin and Saunders 1976) and a number of models have been proposed to describe the population fluctuations. However, in laboratory cultures the oscillations always seem to become damped and, in some cases, steady-state population levels are realised. Dent *et al.* (1976) have concluded that sustained oscillations as predicted by, for example, the Lotka–Volterra model, are not observed because the apparently simple prey–predator mixed culture is 'a rather more complex microbial ecosystem composed of a variety of physiological and ecological interrelationships'. Obviously the ability to predict accurately and to quantify prey–predator interactions in naturally-occurring populations must await the full elucidation of all the interactions.

6 Competition, (− −). The final fundamental type of interaction is competition which results when both populations are limited, either in terms of growth-rate or final population size, by a common dependence on an external factor required for growth. Competition is probably the single most important interaction in nature, providing *inter alia* the selective mechanism for evolution. Competition, as Gause observed, is the basis of the struggle for existence and is discussed in greater detail in the following sections.

3 The kinetics of competition

In this analysis of the kinetic principles of competition a number of simplifying assumptions have been made:
1. The analysis is restricted to competition for a limiting amount of an essential nutrient in a

mixed culture containing only two different microbes. However, the principles which can be elucidated can be applied to free competition in multi-species communities.

2. It is assumed that there are no other interactions between the two component populations. Obviously undefined interactions would modify the dynamics of competition, for example, by eliminating free competition for the limiting substrate (e.g. the yeast–*Lactobacillus* relationship described in section 2).

3. It is assumed that the two populations show a similar response to the prevailing growth conditions. In many natural environments interactions between the organism and its physical environment are probably of major significance, although the kinetic consequences of these interactions are poorly understood and usually ignored. As an example, if one member of a microbial community is able to attach itself to the surfaces present within the environment, the adhering population has a greater chance of being retained within that habitat, particularly if it is an open growth system, and may be able to counter the competitive disadvantage it may have with the other organism for the growth-limiting substrate. These physical interactions may be of crucial significance in modifying and regulating free competition.

3.1 Free competition in a closed environment

If two different organisms, A and B, are grown together in a closed environment (e.g. a batch culture) in which all the nutrients are initially in excess then the growth of the two populations may be written:

$$\frac{dx_A}{dt} = \mu_{MA} x_A \quad \text{and} \quad \frac{dx_B}{dt} = \mu_{MB} x_B$$

where μ_{MA} and μ_{MB} are the maximum specific growth rates for A and B respectively and x_A and x_B are the population sizes.
Hence

$$\ln x_{At} - \ln x_{A0} = \mu_{MA} t$$

and

$$\ln x_{Bt} - \ln x_{B0} = \mu_{MB} t$$

The rate of increase of the two populations under conditions of an unrestricted nutrient supply depends on the maximum growth-rate of the populations and is determined by the genotype of the organisms (see Bull (1975) for further details of the fundamentals of microbial growth).

From equations (1) and (2) we have by subtraction:

$$[\mu_{MA} - \mu_{MB}] t = [\ln x_{At} - \ln x_{A0} - \ln x_{Bt} + \ln x_{B0}]$$

$$= \ln \left[\frac{x_{At}}{x_{Bt}} \right] - \ln \left[\frac{x_{A0}}{x_{B0}} \right]$$

$$\mu_{MA} - \mu_{MB} = \frac{\ln R_t - \ln R_0}{t} \tag{3}$$

where R_t is the ratio of the population size of A to population size B at time t and R_0 is the same ratio at zero time. Assuming that the growth conditions remain constant for the observed period of growth, then the relationship *lnR* against *time* yields a straight line, the slope of which gives the difference between the maximum specific growth rates of the two populations (Fig. 8.5). As μ_{MA} tends towards μ_{MB} then $\ln R/t$ tends towards zero and the degree of competition between the two populations decreases. Conversely as μ_{MA} tends towards $\mu_{MA} \gg \mu_{MB}$, $\ln R/t$ tends towards infinity and competition increases. The difference in the maximum specific growth rates provides a simple measure of the degree of fitness between the two populations and an assessment of their ability to survive and exploit their growth environment. So far we have assumed that all the nutrients are in excess and so, strictly, there is no direct competition for a limiting growth requirement during the growing phase. In the sense, however, that the final population sizes will be dictated by the quantity of the nutrient which is depleted first, then it is clear that a competitive advantage lies with the more rapidly growing population and that the advantage increases with increasing differences between the maximum specific growth rates. The faster growing organisms can use up more of the limiting nutrient during growth than its less avid brethren. In most natural environments fluctuations in nutrient concentrations occur, leading to alternating periods of feast and famine and with successive cycles those populations which can gradually increase their proportion of the total microbial biomass are more competitive and have an advantage in the survival race. The importance of competition in this respect may be gauged from the following theoretical situation (Fig. 8.5). In one mixed culture organisms A and B were grown in a medium in which the concentration of the limiting nutrient was such that the total population was restricted to 203 individuals. For A, $\mu_{MA} = 0.693 \text{ h}^{-1}$ and for B, $\mu_{MB} = 0.347 \text{ h}^{-1}$ and the initial starting population was 3:2 for A:B, respectively. After exactly 6 h, growth ceased at a final population size of 208 (203 new individuals plus the original 5 from the inoculum) of which 92·3 per cent (192 individuals) were organism A and 7·7 per cent (16 individuals) were organism B. The degree of competitiveness, calculated from the slope of the logarithm ratio curve, was 0.347 h^{-1}, which is, of course, the difference between the two maximum specific growth rates of A and B.

In a second mixed culture organism B was replaced by C whose maximum specific growth rate (0.462 h^{-1}) was greater than that of B. Growth was initiated with 5 individuals in the ratio A:C, 3:2. Population A obviously grew at the same rate as in the first experiment whereas population C grew more rapidly. The final population size again of 208 organisms was reached after 5·8 h of growth, showing a slight decrease over the first experiment, with A comprising 85·1 per cent (177 individuals) of the total population and C 14·9 per cent (31 individuals). As expected from these data the rate of change of the population ratios was less than in the previous experiment, giving a value of 0.231 h^{-1} as a measure of the reduced level of competition.

The important point so far as competition is concerned is that in the event of an addition of more of the growth-limiting nutrient, organism A is better placed when it was growing in competition with organism B than with organism C. For the (A + B) mixed culture the new starting ratios would be 24:2 (A:B) compared with 11·4:2 (A:C) for the second mixed culture. An alternative way of looking at this relationship is to say that population C is in a significantly better position to take advantage of the influx of fresh nutrients, compared with organism B, because of its capacity to keep pace with A.

Gause used closed culture systems for all of his competition studies. For example he grew two species of yeast — a budding yeast, *Saccharomyces cerevisiae* and a fission yeast, *Schizosaccharomyces kephir* — in an undefined growth medium composed of glucose and yeast extract. Although the reason for growth cessation was not determined it was assumed to be due to a nutrient limitation (probably glucose) and it is clear that the presence of the yeasts together greatly reduced their maximum population sizes compared with monoculture. For example, in one experiment, *S. cerevisiae* comprised 55 per cent of the total population whilst

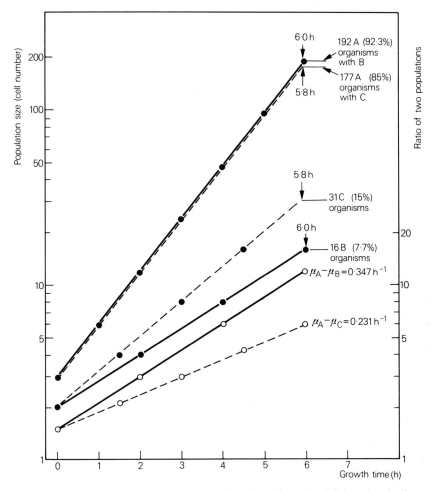

Fig. 8.5 Competition between two pairs of microorganisms, A and B or A and C, in a closed culture system. For both mixed cultures the starting population sizes were 3:2, A to B or A to C. (●———●) growth curves for A and B together where μ_{MA} = 0·693 h^{-1} and μ_{MB} = 0·347 h^{-1}; (○———○) logarithm of the ratio of population A to population B; (●– – – –●) growth curves for A and C together where μ_{MA} = 0·693 h^{-1} and μ_{MC} = 0·462 h^{-1}; (○– – – –○) logarithm of the ratio of population A to population C.

S. kephir constituted the remaining 45 per cent, indicating that *S. cerevisae* was marginally more competitive than *S. kephir*. If the two yeasts under these conditions had identical maximum specific growth rates, the degree of competition between the two populations would have been zero and each would have comprised 50 per cent of the maximum population size.

3.2 Free competition in an open environment

Competition in a closed environment is possibly of limited significance in nature because no particular growth requirement is present at a concentration which limits the growth rate of the populations. Open growth systems come closer to simulating the natural environment for a number of reasons, particularly because growing populations are continuously supplied with a restricted quantity of a required nutrient (see Veldkamp and Jannasch 1972 for further details).

Nutrient limitation reduces the specific growth rates of the component populations of a community to less than maximal values and the organism which is able to make the best use of the available materials and grow most rapidly has the greater competitive advantage. One of the important consequences of such a system is that the organism with the lower growth rate is eventually eliminated from the open culture. The growth rates of the component species are influenced by a second important growth parameter, the saturation constant K_s, which defines the affinity of an organism for the limiting substrate.

In order to understand the mechanism of competition under these conditions it is convenient to consider the case of a mutant organism appearing in a large population of its parent growing in a chemostat. The following kinetic analysis has a general application to any competitive situation occurring between two or more organisms under nutrient-limited conditions in an open environment.

The parent population A is supplied with the limiting substrate at an initial concentration S_R and constant dilution rate D. Microorganism A utilises some of the substrate for growth, thereby reducing its concentration to s and increasing its biomass concentration to x_A, eventually achieving steady-state values of $\bar{x}_A$ and $\bar{s}$ when the population growth rate μ_A equals the selected dilution rate D. If a mutant B arises within this steady-state population, it is possible to predict whether or not it will be more or less competitive than the parent A. Thus at the time of the appearance of mutant B, the overall situation may be summarised as follows:

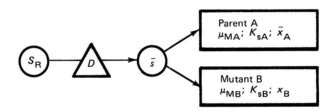

The growth of the two populations is represented by

$$\frac{dx_A}{dt} = (\mu_A - D)x_A \tag{4}$$

and

$$\frac{dx_B}{dt} = (\mu_B - D)x_B \tag{5}$$

where μ_A and μ_B are the sub-maximal specific growth rates determined by the concentration of the growth-limiting substrate. The Monod relationship can be substituted for the individual specific growth rate terms in equations (4) and (5):

$$\frac{dx_A}{dt} = \left[\frac{\mu_{MA} s}{(K_{sA} + s)} - D \right] x_A \tag{6}$$

and

$$\frac{dx_B}{dt} = \left[\frac{\mu_{MB} s}{(K_{sB} + s)} - D \right] x_B \tag{7}$$

When organism A is the major population and the initial steady-state conditions are:

$$\frac{dx_A}{dt} = 0 \quad \text{and} \quad \mu_A = D$$

Substituting in equation (6) and rearranging yields:

$$D = \frac{\mu_{MA}\bar{s}}{(K_{sA} + \bar{s})} \tag{8}$$

For mutant B in a steady-state population with its parent, the mutant initially may represent only one individual in a total population of perhaps 10^{12} organisms. In the first instance mutant B has no measurable effect on the steady-state conditions, including the growth-limiting substrate concentration. Indeed in the event of organism B being more competitive than organism A, it will be some considerable time before population B has increased to sufficient numbers which significantly begin to alter the steady-state conditions and invalidate equation (8). In practical terms, as long as the minor population is at least 10^{-2} times smaller than the dominant population, the steady-state conditions will continue to be dictated by the growth characteristics of the major population.

The important question now is whether or not the mutant population B is more or less competitive than its parent A. A comparison of equations (6) and (7) in the steady-state versions shows that there are three general cases.

Case 1. For the mutant population B to be more competitive than organism A and increase its proportion of the total population, $\mu_B > D = \mu_A$. The greater the difference between μ_B and D, the faster will be the rate of displacement of population A by population B. Now for $\mu_B > \mu_A$ equations (6) and (7) yield:

$$\frac{\mu_{MB}\bar{s}}{(K_{sB} + \bar{s})} > \frac{\mu_{MA}\bar{s}}{(K_{sA} + \bar{s})}$$

and so dx_B/dt will be positive. Since the growth-limiting substrate concentration remains constant so long as $x_B \leqslant 10^{-2}x_A$, dx_B/dt can only be positive if

either $\mu_{MB} > \mu_{MA}$ (see Fig. 8.6 (a))
or $K_{sB} < K_{sA}$ (see Fig. 8.6 (b))

In both these cases, whatever dilution rate is selected, as long as it is less than the maximum specific growth rate of either organism, the mutant population B is always the most competitive and eventually succeeds in eliminating A from the mixture. For example, if a dilution rate D is chosen (Fig. 8.6 (a) and (b)) then for a monoculture of organism A the steady-state, growth-limiting substrate concentration is $\bar{s}_1$. For both cases at this substrate concentration mutant B can be supported at a greater growth rate than the parent population and so dx_B/dt is positive and the ratio of A to B begins to decline. Ultimately the size of population B becomes a significant fraction of the total population and the further growth of population B results in a reduction in the concentration of the growth-limiting substrate concentration. As soon as s_1 tends towards s_2 (Fig. 8.6 (a)) or s_3 (Fig. 8.6 (b)) dx_A/dt decreases and becomes less than the selected dilution rate. When these conditions obtain organism A begins to wash out of the growth vessel. Whilst the elimination of organism A is occurring, dx_B/dt is still greater than the chosen dilution rate and population B continues to increase. Eventually the biomass of organism B tends towards its new steady-state value where $dx_B/dt = 0$ and the new steady-state growth-limiting substrate concentrations $\bar{s}_2$ and $\bar{s}_3$ become established.

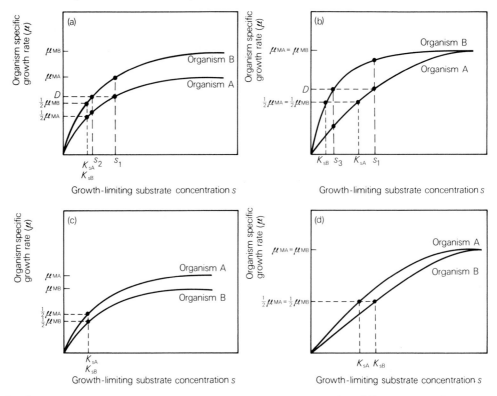

Fig. 8.6 The Monod relationships for a wild-type parent organism A and four different mutants B. (a) $K_{sA} = K_{sB}$, but $\mu_{MA} < \mu_{MB}$. At all growth-limiting substrate concentrations B has a competitive advantage over A. (b) $K_{sA} > K_{sB}$, but $\mu_{MA} = \mu_{MB}$. At all growth-limiting substrate concentrations B has a competitive advantage over A. (c) $K_{sA} = K_{sB}$, but $\mu_{MA} > \mu_{MB}$. At all growth-limiting substrate concentrations A has a competitive advantage over B. (d) $K_{sA} < K_{sB}$, but $\mu_{MA} = \mu_{MB}$. At all growth-limiting substrate concentrations A has a competitive advantage over B.

Case 2. If the mutant population B is less competitive than organism A we have the situation of $\mu_B < D = \mu_A$. As before, for this to be true, equations (6) and (7) yield:

$$\frac{\mu_{MB}\bar{s}}{(K_{sB} + \bar{s})} < \frac{\mu_{MA}\bar{s}}{(K_{sA} + \bar{s})}$$

and so dx_B/dt will be negative. This case occurs when

either $\mu_{MB} < \mu_{MA}$ (see Fig. 8.6 (c))

or $K_{sB} > K_{sA}$ (see Fig. 8.6 (d))

Clearly the kinetic reasoning for the exclusion of organism B is the reverse of that applying in the first situation. In essence, for any given steady-state culture of organism A, the concentration of the growth-limiting substrate cannot support as fast a growth rate for organism B as it can for organism A.

Case 3. Finally it is possible to conceive of the case in which the population B grows at exactly the same rate as population A. Hence:

$$\frac{\mu_{MB}\bar{s}}{(K_{sB} + \bar{s})} = \frac{\mu_{MA}\bar{s}}{(K_{sA} + \bar{s})}$$

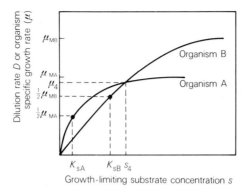

Fig. 8.7　The Monod relationships for two organisms, A and B, in which $K_{sA} < K_{sB}$ and $\mu_{MA} > \mu_{MB}$, showing 'crossover' kinetics. At $\mu_4(s_4)$ in a steady-state culture $\mu_A = \mu_B$ and an equilibrium exists between the two populations. Between $D = 0$ and μ_4 for any substrate concentration $\mu_A > \mu_B$ and A is therefore more competitive than B. When $D > \mu_4$ then $\mu_A < \mu_B$ and B is the more competitive organism.

In this case there is an equilibrium between the two populations with $dx_A/dt = dx_B/dt$ and $\mu_A = \mu_B$. The proportions of the two populations would depend on the initial starting sizes and no selection pressures could operate in favour of either organism. This situation is extremely unlikely in the case of mutant-wild type population interactions, but it is conceivable that in natural environments two different organisms could have identical maximum specific growth rates and saturation constants.

It has been observed with naturally isolated populations that a unique growth-limiting substrate concentration often exists at which $\mu_A = \mu_B$ because of the combined effects of different maximum specific growth rates and saturation constants (Fig. 8.7). These so-called 'crossover' Monod kinetics define a single substrate concentration, s_4, which gives exactly the same specific growth rate μ_4 for both organisms. Experimentally it is not feasible to maintain an exactly constant dilution rate needed to achieve the balanced growth of the two populations, since invariably the average dilution rate will be slightly above or below the value μ_4.

This commonly observed situation illustrates one further point, i.e. the outcome of the competition between two organisms A and B, particularly in enrichment experiments, depends critically on the imposed dilution rate of the continuous-flow culture. If a dilution rate between $D = 0$ and μ_4 was chosen, organism A would dominate because it has the greater affinity for the growth-limiting substrate and hence the more rapid growth rate. Alternatively if a dilution rate greater than μ_4 was set organism B would be selectively enriched.

3.3　Examples of free microbial competition in an open environment

3.3.1　Limiting nutrient bypass competition

The simplest form of competition occurs when one member of a two-membered mixed culture can bypass the growth-limiting substrate restricting the growth rates of the two populations. T. G. Mason and J. H. Slater (unpublished observations) have examined this type of competition between mutants of *E. coli*. A tyrosine auxotroph was grown in a tyrosine-limited chemostat at a fixed dilution rate and after 85 h of steady-state growth a few revertant prototrophs were detected in the culture (Fig. 8.8). This population grew rapidly, at a rate close to the maximum specific growth rate because the new population's growth rate was no longer dictated by the prevailing tyrosine concentration. After 100 h growth the prototroph com-

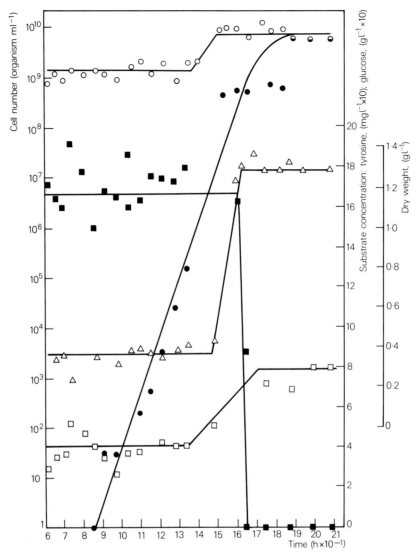

Fig. 8.8 The selection of a tyrosine prototroph of *Escherichia coli* from a steady-state population of tyrosine auxotrophs growing in a tyrosine-limited chemostat. (o———o) total viable population; (●———●) the prototrophic population; (△———△) the biomass concentration; (■———■) the glucose concentration; (□———□) the tyrosine concentration.

pletely replaced the tyrosine auxotroph and established a different steady-state population which was limited by a second nutrient, glucose, indicated by the sharp decline in glucose concentration and concomitant increase in the tyrosine concentration. The overall population concentration also increased because it had been so arranged that the glucose concentration in the inflowing medium sustained a higher biomass concentration when it was the limiting nutrient than when tyrosine was the limiting nutrient.

3.3.2 Competition between microbes with different maximum specific growth rates but identical saturation constants

This case, theoretically illustrated in Fig. 8.6 (a) and (c), can be observed in mutant-wild type interactions when the mutation involves the cellular concentration of the growth rate-limiting enzyme. Mutations which increase the concentration of that enzyme, increase the rate at which the growth-limiting substrate may be converted to biomass but have no effect on the enzyme's affinity for the substrate. Frequently this situation occurs when a mutation alters the regulatory mechanisms controlling the synthesis of the enzyme in question, and since the mutation has not involved a change in the structure of the enzyme the Michaelis constant, which in this case may be related to the saturation constant, remains unaltered.

Some experiments designed to examine the fate of plasmid-containing strains of *E. coli* in lactose-limited chemostats are interesting in this context (unpublished experiments of I. M. Cairns, K. G. Hardy and J. H. Slater, quoted in Bull and Slater 1976). A strain of *E. coli* was studied which contained an F^+ factor also coding for the enzyme β-galactosidase which is the primary catabolic enzyme in the metabolism of lactose and was the growth rate-limiting enzyme. As a consequence of a deletion mutation the chromosome lacked the β-galactosidase structural gene and so the experiments were not complicated by the problem of reversion to a functional *lac* operon on the chromosome. During the first 5 days of growth, the steady-state population contained 100 per cent plasmid-plus individuals and showed a constant β-galactosidase activity of 0·75 units. At day 5 a few plasmid-minus (designated F^-) organisms were detected in the culture and these increased in number to form over 90 per cent of the population during the next 72 h of growth. At the same time there was a 4-fold increase in the β-galactosidase specific activity. Other experiments showed that, eventually, the population becomes completely F^- as a result of the total loss of the plasmid from the population.

There are two points of interest arising from these experiments. First, the F^- population could only succeed in growing on lactose, the sole carbon source, if the functional β-galactosidase structural gene, initially located on the F factor plasmid, had been transferred to the bacterial chromosome via a genetic recombination event, probably associated with the loss of the plasmid. Secondly, it appears that the positioning of the β-galactosidase gene on the chromosome resulted in a significant increase in the specific activity of the enzyme. It was the increased concentration of β-galactosidase in the F^- population which resulted in a mutant population with a strong competitive advantage over the parent F^+ population. Indeed using an equation based on equation (3) (modified to apply to continuous-flow culture competition) it was calculated that there was a 14·5 per cent difference between the F^+ and F^- growth rates, in favour of the F^- organisms. There may be one further competitive advantage gained by the loss of the plasmid: in the F^- organisms there was significantly less DNA per organism which had to be replicated, and under conditions of carbon and energy limitation this may make a considerable saving on the resources available to the F^- organisms, releasing more of the substrate for biomass production.

Hartley *et al.* (1972) have observed the competitive advantage gained by mutants of *K. aerogenes* with increased specific activities for ribitol dehydrogenase. Normally ribitol is the substrate for this enzyme but it also possesses slight activity towards an analogue of ribitol, xylitol. In one series of experiments the parent organism, *K. aerogenes* strain A, was grown in a chemostat at a dilution rate of $0·2\ h^{-1}$ with xylitol as the growth-limiting nutrient. In time a sequence of two mutants appeared replacing the previous population (Table 8.2). The first new population, strain A1, showed a 21 per cent increase in the maximum specific growth rate and a 4·5-fold increase in ribitol dehydrogenase specific activity. The ratio of xylitol to ribitol dehydrogenase activity did not change significantly indicating that there had been no change in the active site of the enzyme and hence its affinity for the substrate remained the same. A second mutant population, strain A11, further improved its maximum specific growth rate in parallel with a 3-fold increase in ribitol dehydrogenase activity. The reason for this dramatic

Table 8.2   Changes in pentitol dehydrogenase activities in chemostat
cultures of *Klebsiella aerogenes*

Klebsiella aerogenes strain	*Maximum specific growth rate (h^{-1})*	*Percentage increase in μ_{max}*	*Ratio of maximum xylitol: ribitol dehydrogenase activity*	*Ribitol dehydrogenase specific activity*	*Percentage of total cell protein in ribitol dehydrogenase*
A	0·53	—	0·24	1·0	1·2
A1	0·64	21	0·21	4·5	6·0
A11	0·82	48	0·17	15·8	18·0

increase in ribitol dehydrogenase activity, accompanied by a related increase in enzyme protein per organism, is quite clear; for an enzyme with a low rate of substrate utilisation any mutation which increases this rate will provide more carbon precursors for biomass production. The net effect of an improvement in the potential for biomass production is to produce a strain with a strong selective advantage.

Equally interesting are the studies of Cox and his co-workers (Gibson *et al.* 1970, Cox and Gibson 1974) in which the competitive advantage of certain mutants of *E. coli* affecting the organism's mutation rate have been analysed. 'Mutator' mutants, designated *E. coli mutT1*, have greatly increased spontaneous mutation rates which, under appropriate conditions, grow more competitively than their isogenic parents. For example, *E. coli W3110 mut*[+] wild-type was grown in a glucose-limited chemostat with its mutator mutant *E. coli W3110 mutT1*. Within approximately 350 h of continuous growth there was over a 100-fold change in the ratio of the two populations in favour of the mutant. The mutator *mutT1* growth rate was calculated as 0·4675 h^{-1} which was an increase of 0·0055 h^{-1} over the dilution rate and the specific growth rate of the parent *mut*[+]. Initially the mutator population was in the minority and for much of the takeover period was by at least two orders of magnitude the smaller population. Thus as we have discussed previously the conditions were set by the parent *mut*[+] population and under these circumstances the mutator population had a 1·2 per cent higher growth rate.

Two different reasons can be offered to explain the advantage which mutator mutants have over wild-type organisms. First, it could be due to some intrinsic property of the mutator gene product such that, for whatever reason, a change in this protein could produce a strain with an increased growth rate. In the most extreme case the mutator gene change might result in the protein product not being synthesised and consequently effect a saving in carbon resources (by not diverting amino acids into this protein) or in energy terms (by saving the cellular energy needed to synthesise that particular protein). The second, and the more plausible of the two explanations, is that an increase in the mutation rate, mediated by the product of the *mut* gene, means that other mutations occur elsewhere on the genome more frequently. These secondary changes may then be the immediate cause in producing organisms better fitted for growth under the imposed conditions. Indeed, Cox *et al.* were able to show that secondary mutations had occurred. For example in many cases in the *mutT1* populations, organisms appeared with an enhanced propensity to adhere to the surfaces of the growth vessel. Furthermore it was later shown that in the 'sticky' *mutT1* mutator strains, the *mutT1* gene could be genetically crossed out by mating with a 'non-sticky' *mut*[+] parent strain. The resulting hybrid organism retained its competitive advantage in a chemostat over the parent organism, despite the absence of the *mutT1* gene, and dominated the culture at exactly the same rate as the original *mutT1* organism. Further evidence that the *mutT1* advantage lies in the fact that it speeds up the rate of appearance of other beneficial mutations comes from observations that in some chemostats there were prolonged periods when the ratio of *mutT1* to *mut*[+] organisms did not change. In these cases presumably no advantageous, secondary mutations had yet occurred and so no selection pressure could operate in favour of either of the two populations.

3.3.3 Competition between microbes with different saturation constants but identical maximum specific growth rates

One of the most elegant examples of this category of competition again comes from the work of Hartley *et al.* (1972) on the pentitol dehydrogenases of *K. aerogenes.* In the parent strain the dehydrogenase has a very poor affinity for an 'unnatural' substrate, xylitol, showing an apparent K_m of approximately 1 M. However, a mutant was selected which had a structural change in the ribitol dehydrogenase thereby improving its affinity for xylitol and giving an apparent K_m of 5 mM. The low affinity parent organism was grown in monoculture with xylitol as the growth-limiting nutrient and at a constant dilution rate. Into a steady-state population of 10^{12} individuals, 20 cells of the high xylitol affinity mutant were introduced and after 120 h of growth the high affinity mutant succeeded in ousting the parent organism. Concomitantly there was the expected decrease in xylitol concentration and increase in the total biomass concentration as a result of the significantly improved xylitol utilisation by the mutant.

A further excellent example of the role of different affinities for the limiting substrate in competitive situations comes from the work of Francis and Hansche (1972 and 1973) who studied the acid phosphatase of *Saccharomyces cerevisiae.* This work is discussed in section 4.

3.3.4 Competition between microbes with different maximum specific growth rates and saturation constants

In natural environments mixed populations of microorganisms tend to show the more complex crossover substrate versus growth rate kinetics of the type described in Fig. 8.7. Jannasch (1968) was amongst the first to show that the outcome of competition depended to some extent on the growth rate of the enrichment culture. He grew a marine *Spirillum* species and *E. coli* together in lactate-limited sea-water in an open enrichment system. At specific growth rates in excess of 0.29 h[-1] (which corresponded to lactate concentrations in excess of 5.0 mg lactate l[-1]), the coliform was the faster growing organism and hence the more competitive. Conversely, the *Spirillum* sp. was more successful at the growth rates between zero and 0.29 h[-1] and hence at the lower lactate concentrations. The ecological significance of these observations is quite obvious: the naturally-occurring marine organism, the *Spirillum,* normally inhabits an environment containing extremely dilute concentrations of carbon compounds. The competitive success of this organism has necessitated the evolution of high substrate affinity (low K_s) mechanisms in order to maximise their rate of growth in nutritionally poor environments. On the other hand the coliform normally occupies a nutritionally richer environment and therefore does not need such elaborate, high substrate affinity mechanisms.

More recently Kuenen *et al.* (1977) have found the same phenomenon in aerobic enrichment cultures of organisms competing for limiting quantities of inorganic phosphate. Two parallel enrichments were made using the same limiting medium with phosphate at 7×10^{-4} per cent (w/v) at two dilution rates, 0.03 h[-1] and 0.3 h[-1]. At the low dilution rate after five complete volume changes a *Spirillum* sp. was dominant whereas at the higher dilution rate an unidentified rod-shaped organism predominated. In monoculture experiments it was shown that the rod-shaped organism had a maximum specific growth rate of $0.48 \, h^{-1}$ and a saturation constant for inorganic phosphate of 6.6×10^{-8} M. On the other hand the *Spirillum* sp. had a μ_{max} of $0.24 \, h^{-1}$ and a K_s of 2.7×10^{-8} M. Thus the form of their growth rate versus limiting substrate concentration curves would be the same as in Fig. 8.7 in which B would represent the rod-shaped organism and A the *Spirillum* sp.

4 Competition as a mechanism in enzyme evolution

One way of explaining the diversity of microbial forms and species is to argue that the process of natural selection has operated in all types of environment to produce microbes which are best suited for growth in that environment. The basic mechanism behind natural selection and organism evolution is the process of competition. Clearly those organisms which acquire a particular growth advantage under a given set of environmental conditions will be selected by competing against their less well-endowed neighbours. Experimental enzyme evolution is a particularly exciting field at the present time and microbial enzymes provide convenient experimental systems in which the principles of evolution and natural selection may be further understood.

An elegant demonstration of the use of continuous-flow culture techniques in this context has been provided by Francis and Hansche (1972 and 1973) working with acid phosphatases in *Saccharomyces cerevisiae*. This enzyme plays a major role in furnishing growing cells with phosphate when the medium contains only organic phosphates, the enzyme being able to release inorganic phosphate. In the experiments of Francis and Hansche the yeast was grown at a constant dilution rate $(0.12 \, h^{-1})$ under β-glycerophosphate limitation. The culture was controlled at pH 6.0 whereas the optimum for the original wild-type acid phosphatase was pH 4.2. At the higher pH value the acid phosphatase activity was reduced to approximately 25 per cent of its maximum at pH 4.2. Under these conditions, therefore, two major selection pressures operated on the growing yeast population. First, the conditions favoured the selection of mutant yeasts with an improved substrate affinity and, secondly, yeasts with a modified activity in response to the unfavourable pH conditions.

In one experiment the parent strain S288C maintained a steady-state population for about 1 000 h of continuous growth before it was displaced by a mutant, M1. This mutant was selected because of a decrease in the saturation constant for β-glycerophosphate (S288C had a K_s of 4.58 mg β-glycerophosphate (β-GP) l^{-1} compared with M1's K_s of 1.86 mg β-GP l^{-1}). There was no change in the maximum specific growth rate and it seems likely that the change of K_s value in the mutant was due to an alteration in the acid phosphatase K_m for β-glycero-phosphate (Table 8.3) rather than a change in the enzyme's pH profile. The mutant M1 population proved to be stable for a further 1 300 h of growth until, in its turn, it was replaced by mutant M2. The second mutational event appeared to be more extensive than in the first mutant since M2 was altered in both its substrate affinity, which was increased with a new K_s of 0.25 mg β-GP l^{-1}, and its maximum specific growth rate, which was increased to $0.261 \, h^{-1}$ (Table 8.3). These changes were concomitant with a change in the pH optimum of the acid phosphatase which increased to pH 4.8 but there was no change in the enzyme's affinity for β-glycerophosphate. Although the precise molecular-biological details have not been elucidated, these studies show the role of competition in selecting for an organism better suited for growth in a particular environment.

A second example of competition at work in enzyme evolution comes from our work on the enzyme dehalogenase in *Pseudomonas putida*. Senior *et al.* (1976) isolated a stable microbial

Table 8.3 Changes in the growth and enzyme constants of an acid phosphatase-producing yeast

Strain	Growth time before new strain appeared (h)	Organism Characteristics			Acid Phosphatase Characteristics		
		μ_{max} (h^{-1})	growth rate at 9.1 mg β-glycerophosphate ℓ^{-1}	K_s, (mg β-glycerophosphate ℓ^{-1})	V_{max} (μmoles β-glycerophosphate min^{-1} $cell \times 10^{-10}$)	$K_m \times 10^{-5}$ $(M \times 10^{-5})$	pH optimum
Parent S288C	—	0·227	0·151	4·58	6·72	1·54	4·2
M1	1 040	0·230	0·191	1·86	5·51	1·25	4·2
M2	2 310	0·261	0·254	0·25	10·65	1·33	4·8

community composed of seven microorgansims which could grow on the herbicide Dalapon (22DCPA, 2,2'-dichloropropionic acid) as its sole carbon and energy source. Within this complex community a strain of *P. putida* S3 was present which initially could not grow on 22DCPA. However, after about 2 900 h of chemostat growth a mutant, strain P3, appeared which could use 22DCPA as its growth substrate. Subsequent experiments with monocultures of *P. putida* S3 showed that the selection of a strain with an altered dehalogenase having either a novel specificity towards 22DCPA or greatly elevated levels of the enzyme with a poor 22DCPA specificity, was the basis of this population change. In these experiments the parent S3 strain was grown in a carbon-limited chemostat in which the carbon source was a 1:5 mixture of sodium propionate and 22DCPA. Initially *P. putida* S3 could only grow on the propionate and formed a steady-state population with a low biomass concentration (Fig. 8.9). The propionate functioned as a 'carrying' substrate to enable the pseudomonad to grow in the open culture and simultaneously to be exposed to the non-metabolisable substrate. At this time the S3 population did not have (or possibly had at exceedingly low levels) a dehalogenase with activity towards 22DCPA and could not utilise the major carbon source, and so it is evident that a mutant which possessed 22DCPA-specific dehalogenase activity would have a considerable selective advantage. After 3 600 h of growth such a mutant appeared and the new P3 strain rapidly succeeded in dominating the culture and displacing the S3 parent. At the same time as the change in populations there was a reduction in the unused Dalapon concentration and a fall in the culture pH due to HCl formation from the dechlorination of Dalapon.

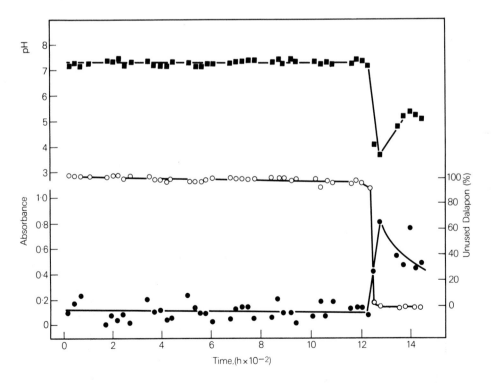

Fig. 8.9 Competition between *Pseudomonas putida* S3 and P3 in a mixed carbon (propionate- and 2,2'-dichloropropionate)-limited chemostat. (●———●) culture absorbance; (○———○) unused 2,2'-dichloropropionate as a percentage of the concentration in the inflowing medium; (■———■) culture pH.

5 Microbial communities

We began this chapter by observing that communities of interrelated organisms probably occur frequently in natural environments, whether it is at the relatively simple level of food chains or webs, or at the more complex level of groups of microbes interacting in a manner which maximises their corporate chance of survival. The systematic study of the biology and biochemistry of microbial communities is only just beginning and to conclude this chapter we wish to describe briefly some of the more complex assemblages which have recently been isolated and characterised.

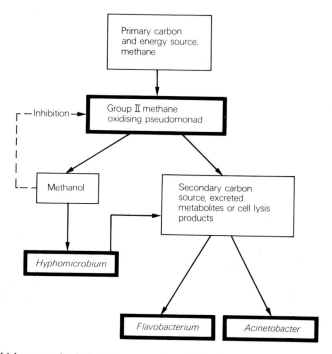

Fig. 8.10 A microbial community isolated from a methane-limited continuous-flow culture enrichment.

Wilkinson *et al.* (1974) found a stable microbial community in a culture enriched on methane as the sole carbon and energy source. There were at least four recognisable species, only one of which was able to oxidise and utilise methane for energy generation and biomass production (Fig. 8.10). This organism remains unidentified but the other three were shown to be species of *Hyphomicrobium, Flavobacterium* and *Acinetobacter.* The metabolic core of this mixture, certainly when the community was grown on methane, revolved around the methane oxidiser and the *Hyphomicrobium*; together these two organisms always comprised 98–99 per cent of the total population, with approximately ten times more methane utilisers than *Hyphomicrobium.* It turned out that the *Hyphomicrobium* had a beneficial effect on the growth of the methane oxidiser because it metabolised the methanol excreted into the growth medium by the methane utiliser. In the absence of *Hyphomicrobium,* methanol accumulated and quickly inhibited the growth of the producing organism. If a slug of methanol was added to the culture the proportion of *Hyphomicrobium* increased rapidly, reaching perhaps 70 per cent of the total population at the expense of the methane oxidiser. Gradually as the methanol was either oxidised or washed out of the culture, the populations reverted to their original equilibrium sizes. The precise role of the *Flavobacterium* and *Acinetobacter* species remains obscure. They are both heterotrophs and must be growing on organic compounds excreted by the major populations, or on cell-lysis products resulting from the death of the primary organisms. It would be interesting to know if they are solely gratuitous 'hangers on' or if they confer some direct benefit on the major populations in return for their supply of growth nutrients.

In our laboratory we have analysed two microbial communities of varying complexities (Osman *et al.* 1976, Senior *et al.* 1976). One of the communities is relatively simple, containing as a stable group three of bacteria growing on orcinol. One organism, *Pseudomonas stutzeri*, is able to metabolise orcinol, while the other two secondary populations, composed of a *Curtobacterium* species and *Brevibacterium linens,* lack the capability to cleave the aromatic ring of orcinol. In this mixture the exact role of the two secondary organisms is uncertain, certainly in terms of any possible nutritional relationships. However, the reason for the three-membered community's stability may have a kinetic explanation, since the presence of the secondary utilisers significantly reduces the apparent saturation constant for orcinol for the complete community. In monoculture the primary orcinol-utiliser, *P. stutzeri*, had a K_s of 100 mg orcinol l^{-1} whereas in the three-membered mixture the K_s was reduced to 71 mg orcinol l^{-1}. In this case we seem to be witnessing the selection of a group of organisms, as opposed to a single organism, along exactly the same lines as outlined in section 3; i.e. the individual or community with the greatest affinity for the growth-limiting substrate will be the most competitive.

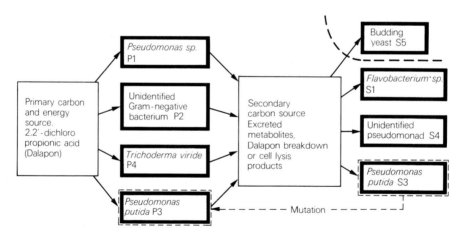

Fig. 8.11 A microbial community composed of seven different species growing on the herbicide Dalapon as its sole carbon and energy source.

The second microbial community is much more complex (Fig. 8.11), containing six or seven microorganisms growing on the herbicide Dalapon (22DCPA). After some preliminary enrichments, seven organisms were identified and these were divided into two basic groups. First, there were three primary Dalapon utilisers which could grow on Dalapon in monoculture; two were pseudomonad-type bacteria and the third was a fungus *Trichoderma viride*. The second group of organisms contained *Pseudomonas putida,* an unidentified pseudomonad, a *Flavobacterium* sp. and a pink budding yeast. In the original enrichment culture these microbes were unable to metabolise Dalapon and so were termed secondary organisms. The yeast was not very tightly integrated into the community and was quickly eliminated from the enrichment chemostat when the culture growth rate was increased. The remaining six-membered group was extremely stable and has been maintained for over two years as a continuously growing culture, despite a number of deliberate, severe perturbations in the growth conditions. The percentage of the individual populations varied with changes in growth rate and in overall terms the bacterium P2 was the dominant organism at low growth rates, whilst the pseudomonad P1 dominated at higher growth rates. The precise nutritional relationships within the community have not yet been elucidated.

We have already mentioned the interesting change in the nutritional status of *Pseudomonas putida* S3 which occurred in this community (section 4). Two other points, however, are illustrated by the community and act as salutary reminders that the simple kinetic models developed for competition in homogeneous laboratory cultures may have limited relevance to natural ecosystems. First, the Dalapon microbial community originally contained three, competing primary populations and, indeed, a fourth appeared later, presumably increasing still further the degree of competition for the limiting substrate. The simplest rationalisation of this kinetically disturbing situation may be that one of the original assumptions we made (section 3) that no other interactions should occur between the competing populations, is invalid. A network of additional, and so far unrecognised, interactions could provide a stabilising influence, sufficient to eliminate strictly free competition. Jost *et al.* (1973) have demonstrated this point in some of their model community studies. For example, competition between two bacteria in a glucose-limited chemostat culture swiftly resulted in the elimination of one of the bacteria. However, if a predator *Tetrahymena pyriformis* was added to make a third member, then paradoxically free competition between the two bacteria was attenuated and a stable three-membered community became established.

Secondly, the long-term stability of the seven-membered Dalapon community was critically dependant on other interactions to preserve its structure under certain growth conditions. At higher growth rates the culture pH (normally about pH 5·0 at low dilution rates) increased to about pH 6·0–7·0. At this pH the filamentous fungus grew poorly and would normally have been expected to disappear from the community. However, the *Trichoderma viride* responded to the unfavourable conditions by extensive colonisation of the walls of the chemostat, thereby effectively reducing the rate at which it was washed out of the growth vessel. It also seems probable that the wall growth provided a microenvironment within which more favourable growth conditions were established and maintained by the fungus. Little is known about microbe–physical environment interactions of this sort but it seems extremely likely that they are of major importance in nature.

6 Conclusions

The study of mixed cultures and microbial communities is the basis of microbial ecology and the current, renewed interest in one sense brings microbiology full circle, since one of its pioneering developments was the introduction of techniques which allowed the isolation of pure microbial populations. It is now clear that the need for such simplifying techniques tacitly acknowledged the complexity of microbial systems in natural environments and it is only now that our basic understanding of microorganisms and our technical developments have reached the stage when worthwhile studies of natural populations may be undertaken.

It is also a welcome development that microbial ecology is rapidly moving from the era of simple, qualitative descriptions – e.g. listing all the various species found in a particular ecosystem and little else – towards a much deeper understanding of the dynamic and ecophysiological principles involved in the structure of the microflora in a particular ecological niche. One facet of these studies is to develop accurate mathematical models, based on an understanding of all the biological factors implicated, which describe and predict the behaviour of one or more populations in their natural environment. In many cases the mathematical principles involved in such models are relatively straightforward and the restrictions lie in the problems involved in determining the biological parameters of the interactions. For example, in the Dalapon microbial community it is at present not worth attempting the construction of a community kinetic model because we do not know the nature or magnitude of any of the postulated interactions involved in such a stable group of organisms.

The behaviour of microbial communities in biological terms in response to a variety of environmental stresses is another developing area of microbial ecology. These may be entirely natural, as for example in the case of the microflora in an estuarine environment; here mixed populations are subjected to regular alterations in temperature, salinity and nutrient concentration, as a result of tidal movements and at present little is known about the effect of these changes in environmental conditions. Alternatively stresses may be due to novel, unnatural factors resulting in the main from man's interference with the biosphere. For example, the effects of xenobiotic materials, such as herbicides and pesticides, are very much unknown factors. A great deal is known about the phenotypic and genetic alterations in pure cultures of microbes grown in the laboratory, but the question of how these mechanisms operate in natural situations remains largely unanswered.

Finally, but by no means least, methods are becoming available which permit the simulation of many natural environments in laboratory growth systems. Particularly important in this respect is the development of a range of different continuous-flow culture systems with particular features of importance in the natural environment. For example, developments have been made to include solids, such as sand or clay particles, in chemostat systems, thereby providing an environment which is much more akin to the aquatic situation. Enrichment techniques have been developed to enable whole, integrated communities to be isolated as a unit, which may then be studied under defined and controlled conditions. In both these cases these studies become exceedingly difficult *in situ*.

General references

BAZIN, M. J., SAUNDERS, P. T. and PROSSER, J. M. (1976) 'Models of microbial interaction in the soil', *CRC Crit. Rev. Microbiol.*, May, 463–98.

MEERS, J. L. (1973) 'Growth of bacteria in mixed cultures', *CRC Crit. Rev. Microbiol.*, January, 139–84.

VELDKAMP, H. and JANNASCH, H. W. (1972) 'Mixed culture studies with the chemostat', *J. Appl. Chem. Biotechnol.*, **22**, 105–23.

VELDKAMP, H. (1977) 'Mixed culture studies with the chemostat', *Sixth International Symposium on Continuous Culture* (Eds. A. C. R. Dean, D. C. Ellwood, C. G. T. Evans and J. Melling), 315–28.

Text References

BULL, A. T. (1974) 'Microbial growth', in *Companion to Biochemistry* (Eds. A. T. Bull, J. R. Lagnado, J. O. Thomas and K. F. Tipton), pp. 415–42. Longman, London.

BULL, A. T. and SLATER, J. H. (1976) 'The teaching of continuous culture', *Sixth International Symposium on Continuous Culture* (Eds. A. C. R. Dean, D. C. Ellwood, C. G. T. Evans and J. Melling), pp. 49-68.

COX, E. C. and GIBSON, T. C. (1974) 'Selection for high mutation rates in chemostats', *Genetics,* **77**, 169–84.

DENT, V. E., BAZIN, M. J. and SAUNDERS, P. T. (1976) 'Behaviour of *Dictyostelium discoideum* amoeba and *Escherichia coli* grown together in chemostat culture', *Arch. Microbiol.,* **109**, 187–94.

FRANCIS, J. C. and HANSCHE, P. E. (1972) 'Directed evolution of metabolic pathways in microbial populations. I. Modification of the acid phosphatase pH optimum in S. *cerevisiae*', *Genetics,* **70**, 59–73.

FRANCIS, J. C. and HANSCHE, P. E. (1973) 'Directed evolution of metabolic pathways in microbial populations. II. A repeatable adaptation in *Saccharomyces cerevisiae*', *Genetics,* **74**, 259–65.

GAUSE, G. F. (1934) *The Struggle for Existence.* Dover Publications, New York.

GIBSON, T. C., SCHEPPE, M. L. and COX, E. C. (1970) 'Fitness of an *Escherichia coli* mutator gene', *Science,* **169**, 686–8.

GRAY, B. H., FOWLER, C. F., NUGENT, N. A., RIGOPOULOS, N. and FULLER, R. C. (1973) 'Reevaluation of *Chloropseudomonas ethylica* 2-K', *Int. J. System. Bacteriol.,* **23**, 256–64.

HARTLEY, B. S., BURLEIGH, B. D., MIDWINTER, G. G., MOORE, C., MORRIS, H. R., RIGBY, P. W. J., SMITH, M. J. and TAYLOR, S. S. (1972) 'Where do new enzymes come from?', in *Enzymes: Structure and Function,* 8th FEBS Meeting (Eds. J. Drenth, R. A. Oosterbaan and C. Veeger), Vol. 29, pp. 151–76.

JANNASCH, H. W. (1968) 'Competitive elimination of *Enterobacteriaceae* from seawater', *Appl. Microbiol.,* **16**, 1616–18.

JENSEN. H. L. (1957) 'Decomposition of chloro-substituted aliphatic acids by soil bacteria', *Canad. J. Microbiol.,* **3**, 151–64.

JOST, J. L., DRAKE, J. F., FREDRICKSON, A. G. and TSUCHIYA, H. M. (1973) 'Interactions of *Tetrahymena pyriformis, Escherichia coli, Azotobacter vinelandii* and glucose in a minimal medium', *J. Bacteriol.,* **113**, 834–40.

KUENEN, J. G., BOONSTRA, J., SCHRODER, H. G. J. and VELDKAMP, H. (1977) 'Competition for inorganic substrates among chemoorganotrophic and chemolithotrophic bacteria', *Microbial Ecology,* **3**, 119–30.

LEWIS, P. M. (1967) 'A note on the continuous culture of mixed populations of *Lactobacilli* and *Streptococci*', *J. Appl. Bacteriol.,* **30**, 406–9.

MEGEE, R. D., DRAKE, J. F., FREDRICKSON, A. G. and TSUCHIYA, H. M. (1972) 'Studies in intermicrobial symbiosis: S. *cerevisiae* and L. *casei*', *Canad. J. Microbiol.,* **18**, 1733–42.

OSMAN, A., BULL, A. T. and SLATER, J. H. (1976) 'Growth of mixed microbial populations on orcinol in continuous culture', *Abstracts of the Fifth International Symposium on Fermentation* (Ed. H. Dellweg), p. 124, Berlin.

SCHOPFER, W. H. (1943) in *Plants and Vitamins.* Chronica Botanica Co., Waltham, Mass., USA.

SENIOR, E., BULL, A. T. and SLATER, J. H. (1976) 'Enzyme evolution in a microbial community growing on the herbicide Dalapon', *Nature,* **263**, 476–9.

TSUCHIYA, H. M., DRAKE, J. F., JOST, J. L. and FREDRICKSON, A. G. (1972) 'Prey– predator interactions of *Dictyostelium discoideum* and *Escherichia coli* in continuous culture', *J. Bacteriol.,* **110**, 1147–53.

WILKINSON, T. G., TOPIWALA, H. H. and HAMER, G. (1974) 'Interactions in a mixed bacterial population growing on methane in continuous culture', *Biotechnol. Bioengn.*, **16**, 41–7.

YEOH, H. T., BUNGAY, H. R. and KRIEG, N. R. (1968) 'A microbial interaction involving a combined mutualism and inhibition', *Canad. J. Microbiol.*, **14**, 491–2.

9
Microbial hormones

G. W. Gooday
Department of Microbiology, University of Aberdeen

1 Introduction

The hormones discussed in this chapter are very powerful metabolites that control fundamental processes in microorganisms. They are the means of communication between individuals of the same species. Only those that have been chemically characterised will be discussed in detail; in all cases they are concerned with bringing cells together and with diverting cells from activities concerned with growth to those concerned with reproduction. Once the cells have met, surface recognition phenomena come into play, such as agglutinins on the amoebae of *Dictyostelium discoideum,* and on the mating cells of *Saccharomyces.* This phenomenon is not discussed here.

Cells are clearly aware of each other's presence. This is most easily seen among eukaryotic microbes but undoubtedly occurs amongst prokaryotes as well. Amoebae of *Dictyostelium discoideum* space themselves apart while feeding; hyphae of a fungal colony grow apart so that the mycelium grows out as a circle, with the hyphal branches neatly filling in the spaces between the main hyphae. These are examples of the negative autotaxes and autotropisms that we can interpret as ensuring, by means of a primitive territorial system, maximum exploitation of the substrate by a species. They are examples of chemical communication between cells, though are probably not due to specific chemicals but instead to gradients of oxygen or carbon dioxide around the actively growing cells.

However, the use of specific chemicals is evident when cells of the same species need to cooperate with each other. For example, a hypha of an Ascomycete or a Basidiomycete will commonly grow towards and fuse with a neighbouring hypha of the same species whether or not it is from the same parent mycelium. The resultant anastamoses allow the coordinate development of structures such as fruit bodies from large areas of mycelia, by providing greatly increased passage of material through the fusing hyphae. The chemicals controlling these positive autotropisms are species-specific and although they have not been characterised, there is evidence that suggests that they are unstable macromolecules, having a radius of action of about $10\,\mu m$ around a hyphal apex. There is one autotactic hormone that has been fully characterised, the acrasin of *Dictyostelium,* which is discussed later, but certainly there are many more such agents throughout the microbial world that remain to be discovered.

There are now a handful of microbial sex hormones that have been fully characterised. There are others, such as those from the alga *Oedigonium* and the fungus *Tremella* that are very close to being characterised, and again there is evidence that there are many more waiting to be described. Sexual reproduction is a diversion of energy and materials from vegetative growth and asexual reproduction, and so the successful mating of gametes is too important a process to be left to chance. It must be controlled spatially and temporally. In all cases the sex hormones greatly add to its efficiency, by ensuring that sexual cells are only produced when there is a compatible mating partner in the vicinity, or by attracting the sexual cells of a mating partner. We see the same sorts of controls for example amongst animals, where we commonly find communication between members of the same species by chemical, optical, vibratory and mechanical signals. Amongst the microbes we only find the use of chemical signals, but these can be very sophisticated. Modern biochemistry has shown how a macromolecule such as an enzyme can have the property of recognising and binding a particular molecule with a very high degree of specificity. A slight change in a functional group of the ligand can completely alter its affinity for the binding site. In all cases we see high specificity in the microbial hormones, and we can predict that each hormone has its own receptor molecule.

As a semantic digression, the word hormone is clearly justified for the substances discussed here as they all have the following properties: they are produced in very small amounts; they are biologically active in very small amounts, typically 10^{-8} to 10^{-12} molar; they diffuse from their producing organisms, either to another part of the same organism, or to other organisms of the same species; they tend to be destroyed by the recipient cells; and the recipient cells show profound responses of movement and/or morphogenesis. They are all very specific, as their production and their effect is confined to one species or a few closely related species. As defined by Raper, the sex hormones all induce the formation of, regulate the development of, or elicit response from sexual organs or sexual cells. Qualifying terms for particular types of hormone are used elsewhere, such as pheromone (ectohormone) or gamone (sex hormone) but I think their use is unnecessary for microbial hormones.

2 Autotactic attractants

2.1 The aggregation of *Dictyostelium*

Dictyostelium discoideum is a cellular slime mould, a member of the Acrasiales. Its vegetative cells are amoebae, which when starved undergo a process of endotrophic development, leading without cell division to the formation of a fruiting structure consisting of a mass of spores on top of a long stalk. What concerns us here is the initial stage in this sequence, the aggregation of the amoebae to form a pseudoplasmodium. They move towards a centre, often forming into streams as they go. Numerous experiments convincingly showed that aggregation is due to chemotactic attraction of the amoebae to an agent emanating from the centre of the aggregate. Bonner named this agent 'acrasin'. After a great deal of work, and with a fair dose of serendipity, acrasin was identified as adenosine 3′,5′-cyclic monophosphate (cyclic AMP).

This was a remarkable discovery, as cyclic AMP is a universal *intracellular* regulator of metabolism; for example in bacteria it controls transcription on release from catabolite repression, and in animals it is the 'second messenger', mediating the response to hormones such as adrenaline and glucagon. However, in *Dictyostelium* species it is clearly an *intercellular* hormone. It is specific as it has not been found to attract any other cells, even amoebae of the related slime moulds, *Polysphondylium* spp. The acrasin bioassay is sensitive to 10^{-9} molar cyclic AMP, and individual amoebae respond to less than 10^{-12} molar.

The chemotactic signal that is sensed during aggregation of a colony of amoebae is not a concentration gradient, but a propagating wave of cyclic AMP. When a pulse of cyclic AMP diffuses into contact with an amoeba a series of events occurs. The cyclic AMP is bound to cell-surface receptors; very quickly a surface phosphodiesterase hydrolyses the cyclic AMP to the chemotactically inactive 5'-AMP; at the same time the cell elongates, by pushing out a pseudopodium at the point that first received the cyclic AMP, i.e. towards its source; the cell loses the ability to respond chemotactically to cyclic AMP for a definite time — the refractory period; it releases its own amplified pulse of cyclic AMP, in part or completely newly synthesised by the enzyme adenylate cyclase; and extracellular phosphodiesterase and/or the cell-surface phosphodiesterase break down any cyclic AMP in the vicinity; the cell recovers from its refractory period and 'awaits' the next pulse of cyclic AMP. Meanwhile the cell behind it is responding to the pulse of cyclic AMP that has just been released.

The refractory period ensures the unidirectional propagation of the wave. The exact kinetic properties of all of the responses — e.g. the speeds for the tactic response — for the release of cyclic AMP, and for breaking it down, and the duration of the refractory period, will exactly determine the speed of propagation and response of the cells during aggregation under any set of conditions, but these will also depend on such factors as density of cells, and the particular strain and species under observation. The cells aggregate to a centre at which there must be a cell that initiates the waves of cyclic AMP. It is not clear yet how this central cell differs from the other amoebae.

Amoebae of the related *Polysphondylium* species appear to have their own acrasin, which has not been identified. They show very similar aggregation patterns to those of *D. discoideum*, but amoebae from the two genera will sort themselves out and form separate aggregates after being mixed together.

3 Sperm attractants

3.1 The marine algal sperm attractants

Amongst the brown algae there are examples where sexual fusion takes place between a motile male gamete and a non-motile egg, as in *Fucus,* or between a motile male gamete and a settled female gamete, as in *Ectocarpus* and *Cutleria*. In these three cases the sperm are actively attracted by the female gametes, and cluster around them.

Cook and Elvidge in 1951 realised that the attractant produced by eggs of *Fucus serratus* and *Fucus vesiculosus* is volatile. They collected an active fraction by bubbling a suspension of eggs with nitrogen or hydrogen, and collecting the gaseous fractions by condensation in a liquid air trap. Bioassays were done by placing capillary tubes containing the collected fractions into suspensions of sperms in sea-water. The active material strongly attracted the sperms to the mouth of the capillary tube. However, at that time even the most sophisticated techniques failed to identify the active material. Nevertheless Cook and Elvidge suggested that the attractant was a hydrocarbon. This has proved to be the case. It has been identified as *trans, cis* -1,3,5—octatriene, 'fucoserratene' (Fig. 9.1). A yield of about 3 μg is obtained from 1 kg of fresh female receptacles. As it is so volatile, it has been impossible to measure the concentration required for activity. Biological activity is detected by observing the clustering of sperm around a spot of Vaseline that has been exposed to a stream of air blown through the sample being tested.

The attractants produced by female gametes of *Ectocarpus siliculosus* and *Cutleria multifida* are also hydrocarbons (Fig. 9.1). Ectocarpene has a 7-carbon ring, 6-(*cis*-but-1-enyl)-cyclohepta-1,4-diene; multifidene has a 5-carbon ring, *cis*-3-(*cis*-but-1-enyl)-4-vinylcyclopentene. Yields are about 80 mg and 2·5 mg respectively from 1 kg of fresh gametophyte seaweed. The biological activity is assessed by using the Vaseline drop technique as for fucoserratene, and also by observing clustering of gametes around droplets of mineral oil or glass beads that have previously been exposed to the test solution. Radioactive ectocarpene is rapidly taken up by male gametes but is also quickly metabolised. The gametes are heterokont, i.e. they have one flagellum pointing forwards that is covered with two rows of hairs, known as flimmer, and a second flagellum pointing backwards. Autoradiographs of gametes exposed to tritiated ectocarpene indicate that radioactivity rapidly becomes associated with the anterior flagellum, but later becomes associated instead with the body of the cell. During mating this anterior flagellum will receive the highest concentration of ectocarpene as the cell swims towards the female gamete, and initial sexual fusion is between the tip of this flagellum and the female cell.

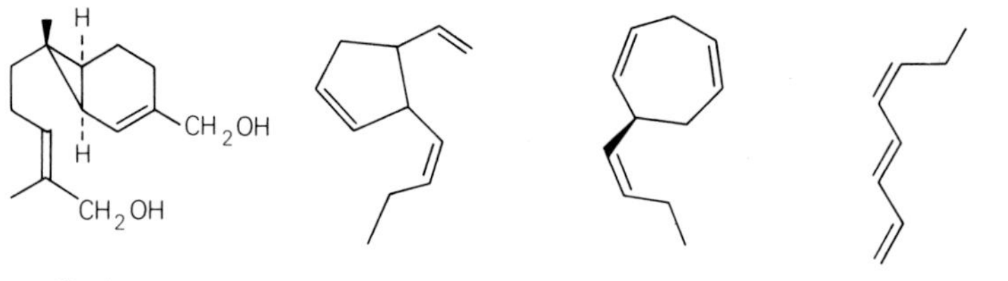

Fig. 9.1 Structures of microbial sperm attractants.

Jaenicke has suggested that all three hydrocarbon hormones could share a common biosynthetic origin, from a polyunsaturated fatty acid such as linolenic acid. β-Oxidation and oxidative decarboxylation could give rise to a C_{11} triene-ol, *cis,cis*-undeca-1,5,8-triene-3-ol, which in turn could give rise to a C_{11} carbonium ion which could cyclise to give ectocarpene and multifidene, according to the species of seaweed. In *Fucus,* the C_{11} triene-ol could be hydroxylated and degraded to fucoserratene. The three hormones are all specific, both in production and effect. They are only produced by the female gametes, in the case of ectocarpene only after these have settled; they are only active on spermatozoa of the same species.

3.2 Mating in *Allomyces* — sirenin

Species of *Allomyces* are water moulds, growing saprophytically on decaying vegetable matter. When triggered by a drop in nutrient level, the haploid plants produce male and female gametangia, usually in pairs on the same hypha. The gametes of both sexes are uninucleate naked cells, swimming by means of a posterior whiplash flagellum, but the male is smaller, more active, and bright orange in colour with γ-carotene. Machlis showed that these male gametes swim towards and cluster around the more sluggish female cells. This attraction is due to the production of a chemical, as it still occurs when male and female gametes are physically separated by an agar block or a permeable membrane. Machlis called the attractant sirenin after the Sirens who attempted to lure Odysseus. An assay was devised of counting the number of male gametes settling on a cellophane membrane which separated them from the solution under test, and this has allowed the successful purification and characterisation of sirenin. Further refinement of the assay now means that it can detect a 50 picomolar solution of sirenin.

As it is active at such a very low concentration, it is not surprising that it is produced in very small amounts. Optimal conditions for the production of sirenin have yielded a micromolar concentration in the culture filtrate. Male cells inactivate sirenin, and also stop its synthesis by fertilising the female gametes, and so it was necessary to grow cultures of predominantly female interspecific hybrids even to obtain this yield. These hybrids had previously been created and investigated in great detail by Emerson as part of a study of speciation in this genus, but their existence has proved invaluable in the elucidation of the role of sirenin, as has the existence of predominantly male hybrids for the production of gametes for the bioassay. This is a clear example of an academic study having unforeseen rewards.

Sirenin has the unusual cyclopropane ring (Fig. 9.1). It is certainly a sesquiterpene, derived from three isopentenyl residues, and is most probably biosynthesised by cyclisation of farnesyl pyrophosphate, the C_{15} precursor of sterols. It has been chemically synthesised, and only the L-isomer is biologically active. Chelated trace elements and calcium ions are required for maximum response. The male gametes specifically inactivate sirenin, and this may be a mechanism whereby they maintain their sensitivity over the 100 000-fold concentration gradient between 10^{-10} and 10^{-5} molar. The sirenin cannot be extracted back from the male gametes after they have removed it from solution, and so they probably metabolise it to an inactive compound. Perhaps this metabolic step is an essential part of the chemoreception mechanism. The gametes do lose their ability to respond to sirenin for a time after taking it up. For example 45 min. in the absence of sirenin was required before the chemotactic response was regained by cells that had previously been 30 min. in 5 nM sirenin.

Five different types of motile cell are produced during the life cycle of *Allomyces*: haploid zoospores, diploid zoospores, male gametes, female gametes, and zygotes. Of these, only the female gametes produce sirenin, only the male gametes respond to and inactivate sirenin. On the other hand, only the zoospores and zygotes are attracted by amino acids. It is interesting that the response to amino acids shown by the zygotes, which can be interpreted as a taxis towards nutrients, is much less sensitive than that of the gametes to sirenin, namely 400 μg per ml compared to 20 pg per ml.

The molecular mechanism of response to sirenin is unknown, but one can envisage a receptor macromolecule, located in the cell membrane of the male gamete, undergoing a conformational change on binding sirenin, and triggering a polarised change in some property of the membrane to lead to a directional response of the swimming cell.

4 Fungal, algal and protozoal sex hormones

4.1 *Achlya*—antheridiol and oogoniol

The species of *Achlya* are water moulds, living on decaying animal or vegetable matter. They can be isolated by baiting a stream or pond with a suitable substrate such as boiled hemp seed. They are members of the Oomycetes and are therefore phylogenetically quite distinct from most of the fungi; e.g. they have cellulose as the structural component in their walls instead of chitin. Margulis places them in the Kingdom Protista in her five-kingdom classification of organisms instead of in the Kingdom Fungi.

The biology of *Achlya* was investigated in detail by Raper, who elegantly showed that in the heterothallic species the female plants release a hormone into the culture medium that completely alters the course of differentiation of the male plants. Instead of growing vegetatively, the hyphae produce many short narrow branches, the antheridial branches. The number

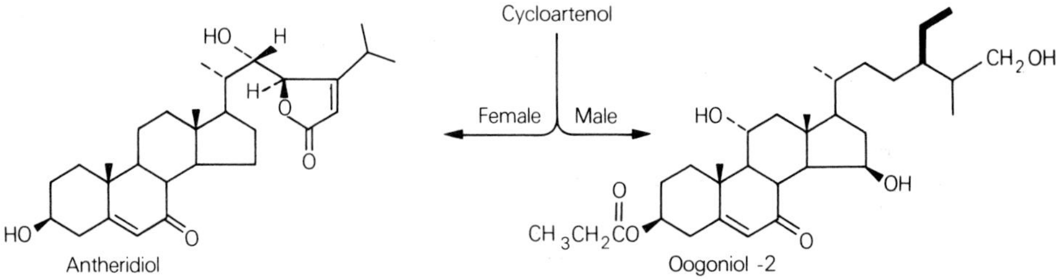

Fig. 9.2 Structures and suggested biogenesis of sex hormones in *Achyla*.

of these branches is directly proportional to the concentration of the hormone, now called 'antheridiol'. This reliable bioassay was crucial in the purification and characterisation of antheridiol. In 1942, Raper and Haagen-Smit, after tedious extractions from 1 440 litres of culture filtrate, managed to obtain 2 mg of active compound, but this was far too little for the contemporary methods of determining chemical structures. However, this original sample was kept and was analysed recently and shown to be substantially pure antheridiol. It was 25 years later that Barksdale and McMorris completely characterised it. It has proved to be a sterol (Fig. 9.2). This is of great interest since steroids have been widely exploited by animals as hormones, but this is the first example outside the animal kingdom. The planar steroid nucleus can be given a very high degree of specificity by the particular functional groups attached to it which will stick above or below it. So it is not surprising that the structural and stereochemical requirements for antheridiol hormone activity are highly specific. Many isomers and derivatives of antheridiol have now been synthesised. Most are inactive; none has an activity anything like that of antheridiol itself. For example, its three C-22,23 stereoisomers have less than 0·1 per cent of its activity, while 7-deoxy-7-dihydroantheridiol has about 5 per cent of its activity.

Antheridiol is active in very low concentrations. The bioassay can detect a 10^{-11} molar solution. Barksdale has shown that antheridiol itself can elicit the complete sequence of male responses that is seen leading up to plasmogamy. She adsorbed antheridiol on to particles of polyvinyl powder (plastic oogonia) which were then sprinkled over male hyphae. Antheridial branches were formed, which grew towards the source of antheridiol, wrapped themselves around it, and delimited antheridia. Thus successively higher concentrations of antheridiol elicit the range of morphogenetic responses of branching, chemotropism, formation of the delimiting septum and meiosis in the resultant antheridium. These responses can occur in the absence of exogenous nutrients, i.e. endotrophically. Low levels of exogenous nutrients will encourage the formation of antheridial branches, but higher levels will cause the branches to 'revert' as vegetative hyphae. The extent of response to antheridiol also varies considerably according to which strain of *Achlya* is being tested. Different isolates from the different species, both homothallic and heterothallic, can be arranged in a sexual series, and it is the strongest male isolates that show the greatest response.

The male cells of *Achlya* efficiently remove antheridiol from a solution added to them. For example, when a solution of about 1 μg ml^{-1} antheridiol was added to strains of heterothallic and homothallic species of *Achlya* they removed it at rates that reached about 10 ng per mg dry weight per minute. The rate and the time taken to achieve the maximum rate were related to the responsiveness of the strains. The availability of radioactive antheridiol has enabled experiments to be done that show that male cells respond to antheridiol by the induction of a system that converts it to inactive metabolites. There is a lag period of 30 to 80 min. for the appearance of this ability, and the addition of the inhibitors of RNA and protein synthesis, actinomycin D and cycloheximide, prevents the formation of the metabolites. So antheridiol induces a system for its own metabolism. Perhaps this is part of the mechanism of reception of the hormone, and perhaps it serves to accentuate the concentration gradient as the male branches grow chemotropically towards the source of antheridiol.

Some of the biochemical responses to the action of antheridiol have been described. In the first 8 h there is an increase in the rate of synthesis of protein and a marked increase in the rate of synthesis of RNA, particularly that rich in polyadenylic acid, a characteristic of messenger RNA. These increases are prevented by the inhibitors of RNA synthesis, actinomycin D and cordycepin, and by cycloheximide. The electrophoretic pattern of proteins synthesised following antheridiol addition shows that one protein in particular is preferentially made during the first few hours of induction, but its role is not yet elucidated. The branching of responding hyphae is accompanied by an increase in activity of the enzyme cellulase, which is released to the medium. This cellulase may have a direct morphogenetic role, in softening the cellulosic hyphal wall to allow the branches to push out; it does not seem to have a nutritional role. Electron micrographs do show accumulations of vesicles, which may contain the cellulase, being released into the wall where the branches are being formed. At the same time a storage polysaccharide, a β-1,3-glucan, disappears from the cytoplasm, presumably providing building blocks or a source of energy for the developing antheridial branches. At the same time also, the rate of oxygen consumption increases.

A further response to antheridiol, as shown by Raper's earliest experiments, is for the male hyphae to produce a new hormone, oogoniol, which diffuses back to the female hyphae and causes them to differentiate to give oogonial initials, which in the continued presence of oogoniol will delimit the oogonia with accompanying meiosis. Oogoniol also probably stimulates the production of antheridiol. Thus, apart from the chemotropic action of antheridiol, the actions of the male and female hormones are closely comparable. The characterisation of oogoniol was much more difficult than that of antheridiol, as it is produced in much smaller quantities, its bioassay is much less sensitive, it is usually only produced in the presence of antheridiol itself, and its chromatographic properties are very similar to those of antheridiol. It was finally extracted in sufficient quantities from the culture filtrate of a strain of the homothallic *A. heterosexualis* that produces it constitutively. Oogoniol is also a sterol, but is esterified at C-3 with isobutyrate, propionate and acetate in oogoniols 1,2 and 3, respectively (Fig. 9.2).

Antheridiol and oogoniol probably share common precursors, but at some point their biosynthetic pathways diverge. The choice of which branch is taken must be one of the manifestations of the genetic control of sexuality. In the homothallic strains, adjacent parts of the genotypically homogenous thallus become phenotypically different to acquire male or female characteristics. The mechanism for the inheritance of sex in the heterothallic strains remains unknown, as the ratios of progeny from crosses do not offer any clear-cut answers such as a two allele or sex chromosome system, or cytoplasmic inheritance. To try to explain this complexity, it has been suggested that the phenotypic sexuality observed in an isolate of *Achlya* is largely a result of the summation of the contributions made by the many genes responsible for the production and response to the hormones.

Sterols are also involved in the regulation of sexual reproduction in the very important plant pathogens, *Phytophthora* and *Pythium*. These are Oomycetes, close relatives of *Achlya*, but they differ from it in apparently being unable to synthesise sterols. They do not require added sterols for vegetative growth, but these are essential for the formation of the sexual oospores. In their natural parasitic mode of life, the host plant would provide such sterols. Could it be that the sterols are metabolised to a hormone analagous to antheridiol, just as insects can convert dietary sterols to ecdysone?

4.2 *Mucor* and trisporic acid

Sexual differentiation in the Mucorales is controlled by the one hormone, trisporic acid. It has been identified from homothallic and heterothollic species for genera considered as taxonomically widely separated, and so any differences observed in the control of its synthesis are probably only of minor significance.

Sexual reproduction in the Mucorales was elucidated by Blakeslee in 1904, who coined the term heterothallism to describe the requirement of the coming together of mycelia of the two mating types, (+) and (−), for the formation of zygospores. In 1924 Burgeff, by growing (+) and (−) *Mucor* separated by a permeable membrane, demonstrated the existence of diffusible hormones; the presence of the opposite mating type on the other side of the membrane caused the formation of sexual zygophores instead of asexual sporangiophores.

We now know that when they are in 'diffusion contact', i.e. close to each other in the air or in an aqueous medium but not necessarily touching, the (+) and (−) mycelia institute a collaborative biosynthesis of trisporic acid. By pre-labelling either one of them with radioactive ^{14}C it can be shown that they both contribute to the final yield of trisporic acid. When (+) and (−) are grown separately, in effect trisporic acid is not produced (there may be just a trace of it); when they are brought together it is made by their cooperative metabolism; when they are separated again its synthesis stops.

Trisporic acid (in fact a mixture of four readily interconvertible metabolites) is formed from β-carotene. This C_{40} compound is probably cleaved to the C_{20} retinal, by a carotene oxygenase, as occurs in the mammalian eye. The retinal is in turn cleaved to a C_{18} compound, perhaps 4-dihydrotrisporin. This is then envisaged at the branch point between the metabolisms of (+) and (−). The (+) hyphae can convert it to methyl 4-dihydrotrisporate, but are virtually or totally unable to metabolise this to trisporic acid; the (−) hyphae convert it to trisporal (or, in some species apparently only to its precursor, trisporin), but are virtually or totally unable to metabolise this is trisporic acid. Thus, in the unmated cultures, each mating type produces very small amounts of a specific metabolite.

However, (−) hyphae can metabolise methyl 4-dihydrotrisporate to trisporic acid; (+) hyphae can metabolise trisporol (and trisporin) to trisporic acid.

So, (−) can oxidise at C-4 through to the ketone, (+) can only oxidise to the hydroxyl group; (−) can readily hydrolyse the methyl ester at C-1, (+) cannot; (+) can oxidise the methyl group or its alcohol at C-1 to the acid, (−) cannot (Fig. 9.3).

Then, when (−) grows near to (+) it can take up the methyl 4-dihydrotrisporate, convert it to trisporic acid, and respond sexually; the (+) meanwhile is taking up the trisporol, converting it to trisporic acid, and responding sexually. However, at this stage there will be large diffusion losses of the precursors, so this mechanism would not be very efficient, were it not for the strong stimulatory effect that trisporic acid has on the production of the precursors. At least some of the enzyme activities responsible for the synthesis and subsequent conversion of the precursors are induced to a much higher level in response to addition of trisporic acid. Thus the collaborative biosynthesis of trisporic acid is self-amplifying. The extent of this self-amplication, and indeed the relative rates of all of the different metabolic conversions involved, will differ between different strains and species. For example, *Blakeslea trispora* (from which trisporic acid was characterised and named before its hormone action was suspected) produces enormous quantities, so that trisporic acid becomes a major metabolite in mated cultures. Homothallic species, although capable of the full biosynthetic pathway, presumably have a phenotypic spatial separation of (+)ness and (−)ness, and so of the complementary biosynthetic attributes, in adjacent hyphae so that zygophores are differentiated and mate together in pairs.

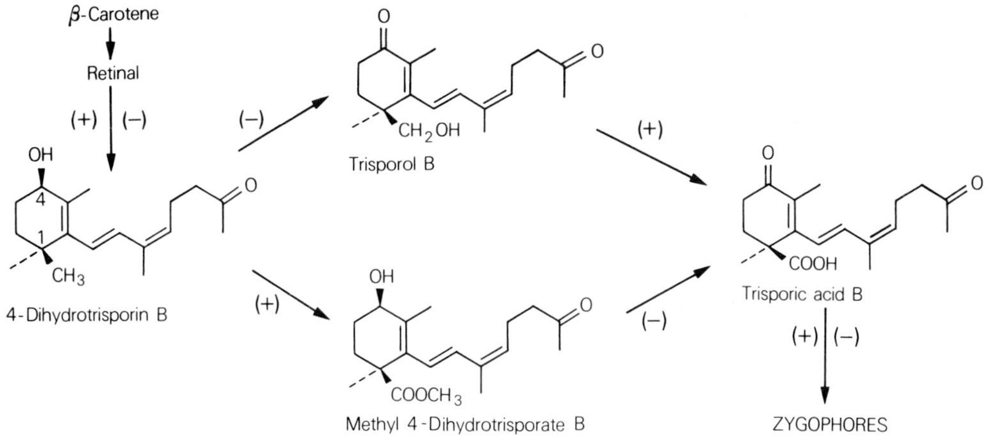

Fig. 9.3 Suggested biogenesis of trisporic acid: collaborative biosynthesis by (+) and (−) mating types.

Some species, for example those of *Rhizopus,* do not have detectable carotenoids. However, we can expect them to have the metabolism through to trisporic acid, even if the intermediates do not accumulate in detectable amounts, as they and many other isolates of the Mucorales readily undergo interspecific sexual interactions, involving the formation of zygophores and leading up to, but not including, plasmogamy.

The major morphogenetic effect of trisporic acid is to cause the formation of the zygophores, the characteristic sexual hyphae. In *Mucor mucedo* the zygophores are readily distinguishable from other aerial hyphae, and so the number produced can be counted as a bioassay. The limit of detection is about 10^{-8} molar trisporic acid. It causes the formation of zygophores in (+) and (−) mycelium. It can also stimulate zygospore production in some strains of homothallic species, and in one case has 'restored' fertility to an isolate of the homothallic *Zygorhynchus moelleri* which was thus presumably sterile because of attenuated hormone production.

The major biochemical effects of trisporic acid that have been investigated are its stimulation of carotene and sterol production and of the metabolism of carotene to the mating type-specific trisporate precursors. These responses are prevented by inhibitors of RNA and protein biosynthesis such as 5-fluorouracil and cycloheximide. However, further biochemical responses are clearly implied by the behaviour of the zygophores. When they come into contact, compatible zygophores immediately fuse. This is not shown by vegetative or asexual hyphae, and so suggests that the zygophores have mating type-specific surface agglutinins, such as have been described for mating cells of the yeast *Hansenula wingei.*

The zygophores of opposite mating type grow towards each other. This phenomenon is shown particularly strongly by *M. mucedo*, and is clearly caused by chemotropism to complementary volatile effectors, released and perceived especially at the zygophore apices. The exciting possibility, invoking Occam's razor, is that the chemicals involved in these mutal attractions are also intimately involved in the control of the biosynthetic pathways to trisporic acid. For there, as described above, we already have a series of chemicals that are specifically produced by one mating type only, and are specifically recognised by the other mating type — just the requirements for the chemotropic effectors.

4.3 *Saccharomyces cerevisiae*

This species, the most well-known eukaryotic microbe, may appear an unlikely source of sex hormones, as we are most familiar with it as a large population in a fermenting vat. However, it has now been clearly shown that haploid cells of opposite mating type, *a* and α, undergo a period of courtship, mediated by two complementary hormones, which leads to cell fusion and the formation of the diploid zygote.

Cells of α mating type constitutively produce the peptide, α-factor. Four peptides with hormone activity can be isolated from the culture medium, but they are very closely related, and the tridecapeptide α1 could give rise to the dodecapeptide α2 by loss of the *N*-terminal tryptophan, possibly by the action of an aminopeptidase, and α1 and α2 could give rise to α3 and α4 by chemical oxidation of the methionine to methionine sulphoxide in the acidic growth medium. It is a hydrophobic peptide of unusual composition, especially with its tryptophan–histidine–tryptophan–grouping:

$$H_2N\text{-Trp-His-Trp-Leu-Gln-Leu-Lys-Pro-Gly-Gln-Pro-Met-Tyr-COOH}$$

α-Factor binds cupric ions, and originally appeared on chromatograms as a blue spot. The inhibitor of eukaryotic protein synthesis, cycloheximide, rapidly and efficiently stops its synthesis. This suggests that α-factor is specified by messenger RNA and synthesised on ribosomes, rather than being specified and synthesised on synthase enzyme complexes as are at least some of the biologically active peptides of similar size produced by prokaryotic cells, such as the antibiotics associated with sporulation in *Bacillus* species. Possibly the α-factor is produced by proteolytic cleavage of a protein precursor, as is insulin in mammals.

α-Factor is only synthesised by α-cells, not by *a*-cells or *a*/α diploids. It is only active on *a*-cells. They respond in several ways: they elongate to become pear-shaped; within one generation they are arrested at the G1 phase of the cell cycle (after cytokinesis) so that nuclear DNA replication and further cell division are blocked; they gain the property of agglutinating with α-cells; and they initiate, or greatly increase, their production of the complementary *a*-factor. Synthesis of mitochondrial DNA, cellular protein and RNA continue unaffected, but altered cell-wall components are produced to give more glucan and less mannan in the elongating wall. *a*-Cells remove α-factor from solution, perhaps by binding it. On the addition of α-factor, it is initiation of DNA synthesis of *a*-cells that is inhibited, as cells in S-phase continue with their round of DNA replication. Washing out α-factor from *a*-cells results in a synchronized burst of DNA synthesis, and so the G1 arrest is reversible.

The *a*-factor, yet to be fully characterised, is probably also a peptide. It acts on α-cells, causing a transient G1 arrest, with inhibition of DNA synthesis and cell division, and causing swelling and elongation of the cells.

These responses of *a* and α cells to α and *a* factors are clearly appropriate preliminaries for mating. When *a* and α cells are mixed, unbudded cells accumulate in the culture during a courtship period, and subsequent pair formation is seen only between single unbudded cells. At the G1 stage of the cell cycle, the haploid cells will each contain a single complement of the genome, as will be required for the resultant karyogamy. Thus the two hormones ensure a mutual synchronization of the cell cycles.

4.4 Induction of sexuality in *Volvox*

Volvox is a colonial green alga. It was first described by Antony van Leeuwenhoek in a letter of 2nd January 1700, with as much wonderment as was in his first description of bacteria some years earlier: 'When I brought these little bodies before the microscope, I saw that they were not simply round, but that their outermost membrane was everywhere beset with many little projecting particles, which seemed to me to be triangular, with the end tapering to a point: and it looked to me as if, in the whole circumference of that little ball, eighty such particles were set, all orderly arranged and at equal distances from one another; so that upon so small a body there did stand a full two thousand of the said projecting particles. This was for me a pleasant sight. . .'

Sexual reproduction in such a colony involves meiosis with the differentiation of sperm or eggs. The activity of a hormone was suggested when it was observed that sexual development in cultures occurred following the appearance of a few male colonies. The male colonies have been found to secrete a substance that induces sexual differentiation in vegetative colonies. This hormone is not sex-specific, as it can cause either male or female colonies to be produced. However, it is species-specific, as it has no action on closely related *Volvox* species. Its bioassay consists of scoring the relative proportions of female and asexual colonies in a series of tubes containing growth medium plus serial dilutions of test solution that have been inoculated with a set of colonies of a female tester strain. No females at all will be present in the absence of hormone activity.

The hormones from *Volvox carteri* and *V. carteri* var. *nagariensis* are glycoproteins, of molecular weight about 30 000. The latter glycoprotein has about 45 per cent carbohydrate. It is remarkably active. The bioassay can detect a concentration of 10 pg per litre, corresponding to a molar concentration of 3×10^{-16}.

4.5 *Blepharisma* — blepharismone and blepharmone

Blepharisma intermedium is a ciliate protozoan. Its sexual reproduction involves conjugation between the two mating types I and II. Type I cells constitutively release a hormone to the medium, termed blepharmone, which induces type II cells to release, or greatly increases their production of, a complementary hormone, termed blepharismone, which in turn causes an increase in blepharmone production by type I cells. Thus when the two mating types come together, there is positive feedback to increase the production of both hormones, just as we have seen with fungal systems.

However the chemical natures of the two hormones are quite different. Blepharmone is a glycoprotein, of molecular weight about 20 000. The amino-acid composition shows a high tyrosine content, and high contents of aspartate, threonine and serine as expected from a glycoprotein. The carbohydrate is about 5 per cent of the molecule and consists of three residues each of glucosamine and mannose. Blepharismone is the calcium salt of an aromatic acid: calcium -3-(2'-formylamino-5'-hydroxybenzoyl) lactate (Fig. 9.4). It is probably biosynthesised from tryptophan, and L- and D-tryptophan and 5-hydroxytryptophan inhibit its biological activity.

Blepharmone and blepharismone cause type II cells and type I cells, respectively, to agglutinate. These are the bases of the two bioassays. Blepharmone will cause pair formation of type II cells at 60 pg ml^{-1}; blepharismone will cause pair formation of type I cells at 1 ng ml^{-1}. These responses are inhibited by cycloheximide, suggesting that protein synthesis is required, and in fact increased protein synthesis is observed in cells treated with the hormones. About 2 h elapse between adding the hormones and observing the agglutination, and a similar time elapses before conjugant pairs are seen when cells of type I and type II are mixed together.

These results suggest that each hormone induces, or at least greatly increases, the synthesis of a cell-surface agglutinin in the cells of the opposite mating type. The agglutinin probably acts

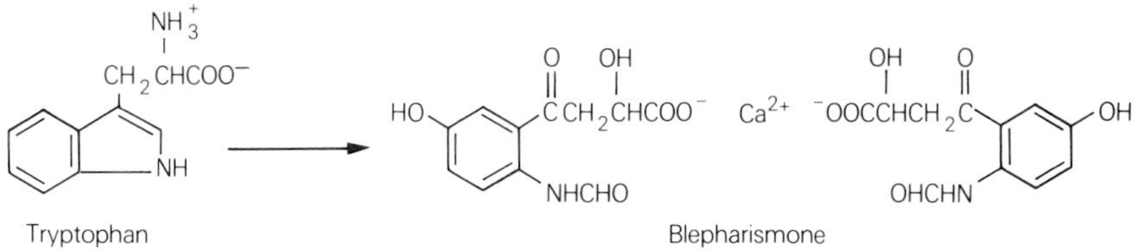

Fig. 9.4 Suggested biogenesis of blepharismone by type II cells of *Blepharisma*.

as a component of the ciliary membrane. As with some other microbial hormone systems, one of these hormones can act as an attractant as well as a morphogen. Type I cells will swim towards a source of blepharismone, and this of course will greatly help them to agglutinate with type II cells rather than with each other.

4.6 Control of the production of the sex hormones

These hormones that regulate sexual differentiation directly leading to mating show the intriguing property of being under the control of mutual positive feedback by the two mating types. This is seen in *Achlya, Mucor, Saccharomyces* and *Blepharisma*. The result is that full production of the sex hormone systems is only initiated when the two mating types come close to each other, and only maintained while they stay close together. Thus the sex hormones are a chemical 'sexual display' analogous to the visual displays of animal courtship that are triggered by the nearness of a potential mate.

5 Conclusions

Chemically, the hormones discussed here are very diverse; in increasing molecular size there are hydrocarbons, an aromatic derivative of an amino acid, a nucleotide, terpenes, peptides, and glycoproteins. Nevertheless they are remarkably similar in their properties. All of them satisfy the requirements defined in the introduction: they are very specific in production and action and very powerful in action.

Each hormone must have a receptor molecule with two properties: it must bind the hormone in order to recognise it; it must transmit this signal to the cell. Progress is now being made in identifying mammalian hormone receptors (see Perry 1974), and now that these microbial hormones have been identified it should be possible to start to characterise their receptors, and to determine their mechanisms of action.

How did the microbial hormones evolve? A hormone system requires the acquisition of three attributes by a species: the ability to make the hormone; the ability to detect it; and the ability to transmit this signal to the cell's metabolism. It is unlikely that such separate properties should arise at the same time independently. Perhaps the organism acquired the receptor molecule as a new protein at any time after becoming able to make the metabolite, which until then would just have been another natural product with no biological activity; or perhaps the receptor molecule arose from a pre-existing enzyme that could bind the metabolite. Such questions are now open to investigation. The evolutionary significance of the existence of microbial hormones is clear. J. B. S. Haldane has pointed out that they were a necessary preliminary for the formation of the Metazoa and higher plants. To become an organism a mass of mutually adherent cells must be able to influence one another and communicate. Thus the microbial hormones, diffusing through the external medium from cell to cell, are the antecedents of the hormones circulating through the bodies of animals and plants.

6 References

This is a brief selection of original papers and reviews which should act as starting points for further reading.

ALACANTARA, F. and MONK, M. (1974). 'Signal propagation during aggregation in the slime mould *Dictyostelium discoideum*', *J. Gen. Microbiol.*, **85**, 321–34.

BARKSDALE, A. W. (1969). 'Sexual hormones of *Achlya* and other fungi', *Science,* **166**, 831–7.
Characterisation of antheridiol.

BU'LOCK, J. D., JONES, B. E. and WINSKILL, N. (1976). 'The apocarotenoid system of sex hormones and prohormones in Mucorales', *Pure and Applied Chemistry*, **47**, 191–202.
Biogenesis of trisporic acid.

CARLILE, M. J. (Ed.) (1975). *Primitive Sensory and Communication Systems.* Academic Press, London.
Contains accounts of acrasin by Konijn, T. M. and fungal and algal attractants by Gooday, G. W.

GOODAY, G. W. (1973). 'Differentiation in the Mucorales', *Soc. Gen. Microbiol. Symp.*, **23**, 269–94.

GOODAY, G. W. (1974). 'Fungal sex hormones', *Ann. Rev. Biochem.*, **43**, 35–49.

JAENICKE, L. (Ed.) (1974). *Biochemistry of Sensory Functions.* Springer-Verlag, Berlin.
Contains accounts of the algal hydrocarbon attractants by Jaenicke, L.; *Blepharisma* hormones by Miyake, A.; and acrasin by Gerisch, G., Malchow, D. and Hess, B.

KONIJN, T. M., CHANG, Y. Y. and BONNER, J. T. (1969). 'Synthesis of cyclic AMP in *Dictyostelium discoideum* and *Polysphondylium pallidum*', *Nature,* **224**, 1211–12.
Cyclic AMP as acrasin.

MACHLIS, L., NUTTING, W. H., WILLIAMS, M. W. and RAPOPORT, H. (1966). 'Production, isolation and characterisation of sirenin', *Biochemistry,* **5**, 2147–52.

MCMORRIS, T. C., SESHADRI, R., WEIHE, G. R., ARSENAULT, G. P. and BARKSDALE, A. W. (1975). 'Structures of oogoniol-1, -2 and -3, steroidal sex hormones of the water mould, *Achlya*', *J. Amer. Chem. Soc.*, **97**, 2544–5.

MESLAND, D. A. M., HUISMAN, J. G. and VAN DEN ENDE, H. (1974). 'Volatile sexual hormones in *Mucor mucedo*', *J. Gen. Microbiol.*, **80**, 111–17.

PERRY, M. C. (1974). 'The hormonal control of metabolism', in *Companion to Biochemistry* (Eds. A. T. Bull, J. R. Lagnado, J. O. Thomas and K. F. Tipton), pp. 587–607. Longman, London.

STARR, R. C. and JAENICKE, L. (1974). 'Purification and characterisation of the hormone initiating sexual morphogenesis in *Volvox carteri f. nagariensis* Iyengar (glycoprotein)', *Proc. Natl. Acad. Sci. USA*, **71**, 1050–4.

STÖTZLER, D., KILTZ, H. and DUNTZE, W. (1976). 'Primary structure of α-factor peptides from *Saccharomyces cerevisiae*', *Eur. J. Biochem.*, **69**, 397–400.

SUTTER, R. P., HARRISON, T. C. and GALASKO, G. (1974). 'Trisporic acid biosynthesis in *Blakeslea trispora* via mating type-specific precursors', *J. Biol. Chem.*, **249**, 2282–4.

10

Morphogenesis in bacteria

R. Whittenbury and **C. S. Dow**
Department of Biological Sciences, University of Warwick, Coventry

1 Introduction

Escherichia coli and many other morphologically unexciting bacteria have dominated the microbiological scene for the past decade or so as vehicles for unravelling the complexities of genetics, molecular biology and biochemistry. Now it seems to be the turn of morphologically complex microbes – such as the stalked and 'budding' bacteria – to be favoured as models, this time in the pursuit of the basic principles governing morphogenesis and differentiation. A major disadvantage in changing to new organisms, of course, is the absence of banks of defined mutants and detailed records of biochemical and molecular biological information which so favour the continued use of *E. coli* and a few other select species. However, as Shapiro and her colleagues in the USA have already demonstrated for *Caulobacter crescentus,* it is possible within a relatively short period to assemble basic information, techniques, mutants, bacteriophages and so on, to permit investigations on development in prokaryotes not attainable by the use of *E. coli* and other morphologically simple bacteria.

For a variety of reasons, establishing the mechanisms and molecular events controlling and/or influencing morphogenesis and differentiation in eukaryotes is proving to be a slow process. Bacteria have proved useful in the past in the elucidation of fundamental processes applicable to all cells and some microbiologists feel that bacteria might also be exploited in the uncovering of principles governing development – if not the actual processes – of eukaryotes. A good deal of effort has already been expended in this direction but it has become obvious that such organisms have their limitations, if only because it has not yet proved possible to synchronize selectively the large quantities of cells necessary for quantitative studies, or because the differentiation event is only occasional and not obligate (endosporeformers) or because 'landmarks' in morphogenesis are so few (*E. coli* has only one, division).

Caulobacter crescentus, a stalked bacterium, undergoes a dimorphic cell cycle and because of this characteristic has been shown to have many useful attributes for the study of morphogenesis. As a consequence of success with this bacterium, attention has turned to other bacteria with complex life cycles in a search for additional models for developmental studies. As yet, only a few have been studied in any detail, and of these the 'budding' bacteria seem to be potentially very useful. Before embarking on this topic, it seems worth while to look at the reason(s) why bacteria may have evolved morphologies more complex than that of a simple rod or coccus. A major reason could be the nature of the environment and its dictates on the survival of the organism; the aquatic environment seems to be such an example.

2 The aquatic environment and the morphology of microbes

Natural bodies of water seem to be the source of many morphologically unusual bacteria. Microbial habitats in such waters will be diverse and constantly changing, depending on temperature, length of day, fluctuating nutrient status, stratification, and so on. Consequently, microbes whose natural habitat is water will be under severe pressure to evolve means of coping with survival in zones and areas suitable for their growth and reproduction. Movement by flagella, coupled to chemotactic and/or phototactic responses, would seem an obvious response to zonal changes through the day (e.g. some large species of photosynthetic sulphide-utilizing bacteria), while the formation of gas vacuoles to modify cell density is another such adaptation (e.g. other species of sulphide-utilizing photosynthetic bacteria).

In many instances, however, microbes seem to have come to terms with the instability of their environment either by: (1) adhesion to surfaces by one means or another (special

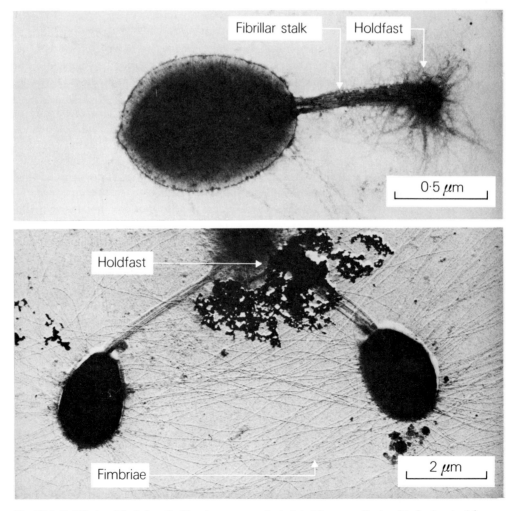

Fig. 10.1 Holdfast and fimbriae of a *Planctomyces* species isolated from an oligotrophic freshwater lake. (Courtesy of Miss A. Lawrence.)

holdfasts, fimbriae (Fig. 10.1) or capsular polysaccharide stickiness), or (2) microcolony formation which may incorporate holdfast anchoring of cells and the initiation of escape mechanisms from a deteriorating environment (production of motile swarm cells). It is the latter group of microbes on which we wish to focus attention, if only to draw attention to the benefits which a genuine colonial habitat may confer upon microbes in a particular environment.

3 The colonial microbes

The term a 'colony of microbes' usually conjures up the image of cell aggregation as seen on agar media; these colonies are, of course, usually artefacts not normally found in the natural environment. The colonies we are concerned with are those in which organisms are structurally linked to each other or confined in some way and seem to have evolved to cope with an aquatic environment. Examples are given below.

3.1 *Sphaerotilus natans*

This bacterium when grown in a nutrient-rich medium is a polarly flagellated, heterotrophic, Gram-negative, aerobic rod, similar in many ways to pseudomonads. In dilute media, and more importantly in the nutritionally dilute aquatic environment, this bacterium and its siblings are found confined within a sheath. Stages in development of a confined colony of this species are shown in Fig. 10.2; major points to note are the initial formation of a holdfast and the pseudo-branching sequences which permit a vast number of bacteria to remain together anchored at the one site. In this form the organism is commonly known as 'sewage fungus'. The macroscopic appearance of these colonies is one of a rust-coloured, fine-stranded mass. The rust colour is a result of ferric hydroxide adsorption to the sheath: this characteristic, unfortunately, has led to this and related species being collectively called 'iron bacteria' with the implied, and certainly unproven, notions that either they use ferrous iron as an energy source (autotrophic growth) or that abnormally high concentrations of iron favour their growth. The sheath is composed of a protein–lipopolysaccharide complex and is cytoplasmic membrane-like in nature. Any role of this sheath, other than containment of the bacteria, remains to be determined. From an environmental point of view, this mode of development seems ideally suited to an aquatic existence; the bacteria are kept end-to-end in a giant colony, presumably anchored in a favourable environment. Local environmental failure, whilst spelling the end for further development of a particular colony, is not lethal to all the members of the colony as single motile cells escape from time to time through breaks in the sheath; such escapees can move, presumably, by chemotactically directed movement of flagella to a favourable environment and set up a new colony as before.

3.2 *Beggiatoa*, *Thiothrix* and *Leucothrix*

These genera of Gram-negative, aerobic, heterotrophic bacteria (thought by some to be capable of autotrophic growth with H_2S as the electron donor) are sometimes referred to as 'colourless blue–green algae' because of their overall morphological resemblance to certain filamentous blue–green algae (cyanobacteria). They occur as filaments which, in the wild, are made up of hundreds of cells which remain firmly attached to each other. This 'colonial' feature is a distinctive characteristic of these species; they do not appear to occur naturally as single cells or

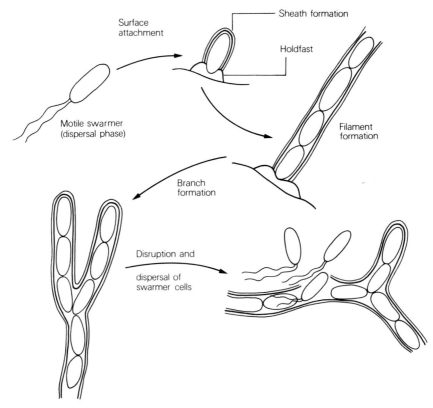

Fig. 10.2 Colonial development of *Sphaerotilus natans*.

pairs, except in the so-called reproductive phase of *Thiothrix* and *Leucothrix* (Fig. 10.3). Reproduction in *Beggiatoa* (increase in number of viable units, not cell division) is accomplished by the 'release' of small lengths of filament (hormogonia) containing a few cells.

In the natural state these bacteria are found as vast conglomerates attached to stones in streams or to stones over which water continually washes. Obviously such bacteria, by their mode of growth and colonial form, are ideally suited to existence in an aquatic environment. Creeping motility (*Leucothrix* and *Thiothrix*) and the release of motile cells or of small lengths of cells, can be seen as escape mechanisms for a portion of the colony from an unfavourable change in the microenvironment.

3.3 *Anabaena, Oscillatoria* and other filamentous cyanobacteria

These photosynthetic prokaryotes can be viewed as 'colonial' microbes in the same sense as the bacteria in the previous groups; all are 'gliders' and are able to increase their viable unit numbers by release of small filaments of cells. However, differentiation of some of the cells in the filaments increases the survival potential of these colonial microbes over that which could be expressed by a single cell of the same species. Selected cells within the filaments become (1) cysts (called akinetes) which are analogous to the cysts formed by some flexibacteria, myxobacters, some azotobacters and methane oxidizers and (2) heterocysts (unique to the cyanobacteria), cells which have differentiated to a biochemical role (nitrogen fixation) in the service of adjacent vegetative cells and have lost the ability to reproduce. The numbers and spatial

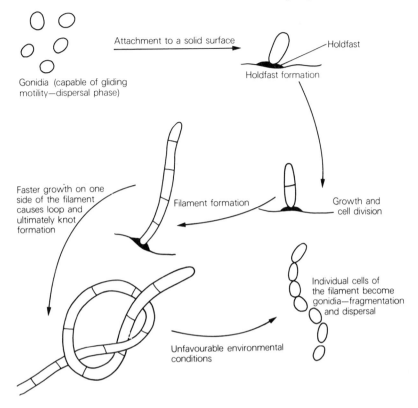

Gonidia (capable of gliding
motility—dispersal phase)

Attachment to a solid surface

Holdfast

Holdfast formation

Growth and
cell division

Filament formation

Faster growth on one
side of the filament
causes loop and
ultimately knot
formation

Individual cells of
the filament become
gonidia—fragmentation
and dispersal

Unfavourable environmental
conditions

Fig. 10.3 Diagrammatic representation of the life cycle of *Leucothrix*.

location within the filament of these differentiated cells varies with the species and the nutrient conditions, but is always less than a one-to-one ratio of differentiated to vegetative cell. In what is essentially a multicellular prokaryote, these differentiated cells confer powers of survival and nutritional virtuosity (fixed nitrogen content of medium appears indirectly to influence the commitment of proheterocysts to nitrogen-fixing heterocysts) on the 'colonial' filament obviously not possible in a single cell.

3.4 Multi-layered sheath or encapsulated bacteria

A number of bacteria (Fig. 10.4), including photosynthetic bacteria and cyanobacteria, are found entrapped within a sheath or large capsule able to adhere to surfaces. Very little is known about the growth and development of such agglomerations, but such cell collections confined in this way can be viewed as colonies adapted to an aquatic existence.

3.5 Bacteria within a polysaccharide matrix

Nevskia ramosa (Fig. 10.5) is probably the best known example of this means of ensuring that cells are kept as a stable colony, less well organized than previously described examples, but nevertheless just as effective it would seem. Other bacteria, as can be seen by sampling water surfaces, are found in masses embedded in a confluent slime. These slimes contain a number of species of bacteria and it is not possible to determine which may or may not be responsible for

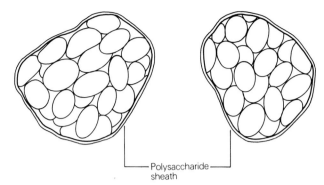

Fig. 10.4 Encapsulation by polysaccharide (?) of large number of cells, e.g. *Thiocystis* and *Thiocapsa*.

forming the polysaccharide matrix within which they are found. However, the effect is that the organisms are concentrated within a colonial matrix which is stable in an aquatic environment.

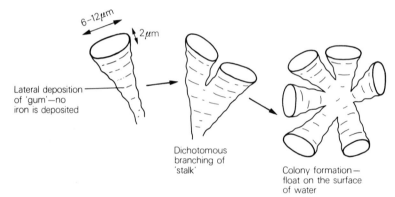

Fig. 10.5 Diagrammatic representation of *Nevskia ramosa*.

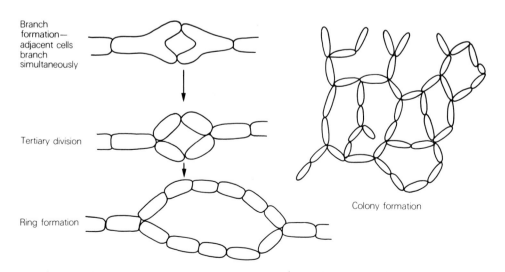

Fig. 10.6 Cell division and colony formation in *Pelodictyon*.

3.6 Lattice formation

A number of aquatic bacteria grow and divide in ways such that hundreds of cells remain in contact, either as a colony in sheets (*Lampropedia*) or rings (*Pelodictyon*). The division process of the latter organisms, tertiary fission, appears to be a unique way of developing a network of interlinked cells (Fig. 10.6) which result in a genuine colony, again stable in an aquatic environment.

3.7 Prosthecate linking

Some budding bacteria – *Rhodomicrobium vannielii* is an example – remain linked by the budding filaments under certain growth conditions (Fig. 10.7). This linking can lead to the formation of massive entangled colonies on the surface of sediments; again an example of colonial development in aquatic situations.

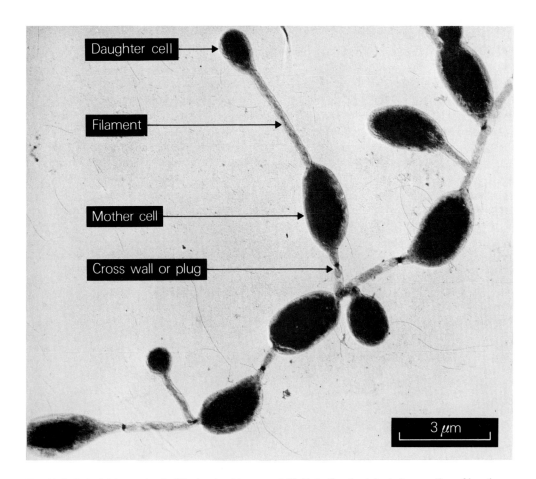

Fig. 10.7 Colonial formation in *Rhodomicrobium vannielii*. Note the physiological separation of 'mother-daughter' cell units by plug or cross wall formation. (Gold/palladium shadowed electron micrograph.)

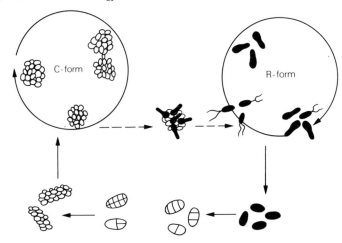

Fig. 10.8 Diagrammatic representation of the life cycle of *Geodermatophilus* strain 22–68. Growth and cell division in the C-form requires the presence of a factor found in tryptose. Absence of this factor induces differentiation to the R-form. Readdition of the tryptose will induce differentiation from the R-form to the C-form. (After Ishiguro and Wolfe, 1970.)

4 Categorisation of cell cycles and related events

The examples above of how bacteria appear to have evolved morphologically to cope with the problems posed by life in an aquatic environment are also examples of organisms which should prove admirable for the investigation of morphogenesis and differentiation. But first it is necessary to give a few simple definitions of terms used in describing cell cycles. It is not always clear how events in one organism compare with supposedly similar events in another, nor whether they are parallel to each other as is implied in the literature. We do not pretend that the definitions are profoundly conceptual; they are still rudimentary and doubts remain about the validity of comparing prokaryotes and eukaryotes in a developmental context.

Three categories are described: cellular events, cell cycles and levels of cell organisation.

Fig. 10.9 *Below Hyphomicrobium* life cycle. *Facing page* Gold palladium shadowed electron micrographs: 1, flagellated swarm cell; 2, swarm cell entering the maturation phase; 3, filament synthesis; 4, bud (daughter cell) formation; 5, asymmetric division.

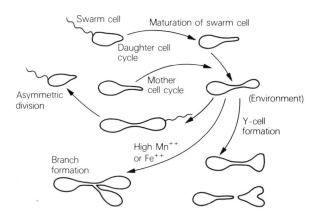

4.1 Cellular events

4.1.1 Morphogenesis – describes changes in external morphology and the internal architecture of the cells occurring during the vegetative cell cycle. The end result is cell multiplication.

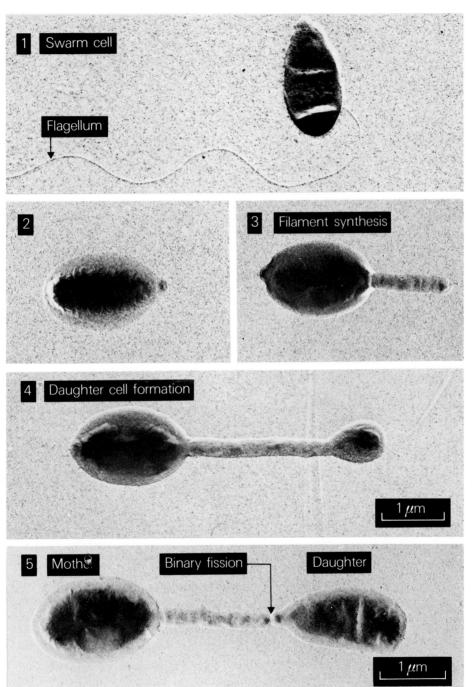

4.1.2 Differentiation — describes events initiated by transcriptional and translational changes in the cell cycle leading to the formation of a new type of cell (e.g. spore or heterocyst) or a modified cell (e.g. cyst). Such differentiated cells may revert to vegetative cells (e.g. spore and cyst germination) or be permanent (e.g. heterocyst).

4.1.3 Development — describes a composite event involving morphogenesis and differentiation under intercellular influence as in some myxobacters and filamentous cyanobacteria where cells are modified to serve a role necessary to the function of a multicellular complex.

4.2 Cell types

4.2.1 Monomorphic vegetative cell type is a description applied to microbes which express only the one phenotype; e.g. *E. coli* is a rod and always manifests itself in rod form.

4.2.2 Polymorphic vegetative cell type describes bacteria which may adopt more than one distinctive morphological form; the particular form expressed being dependent either upon the nutrient concentration of the medium or upon the presence or absence of a particular medium component. Such cell forms often undergo constant and distinctive cell cycles (e.g. *Arthrobacter* species (Fig. 10.24), certain cyanobacteria (Fig. 10.12), *Geodermatophilus* (Fig. 10.8), *Hyphomicrobium* species (Fig. 10.9), *Ancalomicrobium* (Fig. 10.10).

4.2.3 Spore and cyst cycles apply to those cell cycles of microbes which lead to the formation of resting cells. Such species may be of the monomorphic vegetative cell cycle type (e.g. *Bacillus* species (Fig. 10.11(a)) and *Methylosinus* species (Fig. 10.11(b)) or of the polymorphic cell cycle type (e.g. *Rhodomicrobium vannielii* (Fig. 10.33)).

Fig. 10.10 *Below* Phenotypic variation characteristic of *Ancalomicrobium* species. *Facing page* Phase contrast photomicrographs and gold/palladium shadowed electron micrographs of phenotypic variation in an *Ancalomicrobium* species. The cell types have been correlated with those in the diagrammatic representation.

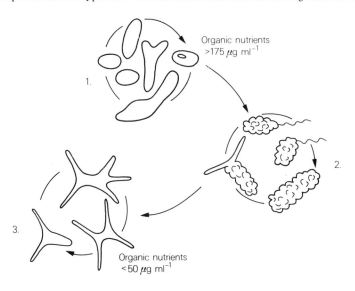

4.3 Levels of organisation

4.3.1 Single celled — the organism is normally independent of its fellows in terms of function and reproduction.

4.3.2 Occasional intercellular co-operation — describes bacteria able to lead a single-celled existence but which can cooperate in an intercellular manner to form a multicellular structure (e.g. fruiting body formation by some species of *Myxobacter*).

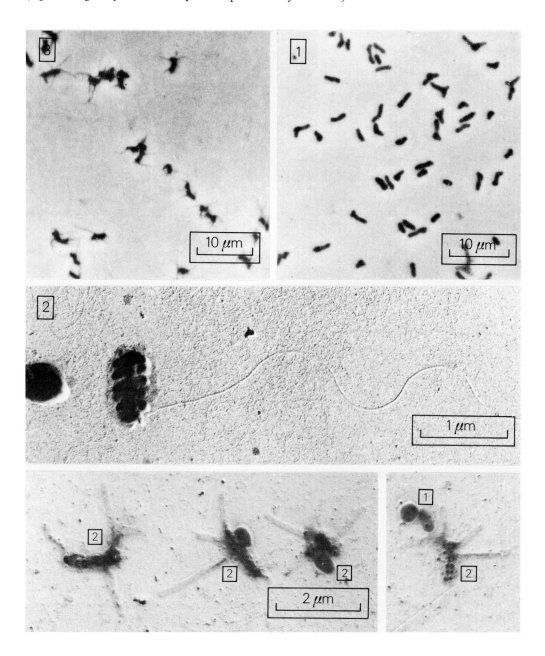

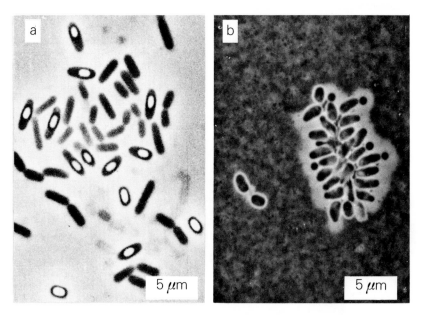

Fig. 10.11 (a) Light micrograph of endospores of *Bacillus subtilis*. (b) Spore formation in *Methylosinus* spp. (Light micrograph of an India ink preparation.)

4.3.3 Multicellular prokaryote – describes those prokaryotes able to differentiate and develop into a multicellular entity which (*a*) expresses an intercellular activity resulting in the differentiation of selected cells of that complex, and (*b*) clearly has survival potential far superior to that of a single cell of that species in the same environment.

The possibility that a multicelled organism composed of more than one type of cell of a prokaryotic nature may exist, and always functions as a multicelled organism, is generally overlooked by microbiologists who normally associate such a phenomenon only with eukaryotic cells. That there are instances of cellular interdependence in prokaryotes (e.g. as in *Myxobacter* species) has been recognised for a long time. However, certain filamentous cyanobacteria (e.g. *Chlorogloea fritschii*) clearly exist naturally and permanently (except in the cyst state) as multicellular complexes. This latter microbe is also unusual in that it has a polymorphic cell cycle, being able to express itself in distinctly different morphological forms, depending on its environment (Fig. 10.12).

Although these definitions may seem rather simple and a matter of common sense we consider that they will be helpful in comparing events in one organism with another. More usefully, perhaps, they highlight instances when two organisms are not strictly comparable. For instance *E. coli* is seen to be quite dissimilar from *R. vannielii* in that the former species is a bacterium with a monomorphic vegetative cell cycle (Fig. 10.18) leading to the formation of two siblings at division, whilst the latter organism is a bacterium with a spore cycle and a polymorphic vegetative cell cycle (Fig. 10.35) resulting in a first generation daughter cell and a mother cell at division (these cells are not siblings – see later). *R. vannielii* also differentiates and develops into a multicellular prokaryote (Fig. 10.7).

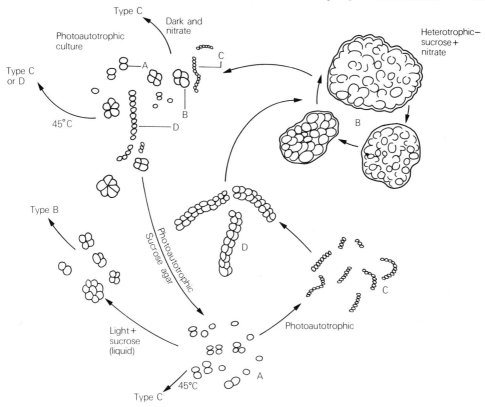

Fig. 10.12 Morphological variations induced in *Chlorogloea fritschii* by different environmental conditions. (After Evans, Foulds and Carr, 1976.) Type A: large granulated cells (2 × 3 μm) existing either singly or as clumps containing two or more cells which arise from division in up to three planes. Type B: found in clumps which combine larger groups of cells surrounded by a mucilaginous sheath. Type C: small cells (1 μm) found in filaments. Type D: cells (1·5 μm) found in filaments in the process of dividing.

5 'Budding' bacteria

Until a few years ago, only two species (*R. vannielii* and *Hyphomicrobium vulgare*) were recognised as 'budding bacteria'. That these two microbes were clearly different in their mode of reproduction from all other known bacteria was obvious (Figs. 10.35 and 10.9). A new cell (daughter cell) is formed *de novo* at the end of a filament, unlike the new cells of *E. coli*, which result from the binary fission of an elongated cell. This apparent morphological distinction between the budding bacteria and *E. coli* seems to have been accepted as sufficient evidence in itself to perpetuate the 'budding bacteria' group and to add to it year by year, as new species have been discovered or existing bacterial species have been found to bud.

Before discussing this group of bacteria further, however, it is necessary to be certain that 'budding' is a real phenomenon and whether or not it is confined only to those organisms now identified as 'budding' bacteria. A point to dispose of straightaway is the often implied assumption that 'binary fission' is a process distinct from 'budding'. Binary fission is as much part of the budding cycle as it is of division processes in all bacteria (Fig. 10.13). Although binary fission may not always be part of the process (Fig. 10.7), plugging of the filament rather than division being the end-point in cell separation, it inevitably has to occur in order to allow

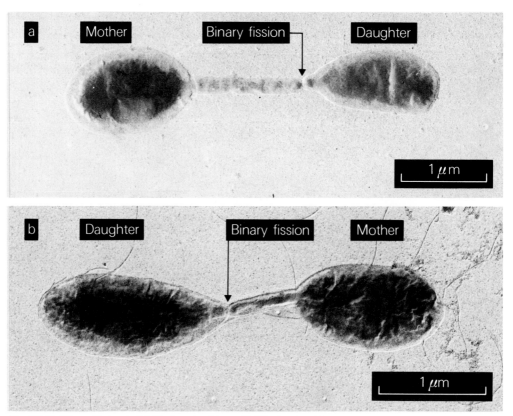

Fig. 10.13 Binary fission in (a) *Hyphomicrobium*, (b) *Rhodomicrobium*.

separate cell units to be released. This occurs in swarm cell production. If binary fission never occurred, the bacterium concerned would never be dispersed, but would remain as one giant colony.

The obvious distinction between the cell cycles of *E. coli* and *H. vulgare* is the symmetry of division of the former and asymmetry of the latter: this asymmetry of growth and division is the one important feature which seems to differentiate the 'budding' bacteria from all others. It reflects the polar mode of growth of 'budding' bacteria; all budding bacteria appear to have a single growth zone or growth point usually situated at one or both poles. Cell growth may proceed from only the one pole (e.g. *Rhodopseudomonas palustris*) or from both poles (*R. vannielii*) or even from the side of the cell ('mushroom-shaped budding bacterium', Fig. 10.14).

The non-budding bacteria appear to differ mainly from the budding bacteria in that envelope extension is a consequence of growth at multiple points along the axis of the cell. However, this is not a real distinction between the two types of bacteria; corynebacteria grow polarly, accounting for their irregular shape, whilst in minimal medium *E. coli* extends only from one pole (Fig. 10.22). In rich medium *E. coli* has multiple growth points. Consequently, there seems to be no absolute distinction between the budding process and the normal cell extension process leading to division as in *E. coli*; and, therefore, no clear-cut reason for categorising budding bacteria as a distinctive group on their own. The only real difference which may warrant such a distinction is that in budding bacteria growth is always from one growth zone, whereas this seems not to be the case with other bacteria.

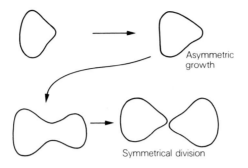

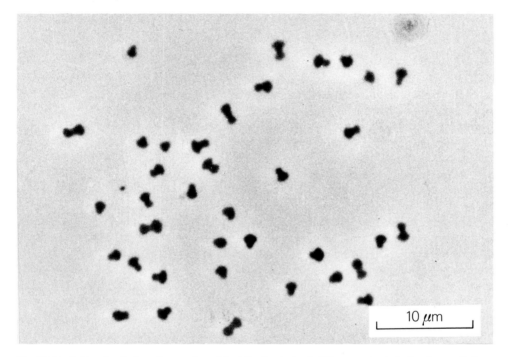

Fig. 10.14 Cell cycle of a 'mushroom-shaped budding bacterium'. (Phase contrast photomicrograph.)

The adoption of a polar rather than a multi-point mode of growth has far-reaching significance and seems to account for the real differences which exist between budding and non-budding bacteria. What we are saying here is that it is not the reproduction process itself which distinguishes between the microbes but rather the consequences of the mode of growth adopted by the organisms. These are best outlined by comparing a bacterium which grows ideally by intercalary processes with a bacterium growing polarly, and considering the consequences for the respective organisms.

5.1 Intercalary growth

If a rod-shaped bacterium is growing exponentially in a steady-state continuous-flow culture by ideal intercalary processes (i.e. old and new structures and cytoplasmic material including DNA are equally shared between dividing cells), then at division the two resultant siblings will be both qualitatively and quantitatively identical. Not only will they be morphologically indistinguishable, they will also be identical in regard to the distribution and ratio of old and new components to the cell which gave rise to them at the moment that cell was itself formed as a result of division. Five major points characterize such an organism in such circumstances:

1 It is potentially immortal.
2 It has an age span which does not exceed division time, i.e. if the generation time is 2 h, the organism will never be older than 2 h, reverting to zero hours at division as it no longer exists.
3 Siblings are identical in all respects and are genuinely siblings.
4. Membranous and other internal structures are likely to be symmetrical or, at least, small enough not to affect division.
5 Such organisms will always have a simple cell cycle and relatively simple or symmetrical internal structure imposed upon them by their mode of reproduction.

5.2 Polar growth

Taking a *Hyphomicrobium* species as an example, the following features are characteristic of its cell cycle in ideal continuous-flow circumstances.

At division the cell divides asymmetrically yielding (*a*) a 'mother' cell, which is a filamented cell constituted of old structural material, originally present at the division leading to its formation, and new structural material in the filament which is formed post-division; and (*b*) a 'daughter' cell which is a flagellated cell composed mainly of new material synthesised in the preceding cell cycle. In the next reproductive round the original 'mother' cell will not be made up of new material synthesised in the preceding cell cycle, but of material originally synthesised in the particular cell cycle leading to its formation. In this sense the mother cell ages with each succeeding round of reproduction.

Major features characterising obligate polar growth are as follows:

1 Organisms are probably mortal, undergoing only a limited number of reproductive cycles (e.g. *R. vannielii* in one phenotypic form, Fig. 10.35, can give rise to only four 'daughter' cells at most).
2 Ageing is a genuine characteristic of such organisms. Cell walls, for instance, are not renewed or intercalated with new material as judged by penicillin studies or by radioisotope uptake.
3 Cell division is frequently asymmetrical (an exception is the mushroom-shaped bacterium, Fig. 10.14).

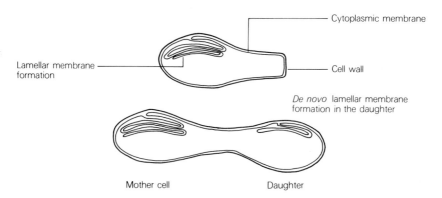

Fig. 10.15 Asymmetric membrane formation in *Rhodopseudomonas palustris*.

4 Internal membrane patterns may be asymmetrically constructed (Fig. 10.15) as constraints imposed by asymmetrical division processes are not present. In such instances, membranous bodies (organelles) are formed *de novo* in the 'daughter' cell.

5 The 'daughter' cell, at division, is immature: new internal and external structures have to be formed (e.g. invaginated membrane unit) or lost (e.g. flagella) before the cell reaches that point in its development when reproduction processes can begin. This difference in maturity between the new cell and the mother cell is also reflected in the increased time of the new cell's first division cycle compared with that of the mature mother cell's division cycle.

5.3 Mother and daughter cell concept

The main purpose of this discussion on polar and intercalary growth is to emphasise the dramatic consequences of polar growth for the organism concerned and, in particular, the likelihood that cell cycles of such organisms will have characteristics not normally associated with, say, *E. coli*.

One of the consequences is the mother/daughter cell relationship which applies not only to budding bacteria, but to others – such as *Caulobacter* and *Asticaccaulus* species – which have dimorphic life cycles but do not necessarily grow polarly. The differences between mother and daughter cells always arise as part of the life-cycle leading to reproduction; the differences may be directly related to the reproduction process (filament synthesis in budding bacteria) or be apparently unrelated to the reproduction process (stalk synthesis in *Caulobacter*).

Whatever the relationship to reproduction, these post-division developments lead to a 'parent/offspring' relationship not a sibling relationship; only the daughter cells of the one mother cell are siblings, and they differ in age relationship unlike the siblings from the intercalary growth type cells like *E. coli* which are always identical in age.

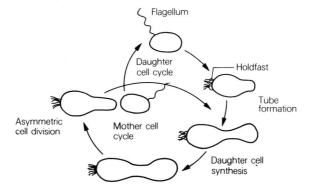

Fig. 10.16 Cell cycle of *Rhodopseudomonas palustris*.

5.4 Polarisation of the cell

A polar mode of growth or specialisation of function at one pole (e.g. *Caulobacter* stalks) leads inevitably to the polarisation of the cell. Such bacteria, in other words, have a 'front' and a 'back' end or a reproductive and non-reproductive pole. This is the case with *Rhodopseudomonas palustris* (Fig. 10.16) which only buds from the one pole and always has the holdfast at the other pole. Swarm cells prove to be mirror images of the mother cell, budding occurring from the pole first exposed at division. This particular organism is also oriented from top to bottom as well as end to end, as indicated by the site of the sub-polar flagella (Fig. 10.16) and the horseshoe-shaped cross-section of the lamella membrane complex. Polarisation is also evident in *Hyphomicrobium* spp. (e.g. swarm cell, Fig. 10.9) as well as other budding and stalked bacteria.

5.5 Obligate polar growth — potential for morphogenetic evolution

By opting at some stage in the cell cycle for obligate polar growth, not necessarily directly concerned with the reproduction of a daughter cell (e.g. *Caulobacter* stalk synthesis) the organism seems to have acquired the potential for morphogenesis unfettered by considerations of cell division. On such an assumption a comparison of budding bacteria can be made. Figure 10.17 demonstrates possible stages of development, as found in existing budding species, in order of increasing morphological development and specialisation. Cell polarisation is evident in the least complex examples (*Rhodopseudomonas acidophila, Rh. palustris* and *Rh. viridis,* and *Nitrobacter winogradskii*) in that only one pole is the reproductive pole and the membrane invagination pattern is asymmetrical. The obligate polarity disappears with the development of budding filaments; such bacteria can use both poles for reproduction.

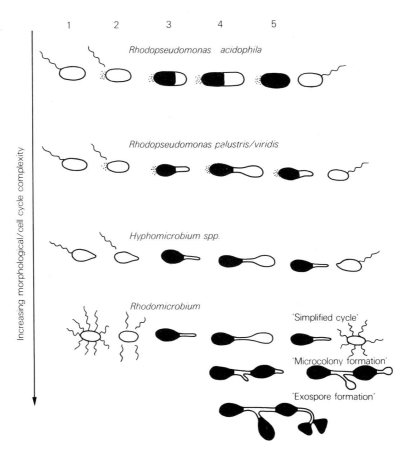

Fig. 10.17 Comparison of the cell cycles of 'budding' bacteria ordered in degrees of morphological complexity: 1, Motile swarm cell; 2, loss of motility — flagella shedding; 3, maturation; 4, daughter cell synthesis by obligate polar growth; 5, asymmetric cell division.

6 Varieties of budding bacteria

Every commentator on budding bacteria produces his or her own list of favourites to be included under the umbrella heading of 'budding bacteria' and usually takes the opportunity to exclude, in a pointed manner, those organisms considered to be falsely described as budding. Usually, this exercise follows a preamble on definition of budding and, sometimes, the sub-division of budding into categories. All this, we propose, is a meaningless activity. Budding, as we have indicated earlier, is a superficial concept. Obligate polar growth and asymmetry of division (not always immediately obvious, e.g. *Rh. acidophila*) are the key features of many so-called budding bacteria which make them such potentially excellent models for the study of development. And such features are not always confined to that group. Obligate polar stalk growth in *Caulobacter* species imposes similar constraints to those observed for budding bacteria growth, though Caulobacters are not obligately polarly growing at the reproductive pole; these constraints lead to dimorphic cell cycles and asymmetric division. Just to emphasise that 'budding' does not necessarily lead to asymmetry of division or cell immaturity (in that the newly synthesised daughter cell lags behind the mother cell in the subsequent division round), we highlight the example of the 'mushroom-shaped bacterium' (Fig. 10.14).

So our proposition is to put aside concepts of budding and to desist from perpetuating classifications based on budding which falsely separate *E. coli* from, say, *Nitrobacter agilis*. If any classification is felt to be a pressing necessity in the context of morphogenesis, then it should be geared to a clear-cut definition of processes leading to the observed differences between organisms. Such definitions would be best rooted, in our opinion, in the concept of cell cycles we have proposed earlier.

Table 10.1 embraces a selection of those bacteria which we feel are sufficiently well described, insofar as their morphological development is known, to be useful in studies on morphogenesis. Superficially, they seem not to be the organisms normally grouped on such a basis. However, our comments in the table indicate the one or more properties they share and also indicate those organisms which are more complex than they might seem at first sight.

Table 10.1   Classification of microbial cell cycles

Organism	Monomorphic	Dimorphic	Polymorphic	Spore/cyst	Multicellular
Escherichia coli	+	–	–	–	–
Arthrobacter	+	+	–	–	–
Bacillus	+	–	–	+	–
Mushroom bacterium	+	–	–	–	–
Caulobacter	–	+	–	–	–
Planctomyces	–	+	–	–	–
Nitrobacter	–	+	–	–	–
Rhodopseudomonas palustris	–	+	–	–	–
Hyphomicrobium	–	+	+	–	–
Rhodomicrobium	–	+	+	+	+
Geodermatophilus	+	+	–	(–)	–
Streptomyces	+	–	–	+	(+)
Myxobacteria	+	–	–	+	+
Blue–green algae (*Anabaena*)	–	+	+	+	+
Sphaerotilus	+	–	–	–	+

Cell cycles of certain of the bacteria, chosen to represent the spectrum of types being studied and the different sorts of information being released, and a summary of the pertinent information emerging are given below. The first of these discussions focuses on the simple, monomorphic cell cycle of *E. coli* and serves as a prelude to the more elaborate morphogenetic systems to be found in prokaryotes. The problem of resolving the regulation of bacterial cell division is analysed in depth by Richard James in Chapter 3 of this book, to which the interested reader is directed.

7 *Escherichia coli* — Molecular biology of the cell cycle

Studies based on the effect of specific inhibitors (e.g. chloramphenicol, rifampicin and nalidixic acid) on synchronous cultures, and on the use of mutants, particularly the TAU bar auxotroph (requiring thymine, arginine and uracil), have led to the following model (Figs. 10.18 and 10.19) being proposed to explain the observed sequential biochemical events occurring during the cell cycle.

Genome replication and cell division in *E. coli* do not, as was originally believed, constitute a linear sequence of events (C + D periods). There are two autonomous, parallel sequential processes linked only at initiation and termination (Fig. 10.19). A cell reaching its initiation mass, $M\hat{\imath}$ (doubling of the initiation mass takes place every mass doubling time, λ min.), initiates both the synthesis of division proteins and replication of the chromosome. Both events are completed after approximately 40 min. (this corresponds to the C period). Termination of DNA replication, i.e. when the growing points reach the chromosome terminus, induces the synthesis of termination protein. Completion of division protein synthesis in turn initiates a sequence of events which takes 20 min. to complete but does not require DNA, RNA or protein synthesis (events of the D period). The events occurring during the D period are indispensible, being required for functions of the division proteins and possibly for the formation at the potential division site of a septum 'primordium'. After this period interaction between the septum 'primordium' and the preformed termination proteins leads to cell division. This model attempts to integrate the observed biochemical and physiological events of DNA replication and cell division into the cell cycle. In the context of cellular morphogenesis, however, *E. coli* has

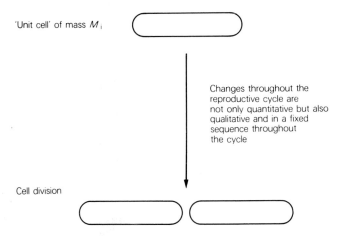

'Unit cell' of mass M_i

Changes throughout the reproductive cycle are not only quantitative but also qualitative and in a fixed sequence throughout the cycle

Cell division

Fig. 10.18 *Escherichia coli* cell cycle.

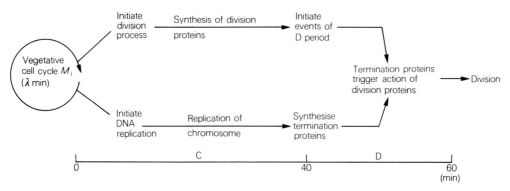

Fig. 10.19 Model of the cell cycle in *Escherichia coli.*

also been of value in studies concerned with cell envelope growth and the location and predisposition of potential division sites.

Initial experiments designed to identify the cell-wall growth point(s) employed both auto-radiography and immunofluorescence techniques (Fig. 10.20). These gave conflicting and

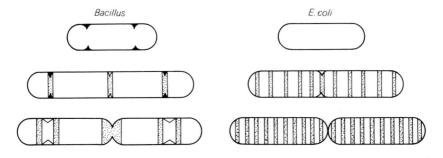

Fig. 10.20 Modes of cell-wall replication as seen by immunofluorescence, i.e. *Escherichia coli*, cell growth by a process equivalent to diffuse intercalation.

confusing results, primarily because of the misconception that the cell envelope, with the exception of the growth points, was a relatively static, inert entity. Later experiments involving the analysis of the distribution of specific cell envelope markers among progeny cells, e.g. β-galactoside permease (Fig. 10.21) and T_6 receptors indicated that cell-wall growth occurs at a limited number of sites rather than by diffuse intercalations. Recent studies on the growth of individual cells relative to fixed latex particles in slide cultures and on the location of the penicillin-sensitive sites (Fig. 10.22) concur with this view.

Higgins and Shockman (Fig. 10.23) have collated the available information on the growth of *E. coli* B/r and have presented a model to explain envelope growth of rods which is compatible with the data on the growth of rods and cocci, the three-dimensional structure of peptidoglycan and which retains the 'unit cell' concept in addition to integrating chromosome replication and segregation with cell division.

7.1 Model for envelope growth of rods (Fig. 10.23)

Consider a cell growing with a doubling time of 60 min., where $I = C + D$. I is the time required

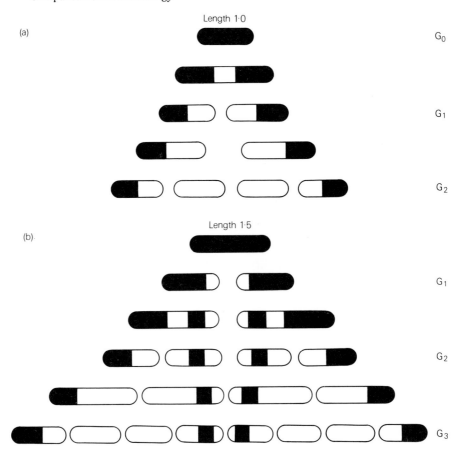

Fig. 10.21 Expected distribution of β-galactoside permease during de-induction among the progeny of an individual bacterial cell assuming that while bacteria increase from relative length one or two there is a single (median) growth point. (a) Bacterium with a relative length of 1·0 at the time of de-induction. (b) Bacterium with a relative length of 1·5 at the time of de-induction.

A non-synchronous population includes individuals whose lengths are continuously distributed between relative lengths one and two. (After Autissier, Jaffe and Kepes, 1971.)

for the accumulation of a sufficient quantity of hypothetical initiator substance to allow the initiation of one round of chromosome replication, C is the time required to replicate the chromosome, and D is the time required after chromosome replication for cell division (these are considered to be fairly constant at $37°C$ for *E. coli* B/r over a range of generation times). The proposed sequential events for such a cell, which would also correspond to a 'unit cell' possessing a single envelope growth point, would be as follows: concomitant with the initiation of a round of chromosome replication a wall elongation site is initiated at the junction between polar and cylindrical wall. A layer of new wall is begun under the old wall, these being linked by electrostatic interaction, hydrogen bonding or covalently by transpeptidation. The old wall is cleaved at this point so that the addition of new wall precursors will allow for the elongation of the surface. While the chromosome is duplicating, approximately a 'unit cell' length of cylindrical wall is unidirectionally fed out towards the nearest pole. Completion of the DNA replication cycle is followed by the formation and segregation of a new chromosome attach-

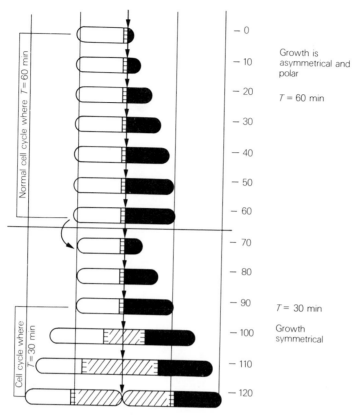

Fig. 10.22 Growth of the cell envelope of *Escherichia coli* according to the unit cell hypothesis; constructed from data on the direction of growth and from the positioning of the penicillin-sensitive sites (equated with potential division sites) at two different growth rates.

A unit cell (the smallest possible cell for that particular bacterium under any growth conditions) is grown for one cell cycle in a medium where the mass doubling time is 60 min. (at 60 min. the daughter cell on the left is transferred to a richer medium where the mass doubling time is 30 min. During such a shift up there will be a 60-min. interval between transfer to the new medium and the next cell division. In consequence the length of cells at this division will be twice the length of dividing cells in old medium. The growth sites are shown as dashed horizontal lines across the cells and the direction of growth at each site is arrowed. Each growth site gives rise to two new sites of opposite polarities when the cell reaches a length of 2 unit cells. The central vertical line corresponds to the spatial location of cell divisions. The triangles show the penicillin-sensitive sites. (After Donachie and Begg, 1970.)

ment site and the division trigger (termination proteins?) reacts with a postulated membrane site (septum primordium?). This interaction converts the unidirectional elongation site into a cross-wall site, i.e. the membrane carrying the wall-synthesis enzymes and precursors invaginates. The cross-wall is completed and strengthened, and the cells separate. A shift up of such a cell to a faster growth rate ($I = \frac{1}{2}(C + D)$) results in the initiator substance(s) being synthesised twice as fast as in the previous case, i.e. a new round of chromosome replication is initiated at approximately the time the previous round is completed. Upon completion of the first round, a wall elongation site is converted to a cross-wall site; upon initiation of the new round, two new wall elongation sites are initiated at the junction of the nascent cross-wall and cylindrical-wall and each of these feeds out new cylindrical wall in opposite directions, towards the nascent cross wall, so mimicking symmetrical cell growth. By the time the cross wall is complete

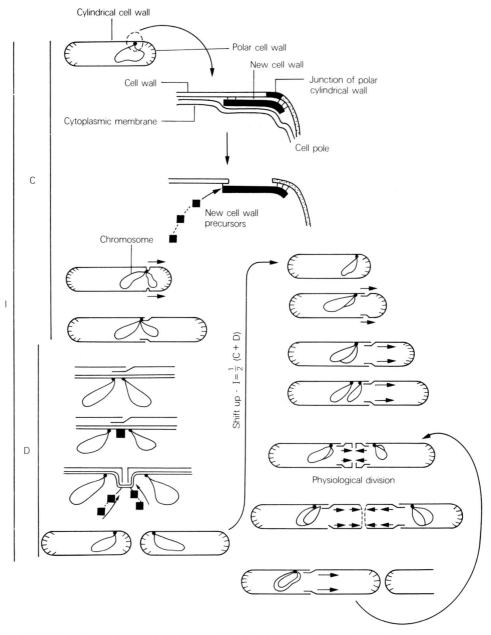

Fig. 10.23 Model for envelope growth of rods. (After Higgins and Shockman, 1971.)

additional wall has been made so enlarging the cell at the time of division. (See Ch. 3, sections 3 and 4 for further details of cell septation and its regulation.)

Studies on *E. coli* have, therefore, permitted the formulation of two compatible models of the cell cycle, one concerned with the sequential biochemical events leading to cell division, the other with envelope growth and chromosome segregation. Consequently, an organism which is morphologically very simple has served to illustrate some of the principle molecular aspects of cellular morphogenesis; nevertheless, as discussed earlier, *E. coli* has severe limitations when considering cellular morphogenesis, which may only be overcome by studying more complex microbial cycles.

8 *Arthrobacter crystallopoietes* — nutritionally-controlled morphogenesis

In complex growth media bacteria of the genus *Arthrobacter* exhibit nutritionally-induced morphogenesis, a sphere to rod transition, which is striking, stable and occurs within a completely vegetative life cycle (Fig. 10.24). Control of this morphogenetic switch can be achieved by growing the organism on chemically defined media (Fig. 10.25). Consequently, the ease of cultivation and the marked morphological variations accompanying the growth cycle make this organism particularly suited to a study of the regulatory mechanisms involved in cellular morphogenesis.

It has been established that sphere—rod morphogenesis is accompanied by detectable changes in cell-wall structure and in the activity of a wall-associated autolytic *N*-acetyl-muramidase (Fig. 10.26). However, it is not known whether these alterations to the cell wall are the cause or a result of morphogenesis. The question as to what causes *A. crystallopoietes* to undergo the sphere—rod—sphere transition remains unanswered, as does elucidation of any advantage(s) conferred on the organism by such a morphogenetic switch.

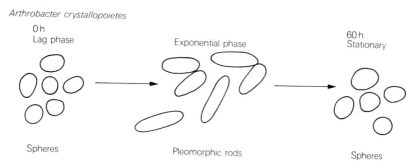

Fig. 10.24 *Arthrobacter* — morphogenetic cell cycle.

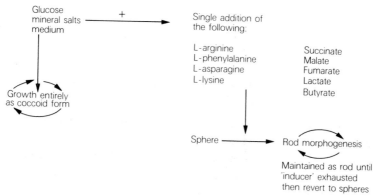

Fig. 10.25 *Arthrobacter* — nutritionally controlled morphogenetic cell cycle.

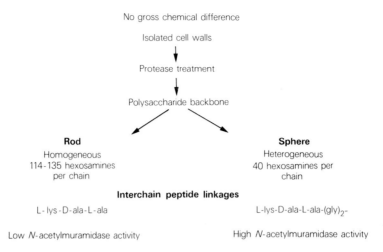

No gross chemical difference

Isolated cell walls

Protease treatment

Polysaccharide backbone

Rod
Homogeneous
114-135 hexosamines
per chain

Sphere
Heterogeneous
40 hexosamines per
chain

Interchain peptide linkages

L-lys-D-ala-L-ala

L-lys-D-ala-L-ala-(gly)$_2$-

Low *N*-acetylmuramidase activity

High *N*-acetylmuramidase activity

Fig. 10.26 Comparison of spherical and rod-shaped cells of *Arthrobacter*.

9 *Bacillus* species — endospore morphogenesis and differentiation; non-obligate differentiation

Although an ordered series of events is observed during the morphogenesis and differentiation of several biological systems, bacterial sporogenesis is a unique event. Regulation of sporogenesis appears to be intermediate in complexity between the production and activity of bacterial enzymes and the differentiation in more complex organisms of specialised cell types. *Bacillus* endospores (Fig. 10.11(a)) provide two model systems for the exploration of such changes:

1 Endospore formation (Fig. 10.27(a)).
2 Endospore outgrowth (Fig. 10.27(b)).

The temporal sequence of morphological and physiological events of endospore formation (Fig. 10.28) and germination (Fig. 10.29) have been well characterised. The literature on sporulation is voluminous and a comprehensive treatment of the process is not within the scope of this chapter. Consequently, we will consider only those aspects of current interest which concentrate on the molecular events controlling the change(s) in cellular expresion.

9.1 Endospore formation — (relief of catabolite repression?)

In the presence of excess carbon and nitrogen, bacilli grow vegetatively and sporulation is repressed. A deficiency of either substrate (amongst others) may induce spore formation (differentiation). A nitrogen-containing metabolite, the intracellular concentration of which depends on the rate of metabolism of both the available carbon and nitrogen sources, has been postulated as the repression agent. As phosphate-limitation also induces sporulation, it is probable that this particular metabolite is phosphorylated. There is unlikely to be an all-or-none response since studies on continuous cultures revealed a low probability that all cells will sporulate even under optimum growth conditions. Moreover, there is a continuous range of probabilities that the incidence of spore formation is a function of the growth rate.

It is generally assumed that repression of sporulation is similar to the catabolite repression of inducible enzymes. Attempts to define the biochemical nature of the sporulation repressor(s) showed that several compounds metabolised via different metabolic routes can suppress

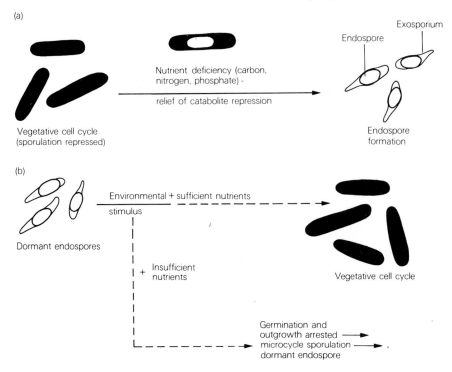

Fig. 10.27 (a) Endospore formation; (b) endospore outgrowth.

sporulation. An active ATP-producing system is required for sporulation, the reducing power being generated by an active tricarboxylic acid cycle. However, at the end of exponential growth (in *B. subtilis*) there is a transient fall in the ATP level within the cell. Thus a change in ATP level (or adenylate energy charge), brought about by starvation of an energy source, may be the signal which initiates sporulation, a possible repression mechanism being the phosphorylation or adenylation of an aporepressor protein. Following this signal the ATP level is restored and sporulation effected.

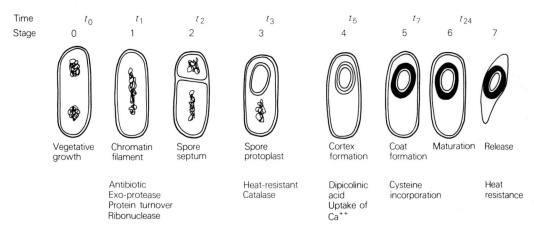

Fig. 10.28 Temporal sequence of morphological and physiological events characterised during endospore formation.

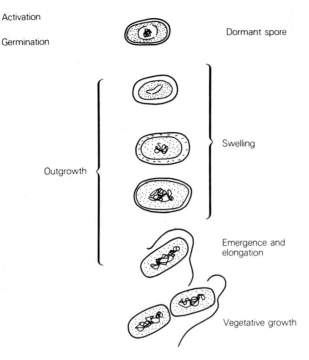

Fig. 10.29 Morphological characteristics of endospore germination.

Within most Gram-negative bacteria catabolite repression of inducible enzymes is intimately linked with 3'5'-cyclic adenosine-5'-monophosphate (cAMP) levels, but all attempts to demonstrate the presence of cAMP in *Bacillus* species have failed. However, the intracellular concentration of the structurally analogous compound 3'5'-cyclic guanosine-5'-monophosphate (cGMP) rises sharply with the initiation of sporulation, prompting the speculation that cGMP plays a role in catabolite repression of sporulation. In addition, a series of unique highly phosphorylated nucleotides has been observed at the start of sporulation. Their precise function is unknown and it is not clear whether these nucleotides accumulate simply as a result of the onset of sporulation, perhaps reflecting a functional change in the translational machinery, or whether they promote the expression of sporulation specific genes.

It would appear, therefore, that the initiation of sporulation is a complex system involving several integrated controls. The available data strongly indicate that initiation is brought about by the relief of repression. How this is effected is still a matter for speculation.

9.2 Transcriptional control of sporulation

It is thought that the events which follow the initiation of sporulation are controlled by sequential transcription of relevant loci and much effort has gone into the elucidation of such controls. It is known that there are specific mRNAs found only in the sporulating cell which implies transcriptional control of gene expression during the differentiation cycle. This, in turn, suggests the regulation of specific operons by regulatory proteins. As yet, there is no direct evidence that this happens during sporulation. However, there is sufficient indirect evidence to permit the postulation of two possible control mechanisms.

(*a*) The first concerns the role of RNA polymerase modification in sporulation, i.e. an analogy may exist between sporulation events and those occurring during bacteriophage T4 infection of a host cell. Studies on isolated RNA polymerases from vegetative and sporulating cells have shown that both possess the same core enzyme. However, there is considerable interference with the binding of the vegetative σ factor with the core enzyme during sporulation (although the vegetative σ factor itself does not disappear) and this modification is thought to be mediated by spore specific protein factors associated with the RNA polymerase core. This is an attractive model primarily because transcriptional control (change in template specificity during the switch from vegetative to sporulation loci) might be achieved by a simple modification or substitution of the σ factor (responsible for recognition of the promoter site), leaving the core enzyme unchanged. Such a regulatory function ascribed to the RNA polymerase undoubtedly has its attractions, particularly when comparing the prokaryotic system with the more complex eukaryote where several different RNA polymerases appear to be active within the differentiating cells. However, the fact that there are a large number of spore-specific loci which are transcribed at a number of different times makes it unlikely that RNA polymerase specificity is the only transcriptional control mechanism. This would require different modifying proteins to be synthesised at several stages during the differentiation cycle.

(*b*) A second possible transcriptional control mechanism involves conformational changes to the DNA and this is known to play a central role in eukaryotic differentiation. Irradiation of sporulating cells yields an unusual photoproduct (a thymine photoproduct rather than a thymine dimer) the appearance of which coincides with a change in the appearance of chromatin in electron micrographs. Two possible effectors of this confirmation change are Ca^{2+} ions, the uptake of which closely parallels the appearance of the 'spore photoproduct', and low molecular weight, spore-specific proteins which are known to bind to DNA and to alter its melting point.

9.3 Peptide antibiotics — (function in sporulation?)

Like other secondary metabolites, peptide antibiotics are produced in conditions which are unfavourable for vegetative growth, i.e. those conditions which induce sporulation. They are produced only by sporeformers and mutational loss of synthesis is often associated with asporogeny. Also, antibiotic production and sporulation respond in parallel to specific inhibitors as well as to changes in culture conditions. The question is — do these observations indicate a causal relationship where peptide antibiotics play an essential function in sporulation at concentrations at which they are produced by sporulating cultures? Certainly they are potent inhibitors of vegetative growth.

The first direct evidence of a regulatory function for a peptide antibiotic comes from the study of the effects of tyrothricin on the life-cycle of the producing organism, *Bacillus brevis*. Addition of tyrothricin to exponentially growing cells at the concentration found in sporulating cells inhibits growth by directly affecting RNA synthesis (Sarker and Paulus 1972). Tyrothricin is a potent inhibitor of the RNA polymerase, *in vitro* and *in vivo*. This finding led to a hypothesis which attempts to explain the shut-off of vegetative genes during sporulation, the mechanics of which is still obscure. The peptide antibiotics may act as selective regulators of gene transcription during the transition from vegetative growth to sporulation by affecting the interaction of the RNA polymerase with those promoters associated with vegetative gene function.

A second peptide antibiotic, edeine A, which is also produced by *B. brevis* has been shown to be a reversible inhibitor of DNA synthesis. A possible correlation, therefore, between antibiotic function and spore formation would be the specific inhibition of DNA replication early in the differentiation process.

Rather than simply being fortuitous products of spore differentiation, there is now evidence to suggest that the peptide antibiotics play specific and well-defined roles in regulating differentiation in endospore formers.

9.4 Genetic aspects of morphogenesis and differentiation

Any model of cellular differentiation must be amenable to genetic analysis. Of the endospore formers, *Bacillus subtilis* is the species best understood at the genetic level, primarily because the genetic exchange systems have been extensively exploited. It is, however, assumed that many of the features of sporulation are common to many, if not all, of the endospore formers. Although considerable doubt exists about the precise role that most sporulation-associated events play in the regulation process, it is now clear that there are at least 30 to 40 sporulation loci which are widely dispersed on the chromosome. Genetic analysis makes it doubtful that there exists a linear dependent sequence of induction. More probably there are several parallel pathways of induction. Obviously, it is not possible to introduce here the wealth of genetic knowledge which has been accumulated on sporulation. Several recent key references to this subject are given at the end of this chapter.

The above system was the first to be developed as a prokaryotic model for the study of cellular differentiation and morphogenesis and, as such, has been very useful. However, problems have now been encountered in that molecular changes associated with the differentiation event may either be a consequence of the metabolic shift down initiating sporulation, or part of the differentiation process. Distinguishing between these possibilities is proving troublesome. This stumbling block has led to a search for alternative models. Consequently, species belonging to genera such as *Caulobacter* and *Rhodomicrobium,* which have an obligate differentiation sequence within their vegetative cell cycle, as opposed to the occasional differential event in spore formers, have been considered.

10 *Caulobacter* species — obligate dimorphism

Bacillus endospore formation is an example of cell-cycle-independent differentiation. *Caulobacter* species, on the other hand, undergo cellular differentiation as an *obligate part of the vegetative cell cycle.* The *Caulobacter* cell cycle (Figs. 10.30 and 10.31) leads to formation of two distinct cell types, a flagellated, non-stalked 'daughter' or swarm cell, and a stalked, non-flagellated 'mother' cell. This cycle is characterised by a well-defined temporal sequence of morphogenetic events which can be linked with distinct molecular events and control systems (Fig. 10.32).

Properties which make this bacterium an attractive model system for the study of cellular morphogenesis and differentiation are:

a A simple, well-defined morphological cell cycle which can be studied in synchronised populations.

b The cells grow on defined media, thereby permitting the correlation of biochemical events with morphological expression.

c The presence of genetic exchange systems (transduction and conjugation) which facilitate the characterisation of morphogenetic mutants.

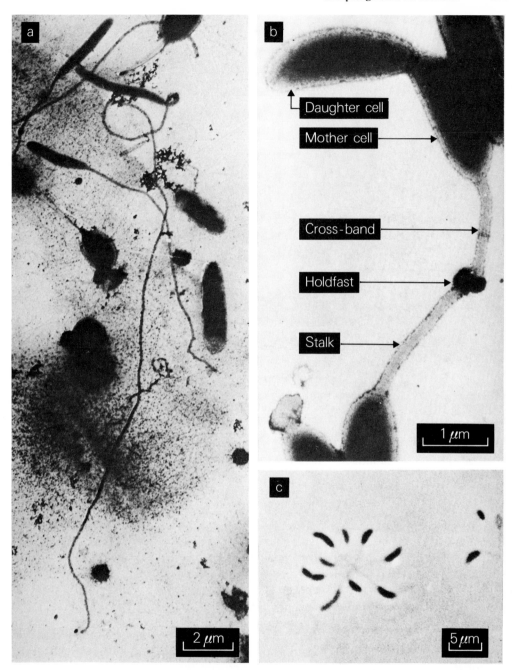

Fig. 10.30 (a) Gold/palladium electron micrograph of an oligotrophic lake sample showing a *Caulobacter* species with extensive stalk formation. (b) Uranyl acetate negative stain of *Caulobacter* showing asymmetric division, holdfast and a cross-band within the stalk. (c) Rosette formation.

As with most morphogenetic experimental models, *Caulobacter* is being exploited in studies concerned with the central problems of differentiation, e.g. the timing of gene expression

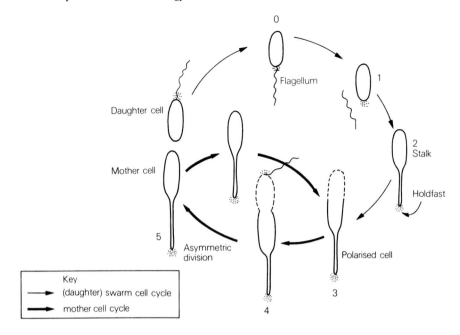

Fig. 10.31 Diagrammatic representation of the *Caulobacter* cell cycle. Temporal sequence of morphological events:

	0	Motile, immature swarm cell.
Maturation	1	Loss of motility, flagella shedding.
differentiation	2	Stalk synthesis.
	3	Initiation of 'daughter' cell synthesis.
	4	Formation of the flagella and holdfast at the pole opposite the stalk.
	5	Asymmetric division.

On release the swarm cell follows the above cycle while the stalked 'mother' cell immediately initiates 'daughter' cell formation, i.e. there is no requirement for a second maturation phase. It has also been reported that the 'mother' cell synthesises a cross-band after each cycle of replication, cross-band formation, acting therefore as a generation marker.

(sequential transcription), the regulation of such gene activity, and the spatial organisation of gene products within the cell. This organism, along with other polarised prokaryotes, e.g. *Rhodopseudomonas palustris* and *Hyphomicrobium* species, possess several unique features both relevant to cellular differentiation and amenable to experimentation.

10.1 Stalk formation and function

The *Caulobacter* stalk (0.2 μm diameter) is an integral cellular extension, i.e. it is enclosed by both the cell membrane and wall. Length of the stalk varies from 1 μm to in excess of 15 μm. Most of the stalk is an inert, non-growing structure with the exception of the growth point situated at the junction with the cell body. Consequently, stalk formation represents temporal- and site-restricted synthesis of cell wall distinct from that occurring at the cell growth points. The cross-bands, which divide the stalk into a series of adjacent compartments, are constructed of concentric rings, probably composed of mucopeptide as they are sensitive to lysozyme. Stalk function is unknown. Stalk length varies depending on environmental conditions. For example, there is a close correlation between phosphate concentration and stalk length, in that low

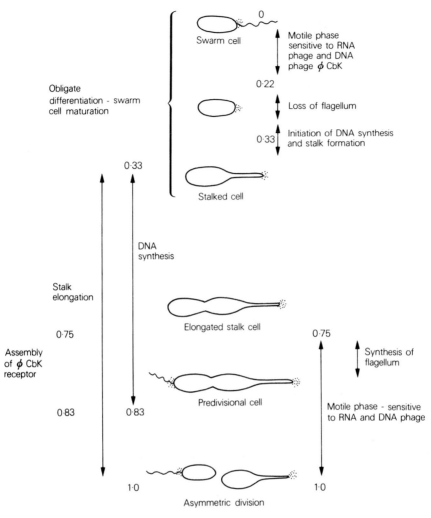

Fig. 10.32 Diagrammatic representation of the *Caulobacter* cell cycle showing the temporally identifiable physiological markers. (As the sequence of events is constant and independent of growth rate they can be considered as fractions of the unit cell cycle.) (After Shapiro, 1976.)

phosphate concentrations (0·1 mM) induce very long stalks while high phosphate concentrations (2 mM) are reflected by very short stalks. This observation has led to the suggestion that the *Caulobacter* stalk is a specialised structure for the uptake and possible concentration of nutrients from low-nutrient environments. (*Caulobacter* species are indigenous to oligotrophic aquatic systems). This hypothesis, although first mooted several years ago, is not supported by experimental evidence. However, stalk synthesis is an obligate part of the cycle in that its formation cannot be repressed, although recently stalkless mutants have been reported.

Stalk regulation can result from any one of three factors: (*a*) changes in nutrient concentration; (*b*) the number of cell-cycles gone through by a particular cell; and (*c*) the availability of cyclic GMP, leading to an increase in stalk length. Present evidence also suggests that the control of stalk initiation is dependent on prior cell division, i.e. it is intimately involved with the sequential events of the cell cycle.

10.2 Swarm cells

The most important feature of *Caulobacter* swarm cells is that they are immature, and must thus differentiate to become reproductive 'mother' cells (obligate cellular differentiation). Two other points are worth emphasing.

(*a*) DNA replication is suppressed initially and begins only when stalk synthesis has been completed. Consequently, initiation of DNA replication in a synchronised swarm cell population will also be synchronous, i.e. there is no requirement to physiologically synchronise the cells, as is the case with *E. coli,* and therefore less likelihood of artefacts being introduced at the molecular level. *Caulobacter* should prove an ideal system for studies on the control of the initiation of DNA synthesis, e.g. membrane attachment, involvement of initiation proteins.

(*b*) Recent studies have revealed that initiation of swarm cell differentiation can be controlled by environmental parameters, i.e. as a batch culture approaches stationary phase and the organisms are subjected to shift down conditions, swarm cell development (not formation) is suppressed. It is tempting to speculate on the presence of a control system in swarm cells which is not activated unless environmental conditions are favourable for growth, which, in turn, suggests that the swarm cell is a dispersal phase and only differentiates when it finds a suitable ecological niche.

11 *Rhodomicrobium vannielii* — summation of microbial diversity

Rhodomicrobium vannielii is a photoheterotrophic budding bacterium of unusual morphology.

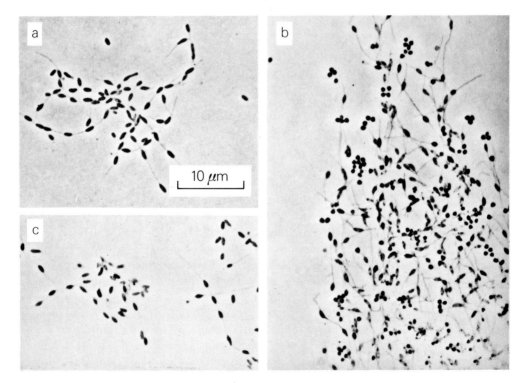

Fig. 10.33 When *Rhodomicrobium* is grown in batch culture (malate salts medium, 30°C, with an incident light intensity of 1 000 lux) the following cell forms are observed: (a) non-filamented motile cells; (b) exospores; (c) cellular matrices.

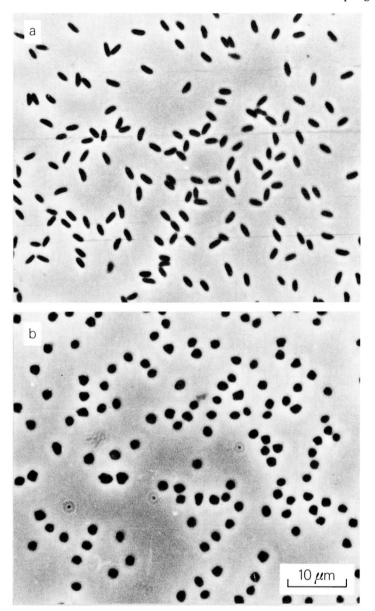

Fig. 10.34 Selective synchronisation of *Rhodomicrobium*. (a) Swarm cells; (b) exospores.

When grown in batch culture, the following cell forms are observed: (*a*) ovoid cells (2 to 3 μm $\times$ 1 μm) linked together by hyphae which may be branched so forming a ramifying group of cells; these we call cellular matrices or microcolonies; (*b*) non-stalked, peritrichously flagellated swarm or 'daughter' cells; and (*c*) non-motile angular cells termed by us exospores (Fig. 10.33). The extreme morphological differences apparent in exponential batch cultures have been exploited in devising a quick and easy method of obtaining homogeneous swarm cell or exospore populations. Simply passing a heterogeneous culture down a glass wool column (Fig. 10.34) allows swarm cells to be selectively synchronised. When such homogeneous swarm

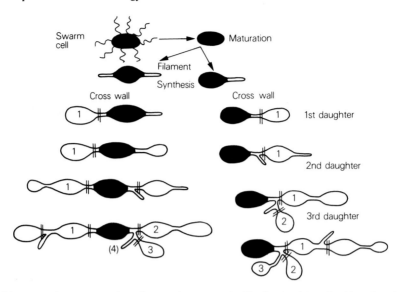

Fig. 10.35 Diagrammatic representation of vegetative growth in *Rhodomicrobium.* In this cell cycle the exact history of each cell may be traced. One mother cell can give rise to a maximum of four daughter cells.

cell populations are monitored, they follow a well-defined and characteristic temporal sequence of morphogenesis in differentiating from an immature, non-reproductive swarm cell to a mature, reproductive, terminally differentiated mother cell (Figs. 10.35 and 10.36) — a sequence identical in principle to that followed by differentiating *Caulobacter* swarm cells.

Synchronised swarm cell populations can be exploited experimentally in the same way as *Caulobacter* swarm cells; the selective synchronisation procedure used with this organism has the advantage that quantities of cells far in excess of those obtained by previous methods — from *Caulobacter* for instance — can be obtained. It is therefore possible to correlate molecular events with specific stages in cellular morphology (Fig. 10.37).

Rhodomicrobium swarm cell differentiation is repressed under conditions of low light intensity (regardless of the nutrient status of the medium), i.e. swarm cells are formed and remain motile but they do not initiate budding filament formation. This differentiation 'delay' would appear to be a fail-safe mechanism in that swarm cells have to move to a favourable environment before differentiation continues to the mother cell state. It now seems likely that *Caulobacter* and *Hyphomicrobium* swarm cells are also subject to the same developmental constraints. Such organisms, therefore, may provide an experimental opportunity to unravel the mechanism of control (at the molecular level) of cellular morphogenesis in response to environmental stimuli, obviously a very important ecological and developmental question.

An important asset of *R. vannielii* is that it forms exospores. Therefore, the potential exists to study sporulation and germination events in a parallel manner to that employed with *Bacillus* species. In contrast to the endospore events, exospore formation (Fig. 10.38) is extremely difficult to follow because of the heterogeneity (cell types) of the culture and the lack of a synchronising procedure. However, germination (a complex affair) can be followed relatively easily in a homogeneous population (obtained by column filtration) (Fig. 10.39).

Rhodomicrobium vannielii, therefore (in contrast to other organisms studied so far), offers three systems for the study of cellular morphogenesis and differentiation: (*a*) swarm cell formation; (*b*) swarm cell differentiation (maturation); and (*c*) exospore formation and germination. In addition there can occur a phenotypic change in the vegetative cell cycle in response

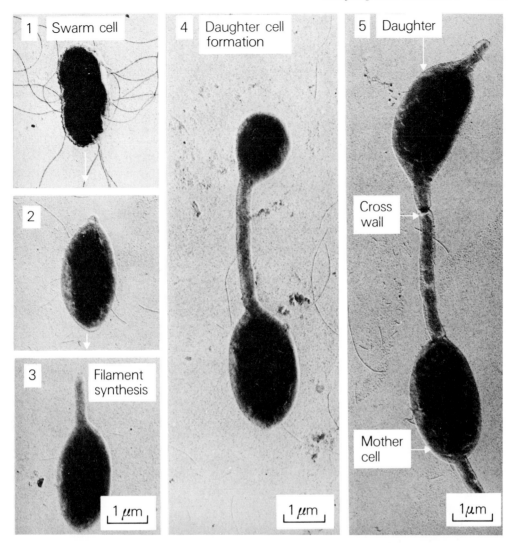

Fig. 10.36 Sequential morphological events occurring during swarm cell maturation. (Gold/palladium shadowed electron micrographs.)

to growth conditions. Under high exogenous CO_2 concentrations, cellular matrices cease to be formed but reappear and dominate on reduction of the CO_2 levels. The point of importance here is that *R. vannielii,* in the vegetative phase, can be maintained in either one of two distinctive cell cycles, dependent upon environmental conditions.

Rhodomicrobium vannielii is therefore an organism of considerable complexity and diversity, both with respect to cellular expression and cellular cycles. It is, nonetheless, very amenable to experimentation and analysis as regards events concerned with cellular morphogenesis and differentiation.

12 Filamentous cyanobacteria — (multicelled prokaryotes?)

The filamentous cyanobacteria (e.g. *Anabaena*) form a variety of differentiated cells and exhibit

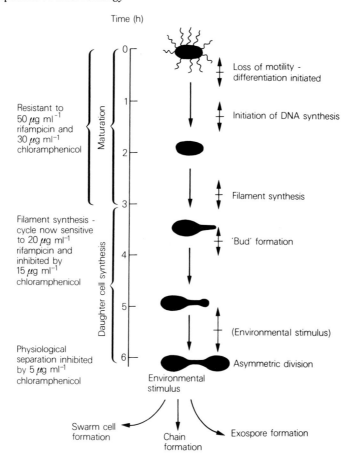

Time (h)

Resistant to
50 μg ml⁻¹
rifampicin and
30 μg ml⁻¹
chloramphenicol

Maturation

Filament synthesis -
cycle now sensitive
to 20 μg ml⁻¹
rifampicin and
inhibited by
15 μg ml⁻¹
chloramphenicol

Daughter cell synthesis

Physiological
separation inhibited
by 5 μg ml⁻¹
chloramphenicol

Loss of motility -
differentiation initiated

Initiation of DNA synthesis

Filament synthesis

'Bud' formation

(Environmental stimulus)

Asymmetric division

Environmental
stimulus

Swarm cell
formation

Chain
formation

Exospore formation

Fig. 10.37 Morphological and physiological events so far identified in the *Rhodomicrobium* cell cycle.

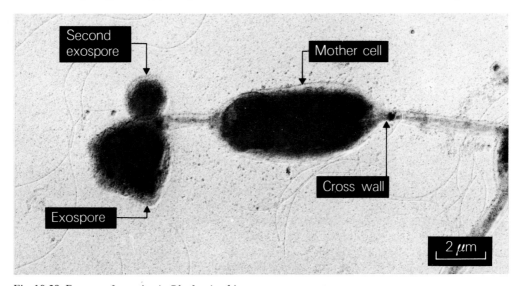

Fig. 10.38 Exospore formation in *Rhodomicrobium*.

Fig. 10.38 (*continued*)

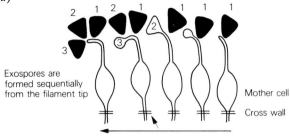

Exospores are
formed sequentially
from the filament tip

Mother cell

Cross wall

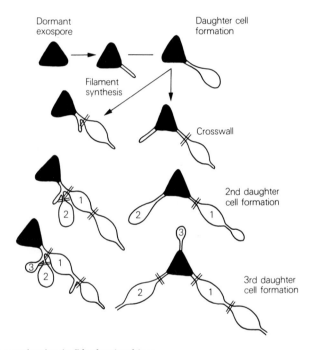

Dormant
exospore

Daughter cell
formation

Filament
synthesis

Crosswall

2nd daughter
cell formation

3rd daughter
cell formation

Fig. 10.39 Exospore germination in *Rhodomicrobium*.

life-cycles of a complexity without parallel in other bacterial systems. As argued at the beginning of this chapter, these characteristics, including the ability of component cells to communicate intercellularly, clearly point to the conclusion that filamentous cyanobacteria are multicelled prokaryotes, a concept usually reserved for eukaryotes.

An *Anabaena* filament can be composed of several cell types:

a Vegetative cells, possessing a complete photosynthetic system, which divide asymmetrically (Fig. 10.40), yielding a large and small cell, the latter being 'immature' (it has a longer generation time than the large cell).

b Heterocysts, formed when fixed nitrogen is limiting, are the primary sites of N_2 fixation and have lost both CO_2-fixing capability and photosystem II. This is the first clear example in prokaryotes of differentiation to a specialised or slave biochemical role to serve neighbouring cells; the ability to reproduce is lost.

c   Akinetes, resting cells (cysts) formed under certain environmental conditions, are often sited close to heterocysts. They contain considerable quantities of cyanophycin granules (a polypeptide containing only aspartic acid and arginine) which may account for their juxtaposition. A large nitrogen store, furnished indirectly by the adjacent heterocysts, is essential for akinete germination in an environment which lacks readily available fixed nitrogen.

d   End cells, which are the pointed cells found at the filament ends. Although ultrastructurally distinguishable from 'normal' vegetative cells, the precise function of end cells remains unknown.

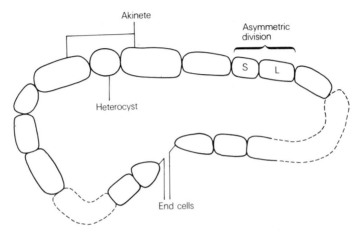

Fig. 10.40 *Anabaena* filament.

A composite filament containing these different types of cells is termed a trichome, which clearly possesses many advantages over a single cell of that species, were such a cell to exist as a separate entity. Such advantages are:

a   The heterocysts' sole function seems to be the fixation of atmospheric nitrogen *under aerobic conditions*. This activity is of benefit to the trichome, but not directly to the heterocyst, which is supported naturally by adjacent cells. This division of labour means that growth and reproduction of the filament can be maintained in the absence of fixed nitrogen under aerobic conditions. This would not be possible if the organism existed as a single cell.

b   Akinete formation, ensuring survival in adverse environmental conditions, has become dependent upon heterocyst activity and would not be possible in a single-celled form of existence in fixed nitrogen deficient conditions.

A major interest in *Anabaena,* from a differentiation point of view, has been pattern formation in relation to the regularity of heterocyst spacing along a filament. A summary of the pattern model proposed is as follows. Mature heterocysts produce a substance which inhibits heterocyst formation by adjacent cells, i.e. there is a zone of inhibition around each heterocyst. To establish an inhibition gradient, the vegetative cells assimilate, dissipate or destroy the inhibitor. This, in turn, has led to the suggestion that the inhibitor may be a nitrogenous compound. Below a certain inhibitor threshold value a vegetative cell (always a 'small' cell) may undergo differentiation to a heterocyst.

Within the filamentous cyanobacteria, therefore, exist further examples of cellular morphogenesis and differentiation but, more importantly, there exists the additional sophistication of multicellular cooperation in a form not evident in other bacteria.

As with *R. vannielii*, the cyanobacteria also offer excellent model systems for the study of the 'molecular interpretation' of environmental stimuli. In this context, *Chlorogloea fritschii* has been shown to undergo specific and characteristic morphological variations in response to defined cultural conditions (Fig. 10.12). Thus, a model system is available, not only for the study of cellular morphogenesis and differentiation, but also for the study of cellular expression as influenced by the environment.

13 Myxobacteria

The myxobacters, which closely resemble in many aspects of their life-cycle the eukaryotic slime moulds, are a further example of what may be considered to be multicellular prokaryotes, in that they go through a life-cycle involving both cellular and colonial morphogenesis (Fig. 10.41).

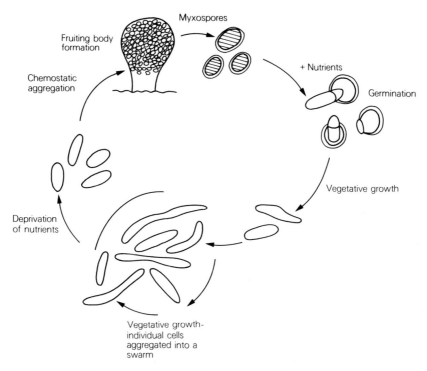

Fig. 10.41 The life cycle of *Myxococcus xanthus*. (After Dworkin, 1973.)

In the vegetative phase, the cells glide over the substrate and remain closely associated within a swarm. This phenomenon is one of the most striking manifestations of the coordinated behaviour of myxobacters and the question as to what restricts or regulates the motility of individual cells is of considerable interest. Present evidence implicates cAMP in this activity. However, there are two particularly intriguing observations concerning the swarm. First, a mutant has been isolated which exhibits motility only when in close proximity to other cells of the clone, i.e. in a swarm all the cells are motile, but when separated these same cells become immotile. This is the first cell interaction mutant ever isolated. Secondly, cells separated from a swarm show no phototactic response, but when reconstituted to form a swarm, they migrate away from high levels of light. Obviously, this model is of considerable importance to the study of coordinated expression. A further example of cellular communication is the assembling of the swarm, in response to a chemotactic stimulus, at a focal point to form a fruiting body. Little has been discovered concerning the developmental processes leading to the formation of fruiting bodies although they are among the most spectacular of microbial structures.

Under conditions of nutrient limitation (e.g. amino acid starvation), myxobacter vegetative cells differentiate to dormant myxospores (equivalent to cysts). This change can also be brought about synchronously by the addition of 0·5 M glycerol. Several other compounds also act as inducers of cyst formation, for instance fully saturated aliphatic compounds of which there is little detectable uptake. This latter observation has prompted the suggestion that the induction signal is mediated by an activation site on the cytoplasmic membrane. In spite of the considerable effort directed at understanding the detailed processes of cyst induction, a satisfactory explanation has yet to be found.

Cysts germinate on complex media. They will also germinate in distilled water containing either a specific germinant (1 to 10 mM phosphate) or when the cell density exceeds 5×10^9 cells ml^{-1}. Throughout the myxobacterial life-cycle this requirement for high cell density is necessary for the successful completion of different phases of intercellular activity.

In summary, the myxobacteria possess several unique characteristics which span the spectrum from cellular morphogenesis and differentiation to cell–cell interactions and the formation of a complex multicellular structure.

References

ANTISSIER, F., JAFFE, A. and KEPES, A. (1971) 'Segregation of galactoside permease, a membrane marker during growth and cell division in *Escherichia coli, Mol. & Gen. Genetics,* **112**, 275–88.

DONACHIE, W. D. and BEGG, K. J. (1970) 'Growth of the bacterial cell', *Nature,* **227**, 1220–4.

DONACHIE, W. D., BEGG, K. J. and VICENTE, M. (1976) 'Cell length, cell growth and cell division', *Nature,* **264**, 328–33.

DWORKIN, M. (1973) 'Cell–cell interactions in the myxobacteria', in *23rd Symposium of the Soc. Gen. Microbiology* (Eds. J. M. Ashworth and J. E. Smith), pp. 125–43. Cambridge University Press, London.

ENSIGN, J. C. and WOLFE, R. S. (1964) 'Nutritional control of morphogenesis in *Arthrobacter crystallopoietes', J. Bacteriol.,* **87**, 924–32.

EVANS, E. H., FOULDS, I. and CARR, N. G. (1976) 'Environmental conditions and morphological variation in the blue–green alga *Chlorogloea fritschii', J. Gen. Microbiol.,* **92**, 147–55.

HANSEN, J. N., SPIEGELMAN, G. and HALVORSON, H. O. (1970) 'Bacterial spore outgrowth: its regulation', *Science,* 1291–8.

HIGGINS, M. L. and SHOCKMAN, G. D. (1971) 'Prokaryotic cell division with respect to wall and membranes', *Crit. Rev. Microbiol.,* **29**, 29–73.

HIRSCH, P. (1974) 'Budding bacteria', *Ann. Rev. Microbiol.,* **28**, 392–444.

ISHIGURO, E. E. and WOLFE, R. S. (1970) 'Control of morphogenesis in *Geodermatophilus:* ultrastructural studies', *J. Bacteriol.,* **104**, 566–80.

KURN, N. and SHAPIRO, L. (1975) 'Regulation of the *Caulobacter* cell cycle', in *Current Topics in Cellular Regulation* (Eds. B. L. Horecker and E. R. Stadtman), Vol 9, pp. 41–62.

KURYLO-BOROWSKA, Z. and SZER, W. (1971) 'Inhibition of bacterial DNA synthesis by edeine', *Biochim. Biophys. Acta,* **287**, 236–45.

LEWIN, B. (1974) 'Gene Expression – 1', in *The Cell Division Cycle,* pp. 550–76. John Wiley, New York.

PARISH, J. H., WEDGWOOD, K. R. and HERRIES, D. G. (1976) 'Morphogenesis in *Myxococcus xanthus* and *Myxococcus virescens'*, *Arch. Mikrobiol.,* **107**, 343–51.

PIGGOT, P. J. and COOTE, J. G. (1976) 'Genetic aspects of bacterial endospore formation', *Bacteriol. Rev.,* **40**, 908–62.

SARKER, N. and PAULUS, H. (1972) 'Function of peptide antibiotics in sporulation', *Nature,* **239**, 228–30.

SCHMIDT, J. M. (1971) 'Prosthecate bacteria', *Ann. Rev. Microbiol.,* **25**, 92–110.

SHAPIRO, L. (1976) 'Differentiation in the *Caulobacter* cell cycle', *Ann. Rev. Microbiol.,* **30**, 377–402.

SHAPIRO, L., AGABIAN-KESHISHIAN, N. and BENDIS, I. (1971) 'Bacterial differentiation', *Science,* **173**, 884–92.

SIMON, R. D. (1976) 'Sporulation in the filamentous cyanobacterium *Anabaena cylindrica'*, *Arch. Mikrobiol.,* **111**, 283–8.

STALEY, J. T. (1974) 'Budding and prosthecate bacteria', in *Handbook of Microbiology* (Eds. A. I. Laskin and H. A. Lechevalier), pp. 29–49. CRC Press Inc.

STARR, M. P. and SKERMAN, V. B. D. (1965) 'Bacterial diversity', *Ann. Rev. Microbiol.,* **19**, 407–54.

WHITTENBURY, R. and DOW, C. S. (1977) 'Morphogenesis and differentiation in *Rhodomicrobium vannielii* and other budding prosthecate bacteria', *Bacteriol. Rev.,* **41**, 754–808.

WILCOX, M., MITCHISON, G. J. and SMITH, R. J. (1973) 'Pattern formation in the blue–green alga, *Anabaena'*, *J. Cell Science,* **13**, 637–49.

ZAVARZIN, G. A. (1961) 'Budding bacteria', *Microbiology USSR* (Eng. trans.), **30**, 774.

11
Membrane fluidity in microorganisms

D. J. ELLAR

Department of Biochemistry, University of Cambridge

1 Introduction

In many respects the relative structural simplicity of prokaryotes is an advantage to the investigator. For the study of membrane structure and function, it is certainly true that the lysozyme-sensitive Gram-positive bacteria have allowed us to isolate and purify the plasma membrane easily and rapidly. As a result, a body of reliable data has been accumulated describing the membrane biochemistry of these organisms. The Mycoplasmas, which lack any form of cell wall outside their plasma membrane, are another prokaryotic form from which homogenous membrane preparations can be obtained quickly and with only minimal manipulation. Gram-negative bacteria have proved more refractory as a source of plasma membrane. This is a result of the greater complexity of their extracellular envelope, which comprises not only a layer of peptidoglycan with an associated lipoprotein layer, but also an additional outer membrane containing lipid, proteins and polysaccharide. (For comprehensive information on prokaryotic envelopes see Meadow 1974, Costerton *et al.* 1974, Braun and Hantke 1974, and Bayer 1975.) Despite these difficulties, techniques have now been developed which permit the separation of Gram-negative envelopes from several organisms into acceptably homogeneous preparations of plasma and outer membranes. These techniques, which essentially involve density gradient centrifugation, exploit the observation that the carbohydrate content and hence the density of outer membrane is significantly different from that of the plasma membrane. Localisation of the characteristic Gram-negative lipopolysaccharide in the outer membrane is the basis for this difference.

The advantages of a rapid and relatively simple isolation procedure for prokaryotic plasma membranes have to be set against the complexity of the final product. Because of the paucity of internal membranes in these organisms, the variety of functions distributed among the specialised organelles of eukaryotic cells is of necessity generally concentrated in a single membrane in bacteria. Among these functions are the synthesis of peptidoglycan, teichoic acids and other cell-envelope constituents; electron transport and oxidative phosphorylation; transport of ions and nutrients; synthesis of phospholipids; DNA attachment and control of chemotaxis. This multiplicity of membrane functions appears daunting when compared to the limited responsibilities of the erythrocyte membrane. Nevertheless, as will be seen below, bacteria and mycoplasmas offer the investigator unique opportunities for genetic and nutritional manipulation of membrane synthesis and function which have proved invaluable for studying the activity of membrane enzymes and probing the mechanisms of membrane assembly.

The current literature contains some excellent general and specialised reviews dealing with many aspects of prokaryote membrane structure and function. A number of these reviews are grouped together in the reference list at the end of this chapter and cover such topics as membrane structure, membrane isolation and analysis, membrane lipids, genetics of membrane components, membranes of photosynthetic bacteria, membrane variability and membrane biosynthesis. It should be emphasised here that the attractive features of bacteria to membrane biochemists were only lately recognised in the history of membrane research, and have in no way inhibited the growth of eukaryote membrane research. On the contrary, pioneering investigations of such eukaryote membranes as myelin, retinal rods and mitochondria, with their own unique features, form the basis for much of our current view of membranes in general as this chapter will show. For this reason, and because of the similarities in membrane architecture between the two cell types (see section 3), any useful list of references must include selected reviews on the eukaryotic cell membrane. The experimental study of artificial lipid bilayers is another major contributor to our knowledge of membranes in general and an introduction to the literature on this topic is also essential.

Against this background information, it is the purpose of this chapter to explore the impact of new knowledge about the physical and chemical nature of membranes on our concept of membrane biosynthesis and differentiation. In particular, to consider data which reveal the fluid-dynamic state of membrane lipids and to examine the implications of this data for the assembly and function of membrane proteins. Some questions which might usefully be asked along these lines are as follows:

1 Is the current working model of membrane structure (fluid mosaic model) compatible with our knowledge of plasma membranes from a variety of bacteria?

2 Does this current model of membrane structure accommodate the changes in bacterial membrane composition which can occur in response to environmental changes?

3 To what extent is the prokaryotic plasma membrane asymmetric; What are the implications of membrane asymmetry for bacterial membrane biosynthesis?

4 To what extent do bacterial membrane components show lateral or rotational mobility?

5 What is known about the biosynthesis of bacterial membrane components and the mechanisms of membrane assembly? How is our understanding of these processes influenced by considerations of membrane fluidity?

6 How do we reconcile the existence of membrane phospholipids as a viscous fluid with the maintenance of fixed sites on the bacterial surface?

These and other questions form the framework of this chapter. It will be apparent that the choice of questions is both idiosyncratic and restricted, influenced in part by the growing literature on the implications of membrane fluidity for the growth and function of eukaryotic cells. For the sins of omission which result from this approach the author apologises in advance and directs the reader to the list of references.

2 Membrane isolation and analysis

In attempting to expand the theme of this chapter in the space available, it is not possible to enter into a detailed description of methods for the isolation and analysis of membranes or to deal comprehensively with the chemistry of lipids from diverse microorganisms. Fortunately these topics have recently been discussed thoroughly in reviews. Certain cautionary notes about membrane purification will be repeated here because of their special relevance to considerations of membrane fluidity and variability. Again, it will be necessary in the sections which follow to discuss in some detail the properties of membrane lipids and selected membrane proteins in order to explore the relationship between membrane composition and membrane fluidity.

With some notable exceptions, prokaryote membranes are chemically and structurally similar to those of higher organisms. As expected from its multifunctional role, the protein-to-lipid ratio of the prokaryote membrane is generally greater than that of most animal cell plasma membranes and more resembles the 3:1 ratio found for mitochondrial inner membranes or the 2:1 ratio found in sarcoplasmic reticulum. However, operational pitfalls in the isolation and handling of membranes make for caution even in this simple comparison. Recent studies have alerted us to the insidious effects of proteolytic and lipolytic enzymes during membrane purification and the need for care in the selection of washing buffers. In the former case, lytic enzymes, which may be endogenous membrane constituents or be adsorbed from the cytoplasm upon cell disruption, can have dramatic effects on the composition and structure of the final purified product if steps are not taken to minimise their effect. Precautions which should be routinely employed wherever possible include the addition of protease inhibitors such as PMSF (phenylmethylsulphonyl fluoride) and EDTA (ethylenediaminetetraacetic acid) and the maintenance of low temperatures (4°C) throughout purification. Without these safeguards, proteolytic activity persisting during membrane isolation can cause further problems in the analysis of membrane polypeptides by SDS-polyacrylamide gel electrophoresis. In this technique and associated methods such as crossed immunoelectrophoresis (see Weber and Osborn 1975, and Salton and Owen 1976) membrane polypeptides are wholly or partly denatured by exposure to detergents and reducing agents. Denatured in this way, proteins often become more susceptible to proteolysis. In the case of membrane lipids, cell disruption may well trigger the onset of turnover brought about by membrane bound phospholipases. During the lysozyme-induced formation of protoplasts from a variety of Gram-positive bacteria, membrane phospholipids were found to be degraded by endogenous lipase with the liberation of vesicles enriched in diglyceride (Kusaka 1975).

The choice of conditions for washing membranes following cell breakage (type of buffer, ionic strength, pH, inclusion of cations, etc.) is the second major way in which the nature of the purified product may be found to vary. This derives from the fact that integration of a significant proportion of proteins into the membrane matrix is achieved through relatively weak interactions such as ionic bonding, hydrogen bonding and London–Van der Waals forces. As a result, these membrane proteins which have been designated extrinsic or peripheral proteins (see section 3) may be dissociated from the membrane if it is exposed to conditions capable of disrupting these weak bonds. Manipulation of the ionic strength and the pH of washing buffers, addition of divalent cations or their removal with chelating agents and mechanical agitation – are all factors which may deplete the membrane complement of extrinsic proteins. (An extended discussion of these effects can be found in Steck and Fox 1972, Steck 1972 and Salton 1971.) In practice the controlled use of these factors has been exploited as a means of selectively detaching and purifying a number of these proteins. Nevertheless, in experiments designed to investigate questions of membrane topography and asymmetry it is as well to consider the possibility that the choice of certain purification conditions may lead to artefacts by rearrangement of protein–protein or protein–lipid interactions.

The dilemma facing the membrane biochemist in his desire to isolate the membrane in as 'native' a state as possible is admirably illustrated in the case of the chelating agent EDTA. We have already seen that the presence of this compound is potentially valuable during membrane purification as an inhibitor of metal-requiring proteases, but on the other hand EDTA is extremely effective in stripping extrinsic proteins from membranes, through its ability to sequester metal ions and thereby disrupt weak ionic interactions. Even the ostensibly wise precaution of maintaining a temperature of 4°C during membrane isolation may lead to perturbation or rigidification of the native membrane structure as a result of irreversible temperature-dependent effects. In view of a recent report (Higgins *et al.* 1976) it may be that even the most carefully designed isolation procedures are inadequate to prevent serious membrane perturbation in some instances. In their pursuit of the origins of the bacterial mesosome, these workers have observed that the typical tubular/vesicular bag-like invagination seen in the cytoplasm is an artefact generated by fixation or any other manipulation which interrupts normal growth of the organism such as centrifugation or filtration. Clearly the mesosomal material represents a special population of membrane components, but its true location and function *in vivo* remains obscure. The suggestion by Salton and Owen (1976) that mesosomes originate from vesicularisation of the outer half of the membrane bilayer could conceivably be tested by the use of non-permeant chemical labelling as discussed in the next section and in the review by Carraway (1975). The above considerations serve to alert us to the probability that the purified membrane is no longer native and can only be defined in terms of the operations employed in its preparation. Where the experimental objective is the purification of a single membrane component, as distinct from comprehensive analysis of the native membrane as a whole, these considerations are less important. While recognising the experimental difficulties, it is possible to obtain acceptable data on the chemical composition of microbial membranes and this can be found in the cited reviews.

Since much of this chapter explores the implications of the properties of membrane lipids, a brief outline of their chemistry is included here.

In general terms, the types of phospholipid found in microbial membranes are similar to those of higher organisms and include phosphatidic acid, phosphatidyl glycerol, cardiolipin (diphosphatidyl glycerol), phosphatidyl ethanolamine, phosphatidyl serine and phosphatidyl inositol. The structures of some of these phospholipids are shown in Fig. 11.1. One exception concerns phosphatidyl choline which is a major component of plant and animal membranes but is less common in bacteria. In Gram-positive bacteria and mycoplasma a number of aminoacyl derivatives of phosphatidyl glycerol have been identified. The amino acids, which are commonly found esterified to the glycerol moiety in these lipids, include lysine, alanine, ornithine and glycine. Another related phospholipid which has been identified in the membrane of both a Gram-positive and Gram-negative bacterium is glucosaminyl phosphatidyl glycerol (Fig. 11.1). Because of the wide species-diversity of prokaryotes, it is not surprising that this general picture does not apply to all organisms. Thus sphingolipids appear to be characteristic of *Bacteroides* and *N*-methylated derivatives of phosphatidyl ethanolamine are commonly found in *Thiobacilli, Clostridia* and *Rhizobia.*

Glycolipids and glycophospholipids are receiving increased attention as components of bacterial membranes, where they are confined mainly to Gram-positive forms (see Shaw 1975). The two major types of glycolipid are the glycosyl diglycerides and the acylated sugars (Fig. 11.2). The most common glycosyl diglycerides are those in which a disaccharide moiety is linked to the 3-position of glycerol, but derivatives containing one, three, four and five sugar residues have been identified. The carbohydrate components of these lipids include mannose, glucose, galactose, rhamnose, mannoheptose and glucuronic acid. In the acylated sugars which also occur in Gram-negative bacteria, no glycerol is present and the acyl residues are linked directly to the carbohydrate. The glycophospholipids include the example of glucosaminyl phosphatidyl glycerol which was mentioned above. A group of carbohydrate-containing lipids found in bacteria and mycoplasmas which have been designated phosphoglycolipids, include

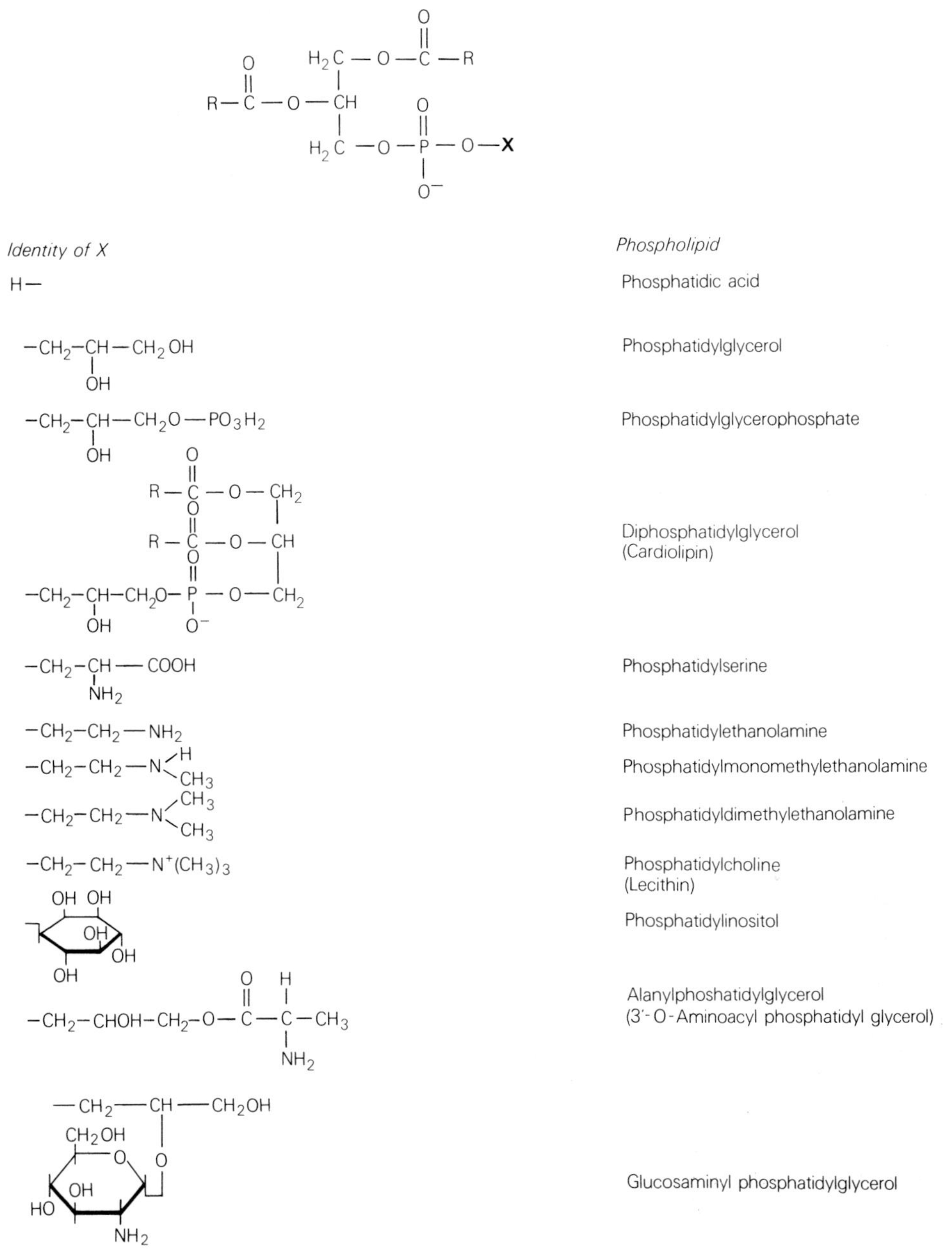

Fig. 11.1 Structures of the major types of microbial phospholipids. The R groups represent fatty acid substituents.

various glycerophosphate derivatives of diglucosyl diglyceride (see Shaw 1975). Teichoic acids are additional important constituents of Gram-positive bacterial membranes (Archibald 1974). Although teichoic acids of differing structure are also found in the Gram-positive cell wall, the teichoic acids found in membranes (lipoteichoic acids) are invariably linear polymers of glycerol phosphate residues with a glycolipid attached to one end through the terminal phosphate.

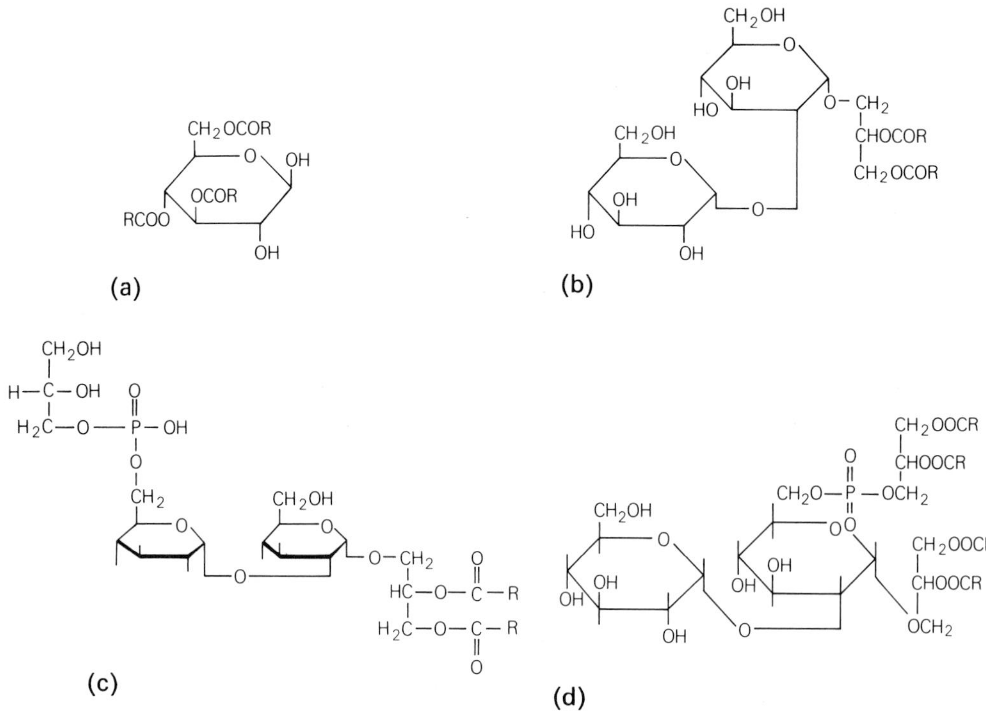

Fig. 11.2 Representative types of glycolipids: (a) acylated glucose; (b) α-diglucosyl diglyceride; (c) glycerylphosphoryl diglucosyl diglyceride; (d) phosphatidylkojibiosyl diglyceride. The R groups represent fatty acid substituents.

The fatty acids of bacterial phospholipids are distinguished by an almost complete absence of polyunsaturated fatty acids and a scarcity of fatty acids with more than 20 carbons in the chain. Distinct differences can be seen between the two major bacterial groups. Thus Gram-positive organisms contain large amounts of iso- and anteiso-branched chain fatty acids together with straight chain saturated forms and only small amounts of the unsaturated fatty acids. In Gram-negative bacteria, branched chain fatty acids are usually absent and their place is taken by saturated and mono-unsaturated fatty acids, together with the characteristic cyclopropane fatty acids (Fig. 11.3). The reader should be aware of certain important exceptions to this over-simplified picture, such as the presence of cyclopropane and unsaturated fatty acids in some Gram-positive bacteria including *Clostridia* and *Lactobacilli*. The 3-hydroxy fatty acids found in Gram-negative bacteria form part of the lipid-A moiety of the outer membrane lipopoly-saccharide and do not occur in the inner membrane phospholipids.

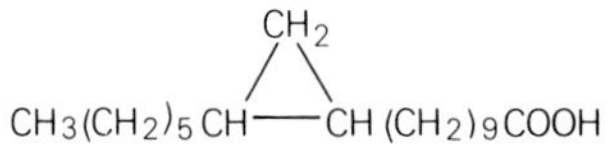

Fig. 11.3 Cyclopropane fatty acid, lactobacillic acid, *cis*-11,12-methylenoctadecanoic acid.

The membrane of obligate halophilic bacteria is another exception to this general pattern. These organisms are being intensively studied because of their unique photophosphorylation mechanism which will be discussed in section 3. Fatty acids are absent from the phospholipids of these organisms and are replaced by long-chain alkyl groups (dihydrophytyl) linked to glycerol by ether rather than ester linkages (Fig. 11.4). The implications of this structure for the structure and properties of halophile membranes will be examined later. Interestingly the stereochemical configuration of the glycerol in these lipids has been shown to be L, contrasting with the D configuration found in the corresponding aminoacyl esters. Like many other microorganisms the halophile membrane typically contains carotenoid pigments, but these organisms are also found to contain significant amounts of squalene and its reduced derivatives.

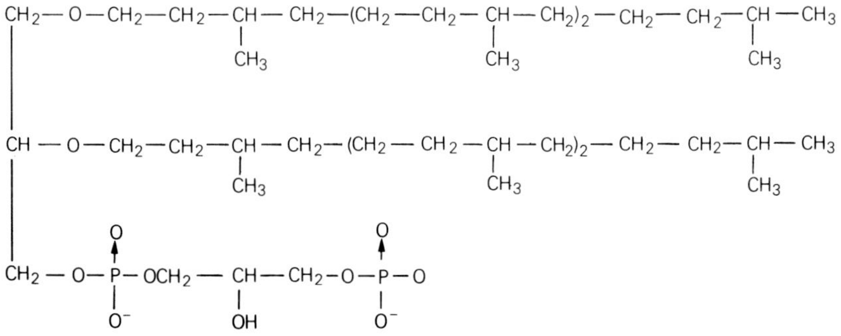

$$CH_2-O-CH_2-CH_2-CH-CH_2-(CH_2-CH_2-CH-CH_2)_2-CH_2-CH_2-CH-CH_3$$

with methyl branches (CH_3).

$$CH-O-CH_2-CH_2-CH-CH_2-(CH_2-CH_2-CH-CH_2)_2-CH_2-CH_2-CH-CH_3$$

$$CH_2-O-\overset{O}{\underset{O^-}{P}}-OCH_2-\underset{OH}{CH}-CH_2-O-\overset{O}{\underset{O^-}{P}}-O$$

Fig. 11.4 The dihydrophytyl ether form of phosphatidylglycerophosphate.

Sulfolobus acidocaldarius, an extremely thermophilic acidophile, has also recently been found to have ether-linked rather than ester-linked lipids in its membrane. In this case the lipids are based on a cyclic glycerol diether and a bifunctional saturated C_{40} isoprenoid residue. A number of bacterial species, particularly those colonising the rumen, have been found to contain plasmalogens. These are phospholipids with alkyl groups linked to glycerol by a vinyl ether bond (Fig. 11.5).

Although it now appears that growing bacteria may incorporate small amounts of exogenous sterol into their membrane (Razin 1975), steroids are not normally found in bacterial membranes. In this respect mycoplasmas are exceptional among prokaryotes. While these organisms are unable to synthesise cholesterol, many of them require its presence in the medium for normal growth and will incorporate significant amounts into their membranes.

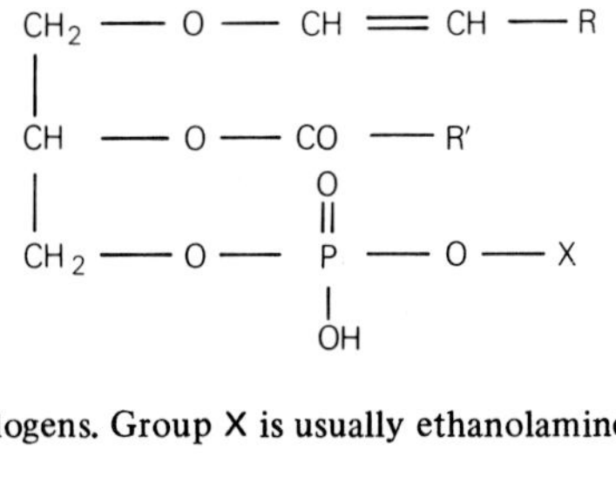

$$CH_2-O-CH=CH-R$$
$$CH-O-CO-R'$$
$$CH_2-O-\overset{O}{\underset{OH}{P}}-O-X$$

Fig. 11.5 The structure of plasmalogens. Group X is usually ethanolamine or choline, but serine is also found.

3 Membranes as an asymmetric fluid mosaic

The major features of the currently popular fluid mosaic model of membrane structure as described by Singer and Nicolson (1972) are illustrated in Fig. 11.6. Much of the evidence in

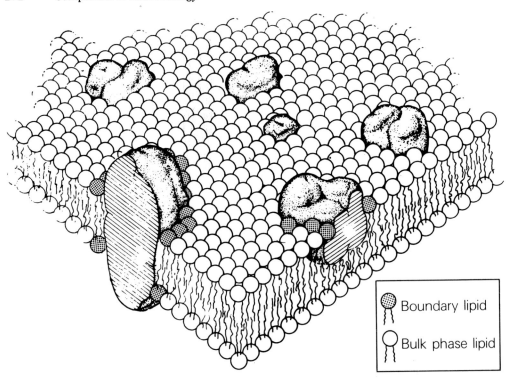

Fig. 11.6 Diagrammatic representation of membrane structure envisaged by the fluid mosaic model, as described by Singer and Nicholson (1972), copyright 1972 by the American Association of the Advancement of Science.

support of this model is drawn from studies of eukaryotes but consideration will be given here to its applicability to microorganisms. Detailed discussion of the evolution of the model can be found in the reference list. In examining the model, evidence will first be presented that bacterial membranes possess a fluid component, namely a lipid bilayer matrix and second that the disposition of proteins and lipids in the membrane is compatible with the ideas implied by the term 'mosaic'.

The fluid mosaic model envisages the membrane as an oriented two-dimensional structure in which proteins are inserted in a fluid lipid matrix. In this context the term 'fluid' is used to denote the fact that under certain conditions the acyl chains of membrane phospholipids are free to flex and rotate and that phospholipid molecules within each half of the bilayer are themselves able to undergo rapid lateral motion. That lipids may be fluid or mobile in this way and yet organised into a stable bilayer surrounding the cell, is entirely feasible on thermo-dynamic and chemical grounds. Pure phospholipids extracted from membranes and dispersed in saline, spontaneously assemble into stable bilayers with their polar head groups in contact with the aqueous phase. Since the provision of a barrier to diffusion is a major requirement for almost all natural membranes, such a stable bilayer is clearly of prime importance to the cell.

The extent of this lipid fluidity or molecular disorder is not surprisingly temperature-dependent. When a phospholipid with a true melting point above 200°C is heated from the solid state, it undergoes a reversible endothermic transition at a temperature well below the melting point. At this *transition temperature* there is a change in state of the phospholipids

Below transition temperature

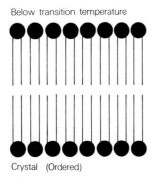

Crystal (Ordered)

Above transition temperature

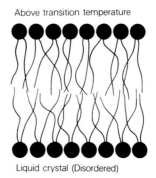

Liquid crystal (Disordered)

Fig. 11.7 Conformation of phospholipid fatty acyl chains in a lipid bilayer.

from the highly ordered crystalline structure to a structure which is intermediate between this and the highly disordered liquid phospholipid. This intermediate state is appropriately termed the liquid crystal form. Below the transition temperature, the lipids are in the solid state with their fatty acids fully extended and oriented at right-angles to the plane of the membrane (Fig. 11.7). In this condition the acyl chains are closely packed in highly ordered hexagonal arrays and the opportunities for molecular motion are thereby severely restricted. As a consequence, bilayers in this ordered or solid state show a minimal cross-sectional area per phospholipid molecule and a maximal thickness. Both these effects will tend to decrease the bilayer permeability. As the temperature is increased, the dynamic freedom of the fatty acyl chains increases until at the transition temperature, they undergo cooperative melting to form the liquid crystal phase (Fig. 11.7). This selective melting of the acyl portions of the lipids is not accompanied by any major alteration of the overall bilayer structure, so that the bilayer configuration can exist both above and below the transition temperature. However the reduction in close-packing and the increased conformational freedom of the lipids in the liquid crystal state, results in a thinner bilayer with a greater surface area occupied by lipid and a greater relative permeability.

The demonstration of this solid-to-fluid thermal transition in intact microorganisms and suspensions of membranes was strong support for the existence of the fluid component of the fluid mosaic model. A number of different techniques have been used to determine the temperature or range of temperatures at which the transition occurs, including differential calorimetry, X-ray diffraction, fluorescence measurements and spin labelling techniques. The application of these methods to bacteria has recently been reviewed (Cronan and Gelmann 1975). Another approach which has been extensively used is the measurement of the temperature-dependence of membrane transport rates. When the logarithm of the activity of a membrane bound enzyme is plotted against the reciprocal of the absolute temperature (Arrhenius plot) a discontinuity is frequently observed, indicating a phase transition and resulting in a change in the activation energy of the process. The interpretation of results from this technique is often difficult and the method will be examined more closely in section 5. Nevertheless it is clear that the correct degree of lipid fluidity or disorder is essential for the activity of a number of membrane-associated functions in bacteria, such as carbohydrate transport, respiration and cell growth, because these functions are severely inhibited at temperatures below the transition temperature. As will be seen later, the temperature dependence of a membrane-bound function has been explored in detail with the aid of *Escherichia coli* fatty acid auxotrophs.

For any species of pure phospholipid the temperature at which the phase transition occurs will be markedly influenced by its constituent fatty acids. Because the temperature of hydrocarbon chain melting will itself be determined by the interaction between adjacent methylene groups and the distance between neighbouring acyl chains, the geometry of the fatty acids will be a crucial factor. The linear rod-like configuration of straight-chain saturated fatty acids will favour close-packing. On the other hand, the presence of a *cis*-double bond in a mono-unsaturated fatty acid results in maximal disruption of hexagonal close-packing. The effect of these different fatty acids is seen from a comparison of the transition temperature of distearyl phosphatidyl choline (60°C) with that of dioleyl phosphatidyl choline (−22°C). The influence of other types of fatty acid on the solid—liquid crystal transition of membrane lipids will be returned to in sections 4 and 5. The nature of the phospholipid headgroup may also affect the phase transition, as seen from the observation that the transition temperature of dipalmitoyl phosphatidyl choline is 20°C less than that of dipalmitoyl phosphatidyl glycerol.

It is important to note several differences between the lipid phase transitions observed with single pure lipids and those of biological membranes. Firstly, the transition in membranes does not generally occur at a precise temperature, but over a temperature range. This is to be expected with most organisms, because of heterogeneity of the polar groups and fatty acids in the membrane phospholipids. Secondly, the amplitude of the phase transition in membranes is less than that observed with the pure extracted lipids. This latter finding suggests that in the native membrane not all of the lipid is in the bulk phase. Such a result would be expected if interaction between membrane protein and lipid withdraws a proportion of lipid from the bulk phase. Measurements of the phase transition by calorimetry, spin labelling and other techniques suggest that in both prokaryotes and eukaryotes, some 15—25 per cent of the bulk phase membrane lipid may be interacting with protein in this way. In the fluid mosaic model this finding is embodied in the proposal that a proportion of membrane protein penetrates into the lipid bilayer or traverses it completely (integral proteins). As a result of this lipid—protein interaction, the molecular motion of those lipid molecules in the vicinity of these integral proteins would be decreased and the rigidity of the lipid phase slightly increased. By means of spin labelling experiments (Vignais and Devaus 1976, Keith *et al.* 1973), the existence of just such a shell of apparently immobilised 'boundary lipid' has been detected both in natural membranes and in artificial systems in which a single purified membrane protein is inserted into a defined lipid bilayer. In another approach using fluorescence probes, Trauble and Overath (1973) showed that in *E. coli* inner membrane, 80 ± 6 per cent of the membrane lipids are capable of taking part in the phase transition. The remaining lipid was associated with integral proteins which were calculated to occupy 10 per cent of the membrane matrix. Their data suggested that each integral protein was surrounded by about 130 lipid molecules which formed a closely coupled shell of relatively immobile or ordered lipid. They also concluded on the basis of their calculations, that approximately half of the 80 ± 6 per cent lipid in the fluid bilayer form was probably associated with a second class of membrane proteins known as *peripheral* or *extrinsic* proteins. In the fluid mosaic model, these are the proteins which bind to membrane surfaces through relatively weak interactions with lipid polar groups. Trauble and Overath take the view that only one-half of the total surface area of the *E. coli* inner membrane is covered with peripheral proteins.

In this discussion of the importance of lipid—protein interactions, we have drawn attention to the two classes of membrane protein which are accommodated in the fluid mosaic model, viz. integral (intrinsic) and peripheral (extrinsic) proteins. These classes are based upon the experimental observation that the membranes of bacteria and higher organisms contain two major types of protein with the properties shown in Table 11.1. Thus while peripheral proteins are released from membranes by relatively mild techniques, integral proteins can only be

Table 11.1   Empirical properties of peripheral and integral proteins

Peripheral protein	*Integral protein*
Released from membrane by relatively mild treatments (high and low ionic strength washing, metal chelating agents)	Only detached from membrane after treatment with detergents, organic solvents, urea, etc.
Usually solubilised free of lipids	Usually associated with lipids after removal from membrane
Usually dissolve readily in neutral aqueous buffers	Frequently insoluble or aggregated in neutral aqueous buffers.
No general lipid requirement for catalytic activity of the solubilised protein	Lipid frequently stimulates or is essential for activity. Order of addition of lipid, protein, substrate, etc. is often critical in reactivating the enzyme
Examples: (prokaryotes) ATP'ase, NADH-dehydrogenase, malate dehydrogenase	Examples: (prokaryotes) succinic dehydrogenase, cytochrome b, isoprenoid alcohol phosphokinase

solubilised by detergents or other reagents which perturb the basic membrane structure. In terms of understanding membrane architecture, it is the different behaviour of these proteins in water which is probably most significant. After release from the membrane, peripheral proteins behave as typical water-soluble proteins and exhibit no special affinity for lipids. In contrast, integral proteins are often extremely hydrophobic and require the presence of detergent to bring them into solution. Furthermore, upon extraction, integral proteins are frequently found to be associated with firmly bound lipid and/or show a dependence on lipid for activation (Machtiger and Fox 1973). On the basis of these and other properties, the fluid mosaic model specifies that integral proteins are responsible for the interruptions in an otherwise continuous bilayer. They are inserted either into or across the bilayer and their position stabilised by hydrophobic interactions between apolar amino acid side chains in the protein and fatty acyl chains in the phospholipids. Conceivably, polar interactions between the lipid headgroup and protein residues could add to this stabilisation. Those neighbouring lipid molecules which participate in this type of interaction constitute the relatively ordered boundary layer which was described above. Although there is as yet little information on the phospholipid composition of such boundary lipid, it is clear that the model would allow for some specificity in the lipid-binding properties of integral proteins. Because they are embedded within the lipid matrix, integral proteins might be expected to be amphipathic, i.e. with binding sites for both polar and non-polar membrane regions. This expectation has been confirmed for cytochrome b_5 of microsomal membranes with the demonstration of a hydrophobic segment in the protein, 40 amino acids in length, which is thought to be the site of interaction with the lipid bilayer (Spatz and Strittmatter 1971). Cytochrome b_5 reductase appears to be another integral protein which is specialised in this way (Spatz and Strittmatter 1973). In eukaryote cells, chemical and immune labelling has indicated that some integral proteins may span the entire membrane thickness. With these techniques, the protein and lipid which is exposed and reactive on either side of the membrane may be labelled by reaction with reagents which do not penetrate the membrane or do so to a very limited extent (Carraway 1975). In this way a major erythrocyte membrane glycoprotein (glycophorin or MN glycoprotein) has been shown to have exposed regions on both sides of the membrane. Its amino terminal end, bearing all the carbohydrate in the glycoprotein, can be detected at the membrane outer surface, while the amino acids at its hydrophilic carboxyl terminal are exposed on the cytoplasmic side. Between these two regions is an α-helical segment consisting of 23 predominantly hydrophobic amino acids. It is proposed that this hydrophobic segment which would be about 35 Å in length, is the site for interaction between this integral protein and the lipid within the bilayer. Although the division between integral and peripheral proteins is an empirical one and does not necessarily imply any invariant and predictable structural differences, the criteria in Table 11.1 allow us to identify several bacterial membrane proteins as candidates for integral proteins. These include some electron transport proteins, and phospholipid synthesising enzymes, as well as proteins such as isoprenoid alcohol phosphokinase and Enzyme II of the phosphoenol pyruvate phosphotransferase system (see Ch. 12) which demonstrate a requirement for phospholipid.

In considering the overall dynamic freedom of phospholipids in the membrane it is now apparent that there is a fluidity gradient within the lipid molecule. NMR and ESR studies have shown that the average motion about C–C bonds in the fatty acid chain increases from the phospholipid headgroup region towards the terminal methyl group. Consequently the central region of the liquid crystal bilayer is highly disordered, with a microviscosity similar to that of a liquid hydrocarbon such as n-decane. These results indicate that the main permeability barrier in the membrane is located in the relatively tightly packed region near the glycerol backbone.

Yet another important aspect of membrane fluidity which has been revealed through the use of X-ray diffraction, spin labelled lipid probes and proton NMR is the capacity of lipid molecules in the liquid crystal state rapidly to diffuse laterally within each half of the bilayer. Measurements with artificial lipid bilayers, sarcoplasmic reticulum and *E. coli* membranes yield a diffusion constant of about 10^{-8} cm s^{-1} for this lateral motion. In other studies, the reported value may be one or more orders of magnitude removed from this figure and there are indications that the size of the reporter group in the case of spin labelled probes, may influence the result. Lateral mobility of membrane phospholipids greatly influences possible mechanisms of membrane biosynthesis, as will be seen later. If such a rapid rate of diffusion were to occur in *E. coli*, it would allow one phospholipid to travel the length of the organism in seconds. This is obviously an oversimplified view and we need to know something of the constraints and barriers to this rapid diffusion which might be expected in the native membrane. Clearly such factors as temperature, fatty acid geometry and headgroup structure, which affect the dynamic freedom of lipids, will also be influential *in vivo*, as well as additional factors such as cation promoted lipid clustering (see later).

Of the possible gymnastic feats which phospholipids could conceivably perform, one which appears to be either extremely rare or non-existent, is rotation about an axis perpendicular to the long axis of the phospholipid. This rotation would allow the phospholipid to 'flip-flop' from one side of the bilayer to another. The observed infrequency of this rotation is perhaps not surprising since the activation energies involved in inserting the charged headgroups through the hydrophobic region of the bilayer are probably restrictive. It has recently been reported however (Rothman and Lenard 1977) that the rate of transmembrane movement of phosphatidyl ethanolamine in *Bacillus megaterium* is at least five orders of magnitude faster than the 'flip-flop' of the same lipid in a lipid bilayer. The authors propose that this rapid rate of movement is one feature of a special mechanism of membrane assembly in organisms capable of rapid growth which requires the concerted action of membrane proteins to facilitate vectorial lipid transfer. Such a mechanism could overcome the energy barrier associated with the simple 'flip-flop' process.

The accumulated evidence for membrane lipid fluidity raises questions about the likelihood of membrane protein fluidity. This will be discussed later in relation to bacterial membrane biogenesis but it should be noted here that numerous experiments, principally with eukaryote cells, have confirmed that lateral movement of membrane proteins and receptors does occur and moreover that the rate and extent of this movement is influenced by membrane lipid fluidity. From the recent demonstration of lateral mobility of lipopolysaccharide in *Salmonella typhimurium,* this also seems to apply to the Gram-negative outer membrane (Muhlradt and Menzel 1974).

The fact that membrane proteins are asymmetrically disposed in prokaryote and eukaryote plasma membranes has been established by a variety of experiments employing non-permeant substrates, chemical and immune labelling, enzyme digestion and cytochemistry (Carraway 1975, Salton and Owen 1976). In the case of erythrocytes, recent work has shown that membrane lipids may also be asymmetrically distributed, with phosphatidyl choline and sphingomyelin concentrated in the outer half of the bilayer and the amino lipids phosphatidyl serine and phosphatidyl ethanolamine contained in the inner (cytoplasmic) half. For this lipid asymmetry to be maintained, the very slow rate of phospholipid 'flip-flop' mentioned above would be important. Alternatively, if those phospholipids in the outer half of the bilayer are able to flip to the inner half, they must flip back at a faster rate to preserve the asymmetry. The first report of lipid asymmetry in bacterial membranes has now appeared (Rothman and Kennedy 1977). This occurs in the same strain of *B. megaterium* which was seen above to be characterised by an unusually rapid rate of transmembrane phospholipid movement.

The term 'mosaic' in the fluid mosaic model refers to the possibility that membrane components may cluster together forming aggregates within or across the membrane. Clustering may occur with heterogenous lipid mixtures of anionic and neutral phospholipids for example. If a membrane with this lipid composition is exposed to divalent cations, it might be expected that the anionic lipids would show a tendency to aggregate. Preferential association of certain lipid species with particular membrane proteins could also lead to a heterogenous distribution of membrane lipids within the bilayer. Similarly, temperature could affect this lipid clustering, if the membrane lipids were sufficiently heterogenous. As we have seen above, variation in fatty acyl chains and lipid headgroups are major factors in determining lipid mobility at any given temperature. In mixtures of dissimilar lipids such as exist in many membranes, it has been shown that each lipid species may segregate into discrete zones below the transition temperature. As the temperature is raised, each species will melt, so that for a particular lipid mixture at a given temperature, the membrane can contain regions of crystalline homogenous lipid coexisting with regions of homogenous liquid crystalline lipid and regions of heterogenous liquid crystalline lipid mixtures. This phenomenon, in which molecules diffuse laterally to form clusters or mosaics of structurally similar lipids, is termed *phase separation.* It is important to note the difference between phase separations of this type and the solid–liquid crystal *phase transition* described earlier.

Although data from spin labelling studies of bacterial membranes have been interpreted as revealing the existence of these phase separations involving dissimilar lipids, Cronan and Gelman (1975) take the view that large-scale phase separations are unlikely in organisms with a homogenous lipid composition. Additional difficulties centre on the possibility that spin labels preferentially locate in certain membrane regions, or alternatively, perturb the sites at which they locate. The situation is even more complicated if bacterial lipids are arranged asymmetrically in the bilayer. Finally, spin labels may detect changes in lipid–protein rather than lipid–lipid interactions.

A different explanation for lipid clustering was suggested by Lee *et al.* (1974). On the basis of their experiments with bilayers constructed entirely from dioleyllecithin, they proposed that quasicrystalline aggregates of more densely packed lipids may form in these bilayers at temperatures 50°C above the solid–liquid crystalline phase transition. When examined by differential calorimetry, animal cell membranes typically show a solid–liquid crystal transition at or below 0°C and yet spin labelling studies reveal that some alterations are occurring in the lipids at temperatures well above 0°C. Similarly, the calorimetrically determined phase transition of *Acholeplasma laidlawii* grown on oleic acid is −20°C, but once again, spin labels incorporated into the membrane indicate lipid changes occurring at temperatures of 20°C. Against this background Lee *et al.* have therefore proposed that short-lived metastable clusters of adjacent entangled molecules may be a feature of membrane lipid mobility and could provide an explanation for some of the results from spin labelling studies. Whatever its molecular basis, a heterogenous organisation of membrane lipids would offer important advantages to an organism. The existence of mosaics of membrane lipid could provide a diverse matrix for membrane proteins with distinct and different lipid requirements as well as discrete foci for the insertion of membrane components during growth. The recent report by Morriset *et al.* (1975) that in *E. coli* K12 the membrane enzymes NADH oxidase and D-lactate oxidase differ in their sensitivity to the extent of lipid fluidity at a given temperature, may be an illustration of the former.

One of the best demonstrations of the existence of membrane mosaicism is provided by the obligate halophilic bacterium *Halobacterium halobium* (Oesterhelt 1976). Under appropriate growth conditions, the membrane of this organism contains visible patches or aggregates. These patches have been separated from the rest of the membrane and shown to consist of a single protein arranged in regular aggregates. This protein, to which retinal is attached, functions as part of a novel photophosphorylation mechanism; it appears to traverse the membrane thickness and could therefore be classed as an integral protein. About 75 per cent of the weight of the patches is made up of this protein (bacteriorhodopsin) which has a molecular weight of about 26 000 daltons. The rest of the patch consists of lipid with an estimated 10 lipid molecules to every molecule of bacteriorhodopsin. An ingenious application of electron microscopy by Henderson and Unwin (1975) has revealed that within the mosaic, bacteriorhodopsin exists in a two-dimensional hexagonal crystalline lattice with the molecules arranged in threes. Each protein molecule contains seven closely packed α-helical segments oriented roughly perpendicular to the plane of the membrane for most of its width. Lipid in bilayer form is believed to occupy the interstices of the protein lattice. The individual molecules have dimensions of 25 $\times$ 35 $\times$ 45 Å and are arranged with the longest axis traversing the membrane width. This description represents one of the best currently available on the structure and disposition of an integral membrane protein. (A fuller discussion of the synthesis and function of bacteriorhodopsin can be found in Oesterhelt 1976). At the moment it is not known whether the lateral association of bacteriorhodopsin molecules to form these mosaics is the result solely of protein–protein interactions or whether it involves interaction of the protein with membrane lipids. Eukaryotic cells also furnish examples of mosaicism in the arrangement of membrane proteins, as for example, the arrangement of proteins in the acetyl choline receptor at the neuromuscular junction. The reader is referred to the reference list for a discussion of these examples and for an introduction to the recent literature implicating eukaryote microtubules and microfilaments in this question of the lateral mobility of membrane components.

An appreciation of the asymmetry and fluid mosaic pattern of membrane structure can thus provide us with a base from which to re-examine many aspects of prokaryote membrane structure and function. The possibility of phase separations for example should now be considered alongside the observations of Higgins *et al.* (1976) on the appearance of mesosomes. Modification of one or more membrane lipid components induced by temperature, autolysis or alterations in the cell's ion balance could lead to selective loss of components from the matrix and their appearance in the form of mesosome vesicles. The widespread occurrence of glyco-lipids in prokaryote membranes raises questions concerning their location in the bilayer. Salton and Owen (1976) have suggested that mesosomes may arise by vesicularisation of the outer half of the bilayer. If glycolipids were concentrated in this half, as has been suggested for animal cells, this might explain the reports of a high proportion of glycolipid in mesosomal preparations.

4 Membrane variability and the role of fluidity

Variation in membrane composition in response to changing conditions is a feature of both prokaryotes and eukaryotes, but the most extensive modifications occur in the former. Considering the extraordinarily diverse range of environments which microorganisms can colonise, it is not surprising that part of their adaptive strategy involves alterations in membrane composition. At this point it is necessary to distinguish between two types of adaptive mechanism: first, a long-term irreversible evolutionary adaptation allowing microorganisms to exploit and become permanent residents of an extreme environment, and second, a short-term reversible adaptive change by which an organism responds to a sudden alteration in its normal habitat. The obligate halophilic bacteria are examples of the first strategy. The long-term genetic modifications which have equipped these organisms to grow optimally in saturated salt solutions have produced novel phytyl ether lipids, an unusual photophosphorylation mechanism involving bacteriorhodopsin, and enzyme variants requiring high salt concentrations for optimal activity. Obligate thermophilic and psychrophilic microorganisms are two additional groups which have adapted by this route. The second strategy which requires no change in the cell's information content, is seen in all other microorganisms and constitutes a general defence mechanism to compensate for environmental changes, allowing organisms to adjust to sudden shifts in growth temperature, osmolality, nutrient concentration, pH, etc. A dramatic example of this is the observation by Minniken *et al.* (1974) that when the growth of *Pseudomonas diminuta* in a chemostat is limited by phosphate, the membrane phospholipids decrease almost to zero and their place is taken by non-phosphate-containing acidic and neutral glycolipids. The range of environmental insults which microorganisms will tolerate in this way provides unique opportunities for exploring the effect of compositional changes on membrane structure and function. Later in this section we will look in detail at some examples of this approach. The relative ease with which the genetic constitution of prokaryotes can be manipulated offers a third approach to the study of membrane biosynthesis and variability. Mutagenised cultures can yield fatty acid and glycerol auxotrophs as well as strains deficient in specific membrane proteins. Examples of the use of some of these mutants will be given later.

Long before the pioneering work of Gorter, Grendel, Danielli, Davson and Harvey, which in the period from 1925 to 1938 revealed the importance of a lipid bilayer in membrane structure, it had been observed that the fatty acid composition of cell phospholipids varied in response to growth temperature. In what could perhaps be called the 'pig in sheep's clothing' experiment, Henriques and Hansen showed in 1901 that the content of unsaturated fatty acids in the outermost back fat of pigs increased in animals kept at $0°C$ compared to animals maintained at $30-35°C$. The evocative choice of a title for this experiment derives from their control experiment in which a pig kept at $0°C$ but covered with a sheepskin coat, failed to show this increase in unsaturated fatty acids. Since this observation, numerous studies have confirmed that fatty acids in the membranes of microorganisms, insects, fish, plants and animals vary markedly in response to changes in the environmental temperature. As expected, these changes are most pronounced in poikilothermic organisms, but they are also detectable in homeothermic animals during hibernation or when a part of the body is persistently exposed to abnormal temperatures. Thus in the leg of the reindeer, it has been found that the proportion of unsaturated acids increases closer to the hoof. The manner in which fatty acid composition responds to temperature in these organisms strongly reinforces the role of fluidity in membrane structure and function. In this section the discussion will be limited to an examination of the relationship which exists between the growth conditions of microorganisms and their apparent need to preserve the physical state of membrane lipids in a form compatible with a fluid mosaic matrix. To do this, we will look at examples of both long-term adaptation as well as short-term membrane modification in response to fluctuations in the environment.

The importance of fatty acid geometry in determining the phase transition or degree of

Table 11.2   Properties of fatty acids and phospholipids

Fatty acid	Chain length	Position of double bond	Melting point ($^{\circ}$C)	Transition temperature (diacyl phosphatidyl choline derivative) ($^{\circ}$C)
Saturated				
Decanoic	10	—	32	—
Lauric	12	—	44	−2
Myristic	14	—	54	24
Palmitic	16	—	63	42
Stearic	18	—	70	58
Arachidic	20	—	75	—
Behenic	22	—	80	—
Branched				
Methyltetradecanoic	15a	—	23	—
Methylhexadecanoic	17a	—	37	—
Methylpentadecanoic	16i	—	62	—
Methylheptadecanoic	18i	—	69	—
Mono-unsaturated				
Oleic	18	9c	13	−22
Elaidic	18	9t	45	12
Vaccenic	18	11c	13	—
Di-unsaturated				
Linoleic	18	9c,12c	−5	—*
Tri-unsaturated				
α-Linolenic	18	9c,12c,15c	−11	—*
Tetra-unsaturated				
Arachidonic	20	5c,9c,11c,14c	−50	—*
Cyclopropane				
Lactobacillic	28	—	28	—

Anteiso- and iso-branched chain fatty acids are denoted by a and i, respectively; c and t are used to denote *cis* and *trans* configurations of the double bond in unsaturated fatty acids.

*For the di- and poly-unsaturated fatty acids, transition temperatures are likely to be below −50°C (Cronan and Gelmann 1975).

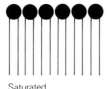

Saturated

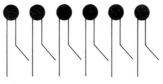

cis -Monounsaturated

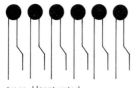

trans -Unsaturated

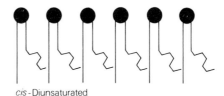

cis -Diunsaturated

Fig. 11.8 Diagrammatic representation of the chain geometry of phospholipids esterified with various fatty acids.

disorder (fluidity) of membrane lipids was seen in the previous section. Table 11.2 lists the melting points of different fatty acids as well as the phase transition temperatures of some of the corresponding diacyl derivatives of phosphatidyl choline. These data re-emphasise the influence of fatty acid geometry on both these parameters. In bilayers composed entirely of straight-chain fatty acids the acyl chains experience no steric hindrance and can therefore pack closely together. These fatty acids have the highest melting point and give bilayers with a relatively high transition temperature. Both the melting point and the transition temperature are directly dependent on the chain length of the saturated acids. The melting point of mono-unsaturated fatty acids is much lower than that of the saturated forms, and this is paralleled by a marked decrease in the transition temperature of the corresponding bilayer. The presence of the single double bond in the mono-unsaturated forms introduces a kink into the molecule (Fig. 11.8) and the resulting restriction on close-packing yields a bilayer which remains disordered (fluid) even at 0°C or below. Introduction of additional double bonds in the di- and poly-unsaturated fatty acids reduces both the melting point and transition temperature still further. Of the two isomers of a mono-unsaturated acid, the *cis* form which is the one generally found in membranes, has the lower melting point. This is reflected in the much lower transition temperature of lipids containing oleic acid compared to the *trans* isomer elaidic acid (Table 11.2). The *trans* double bond causes less interference with close-packing than the *cis* form (Fig. 11.8). Branched chain fatty acids, which are common in the lipids of a variety of microorganisms, are of two types, iso and anteiso, depending on the position of the methyl branch:

$$CH_3-CH-(CH_2)_n-COOH$$
with CH_3 branch on the CH

iso

$$CH_3-CH_2-CH-(CH_2)_n-COOH$$
with CH_3 branch on the CH

anteiso

The fluid properties of the lipid are markedly affected by the position of this branch. Whereas a methyl branch in the iso position causes only minimal interference with close-packing, a methyl group in the anteiso position is much more disordering. Although membrane fatty acid composition is a valuable indicator of the possible state of membrane lipids at a particular temperature, it must be stressed that other factors including the types and heterogeneity of lipid headgroups, protein–lipid interactions and cation–lipid interactions are also important determinants of bilayer fluidity.

In studies to date, mesophilic microorganisms have been found to respond to temperature changes by modifying their fatty acids to produce a membrane whose fluidity is unchanged at the new temperature. This phenomenon has been called homeoviscous adaptation (Sinensky 1974). The types of modification which could be employed by different microorganisms to achieve this constant fluidity are those that might be predicted from the previous discussion of fatty acid geometry. Thus an increase in the fatty acid average chain length, a shift from anteiso to iso branched chain fatty acids or a decrease in the ratio of unsaturated to saturated fatty acids, would all be expected to increase the phase transition temperature. One or more of these effects is indeed found to occur when the growth temperature of a wide range of microorganisms is increased. As expected, decreasing the growth temperature of a mesophile produces changes in the opposite direction. In *E. coli* K12, for example, decreasing growth temperature results in an increase in the proportion of unsaturated fatty acids in the lipids (Table 11.3). In Bacilli and Clostridia which contain principally branched chain fatty acids, the response to temperature may involve either a change in the relative amount of monounsaturated fatty acid or a change in the position of the methyl group in the branched acids. When grown at 20°C, *Bacillus licheniformis* contained 43 per cent by weight of unsaturated acids compared to only 15 per cent at 35°C (Fulco 1970). In another study of a facultatively thermophilic bacillus (Chan *et al.* 1973) the amount of anteiso fatty acids increased from 27·3 per cent of the total at 55°C to 43·9 per cent when the organisms was grown at 37°C. Thus, in summary, a reduction in the growth temperature of mesophilic microorganisms results in an increased membrane content of those fatty acids which produce disorder in the bulk phase

Table 11.3   Fatty acid composition of phospholipids from *E. coli* K12 grown at different temperatures in glucose minimal media (from Haest, De Gier and Van Deenen 1969)

Fatty acid*	*Temperature of growth (°C)*		
	20	*30*	*40*
14:0	1·7	2·3	2·7
16:0	21·0	26·5	36·7
16:1	12·8	6·7	7·3
17:0	8·4	20·0	23·0
18:1	49·0	35·5	15·2
19:0	7·0	9·0	15·1
Total straight-chain saturated acids	22·7	28·8	39·4
Total mono-unsaturated plus cyclopropane fatty acids	77·2	71·2	60·6

*Fatty acids are designated by the number of carbon atoms followed by the number of double bonds, if any.

lipid, with the reverse occurring when the temperature is increased. These effects will be accompanied by corresponding changes in the solid–liquid crystal transition temperature, so that at the new temperature, lipid fluidity remains constant. Table 11.4 illustrates this adaptive response in *E. coli*. While the phase transition temperature of the lipid extracts increases with increasing temperature, the viscosity of the lipids measured at the growth temperature, remains constant. Essentially the same pattern has been observed in *Bacillus stearothermophilus* and *Acholeplasma laidlawii*.

Table 11.4   Phase transition and viscosity of *E. coli* lipid
extracts from cells grown at different temperatures
(from Sinensky 1974)

Temperature of growth ($^\circ$C)	Phase transition temperature ($^\circ$C)	Viscosity* (poise)
15	−1 (± 1)	1·8
30	16 (± 2)	1·9
43	27 (± 1)	2·0

*In each case the coefficient of viscosity was measured at the growth temperature. (A value of 15 poise was obtained when the viscosity measurements were made at 15°C on lipids extracted from cells grown at 43°C).

In addition to confirming the importance of correct lipid fluidity in membrane function, this homeostatic mechanism enables microorganisms to grow over a broader temperature range than would otherwise be possible. From the nature of the fatty acid changes in mesophiles, we might expect that during the course of evolution, those microorganisms which grow obligately at high or low temperatures would have extended the scope of this mechanism to ensure membrane lipids with the correct fluidity in these extreme conditions. Just as it is important to keep in mind that fatty acid composition is not the sole determinant of lipid fluidity, it is equally apparent that membrane fluidity is not necessarily the major factor in determining maximum and minimum growth temperatures. However, examination of a range of thermophilic and psychrophilic microorganisms has shown that by comparison with mesophilic members of the same genus the thermophiles contain a higher proportion of straight-chain saturated and iso branched fatty acids, whereas the psychrophiles possess shorter chain length saturated acids or more unsaturated and anteiso branched chain fatty acids.

While accepting that changes in membrane fatty acids *per se* are not sufficient to convert mesophiles into psychrophiles or thermophiles, it is reasonable to propose that the physical state of membrane lipids is an important factor in determining the growth temperature range of microorganisms in general. This relationship has been explored with several organisms, notably *Acholeplasma laidlawii* and unsaturated fatty acid auxotrophs of *E. coli*. The capacity of *A. laidlawii* for fatty acid synthesis is limited to saturated fatty acids with 12 to 18 carbons. However this organism will incorporate a wide range of saturated and unsaturated fatty acids into its membrane if they are supplied in the growth medium. Thus by altering the fatty acid supplement and the growth temperature, the effect of specific lipid changes on cell growth and membrane function can be studied. Experiments of this type (McElhaney 1976) have shown that under certain circumstances the minimum growth temperature is dependent on the

Table 11.5   Maximum, minimum and optimum growth temperatures and lipid phase transition temperatures of *Acholeplasma laidlawii* B cells grown with different fatty acid supplements (from McElhaney 1976)

Fatty acid added*	Minimum	Optimum	Maximum	Transition midpoint	Transition range
18:0	28	38	44	41	25–55
16:0	22	36	44	38	20–50
None	20	36	44	34	18–45
18:0i	18	36	44	32	18–42
18:1*t*	10	36	44	21	5–32
17:0a	8	36	44	7	0–15†
18:1*c*	8	34	40	−13	−22– −4†
18:2*c*	8	32	38	−19	−30– −10

*Fatty acids are designated by the number of carbon atoms, followed by the number of double bonds. a and i refer to anteiso- and iso-branched fatty acids; *c* and *t* denote *cis* and *trans* isomers, respectively.
†Estimated values.

physical state of membrane lipids. Table 11.5 shows the maximum, minimum and optimum growth temperatures of *A. laidlawii* when grown with different fatty acids and also the mid-point of the transition temperature range. (Because of heterogeneity in the fatty acid composition, the phase transition occurs over a range of temperatures in this organism.) The table reveals that the minimum growth temperature is not determined by the fatty acid composition, provided the membrane is enriched with *cis* unsaturated fatty acids. In this case the transition mid-point is more than 20°C below the minimum growth temperature (8°C). Presumably additional effects such as protein inactivation become growth-limiting in these cells below 8°C. However, when it is grown with fatty acid supplements which yield a higher membrane transition mid-point, the organism displays a new and elevated minimum growth temperature, which appears to be determined by the particular fatty acid supplement used. It is noticeable that changing the membrane fatty acids has very little effect on the optimum or maximum growth temperatures of this organism. Studies in which the fatty acid content of *E. coli* fatty acid auxotrophs was varied have also indicated that cell growth ceases when lipid fluidity is reduced below a certain level (see Cronan and Gelmann 1975). For growth at 37°C, one *E. coli* auxotroph required 15–20 per cent unsaturated fatty acids in its membrane. At 45°C this figure decreased to 11 per cent, rising to 32 per cent for growth at 27°C. Wild type *E. coli* grown at 37°C normally contains 50 per cent unsaturated fatty acid and 50 per cent saturated acid in its membrane and the observed phase transition extends over a range from below 0°C to the growth temperature. For an auxotroph containing 26 per cent unsaturated fatty acid, the phase transition only commences above 35°C. Cronan and Gelmann (1975) have noted that these data indicate that wild type *E. coli* contains considerably more unsaturated fatty acid than is apparently needed for growth at 37°C. They suggest that this excess fluidity might allow the organism to adapt more readily to sudden temperature decreases in its natural habitat. With only the minimum unsaturated fatty acids needed for growth at this temperature (15–20 per cent), *E. coli* would have less scope for adaptation as the temperature decreased. In other experiments, it has been found that an organism can exhibit approximately the same growth rate at a given temperature despite marked fluctuations in its fatty acids, provided that the proportion of ordering or disordering fatty acids does not exceed certain limits. For a number of organisms, growth ceases when the unsaturated (or equivalent) fatty acid content falls below about 15 per cent. A similar value for saturated (or equivalent) fatty acids has been proposed. Homeoviscous adaptation is therefore a mechanism by which organisms attempt to avoid excessive rigidification or fluidisation of membrane lipids as the temperature changes. It will be interesting to see if, in addition to temperature, other environmental factors which can affect the fluidity of membrane lipids, e.g. pressure, can also trigger homeoviscous adaptation. In the absence of this homeostatic mechanism, membrane lipids would presumably become increasingly disordered as the growth temperature increased, resulting in cell leakiness and eventual collapse of the permeability barrier.

In the space available it is not possible to discuss the biochemical basis of the homeoviscous adaptation process, but the reader can find an excellent treatment of this topic in Cronan and Gelmann (1975). In brief, the available evidence indicates that adaptation is achieved by a combination of temperature-dependent changes in the specificity of acyl transferases at the level of phospholipid synthesis, together with alterations in the activity of the fatty acid synthetase system. Again it must be stressed that adaptive changes in membrane lipids are likely to be only one of the factors determining the growth temperature range of a particular microorganism. Since the nature of the phospholipid headgroups and the overall protein-to-lipid ratio are also likely to influence membrane fluidity, the ability to manipulate these factors experimentally can yield additional information on membrane structure and function. This can now be achieved in several ways. We have already noted that the use of phosphate limitation can eliminate phospholipids from the membrane of *P. diminuta* and Shaw (1975) has drawn attention to other possibilities for modifying bacterial glycolipid content. This approach can be supplemented by the use of mutants such as glycerol auxotrophs, which cease net phospholipid synthesis when deprived of glycerol (Mindich 1975), or temperature-sensitive mutants unable to synthesise phosphatidyl ethanolamine owing to phosphatidyl serine decarboxylase deficiency (Hawroot and Kennedy 1975). A range of other mutants is described by Machtiger and Fox (1973).

Before leaving this discussion of membrane variability and fluidity, mention should be made of a possibly anomalous situation in the membrane of the obligate halophilic bacterium *Halobacterium cutirubrum*. This organism, which lacks a peptidoglycan rigid layer and lyses when the sodium chloride concentration is reduced below 3·5 molar, contains the unusual phytyl ether lipids described earlier. Using spin labelled probes, it was found that the lipids in *H. cutirubrum* membranes are extensively immobilised and no temperature-induced phase transition could be detected above $-15°C$. Phase transitions were observed however, in vesicles prepared from extracted lipids (Esser and Lanyi 1973). It was concluded that the presence of membrane proteins somehow resulted in the immobilisation of the lipid phase. The structure of the phytyl side chains with their four branched methyl groups, is thought to facilitate close-packing of the membrane lipids, producing a highly structured bilayer in this organism. If the rotational mobility of these lipids in response to increasing temperature is dominated by chain-bending at the four branch points, then this bilayer should be a highly cooperative structure. For straight-chain fatty acids, as was seen earlier, increased mobility is achieved by the flexing of C–C bonds, which increases towards the centre of the bilayer. In the case of the phytyl lipids, increased motion is likely to occur in successive four carbon segments of the chain by flexing at the four branch points. Because of the highly cooperative nature of this structure, the insertion of hydrophobic membrane protein residues into the lipid matrix could produce a much greater immobilisation of membrane lipids than is the case in bilayers containing typical fatty acids. It would therefore appear that the membrane of this obligate halophile is unusually rigid (Esser and Lanyi 1973).

5 Membrane biosynthesis and the fluid mosaic model

In this last section we will explore the impact of new knowledge about membrane structure on our view of how membranes are assembled *in vivo*. As with the topics of previous sections, no attempt will be made to review all aspects of membrane biosynthesis, but rather to concentrate on how such membrane features as lipid fluidity, lateral and transverse asymmetry and compositional variability are compatible with experimental observations and how they may be accommodated in suggested mechanisms for membrane assembly. Readers will find less specialised reviews of membrane biosynthesis in the reference list.

At the outset then we should identify those membrane properties which are of special interest in this context.

a Membranes are dynamic structures in which lipids and proteins may enjoy considerable freedom of movement in certain directions. Within each half of the bilayer, lipids may rotate about an axis perpendicular to the membrane surface or diffuse laterally at high speed. Spontaneous transfer of phospholipids from one monolayer to the other probably does not occur.

b Membranes show transverse asymmetry with respect to the location of membrane proteins. Most membrane proteins are exposed on the inner (cytoplasmic) surface. Membrane phospholipids are also asymmetrically located in certain eukaryote membranes.

c Membranes can also possess lateral asymmetry. Since proteins and lipids can move laterally in each membrane surface, this allows lipid–lipid, lipid–protein or protein–protein aggregation to occur to produce discrete specialised areas or mosaics. Once established, such mosaics may serve as insertion zones for new membrane components.

d Prokaryote membranes can tolerate considerable variation in their proteins and lipids and yet retain typical membrane functions.

e The correct degree of lipid fluidity is essential for the function of many membrane proteins. This degree of fluidity appears to involve the coexistence of fluid and less fluid areas in many membranes.

f Evidence indicates that some membrane proteins exist in association with lipid molecules whose mobility is less than that of the bulk phase lipid.

g Some, but not all membrane proteins have the capacity to bind lipid and often display a lipid requirement for optimum activity. With this summary of the background information we can now look briefly at the following questions:

1 How are membrane proteins specified?
2 How is asymmetry of lipid and protein established and maintained in the membrane?
3 Where are newly synthesised lipids and proteins inserted into the growing prokaryote membrane?
4 To what extent is membrane synthesis regulated by the cell cycle in prokaryotes?

Inevitably the answer to the first question is that the ultimate location of a membrane protein is genetically determined, since its nearest neighbour preferences are governed by its three-dimensional structure. From the properties of the two extreme types of membrane protein (Table 11.1 and section 3) it is reasonable to assume that peripheral proteins (here restricted to those on the membrane inner surface) are synthesised on soluble ribosomes and subsequently diffuse to the membrane. There they eventually bind directly or through ionic bridges to phospholipid polar groups or associate with residues in other membrane proteins. If these binding sites are randomly distributed over the membrane surface, the peripheral proteins should be similarly arranged. Segregation of peripheral proteins into mosaics is not ruled out however, if they are bound to the membrane matrix through other components which are responsive to lipid fluidity. The special structure of integral membrane proteins must equip them to penetrate into the bilayer, or even through it, in the case of proteins exposed on both membrane surfaces. The large hydrophobic peptides in cytochrome b_5 and cytochrome b_5 reductase (section 3) illustrate this type of positional specificity. These peptide segments would associate with the hydrophobic bilayer to anchor the protein in place. By analogy, with our knowledge of glycophorin (section 3) those proteins which traverse the bilayer might enter the hydrophobic phase in the same way, by initial penetration of hydrophobic or neutral NH_2– terminal segments and then be held in position perhaps by ionic interaction between their COOH–terminal regions and ionic residues on the membrane inner surface. An indication of the molecular basis of such an interaction is given in the recent report of the preferential association of glycophorin with negatively charged phospholipids (Van Zoelen *et al.* 1977). In the sequence of this protein, two arginine and two lysine residues are believed to be sited at the point where the helical region of glycophorin leaves the membrane inner surface. It is suggested that this structural feature would allow a salt bridge to form between polar head groups of anionic phospholipids in the membrane inner half and the protein.

The suggestion that the positioning of integral membrane proteins is the result of synthesis on membrane bound ribosomes, followed by insertion into the membrane of nascent poly-peptides, has been criticised for several reasons (Bretscher 1973). At least in the case of the lactose permease protein, the evidence is that this integral membrane protein is synthesised on soluble ribosomes. The lactose operon is known to be transcribed as a polycistronic message for the three constituent proteins (β-galactosidase, permease, transacetylase). The first and third of these are soluble proteins and since data confirm that the three genes are translated sequentially by the same ribosomes, it would seem that the permease is also synthesised in the cytoplasm. The preference of this member of the trio for a membrane location may again be revealed when its detailed structure is known.

In prokaryotes there are three other classes of protein which probably have a transient existence in the plasma membrane. These are the periplasmic proteins of Gram-negative bacteria, the proteins of the Gram-negative outer membrane and proteins such as exoenzymes which are secreted by the organism. To facilitate exit through the plasma membrane, we might expect some structural specialisation in these proteins. In the case of penicillinase excreted by *Bacillus licheniformis,* the work of Lampen and colleagues (Yamamoto and Lampen 1976) has demonstrated an additional way in which integral membrane proteins may be specialised for their location. While penicillinase is typically excreted during growth of this organism, it also exists as a membrane-bound form with all the characteristics of an integral protein. Treatment of membranes with trypsin released a form of the enzyme which differed from the exopenicil-linase only by the absence of the normal NH_2-terminal lysine, together with a phospholipo-peptide. The partial sequence of this peptide is phosphatidylserine–$(Ser_3, Glx_5, Asx_7, Gly_5)$– Asp–Gln–Ser–Lys–COOH with the lysine being the NH_2–terminal residue of the form of the enzyme which is secreted. It appears therefore that the membrane-bound precursor form possesses an additional hydrophobic phospholipopeptide segment which anchors the enzyme in the membrane and also orients it towards the outer membrane surface in readiness for export. It will be most interesting to see if this strategy of linking protein and lipid covalently is used for specifying and orienting other membrane proteins. A similar mechanism could conceivably guide the export of periplasmic or outer membrane proteins. However, recent experiments with alkaline phosphatase synthesis (Beacham *et al.* 1976), have shown that secretion of this periplasmic protein can occur in the absence of net phospholipid synthesis. Unless a pre-existing pool of phospholipid continues to support secretion under these conditions, a phospholipo-protein intermediate would not appear to be involved in this instance. As an alternative suggestion to explain secretion of this type of protein, the molecule could be transiently anchored by a combination of a neutral NH_2–terminal region, plus salt bridges (as discussed above for glycophorin) at the inner membrane surface. Cleavage of a COOH–terminal peptide containing the ionic residues participating in the salt linkages by an endogenous peptidase, would reduce the charge on the protein still further and might thereby trigger its release to the exterior. Movement in this direction rather than backwards into the cell would be favoured if the neutral NH_2–terminal portion was already partially exposed in the external aqueous environment or alternatively, by the conformational changes which probably occur when the COOH–terminal portion is excised. Retention of the protein within the periplasmic space might then be achieved by aggregation of monomers released in this way into oligomers. In the case of alkaline phosphatase the secreted monomers have been observed to form dimers in the periplasmic space.

This chapter has not dealt in detail with the biosynthesis and assembly of Gram-negative outer membranes, but recent summaries of this rapidly advancing area can be found in reviews by Inouye (1975) and Bayer (1975). Of special interest to the question of membrane protein specialisation is the outer membrane lipoprotein discovered by Braun and his colleagues, which appears to form a link between peptidoglycan and the outer membrane.

The second question concerning the maintenance of protein and lipid asymmetry in a fluid membrane has been answered in part in our consideration of the first question. Genetically programmed protein conformation can specify membrane localisation by determining the extent of lipid solubility and providing loci for lipid-protein and protein–protein interaction. The preponderance of membrane proteins on the membrane inner surface may indicate that most membrane proteins are of the peripheral type. Certainly in prokaryotes, a substantial, if not major proportion of membrane proteins can be detached by procedures such as low ionic strength buffer washing and the use of chelating agents, which do not extract integral proteins. The lipid insolubility of these proteins may restrict them to peripheral interactions with other polypeptides or with lipids in the inner half of the bilayer. In this context, the following observation emphasises the need to examine the role of each polypeptide in an oligomeric protein. Recent work with the ATPase of *E. coli* (Futai *et al.* 1974) has shown that one of the five polypeptides in this protein is involved in binding the catalytic site to the membrane. This illustrates the inadequacy of a simple division into integral and peripheral 'proteins'. In this enzyme we appear to have a bifunctional integral 'polypeptide' with both an affinity for the membrane and an ability to bind one or more of the other ATPase subunits.

An interesting observation which has reinforced the importance of protein structure and lipid fluidity in determining asymmetry, is the case of M13 phage coat protein in *E. coli*. (Wickner 1976). This protein was shown to occur in the host cell plasma membrane with its NH_2–terminus exposed on the outer surface. When purified coat protein was incorporated into lipid bilayer vesicles composed of dilauroyl or dimyristoyl lecithin, its NH_2–terminus was invariably exposed to the exterior, provided that the protein was incorporated into the vesicles at temperatures near the phase transition temperature of the particular lecithin. In contrast, both ends of the coat protein were found to be exposed at the exterior when the incorporation was performed with dimyristoyl lecithin at a temperature $23°C$ below its transisition temperature. This result strongly supports the view that, in this model system at least, protein orientation is determined adequately by protein structure and the extent of lipid fluidity.

In addition to the recent report of phospholipid asymmetry in *B. megaterium* membranes (section 3 page 276), Nikaido *et al.* (1971) have suggested that lipid asymmetry may be a feature of the Gram-negative outer membrane. From the results of experiments using spin labelled fatty acid probes, these authors conclude that distribution of lipopolysaccharide and phospholipids in the outer membrane is asymmetric, with most of the lipopolysaccharide confined to the outer half of the bilayer and the majority of phospholipid restricted to the inner half. The authors point out that segregation of lipopolysaccharide from phospholipid in this way, might place the former in a less fluid environment and thereby account for the observation that the diffusion constant of lipopolysaccharide (3×10^{-13} cm^2 s^{-1}) (section 3; Muhlradt and Menzel, 1974) is four to five orders of magnitude slower than phospholipid diffusion rates (10^{-8} cm^2 $^{-1}$). The existence of this asymmetry raises questions about its origin and main-tenance, including the possible existence of enzymes (flippases) which might catalyse trans-location of molecules from one monolayer to another in the bilayer. However, in the case of lipopolysaccharide there is a body of evidence which suggests that insertion of newly syn-thesised molecules occurs at a limited number of sites (adhesion sites), at which a connection between inner and outer membrane is visible (Bayer 1975). An interesting additional observa-tion in the report by Nikaido *et al.* is the claim that the fatty acid composition of the lipopolysaccharide does not show the typical homeoviscous response to temperature shifts (section 4).

The third and fourth questions which were asked at the beginning of this section concerning sites of biosynthesis and coordination with the cell cycle merit a separate essay to themselves. Here all that is intended is to apprise the reader of the ways in which an awareness of membrane fluidity is influencing the design and interpretation of experiments aimed at answering these questions. In discussing ways in which discrete sites may be maintained in a fluid membrane, the problem becomes one of discovering how lateral mobility of membrane components may be modulated. One important, if obvious point is that membrane proteins would be expected to diffuse laterally at slower rates than membrane lipids. Protein size is clearly an important factor here, since the diffusion rate of a protein and hence its ability to randomise is likely to decrease as the molecule increases in size. The implications of these differences in lipid and protein diffusion rates have been incorporated into an interesting proposal to account for some features of polarity in eukaryote cells (Bretscher 1976). Our previous discussion of peripheral and integral polypeptides suggests other ways in which lateral mobility may vary. Thus a monomeric integral protein like succinate dehydrogenase might diffuse more readily than an oligomer such as ATPase in which only one constituent poly-peptide is interpolated into the bilayer, the remaining four subunits which contain the catalytic site being located in the cytoplasm. The lateral mobility of a transmembrane protein such as glycophorin is likely to be affected by the simultaneous exposure of different portions of the molecule to two different aqueous environments, viz. cytoplasm and the external medium. We can attempt at the outset to distinguish between two types of membrane sites which may be designated mosaics and predetermined sites. A mosaic merely comprises a region of the membrane surface whose composition is determined by factors other than short-lived random associations of lipids and/or proteins. In this case, secondary considerations such as cation-induced lipid clustering (section 3), lipid–protein or protein–protein interactions increase the stability and hence the lifetime of the molecular aggregate. Once formed in this way, a mosaic would presumably still be able to diffuse in the membrane, albeit slowly. Predetermined sites are those regions in the prokaryote membrane where events such as septum formation and spore formation are initiated. In the former case the site is centrally located and in the latter, displaced to one cell pole. In some models of bacterial growth (see p. 242) a polar growth zone is proposed, which would also be defined as a predetermined site. In the extreme case, a predetermined site could be considered to have no lateral mobility, as for example in the positioning of the polar flagellum in bacteria. In addition to the mosaic-stabilising factors cited above, prokaryotes may constrain lateral mobility in other ways, such as covalent linkage of membrane components to wall polymers and the adhesion between inner and outer membranes seen in Gram-negative organisms.

The question of the site of prokaryote membrane synthesis has been intensively studied. In some areas there is consensus and in others controversy persists. Comprehensive reviews of the attempts to answer this question are included in the reference list. Although under normal conditions the cell coordinates membrane lipid and protein synthesis, experiments with glycerol auxotrophs (Mindich 1975) have shown that protein continues to be incorporated into membranes in the absence of net phospholipid synthesis. After glycerol deprivation, net phospholipid synthesis stops and the cells typically go slowly through one further division cycle. During this period some membrane proteins can be incorporated and seen to function. The consensus therefore is that for a number of membrane proteins simultaneous incorporation of membrane protein and lipid is not essential for development of membrane function. This topic has also been explored with the aid of *E. coli* unsaturated fatty acid auxotrophs (Mindich 1975, Cronan and Gelmann 1975). In brief, the object of these experiments was to determine (*a*) whether newly synthesised lipids are inserted into the membrane at one or a limited number of discrete growth zones, and (*b*) whether newly synthesised membrane proteins are inserted in association with simultaneously synthesised lipid. By changing the unsaturated fatty acid supplement during growth of these auxotrophs, it was possible to change the phase transition temperature of their membranes (section 4). Such a change in the lipid environment is reflected in characteristic alterations in the temperature dependence of the membrane protein responsible for β-galactoside transport. Therefore, by a judicious control of the timing of the fatty acid shift and the *lac* operon induction, it was possible to ask whether newly synthesised transport protein was integrated into the membrane alongside 'old' lipid (made prior to the fatty acid shift) or with newly synthesised lipid (containing the new fatty acid supplement). The answer was that the phase transition responses of β-galactoside transport protein indicated that it always inhabited a mixed environment composed of both newly synthesised and pre-existing lipids, provided that the experiment was performed at a temperature at which the membrane lipids were fluid, i.e. at or above the phase transition mid-point (section 4). These results and others, appear to rule out assembly mechanisms in which newly synthesised lipids and proteins initially associate to form mosaics or growth zones and favour instead a scheme in which the transport protein and new lipids are separately inserted at many sites into the lipid matrix and rapidly randomised by lateral diffusion (Thilo and Overath 1976). Nevertheless, several of the models proposed to account for cell growth specify that membrane synthesis occurs at discrete zones. This possibility has been explored in recent years through a variety of experimental strategies (Kepes and Autissier 1972, Mindich 1975; see Leive 1973 for review). In almost every case the results indicate that insertion of a particular membrane component occurs at random or at multiple sites. (Experiments in which the synthesis of wall components was studied are excluded from this discussion). One exception to this general view can be found in the experiments of Kepes and Autissier (1972) who used a penicillin lysis procedure to distinguish between random and non-random integration of sugar transport systems in *E. coli*. These apparently different results may be explained by making a distinction between cell growth and membrane synthesis. On this basis it is possible to suggest that membrane is synthesised at random or multiple sites, but cell growth occurs as a result of subsequent diffusion and accumulation of new material at special zones. Thus the central and polar sites at which division septa and spore membranes are initiated may represent sinks rather than sources for new membrane components. If such a lateral flow occurs, we need to explain the nature of the barriers which cause accumulation at the septal sites or growth zones and to ask whether they hinder the diffusion of both protein and lipid. Here again the possibility of covalent linkages between wall and membrane components, or wall-membrane adhesions may prove to be important. The artificial restraints which must be employed to restore normal septum formation to protoplasts and the failure of bacillus protoplasts to initiate the polar spore septum, suggest that an investigation of membrane fluidity in protoplasts, regenerating protoplasts and intact cells might be fruitful.

An awareness of membrane fluidity certainly prompts many new questions about prokaryote growth and membrane synthesis. For example, does newly synthesised lipid enter into bulk phase lipid more rapidly than into boundary lipid, because of the faster exchange diffusion rate of the former? If this is the case, then the distribution of lipid insertion sites may be influenced by the amount of ordered lipid and the number and size of any mosaics. Obviously any experimental situation which alters the native membrane fluidity may yield clear-cut but incorrect results. These experimental manipulations may be as apparently innocent as chilling, centrifugation or altering the cell's ion loading by washing. It is noticeable that these operations, which we have already seen are sufficient to cause phase separations, are similar to those which Higgins *et al.* (1976) suggest are responsible for the appearance of mesosomes. It might also be informative if we examined the importance of membrane fluidity in the cell cycle. For example, does the state of order/disorder in membrane lipids vary during the cycle or at very low and very high growth rates? A structural link between DNA and the membrane not unnaturally influences our view of events in the cell cycle. One recent report which is therefore particularly interesting indicates that neither the initiation nor the propagation of DNA replication requires a fluid membrane. (Thilo and Vielmetter 1976). In the *E. coli* unsaturated fatty acid auxotrophs used in these experiments, DNA continued to be synthesised when the temperature was 8°C below the phase transition temperature of the membrane lipids. At this stage the results cannot enable us to decide whether the DNA synthesising machinery is peripherally sited on the membrane with no lipid interaction, or whether it is in fact situated in the cytoplasm.

For the future a major task in the area of cell growth and division may well be to determine how cells can initiate and/or preserve predetermined sites on what appears to be a fluid membrane.

General texts and reviews

Prokaryote envelopes and walls

BRAUN, V. and HANTKE, K. (1974) 'Biochemistry of bacterial cell envelopes', *Ann. Rev. Biochem.,* **43**, 89–110.

COSTERTON, J. W., INGRAM, J. M. and CHENG, K. J. (1974) 'Structure and function of the cell envelope of Gram-negative bacteria', *Bacteriol. Rev.,* **38**, 87–110.

LEIVE, L. (Ed.) (1973) *Bacterial Membranes and Walls,* Vol. 1. Marcel Dekker Inc., New York. Contains nine reviews by individual authors on a comprehensive range of topics.

MEADOW, P. M. (1974) 'Structure and synthesis of bacterial walls', in *Companion to Biochemistry* (Eds. A. T. Bull, J. R. Lagnado, J. O. Thomas and K. F. Tipton), pp. 343–65. Longman, London.

Prokaryote membrane structure, function and biosynthesis

BAYER, M. E. (1975) 'Role of adhesion zones in bacterial cell-surface function and biogenesis', in *Membrane Biogenesis* (Ed. A. Tzagoloff), pp. 393–427. Plenum Press, New York.

FOX, C. F. (Ed.) (1975) *Biochemistry of Cell Walls and Membranes,* M.T.P. Int. Rev. Sci. Biochem., Series 1, Vol. 2. Butterworths, Sevenoaks, Kent.
A sound source of advanced reading on most membrane topics in prokaryotes and eukaryotes.

INOUYE, M. (1975) 'Biosynthesis and assembly of the outer membrane proteins of *Escherichia coli*', in *Membrane Biogenesis* (Ed. A. Tzagoloff), pp. 351–91. Plenum Press, New York.

KEPES, A. and AUTISSIER, F. (1972) 'Topology of membrane growth in bacteria', *Biochim. Biophys. Acta,* **265**, 443–69.

MACHTIGER, N. A. and FOX, C. F. (1973) 'Biochemistry of bacterial membranes', *Ann. Rev. Biochem.,* **42**, 575–600.

MINDICH, L. (1975) 'Studies on bacterial membrane biogenesis using glycerol auxotrophs', in *Membrane Biogenesis* (Ed. A. Tzagoloff), pp. 429–54. Plenum Press, New York.

PARDEE, A. B. and ROZENGURT, E. (1975) 'Role of the surface in production of new cells', *Biochemistry of Cell Walls and Membranes,* M.T.P. Int. Rev. Sci. Biochem. Series 1, Vol. 2 (Ed. C. F. Fox), pp. 123–54. Butterworths, Sevenoaks, Kent.

SALTON, M. R. J. (1971) 'Bacterial membranes', *CRC Crit. Rev. Microbiol.,* **1**, 161–97.

SALTON, M. R. J. and OWEN, P. (1976) 'Bacterial membrane structure', *Ann. Rev. Microbiol.,* **30**, 451–82.

Eukaryote membranes

BRANTON, D. (1971) 'Freeze-etching studies of membrane structures', *Phil. Trans. Roy. Soc. Lond.,B* **261**, 133–8.

BRETSCHER, M. S. (1973) 'Membrane structure: some general principles', *Science,* **181**, 622–9.

BRETSCHER, M. S. and RAFF, M. C. (1975) 'Mammalian plasma membranes', *Nature,* **258**, 43–9.

STECK, T. L. (1972) 'Membrane isolation', in *Membrane Molecular Biology* (Eds. C. F. Fox and A. D. Keith), pp. 74–114. Sinauer Associates, Stanford, Conn.

STECK, T. L. and FOX, C. F. (1972) 'Membrane Proteins', in *Membrane Molecular Biology* (Eds. C. F. Fox and A. D. Keith), pp. 27–75. Sinauer Associates, Standord, Conn.

Membrane lipids and techoic acids

ARCHIBALD, A. R. (1974) 'The structure, biosynthesis and function of techoic acid', *Adv. Microb. Physiol.,* **11**, 53–95.

FINNERTY, W. R. (1975) 'Microbial lipid metabolism', *CRC Crit. Rev. Microbiol.,* **4**, 1–40.

O'LEARY, W. M. (1975) 'The chemistry of microbial lipids', *CRC Crit. Rev. Microbiol.,* **4**, 41–63.

SHAW, N. (1975) 'Bacterial glycolipids and glycophospholipids', *Adv. Microb. Physiol',* **12**, 141–67.

Lipid bilayers

BANGHAM, A. D. (1972) 'Lipid bilayers and biomembranes', *Ann. Rev. Biochem.,* **41**, 753–76.

Fluid mosaic model

EDIDIN, M. (1974) 'Rotational and translational diffusion in membranes', *Ann. Rev. Biophys. Bioeng.*, 3, 179–201.

SINGER, S. J. (1974) 'The molecular organization of membranes', *Ann. Rev. Biochem.*, 43, 805–33.

SINGER, S. J. and NICOLSON, G. L. (1972) 'The fluid mosaic model of the structure of cell membranes', *Science,* 175, 720–31.

Microtubules, microfilaments and membrane fluidity

NICOLSON, G. L. (1976) 'Transmembrane control of the receptors on normal and tumor cells. I. Cytoplasmic influence over cell surface components', *Biochim. Biophys. Acta.,* 457, 57–108.

Glycophorin

MARCHESI, V. T., TILLACK, T. W., JACKSON, R. L., SEGREST, J. P. and SCOTT, R. E. (1972) 'Chemical characterization and surface orientation of the major glycoprotein of the human erythrocyte membrane', *Proc. Natl. Acad. Sci. USA,* 69, 1445–9.

Analytical techniques

CARRAWAY, K. L. (1975) 'Covalent labelling of membranes', *Biochim. Biophys. Acta,* 415, 379–410.

KEITH, A. D., SHARNOFF, M. and COHN, G. E. (1973) 'A summary and evaluation of spin labels used as probes for biological membrane structure', *Biochim. Biophys. Acta,* 300, 379–419.

VIGNAIS, P. M. and DEVAUX, P. F. (1976) 'The use of spin labels to study membrane-bound enzymes, receptors and transport systems', in *Enzymes of Biological Membranes* (Ed. A. Martonosi), Vol. I, pp. 91–117. Wiley, Chichester and New York.

WEBER, K. and OSBORN, M. (1975) 'Proteins and sodium dodecyl sulfate: molecular weight determination on polyacrylamide gels and related procedures', in *The Proteins* (Eds. H. Neurath and R. L. Hill), Vol. I, pp. 179–223. Academic Press, New York.

Prokaryote membrane variability

CRONAN, J. E. and GELMANN, E. P. (1975) 'Physical properties of membrane lipids: biological relevance and regulation', *Bacteriol. Rev.,* 39, 232–56.

HEINRICH, L. (Ed.) (1976) *Extreme Environments: Mechanisms of Microbial Adaptation.* Academic Press, New York.
Contains several valuable reviews of membrane modification in response to environmental changes.

Other references cited

BEACHAM, I. R., TAYLOR, N. S. and YOUELL, M. (1976) 'Enzyme secretion in *Escherichia coli*: synthesis of alkaline phosphatase and acid hexose phosphatase in the absence of phospholipid synthesis', *J. Bacteriol.,* 128, 522–7.

BRETSCHER, M. S. (1976) 'Directed lipid flow in membranes', *Nature*, **260**, 21–3.

CHAN, M., VIRMANI, Y. P., HIMES, R. H. and AKAGI, J. M. (1973) 'Spin-labelling studies on the membrane of a facultatively thermophilic bacillus', *J. Bacteriol.*, **113**, 322–8.

ESSER, A. F. and LANYI, J. K. (1973) 'Structure of the lipid phase in cell envelope vesicles from *Halobacterium cutirubrum*', *Biochem.*, **12**, 1933–9.

FULCO, A. J. (1970) 'The biosynthesis of unsaturated fatty acids by bacilli', *J. Biol. Chem.*, **245**, 2985–90.

FUTAI, M., STERNWEIS, P. C. and HEPPEL, L. A. (1974) 'Purification and properties of reconstitutively active and inactive adenosine triphosphatase from *Escherichia coli*', *Proc. Natl. Acad. Sci. USA*, **71**, 2725–9.

HAEST, C. W. M., DE GIER, J. and VAN DEENEN, L. L. M. (1969) 'Changes in the chemical barrier properties of the membrane lipids of *E. coli* by variation of the temperature of growth', *Chem. Phys. Lipids*, **3**, 413–17.

HAWROOT, E. and KENNEDY, E. P. (1975) 'Biogenesis of membrane lipids: mutants of *Escherichia coli* with temperature-sensitive phosphatidyl serine decarboxylase', *Proc. Natl. Acad. Sci. USA*, **72**, 1112–16.

HENDERSON, R. and UNWIN, P. N. T. (1975) 'Three-dimensional model of purple membrane obtained by electron microscopy', *Nature*, **257**, 28–32.

HIGGINS, M. L., TSIEN, H. C. and DANEO-MOORE, L. (1976) 'Organisation of mesosomes in fixed and unfixed cells', *J. Bacteriol.*, **127**, 1519–23.

KUSAKA, I. (1975) 'Degradation of phospholipid and release of diglyceride-rich membrane vesicles during protoplast formation in certain Gram-positive bacteria', *J. Bacteriol.*, **121**, 1173–9.

LEE, A. G., BIRDSALL, N. J. M., METCALFE, J. C., TOON, P. A. and WARREN, G. B. (1974) 'Clusters in lipid bilayers and the interpretation of thermal effects in biological membranes', *Biochem.*, **13**, 3699–705.

McELHANEY, R. N. (1976) 'The biological significance of alterations in the fatty acid composition of microbial membranes, in response to changes in environmental temperature', in *Extreme Environments* (Ed. M. R. Heinrich), pp. 255–81. Academic Press, New York.

MINNIKEN, D. E., ABDOLRAHIMZADEH, H. and BADDILEY, J. (1974) 'Replacement of acidic phospholipids by acidic glycolipids in *Pseudomonas diminuta*', *Nature*, **249**, 268–9.

MORRISETT, J. D., POWNALL, H. J., PLUMLEE, R. T., SMITH, L. C., ZEHNER, Z. E., ESFAHANI, M. and WAKIL, S. J. (1975) 'Multiple thermotropic phase transitions in *Escherichia coli*', *J. Biol. Chem.*, **250**, 6969–76.

MUHLRADT, P. F. and MENZEL, J. (1974) 'Lateral mobility and surface density of lipopolysaccharide in the outer membrane of *Salmonella typhimurium*', *Eur. J. Biochem.*, **43**, 533–9.

NIKAIDO, H., TAKEUCHI, Y., OHNISHI, S. and NAKAE, T. (1977) 'Outer membrane of *Salmonella typhimurium*. Electron spin resonance studies', *Biochim. Biophys. Acta*, **465**, 152–64.

OESTERHELT, D. (1976) 'Bacteriorhodopsin as an example of a light-driven proton pump', *Angewandte. Chem.* (International Ed. Engl.), **15**, 17–24. (German version: (1976) *Angew. Chem.*, **88**, 16–25.)

RAZIN, S. (1975) 'Cholesterol incorporation into bacterial membranes', *J. Bacteriol.*, **124**, 570–2.

ROTHMAN, J. E. and KENNEDY, E. P. (1977) 'Asymmetrical distribution of phospholipids in the membrane of *Bacillus megaterium*', *J. Mol. Biol.*, **110**, 603–18.

ROTHMAN, J. E. and LENARD, J. (1977) 'Membrane asymmetry', *Science*, **195**, 743–53.

SINENSKY, M. (1974) 'Homeoviscous adaptation – A homeostatic process that regulates the viscosity of membrane lipids in *Escherichia coli*', *Proc. Natl. Acad. Sci. USA*, **71**, 522–5.

SPATZ, L. and STRITTMATTER, P. (1971) 'A form of cytochrome b_5 that contains an additional hydrophobic sequence of 40 amino acid residues', *Proc. Natl. Acad. Sci. USA*, **68**, 1042–6.

SPATZ, L. and STRITTMATTER, P. (1973) 'A form of reduced nicotinamide adenine dinucleotide cytochrome b_5 reductase containing both the catalytic site and an additional hydrophobic membrane-binding segment', *J. Biol. Chem.*, **248**, 793–9.

THILO, L. and OVERATH, P. (1976) 'Randomization of membrane lipids in relation to transport system assembly in *Escherichia coli*', *Biochem.*, **15**, 328–34.

THILO, L. and VIELMETTER, W. (1976) 'Independence of deoxyribonucleic acid replication and initiation from membrane fluidity and the supply of unsaturated fatty acids in *Escherichia coli*', *J. Bacteriol.*, **128**, 130–43.

TRAUBLE, H. and OVERATH, P. (1973) 'The structure of *Escherichia coli* membranes studied by fluorescence measurements of lipid phase transitions', *Biochim. Biophys. Acta*, **307**, 491–512.

VAN ZOELEN, E. J. J., ZWAAL, R. F. A., REUVERS, F. A. M., DEMEL, R. A. and VAN DEENEN, L. L. M. (1977) 'Evidence for the preferential interaction of glycophorin with negatively charged phospholipids', *Biochim. Biophys. Acta*, **464**, 482–92.

WICKNER, W. (1976) 'Asymmetric orientation of phage M13 coat protein in *Escherichia coli* cytoplasmic membranes and in synthetic lipid vesicles', *Proc. Natl. Acad. Sci. USA*, **73**, 1159–63.

YAMAMOTO, S. and LAMPEN, J. O. (1976) 'The hydrophobic membrane penicillinase of *Bacillus licheniformis* 749/C', *J. Biol. Chem.*, **251**, 4102–10.

12
Transport of organic solutes by bacteria

H. Tristram

Department of Botany and Microbiology, University College London

1 Introduction

Transport of solutes into bacteria and other organisms raises two main problems. First, the translocation of hydrophilic substances through the plasma membrane, a predominantly hydrophobic barrier (see Ch. 11) and secondly, the nature of the coupling between metabolism and transport. Transport systems are relatively specific, suggesting the involvement of recognition sites in translocating a solute into a cell. A concept, the development of which is discussed by Christensen (1975), which has greatly influenced the study of transport is that of protein carriers, responsible for translocation across the membrane. The notion that specific proteins are involved in bacterial transport arose from the study of mutants with impaired transport activity and the realisation that the activity of many transport systems is inducible.

In bacteria the same solute may be transported by different mechanisms in different genera and in a single species the same solute is often transported by several distinct transport systems which may be distinguished by adequate study of the kinetics, solute specificity, and the investigation of mutants deficient in one or more of the transport systems.

1.1 Mechanisms of uptake of solutes

Solutes are taken up by microorganisms by four distinct processes which may be distinguished on kinetic and thermodynamic grounds (Christensen, 1975). These processes are (*a*) passive diffusion, (*b*) facilitated diffusion, (*c*) active transport, and (*d*) group translocation.

A solute passes through a membrane by *passive diffusion* as a result of random molecular movement. Diffusion occurs down a concentration gradient, i.e. towards that side of the membrane on which the solute is at the lower concentration, the rate of diffusion being proportional to the difference in solute concentration on the two sides of the membrane (Fick's first law). At equilibrium the solute concentration is the same on each side of the membrane. Passive diffusion is not stereospecific. Furthermore solutes and their structurally-related analogues do not compete for passage through the membrane.

Facilitated diffusion displays many of the features of passive diffusion. It is, however, highly stereospecific, displaying competition between a solute and its structural analogues. The rate of translocation is not proportional to the concentration difference across the membrane, but displays saturation kinetics, so resembling an enzyme reaction. These characteristics are consistent with the involvement of a carrier catalysing passage across the membrane at a much more rapid rate than occurs by passive diffusion. As in passive diffusion, an equilibrium is reached when the concentration of solute is identical on two sides of the membrane. It has been frequently stated that facilitated diffusion does not require metabolic energy. However, in 1971 it was claimed that stringently starved *Escherichia coli* is unable to effect facilitated diffusion unless provided with an energy source and more recent investigations have supported this claim. Recent developments in our understanding of energy coupling to transport offer an acceptable explanation of these observations (see section 8).

Active transport resembles facilitated diffusion with one important difference. Systems promoting active transport consist of at least two components, namely a solute-specific carrier which acts catalytically (as in facilitated diffusion) and a continuing supply of metabolic energy, so allowing net movement of solute against an electrochemical gradient. Thus a solute may, by active transport, be accumulated within a cell such that the internal concentration is much greater than the external concentration.

In the three processes already mentioned solute remains chemically unchanged during translocation across the membrane. However *group translocation* results in the modification of the solute during translocation, the solute appearing in the cell as a derivative. Providing the membrane is relatively impermeable to the solute derivative and the derivative is not immediately metabolised, it will accumulate within the cell, but there the resemblance with active transport ends, since the membrane component(s) in group translocation is acting as an enzyme(s) with vectorial properties

1.2 Methods of study

1.2.1 Intact organisms

The transport of solutes into intact microorganisms is usually studied under conditions in which their metabolism is arrested, thereby ensuring intracellular accumulation of the solutes, or, in group translocation processes, accumulation of a derivative of a solute. Transport of carbohydrates (by processes other than group translocation) may be studied by utilising non-metabolisable analogues of the solute or by the use of a mutant which lacks the ability to metabolise the solute. For example, methyl thio-β-galactoside (TMG), a non-metabolised analogue of lactose is often used as a substrate for the *E. coli* lactose transport system and *galK⁻* mutants, lacking functional galactokinase, may be used to study galactose accumulation. Similar approaches may be used to study accumulation of di- and tri-carboxylic acids. Uptake of amino acids may be studied in the presence of chloramphenicol to prevent incorporation of the amino acid into protein. An alternative method, introduced by G. F. Ames, consists of measuring incorporation of radioactively labelled amino acid into protein in growing cells, under conditions in which transport of the amino acid is rate-limiting.

Transport is usually estimated by measuring the incorporation of radioactively labelled solute into the organism by withdrawing samples of suspension and rapidly filtering and washing the organisms on Millipore (or similar) bacterial filters. Temperature of incubation and conditions of washing the organisms on filters are important, and have been reviewed by Maloney, Kashnet and Wilson (1975).

Occasionally transport (though not accumulation) of a solute may be studied by choosing a substrate of the transport system which is rapidly converted intracellularly to a product which is readily assayed and choosing experimental conditions in which transport is rate-limiting. For example, in *E. coli* containing β-galactosidase and the β-galactoside transport system, the rate of hydrolysis of *o*-nitrophenyl-β-galactoside (ONPG) by β-galactosidase is several-fold higher than the rate of ONPG transport. Consequently a simple colorimetric assay of the rate of formation of *o*-nitrophenol (yellow) from ONPG (colourless) yields a valid measure of activity of the β-galactoside transport system.

1.2.2 Membrane vesicles

Kaback and his collaborators pioneered the practice of using bacterial membrane vesicles in the study of solute transport. Briefly the technique consists of first converting bacteria to proto-plasts or spheroplasts. The osmotically-fragile bodies are then lysed by suspension in hypertonic buffer solution, usually containing DNAase and RNAase to digest nucleic acids. The membranes, which spontaneously re-seal to form vesicles, are collected, washed and freed of any unbroken spheroplasts and intact organisms by centrifugation. Membranes prepared in this way consist of closed vesicles since they behave like osmometers; they contain less than 5 per cent of the cellular DNA or RNA, very little residual wall material and are virtually devoid of cytoplasmic or periplasmic enzymes (see Kaback 1970).

The main advantage of vesicle preparations for transport studies is their freedom from cytoplasmic metabolism. However during the preparation of vesicles, membranes may undergo changes in topography which at present can neither be completely controlled nor assessed. Kaback and his coworkers maintain that the membrane vesicles prepared by their standard procedure all have the orientation they possessed in the intact cell, i.e. they are all right-side out. It is becoming increasingly clear that membrane vesicles prepared for transport studies by other workers, essentially by the Kaback method, often contain a significant proportion of everted (that is, inside-out) vesicles, especially if the preparations are frozen before use. More disconcerting is the growing evidence which suggests that, even if membrane preparations contain a preponderance of vesicles having the orientation they possessed in the intact cell, some membrane-bound enzymes become displaced so that formerly inwards-facing enzyme molecules become orientated to the outside and are accessible to substrates dissolved in the extra-vesicular milieu. This aspect of membrane vesicle topography, of prime importance in any consideration of energy coupling to transport, is discussed fully by Simoni and Postma (1975), Hamilton (1975), Harold (1972) and Roseman (1972).

1.2.3 Use and detection of transport mutants

Our knowledge of microbial transport processes owes much to genetics; the study of mutants lacking specific transport systems has proved invaluable, especially in understanding uptake of solutes which are transported by more than one system. Mutants impaired in energy coupling have also contributed to an understanding of this aspect of transport. About 50 structural and regulatory genes concerned with transport have been mapped in *E. coli* (see Bachman, Low and Taylor 1976). Diverse selective techniques, reviewed by Slayman (1973) and Cordaro (1976), have been used to obtain transport-defective mutants.

2 Transport of β-galactosides in *Escherichia coli*

2.1 Biochemical and genetical background

Uptake of β-galactosides by *E. coli* shows all the characteristics of a typical active transport process, the kinetics of which have been studied more intensively than in any other system.

This transport system will be discussed as representative of other active transport processes, but before doing so the biochemical and genetical characteristics of the system will be outlined.

Our knowledge of the control of gene expression stems largely from the study of the catabolism of lactose in *E. coli*. When grown on lactose (a β-galactoside), or on media containing a carbon source such as succinate and certain non-metabolisable β-thiogalactosides, three proteins are induced, namely β-galactosidase which hydrolyses lactose to galactose and glucose, a transport system responsible for uptake of β-galactosides and β-thiogalactoside transacetylase, an enzyme which is not essential for either transport or utilisation of β-galactosides. Three contiguous genes on the *E. coli* chromosome, *lacZ,Y,A*, specify the three functions β-galactosidase, transport and transacetylase, respectively. Together with an operator (*lacO*) and another genetic element, the promoter, these genes constitute an operon, the expression of which is controlled by a regulatory gene, *lacI*. In inducible strains (*lacI*+*lacO*+) transport, β-galactosidase and thiogalactoside transacetylase activities are induced coordinately. Constitutive mutants, of which there are two kinds (*lacI*−*lacO*+ or *lacI*+*lacO*c) produce all three activities without the intervention of inducer.

Mutants unable to ferment lactose (Lac−) may lack either a functional β-galactosidase (*lacZ*−*lacY*+) or a functional transport system (*lacZ*+*lacY*−). In the former, the mutant takes up β-galactosides but the substrate cannot be metabolised; the *lacZ*+*lacY*− mutants are cryptic in that they are unable to transport β-galactoside substrates into the cell, though they possess the enzyme necessary for their catabolism.

Non-induced wild type organisms do not accumulate β-galactosides or β-thiogalactosides, whereas organisms which have been grown with an appropriate inducer accumulate chemically unchanged β-thiogalactosides in a free (soluble) form. Chloramphenicol, a potent inhibitor of protein synthesis in prokaryotes, prevents the induction, indicating that a protein component is involved in transport of β-galactosides. Furthermore, a membrane-bound protein specified by the *lacY* gene has been identified (see section 4.1).

The facts that accumulation of β-galactosides is inducible, involves a protein component which acts catalytically, is highly stereospecific and displays Michaelis—Menten kinetics (see section 2.2) led to the introduction of the term 'permease' for the protein component. The original definition of 'permease' was somewhat ambivalent and has been further blurred by usage. The term has been widely criticised (see Mitchell 1970, Kepes 1971) and is now used less frequently than formerly, most workers preferring the term 'transport system'.

2.2 Kinetics of transport

If the initial rate of uptake (v_{in}) is studied as a function of the external concentration (s_{ex}) of transported solute, saturation kinetics are observed. As with enzyme reactions in the relationship v_{in} versus s_{ex} is hyperbolic, uptake being described approximately by equation (1)

$$v_{in} = V_{max}^{in} \frac{s_{ex}}{K_m^{in} + s_{ex}} + D.s_{ex} \tag{1}$$

where V_{max}^{in} is the maximum velocity of influx at saturating concentrations of solute, K_m^{in} is the solute concentration at which $v_{in} = \frac{1}{2}V_{max}^{in}$ and D is a function describing transmembranal movement by diffusion-like processes not catalysed by the specific carrier.* Except for the introduction of the term Ds_{ex} this equation bears a formal relationship to the classic Michaelis–Menten equation relating velocity of an enzyme reaction to substrate concentration. Uptake of one (labelled) β-galactoside is competitively inhibited by another (unlabelled) β-galactoside, suggesting that the transitory binding of solute to a limiting number of specific binding sites is an essential step in transport. Mutants ($lacY^-$) lacking a functional transport system do not show saturation kinetics, entry being mediated mainly by passive diffusion.

Depending on the β-galactoside studied, net accumulation, during which the intracellular concentration may exceed the extracellular concentration by a factor of 50 to 500, ceases after 5 to 20 min. when the intracellular radioactivity reaches a plateau, the magnitude of which, like the rate of uptake, is dependent on the nature and initial external concentration of the solute. This cessation of uptake is only apparent; the plateau represents a dynamic equilibrium in which rates of influx and efflux are equal, as demonstrated by the experiment illustrated in Fig. 12.1. Uptake of [^{14}C] TMG reached the steady state in about 15 min. (curve a). In b and c influx of [^{14}C] TMG occurred at the same rate as in a, though organisms in b and c had already reached the steady state following the uptake of unlabelled TMG. In other words, when net accumulation had ceased influx and, by inference, efflux were still occurring at the same rate as the initial loading with TMG. The steady state, in which the intracellular concentration of

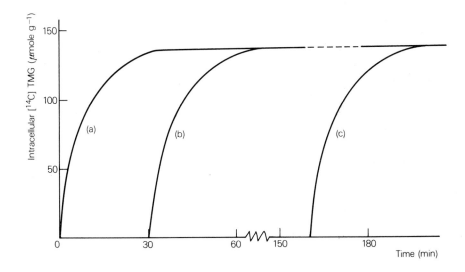

Fig. 12.1 Uptake and exchange of TMG by *E. coli* ML 308. (a) 3 mM [^{14}C] TMG added at zero time; (b) and (c): 3 mM TMG (unlabelled) added at zero time followed by [^{14}C] TMG at 30 min. (curve (b)) or 160 min. (curve (c)). Specific radioactivity was identical in all treatments and weight of [^{14}C] TMG added to (b) and (c) was insufficient to cause any significant change in TMG concentration. (Modified from Kepes, 1971.)

---

*Owing to exchange diffusion and the recapture phenomenon, the Michaelis–Menten equation describes active transport of β-galactosides only approximately. Influx and efflux due to diffusion-like processes are usually small, but can significantly affect the final intracellular steady-state levels under some experimental conditions. In fully induced organisms 'diffusion' accounts for less than 20 per cent of efflux, but contributes about 75 per cent of the total efflux from organisms containing only about 5 per cent of the fully induced level of the β-galactoside-carrier molecules.

β-thiogalactoside is higher than the extracellular concentration, may be expressed by equation (2).

$$V_{max}^{in} \frac{s_{ex}}{K_m^{in} + s_{ex}} + D.s_{ex} = V_{max}^{ef} \frac{s_{in}}{K_m^{ef} + s_{in}} + D.s_{in} \tag{2}$$

where V_{max}^{ef} is the maximum velocity of efflux, K_m^{ef} is the K_m of efflux, s_{in} is the internal concentration of transported solute; other symbols are as in equation (1).

If *E. coli* which has been allowed to accumulate β-thiogalactoside is washed and resuspended in fresh medium lacking β-thiogalactoside, an efflux of radioactive solute is observed. The velocity of efflux v_{ef} is only about 10 to 15 per cent of v_{in} observed during the initial loading stage. This effect is due to the so-called recapture phenomenon, whereby β-thiogalactoside transported outwards through the plasma membrane is trapped in the interstices of the wall and by the presence of the outer membrane, with the result that many molecules are re-translocated back into the cytoplasm without equilibrating with the external fluid. Thus the wall forms a diffusion barrier between the plasma membrane and the extracellular fluid, the effect of which can be alleviated by supplying a high concentration of β-thiogalactoside outside the organism. Under these conditions v_{ef} approximates to v_{in}. The efflux of accumulated solute depends on

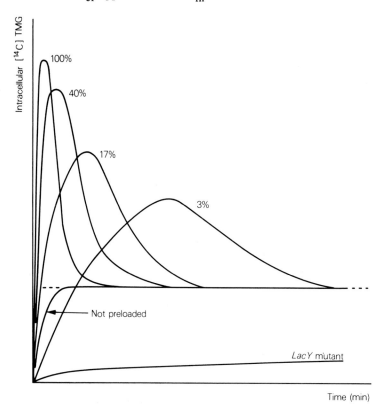

Fig. 12.2 Effect of number of β-galactoside carrier molecules on counterflux in *E. coli* (*lacI⁻ Z⁺Y⁺A⁺* constitutive mutant). In this experiment the number of carriers (expressed as percentage of maximum carrier activity) in each suspension was varied by controlled inactivation with NEM (see section 4.1). The dashed line indicates the concentration of TMG in the medium. Similar results may be obtained by altering the levels of carrier formed due to varying the time of induction in a wild type (*lacI⁺*) inducible strain. For further details, see text. (Modified from Wong and Wilson (1970), *Biochim. Biophys. Acta*, **196**, 336–50.)

the carrier involved in influx since v_{ef} from induced bacteria is greater than that from non-induced organisms and the exit rate from *lacY⁻* mutants, lacking a functional transport system, is very small indeed.

Uncoupling agents such as 2,4-dinitrophenol (DNP) inhibit net accumulation of β-galactoside without significantly affecting influx and efflux. The system no longer behaves as an active transport system, but carries out carrier-mediated facilitated diffusion so that an equilibrium is reached when intracellular and extracellular concentrations of β-galactoside are approximately equal.

Organisms transporting solute by facilitated diffusion also display the phenomenon of counterflux. This refers to transient accumulation against a concentration gradient under conditions in which normal net accumulation of solute is prevented by uncouplers. If bacteria are 'preloaded' by exposure to high concentrations of unlabelled β-thiogalactoside and the organisms are then suspended in medium containing low concentrations of labelled β-thio-galactoside, the latter is rapidly taken up, reaches a peak and subsequently is lost from the organisms. The phenomenon is due to temporary competition for efflux between radioactive β-thiogalactoside (at a low internal concentration) and the high internal concentration of unlabelled β-thiogalactoside. Loss of the preloaded solute relieves the inhibition of efflux of labelled β-thiogalactoside with the result that intracellular radioactivity declines until equilibrium between the two sides of the membrane is reached. The height of the peak is related to the concentration of solute used in the preloading. The rate of influx of radioactivity and hence the time at which the peak is reached is related to the number of functional carrier molecules in the organisms (Fig. 12.2). Counterflux constitutes a powerful test for functional carriers, the presence of which is not easily established from the kinetics of solute movement in facilitated diffusion systems (see Maloney *et al.* 1975).

3 Group translocation

3.1 Phosphoenolpyruvate—sugar phosphotransferase systems

The transport of many carbohydrates in some bacteria by a group translocation process has been elucidated mainly by Roseman, Kundig, Simoni and their collaborators. In 1964 a phosphoenolpyruvate (PEP)-dependent phosphorylation of sugars was demonstrated in *E. coli* and has since been shown to occur in *Salmonella typhimurium, Staphylococcus aureus, Bacillus subtilis* and a number of other bacterial genera (Roseman 1972, Postma and Roseman 1976). Sugars are normally phosphorylated in the 6-position. However the hexitol, mannitol, forms mannitol-1-phosphate and fructose is usually phosphorylated to fructose-1-phosphate, but it too, under some circumstances, may yield fructose-6-phosphate. The process, which is summarised in Fig. 12.3, is known as the PEP—sugar phosphotransferase system. It is complex, normally involving four proteins, three of which are transitorily phosphorylated during the reaction.

Enzyme I and HPr (heat-stable protein) are both located in the cytoplasm and function in the phosphorylation of a wide variety of carbohydrates. During the phosphotransferase reaction, both proteins are phosphorylated at one of the nitrogen atoms of the imidazole ring of a histidine residue. In the presence of Enzyme I, PEP and HPr the latter is phosphorylated to HPr-P and the transfer of its phosphoryl group to glucose, mannose or fructose is promoted by *E. coli* membrane preparations, no other soluble factors being required. The membrane-bound components catalysing the transfer are termed Enzyme II complex which contains proteins IIA

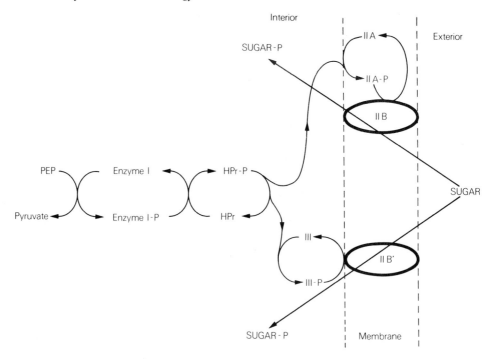

Fig. 12.3 Phosphate transfer and sugar transport by the phosphoenolpyruvate (PEP–sugar phosphotransferase system. (Modified from Postma and Roseman, 1976.)

and IIB, phosphatidylglycerol and a divalent metal. Proteins of the Enzyme II complex, which is specific for glucose, fructose or mannose (and their analogues), have been extracted from membranes and Enzyme IIA has been further fractionated into three proteins, each specific for one of the sugars mentioned. There is evidence that Enzyme IIB is a single protein, functioning in conjunction with the three separate Enzyme IIA's.

In addition to systems involving Enzyme IIA/IIB complexes, other systems in which the membrane-bound IIA is replaced by a soluble (cytoplasmic) protein designated Factor III have been described. Enzyme IIB species which act in conjunction with Factor III's are sometimes designated Enzyme IIB′ to distinguish them from Enzyme IIB associated with Enzyme IIA. Besides the Enzyme IIA/IIB system, the enteric bacteria contain an Enzyme III/IIB′ glucose-phosphorylating system and a III/IIB′ system responsible for phosphorylation of lactose occurs in *Staphylococcus aureus*. The properties and assay of proteins involved in the phosphotransferase reactions are reviewed by Postma and Roseman (1976). To summarise a complex situation, the sugar-specific proteins consist of either Enzymes IIA/IIB (both membrane-bound) or III/IIB′ (only IIB′ being membrane-bound). Both proteins IIA and III are transitorily phosphorylated during transfer of a phosphoryl group from PEP to the sugar.

What is the evidence that these phosphorylating systems are involved in transport of sugars in some bacteria? It was known, even before the discovery of the PEP-dependent phosphoryla-tions, that some sugars are accumulated as phosphate esters in *E. coli* and other bacteria. Furthermore, one or more components of the systems are membrane-bound, suggesting a possible transport role, and finally the study of mutants has provided particularly compelling evidence for a role for PEP–sugar phosphotransferases in the translocation of sugars. Mutants of *S. typhimurium* or *E. coli* which lack a functional Enzyme I (*ptsI* mutants) or HPr (*ptsH* mutants) did not grow on nine different carbohydrates and failed to transport glucose, α-methylglucoside (a glucose analogue), fructose, mannitol, sorbitol, maltose or acetyl-glucosamine. Besides these pleiotropic mutants, others unable to utilise single sugars have been obtained and shown to be deficient in one of the sugar-specific Enzyme II complexes. Similar observations have been made on mutants of *Staphylococcus aureus*. It is now generally accepted that the system first described by Roseman and his colleagues represents a group translocation process whereby a carbohydrate is translocated across the cytoplasmic membrane, becoming phosphorylated in the process. The PEP–sugar phosphotransferase system is therefore a mechanism for coupling translocation directly to an exergonic reaction and accounts for both transport and the initial step in the metabolism of the translocated sugar (see Fig. 12.3).

That phosphorylation of sugar is obligatorily coupled to translocation is suggested by the observations (see above) that mutants lacking one of the phosphotransferase components are unable to transport certain sugars. Furthermore, glucose trapped in *E. coli* membrane vesicles is not phosphorylated, whereas extravesicular glucose accumulates in vesicles as glucose-6-phosphate (Kaback 1970).

3.1.1 Metabolic control and genetics

Probably all sugars are transported into *Staphylococcus aureus* by PEP–sugar phosphotransferase systems, whereas only some sugars are transported in this way by *E. coli*. Most of the sugar-specific Enzyme II complexes are inducible, including the Enzyme IIB′/III (lactose) and IIB′/III(fructose) of *S. aureus* and *E. coli*, respectively. However the Enzyme IIA/IIB(glucose/mannose/fructose) complex of *E. coli* is probably constitutive, as are Enzyme I and HPr. Metabolic control patterns of the PEP–sugar phosphotransferase systems are more complex than this statement suggests, interactions between systems being common. The phosphotransferase systems also modulate the synthesis of other inducible enzymes and (non-phosphotransferase) transport systems (Postma and Roseman 1976, Kornberg and Jones-Mortimer 1977).

The genetics of the PEP–sugar phosphotransferase systems is complex and at present somewhat confused, due partly to the fact that mutants with a genetic lesion affecting a sugar-specific system can usually only be described phenotypically as having a defect in an Enzyme II complex, rather than in a particular component of the complex. Genes *ptsI* and *ptsH*, specifying Enzyme I and HPr respectively, are adjacent and closely linked to a promoter-like element and thus may constitute an operon in *E. coli* and *S. typhimurium*. The genes specifying Enzyme IIB′(lactose) and III(lactose) and the phospho-β-galactosidase of *Staphylococcus aureus* are closely linked to each other and to a regulatory gene involved in the expression of the three structural genes. Similarly, the structural genes for the Enzyme II(mannitol) complex and mannitol-1-phosphate dehydrogenase are linked in *Staphylococcus aureus, S. typhimurium* and *E. coli*. The close linkage and co-regulation of genes specifying a transport protein and the first enzyme of subsequent metabolism of the phosphorylated sugar is of considerable importance in cell economy. Other genes specifying some component of various Enzyme II complexes have also been identified, including those for mannose (*ptsM*), mannitol (*mtlA*), sorbitol (*srlA*), the inducible III/IIB′(fructose) system (*ptsF*) and the III/IIB′(glucose) system (*ptsG*). The map positions and further discussion of these and other genes involved in phosphotransferase systems may be found in reviews by Postma and Roseman (1976), Cordaro (1976) and Bachman, Low and Taylor (1976).

3.2 Other group translocation processes

There is good evidence that nucleic acid bases and nucleosides are also transported by processes involving group translocation (Hochstadt 1974). In *E. coli* exogenous purines and pyrimidines accumulate as nucleoside monophosphates, the mechanism of translocation probably involving the transfer of a 5-phosphoribosyl moiety from phosphoribosyl pyrophosphate (PRPP) to the base by specific membrane-bound transferases. Furthermore, mutants of *S. typhimurium* which lack uracil phosphoribosyl transferase do not transport uracil, suggesting the involvement of this enzyme in uracil uptake. It is reasonably certain that separate enzymes transfer a 5-phosphoribosyl moiety to the pyrimidine uracil, and to the purines guanine, hypoxanthine and xanthine, yielding UMP, GMP, IMP and XMP respectively as the products of translocation. A phosphoribosyl transferase capable of converting cytosine to CMP has never been detected in *E. coli*; presumptive evidence suggests that, prior to uptake, cytosine is deaminated by a periplasmic enzyme (see section 4.2) to uracil which is then translocated by uracil phosphoribosyl transferase. However a study of the uptake of adenine by *E. coli* membrane vesicles revealed several disconcerting anomalies which indicate caution in accepting an obligatory role for phosphoribosylation in uptake of at least this purine.

Purine auxotrophs which also lack nucleoside phosphorylase activity, cannot utilise purine nucleosides to satisfy their purine requirements, but freely utilise the free purine bases. It has been suggested that prior to, or during, translocation purine and pyrimidine nucleosides are cleaved by nucleoside phosphorylase in a group translocation reaction to free base and ribose-1-phosphate. During the cleavage the latter is translocated across the membrane. The free base is liberated into the periplasm and may then be translocated by the appropriate purine or pyrimidine phosphoribosyl transferase, as described above.

In *E. coli,* the capacity to transport fatty acids of chain length C_8 up to about C_{16} is induced concomitantly with the ability to degrade these compounds. Mutants lacking the membrane-bound acyl-CoA synthetase were unable to transport these fatty acids, suggesting that transport may be by a group translocation process involving vectorial acylation as the fatty acid crosses the membrane.

4 Isolation and characterisation of transport proteins

Some progress has been made in isolating proteins involved in transport of solutes across bacterial membranes. The protein components of the PEP—sugar phosphotransferase systems have already been mentioned (see section 3.1). Enzyme I has been highly purified and HPr has been isolated in homogeneous form from *E. coli, S. typhimurium* and *Staphylococcus aureus.* Factor III(lactose) and III(glucose) have been purified to homogeneity from *Staphylococcus aureus* and enteric bacteria respectively. They are all soluble, cytoplasmic proteins and are amenable to conventional techniques of enzyme purification. Their function in transport is, however, of an ancillary nature since they are not directly concerned with translocation of sugars across the membrane. Similarly, some of the proteins possibly involved in transport of purines and pyrimidines (see section 3.2) have been partially purified but their relationship to the plasma membrane is uncertain.

This section is concerned with the membrane-bound or cell envelope-associated proteins which have a recognition site for solute and its analogues and are involved in the transport of that solute.

4.1 Isolation of M protein

Genetic analysis of the lactose catabolic pathway established unequivocally that the product of the *lacY* gene is implicated in the transport of lactose and other β-galactosides (see section 2.1). Transport is inhibited by *N*-ethyl maleimide (NEM) and protected from inhibition by β-galactosyl-1-thio-β-galactoside (TDG). Using a double labelling technique involving treatment of induced and non-induced organisms, an NEM-sensitive protein which was protected by TDG was detected in membranes of induced wild type organisms or constitutive mutants. This protein, designated M (for membrane) protein, was absent from membranes of non-induced wild type organisms of *lacY⁻* mutants (Table 12.1). The observations suggest that the M protein

Table 12.1   Genetic control of the formation of M protein in membranes of *E. coli* mutants

Organism	*Genotype*	*Inducer*	*M protein**
ML30	$lacI^+Z^+Y^+A^+$	−	0
		+	135 ± 3
ML308	$lacI^-Z^+Y^+A^+$	−	111 ± 5
ML3	$lacI^+Z^+Y^-A^+$	+	0
ML35	$lacI^-Z^+Y^-A^+$	−	0
ML308-225	$lacI^-Z^-Y^+A^+$	−	119 ± 5

*Differences (in picomoles/mg protein) in [³H] NEM bound in the presence and absence of TDG. (Modified from Kennedy 1970).

is specified by the *lacY* gene, a conclusion substantiated by the study of a temperature-sensitive revertant of a *lacY* mutant of *E. coli* K-12. The revertant exhibited a Lac⁺ phenotype when grown at 25°C but was Lac⁻ when grown at 42°C. Transport of β-galactosides and binding of TDG to the M protein was similarly temperature-sensitive. Further tests with this revertant, including genetic crosses and measurement of transport activity of recombinants established virtually unambiguously that *lacY* is the structural gene specifying the M protein. The M protein is firmly bound to the membrane fraction, but may be removed by washing with buffer containing the ionic detergent sodium dodecyl sulphate (SDS). This pioneer work on the M protein has been reviewed by Kennedy (1970).

4.2 Periplasmic binding proteins

4.2.1 Nature and location

Heppel and his associates showed that when intact *E. coli* organisms are subjected to osmotic shock about 5 per cent of the total cell protein is liberated into the so-called 'shock fluid' without appreciable loss of viability of the bacteria. The shock fluid contains a number of degradative enzymes, but is devoid of a wide variety of enzymes known to be located in the cytoplasm. The same group of degradative enzymes is liberated during spheroplast formation by the combined action of lysozyme and ethylenediamine tetraacetate (but not during formation of spheroplasts due to the action of penicillin). It was suggested that the proteins liberated by osmotic shock and spheroplast formation are located in the periplasm, between the cytoplasmic membrane and the outer membrane of the Gram-negative cell envelope. This supposition has to some extent been substantiated by histochemical investigations (Heppel 1969).

Among the periplasmic proteins liberated are several for which no enzymic activity is known, but which specifically bind certain solutes and it is widely accepted that these binding proteins are, in some unknown way, involved in transport. Periplasmic binding proteins described include those for galactose, arabinose, ribose, glutamine, cystine, phosphate (all from *E. coli*), sulphate, histidine (*S. typhimurium*) and phenylalanine (*Comomonas*). *E. coli* also contains a protein which specifically binds leucine (LS protein) and another which binds leucine, isoleucine and valine (LIV protein).

The question of the precise location of the so-called periplasmic binding proteins is important to an understanding of their possible role in transport, but it cannot yet be answered unequivocally. Cytochemical methods show that the sulphate binding protein of *S. typhimurium* and the leucine binding protein of *E. coli* occur somewhere outside the cytoplasmic membrane. However, many peripheral (extrinsic) membrane proteins (see Ch. 11) are readily eluted from membranes by very mild treatments. Perhaps the so-called periplasmic binding proteins should be regarded as peripheral membrane proteins, loosely associated with, and readily dissociated from, the outer surface of the cytoplasmic membrane.

4.2.2 Evidence for involvement of binding proteins in transport

Several types of evidence, summarised below, implicate the periplasmic binding proteins in transport (Heppel 1969, Oxender 1972, Boos 1974).

(1) Release of binding protein by osmotic shock is accompanied by a reduction in transport activity, though transport of solutes for which binding proteins are unknown is unimpaired. Attempts to restore transport activity by addition of binding proteins to osmotically-shocked organisms have met with varying degrees of success and have not proved reproducible. In many instances loss of transport activity following osmotic shock (or preparation of membrane vesicles) is only partial, but residual activity can usually be attributed to the presence of multiple transport systems for the solute studied. In the few instances (of which glutamine uptake by *E. coli* is an example), in which transport of a solute is effected by a single system which is dependent on a binding protein, cold osmotic shock treatment of intact cells leads to almost total loss of transport activity.

(2) The specificity of a binding protein usually reflects the specificity of a transport system, as determined by uptake experiments with intact organisms.

(3) The K_m for transport of a particular solute usually approximates to the dissociation constant K_d of binding between that solute and the relevant binding protein.

(4) The most compelling evidence for a role of periplasmic binding proteins in transport comes from the study of mutants. For example, galactose is transported by several transport systems, including the so-called methyl-β-galactoside system (see section 6). Some *E. coli* mutants which failed to accumulate galactose by this system lacked functional galactose binding protein. Selection for revertants with restored transport properties led to restoration of binding protein activity. *Salmonella typhimurium* mutants which lacked a histidine binding protein were also unable to transport histidine by a specific high-affinity system (see section 5.3). Furthermore, a temperature-sensitive revertant of such a mutant regained transport activity at the permissive temperature and formed a histidine binding protein with decreased heat stability. Many examples of the co-regulation, by induction or repression of transport activity and binding protein formation are known. The concomitant appearance of binding and transport activities does not, however, prove an obligatory role for a binding protein in transport (Boos 1974).

4.2.3 Properties and role of periplasmic binding proteins

Periplasmic binding proteins, many of which have been highly purified and some obtained in the crystalline state, are all of moderate size, their molecular weights ranging from 25 000 to 42 000 daltons. Physicochemical studies suggest that most of the periplasmic binding proteins possess only a single solute binding site, though there is some evidence that the galactose binding protein of *E. coli* possesses two such sites (Boos 1974).

Although evidence presented above suggests that the periplasmic binding proteins are implicated in transport, their precise role is still obscure. Many workers regard these proteins as ancillary carriers which facilitate movement of solutes through the periplasm, solute molecules subsequently being transferred to specific, membrane-bound carriers responsible for actual translocation. Singer (1974) has extended the concept that the so-called periplasmic binding proteins are peripheral membrane proteins loosely bound to the outer membrane surface, and has produced a model which visualises a transport role for these proteins. He suggested that they are attached to aggregations of subunits of integral proteins which traverse the membrane, the aggregates enclosing a hydrophilic, water-filled pore through which the solute can be translocated. The solute-binding site of the binding protein is accessible to solute at the outer face of the membrane and, by a conformational change of the protein complex, possibly triggered by an energy-reaction, the binding site is re-oriented to face inwards, the pore is functionally open and solute is translocated across the membrane. Both integral and peripheral (binding) proteins are essential for transport; in the absence of the latter the pore is functionally closed. Whilst there is no direct evidence for the model, a number of transport systems, for example the histidine-specific system in *S. typhimurium* (see section 5.2) and the methyl-β-galactoside transport system of *E. coli* (see section 6) involve two or more proteins, one of which is a periplasmic binding protein.

In addition to their possible role in transport, some of the periplasmic binding proteins also act as chemoreceptors in chemotaxis (see Ch. 13).

4.3 Other membrane-bound transport proteins

Some Enzyme II components of the PEP–sugar phosphotransferase systems (see section 3.1) have been partially purified following extraction from membranes by a variety of detergents and organic solvents. These include Enzyme IIB′(lactose) from *Staphylococcus aureus,* Enzyme IIB′(glucose) and Enzyme IIA/IIB(glucose, fructose, mannose) from *E. coli,* and IIA component of the latter system having been separated into three fractions, each specific for one of the sugars transported.

Using the double-labelling technique previously exploited in the study of the M protein, three membrane-bound proteins have been detected in *B. subtilis* induced for uptake of succinate, fumarate and malate, but were absent from non-induced bacteria. Similarly, osmotic shock released a periplasmic succinate-binding protein and, in addition, two membrane-bound proteins capable of binding succinate have been solubilised from membranes of *E. coli* induced for uptake of succinate, fumarate and malate. Further investigation is required before the precise relationship of these proteins to transport of C_4-dicarboxylic acids is ascertained, especially in view of a claim that *B. subtilis* may contain two inducible systems for transport of these compounds.

5 Transport of amino acids and peptides

The uptake and accumulation of amino acids by *E. coli* appears to be exclusively by active transport, though the nature of the energy coupling may vary in different systems (see section 8). The kinetics of amino acid transport are basically similar to those of other active transport systems as already described in relation to β-galactoside transport (see section 2.2). The amino acid transport systems are fairly specific, though some systems transport groups of structurally-related amino acids. Possibly more important is the observation that several amino acids are transported by more than one system (see reviews by Oxender 1972, Simoni 1972).

Periplasmic binding proteins (see section 4.2) are involved in the uptake of glutamate, glutamine, histidine, cystine, basic amino acids and the branched-chain amino acids (leucine, isoleucine and valine). Most amino acid transport systems are constitutive; however some, especially those involving periplasmic binding proteins, are repressible. In some bacteria inducible systems which appear to duplicate the function of known constitutive systems are occasionally found. For example, *E. coli* forms an inducible system for glutamate and one for tryptophan. Similar inducible systems for glutamate and proline are found in *Pseudomonas aeruginosa* and an inducible system for arginine occurs in *B. subtilis* and other organisms (Slayman 1973). Inducible catabolic pathways, whereby these amino acids may be utilised as sources of carbon and/or nitrogen exist in these organisms and it has been suggested that the function of the inducible amino acid transport systems is the uptake of amino acids for catabolic purposes, rather than for protein biosynthesis.

Structural analogues of amino acids, frequently strongly growth-inhibitory, are usually transported by the system responsible for uptake of the corresponding natural amino acid. The mechanisms responsible for the growth-inhibitory effects of analogues need not concern us, but the effect itself can be turned to good account since analogue-resistant bacterial mutants can be readily selected. Some of these mutants owe their resistance to the ability to exclude analogues from the site (or sites) at which toxicity is exerted as a result of failure to produce a functional transport system. Thus amino acid analogues have proved valuable tools for the selection and genetic mapping of transport-deficient mutants. The review of Slayman (1973) documents the amino acid transport systems and transport mutants of *E. coli, S. typhimurium, P. aeruginosa, Neurospora crassa* and *Saccharomyces cerevisiae*. This account will be confined to discussion of a few representative systems.

5.1 Transport of proline in *E. coli*

The amino acid proline is transported by a highly specific system lacking affinity for any α-amino acids, though a number of proline analogues are transported, as shown by their ability to inhibit proline uptake competitively. Uptake of radioactive proline displays typical Michaelis–Menten kinetics. Some proline analogues, such as 3,4-dehydroproline and azetidine-2-carboxylic acid are strongly growth-inhibitory and selection for analogue-resistance allows the isolation of mutants defective in the ability to transport proline.

Study of the transport of a wide variety of analogues also yields information about the specificity of the amino acid recognition site of a particular transport system. In this way it was concluded that the features which are important for interaction with the recognition site of the proline transport system include, (*a*) the possession of the carbonyl portion of the carboxyl group, though amidation or esterification of the carboxyl group does not destroy activity as a substrate, suggesting a negative charge is unnecessary for activity; (*b*) the spatial position of the carbonyl group; (*c*) the presence of a secondary amine group; (*d*) the size and configuration of the ring; and (*e*) the size, position and polar or non-polar nature of substituent groups in the ring.

5.2 Aromatic amino acids

Uptake of aromatic amino acids is mediated by multiple transport systems. A general aromatic amino acid transport system capable of transporting phenylalanine, tyrosine and tryptophan and many analogues of the aromatic amino acids has been recognised in both *E. coli* and *S. typhimurium,* together with three transport systems, each specific for a single aromatic amino acid. In addition, as already mentioned, *E. coli* forms an inducible tryptophan transport system. Mutants of *E. coli* and *S. typhimurium* deficient in the general aromatic transport system may be obtained by selecting for resistance to azaserine or (in *S. typhimurium*) to some phenyl-alanine analogues. The mutations in both organisms map in a gene designated *aroP* which is presumed to be the structural gene of a protein component of the general aromatic amino acid transport system.

5.3 Histidine

The general aromatic amino acid transport system of *S. typhimurium* also transports histidine with rather low affinity. G. F. Ames has shown that this organism also contains a histidine-specific system. When rate of uptake of histidine is studied as a function of histidine concentration, two K_m values are obtained, one corresponding to the histidine-specific system, the other to the general aromatic amino acid transport system. Complex kinetics of this kind, yielding biphasic Lineweaver–Burk plots, are frequently encountered in transport studies and are usually interpreted as indicative of transport of a solute by two systems, each with a characteristic K_m value. Such a conclusion is justified here since genetic evidence substantiates the view that histidine is transported by the general aromatic system and also by a histidine-specific system, but it has been pointed out that multiphasic transport kinetics are open to alternative explanations (Boos 1974).

By appropriate selection techniques several classes of *S. typhimurium* mutants with impaired or even amplified histidine transport activity have been obtained by Ames and her coworkers (see reviews by Simoni 1972, Slayman 1973). These include (*a*) *hisP* mutants which lack histidine-specific transport activity, but contain a normal complement of the periplasmic histidine binding protein; (*b*) *dhuA* (D-histidine-utilising) mutants which transport histidine more efficiently than the wild type organism and contain elevated levels of the binding protein; and (*c*) *hisJ* mutants which lack the binding protein. Further genetic and biochemical evidence indicates that the *hisP* gene codes for some, as yet, unidentified protein and that a second periplasmic binding protein may be involved in histidine transport. Ames postulated that histidine can be transported via either the binding protein specified by the *hisJ* gene or by the second binding protein (designated protein K), but that the product of the *hisP* gene is involved in both pathways.

5.4 Branched-chain amino acids

Guardiola and his colleagues claim that in *E. coli* not less than six, possibly seven, separate transport systems mediate the uptake of branched-chain amino acids. By kinetic and genetic studies they recognised a very high affinity system (VHA), a high affinity system (HA-1) specified by gene *brnQ* (for branch-chain) and another high affinity system (HA-2) specified by the *brnS* gene. The three systems all transport valine, isoleucine and leucine. A third gene, *brnR*, seems to be involved in the expression of HA-1 and HA-2, but its role is not yet clear. The three genes are not closely linked on the *E. coli* chromosome and the proteins which they specify remain unidentified. In addition to these three general systems there is evidence of three low-affinity systems, each specific for one of the branched-chain amino acids and the possibility of a seventh system, having a high affinity for leucine. This study, which is reviewed by Slayman (1973), illustrates how a genetic approach can assist in unravelling the complexities of multiple transport systems, some having group rather than absolute specificity.

Other workers have isolated a periplasmic binding protein (see section 4.2) which binds leucine, isoleucine and valine (LIV-I system) and also another binding protein which binds leucine specifically (LS system). Production of both these binding proteins is repressed by leucine. Another system, LIV-II, also transports all three branched-chain amino acids, but with rather low affinity. Unlike the LIV-I and LS systems, LIV-II probably does not involve a periplasmic binding protein (Oxender 1972). The LIV-I system can probably be equated with HA-2 and LIV-II with the HA-I of Guardiola and his associates, but the observations of the two groups of workers cannot yet be fully reconciled.

5.5 Transport of peptides

Peptides are transported by systems which are quite distinct from those responsible for uptake of amino acids. *Escherichia coli* possesses one transport system for active uptake of dipeptides of the LL-configuration; the presence of an N-terminal α-amino group and C-terminal carboxyl group are essential for transport of the dipeptide. A separate, single transport system is responsible for uptake of oligopeptides. For transport, an oligopeptide must possess a free N-terminal α-amino group, though absence or substitution of the C-terminal carboxyl group is tolerated. Due probably to a 'sieving' effect of the cell wall only relatively small oligopeptides are transported; transport is limited by a Stokes' radius greater than a certain limiting value. (Payne and Gilvarg 1971).

6 Transport of galactose in *Escherichia coli*

Galactose is transported by several systems in *E. coli* and the study of the uptake of this sugar illustrates the difficulties and complexities which are encountered in investigations of microbial transport. This sugar is a substrate of not less than seven different transport systems (Table 12.2). Without the aid of suitable mutants it would have been virtually impossible to unravel the complexities of galactose transport. By combining a knowledge of the genetic regulation and kinetics of these systems it is possible to discriminate between them (Boos 1974, Kornberg 1976). The methyl-β-galactoside (Mgl) transport system has a very high affinity for galactose. By studying the uptake of 0.5 μM [^{14}C] galactose by a *galK$^-$ mglR$^-$* mutant it is possible to obtain reasonably specific assays of this system. The mutant lacks galactokinase activity, so ensuring that transported galactose accumulates in the organisms and the *mglR$^-$* mutation ensures the constitutive production of the Mgl transport system, thus avoiding the necessity to grow the organisms with galactose or fucose which would also induce other, undesirable, transport systems. Alternatively, by growing a *galK$^-$ lacY$^-$* mutant at 37°C in the presence of an inducer of the Mgl system, formation of the lactose (TMG I) and melibiose (TMG II) systems is avoided, so allowing assay of the Mgl system. Similarly the galactose (GalP) system may be studied in a *galK$^-$ lacY$^-$ Mgl$^-$* mutant grown at 37°C.

Table 12.2   Systems transporting galactose in *E. coli* K-12

Transport system	Some inducers	Structural genes	Regulatory gene	Remarks
Lactose (TMG I)	β-Galactosides β-Thiogalactosides Galactose Melibiose	*lacY*	*lacI*	Induced by galactose in *galK⁻* mutants
Melibiose (TMG II)	Galactose Melibiose	*melB*	?	Induced by galactose in *galK⁻* mutants Formed during growth at 25°C but not at 37°C
Methyl-β-galactoside (Mgl)	Galactose Fucose	*mglA,B,C*	*mglR mglD**	Periplasmic binding protein
Galactose (GalP)	Galactose Fucose TMG	*galP*	*galR*	
Arabinose	L-arabinose	(a) *araF*	*araC*	Periplasmic binding protein. High affinity for arabinose, galactose and fucose
		(b) *araE*	*araC*	No binding protein. Low affinity for arabinose, galactose and fucose
Enzyme II (glucose)† complex of PEP–sugar phosphotransferase				Transports galactose by facilitated diffusion; galactose is not phosphorylated

*Possibly operator gene.
†Probably IIB′ of III/IIB′ (glucose) system (see Postma and Roseman 1976).

7 Transport of hexose phosphates

Cell membranes are normally impermeable to phosphate esters, but *E. coli, S. typhimurium, Staphylococcus aureus* and some other bacteria form an inducible active transport system catalysing uptake of glucose-6-phosphate, 2-deoxyglucose-6-phosphate, mannose-6-phosphate, fructose-6-phosphate, glucose-1-phosphate and fructose-1-phosphate (reviewed by Dietz 1976). Only exogenous glucose-6-phosphate and 2-deoxyglucose-6-phosphate are primary inducers. The mechanism whereby an exogenous (but not endogenous) sugar phosphate induces the system is not understood.

8 The problem of energy-coupling to transport

A currently active field in microbial transport research is directed towards elucidation of the nature of energy-coupling to active transport.

8.1 Permease and redox models

Based on studies of the kinetics of β-galactoside uptake by *E. coli* (see section 2.2), a number of models were proposed, most of which invoked energy-coupling to induce a change in affinity of protein carrier for the transported solute. Many of these so-called 'permease' models envisaged direct phosphorylation of the carrier protein by ATP or a related compound, the phosphorylated and dephosphorylated carrier states possessing different affinities for solute. The models, reviewed by Kepes (1971) have been widely criticised by Hamilton (1975) and others. Direct coupling of ATP to active transport is probably rare in bacteria. Coupling is usually indirect, neither the solute transported nor the carrier itself undergoing chemical modification.

Kaback and his co-workers demonstrated that uptake of solutes by bacterial membrane vesicles can be coupled to oxidation of substrates which differed between different species. They proposed that solute carriers are components of the electron transport chain, participating in electron transport through alternate oxidation and reduction of carrier protein SH groups. Thus electron transport along the chain is regarded as the driving force which energises translocation. In the oxidised state the carrier possesses a high affinity for its specific solute and the solute binding site is accessible to the outer surface of the plasma membrane, whereas in the reduced state the carrier changes its conformation, the affinity of the binding site for ligand is reduced and the solute is released from the inner membrane surface into the cytoplasm. The main feature of this model is the contention that electron transport along the respiratory chain is not only sufficient, but essential to drive accumulation of solutes and that ATP is unable to energise transport across vesicular membranes. Detailed criticisms of the model and the uncertainties which surround the topography of Kaback's vesicles (see section 1.2.2) will be found in the excellent reviews of Harold (1972), Boos (1974), Simoni and Postma (1975) and Hamilton (1975).

8.2 Chemiosmotic theory − role of proton-motive force

There is rapidly growing evidence that energy coupling to transport is best explained by the chemiosmotic theory of Mitchell, originally proposed to explain energy transduction in oxidative phosphorylation (Mitchell 1970). The electron carriers of the respiratory chain are assumed

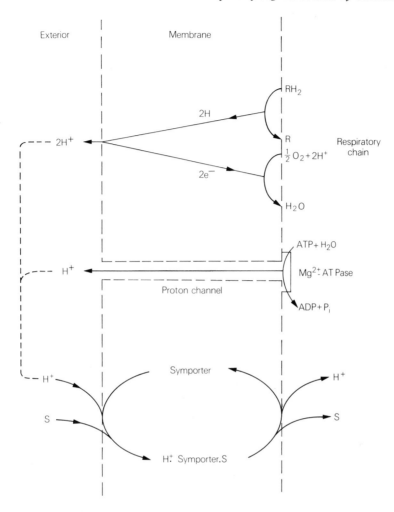

Fig. 12.4 Diagrammatic representation of coupling between proton extrusion and proton:solute (S) symport.

to be arranged across the plasma membrane such that there is an outward flow of hydrogen ions and an inward flow of electrons through each loop (see Fig. 12.4). Translocation of protons across the membrane results in the establishment of a proton-motive force Δp, comprising a membrane potential ($\Delta\psi$, inside negative) and a transmembrane pH gradient (ΔpH, inside alkaline) such that

$$\Delta p = \Delta\psi - Z\,\Delta\text{pH}$$

where Z is a factor converting pH into mV. The proton-motive force Δp may be generated either by respiratory or by photosynthetic electron transport or by ATP hydrolysis. The latter mechanism would be particularly important in bacteria such as *Streptococcus faecalis* which lack a cytochrome system. Furthermore, a membrane potential can be established not only by flux of protons but also by electrogenic movement of ions across the membrane.

Mitchell proposed that transport is coupled to proton movement, and hence to Δp. It was suggested that transport carriers act as bifunctional *symporters,* transferring the ligand specific to the particular transport system and also protons in the same direction across the plasma membrane. The flow of symporter-bound protons down the gradient into the cell is coupled to the transport of symporter-bound ligand. The proposed mechanism is represented in Fig. 12.4. Several other, though similar, possibilities are discussed by Mitchell (1973). Translocation driven by the proton-motive force does not, of necessity, involve symport of solutes, but may involve *uniporters* or *antiporters*. The former involve translocation of a single ligand; the latter results in coupled translocation of two (different) solutes in opposite directions. Examples of such systems are discussed by Mitchell (1970, 1973).

The main evidence supporting the chemiosmotic theory may be summarised as follows:

(1) Using a variety of methods, a membrane potential (interior negative) with values ranging from 50 to 250 mV has been measured in *E. coli, Staphylococcus aureus, Streptococcus faecalis* and other bacteria.

(2) In a number of bacterial species, including *E. coli,* and in membrane vesicles of *E. coli*, proton extrusion and transport of solutes is driven by respiration or by hydrolysis of ATP by Mg^{2+}-ATPase.

(3) In *E. coli*, addition of TMG to anaerobic organisms, in which glycolysis was inhibited by iodoacetate, resulted in uptake of TMG and an influx of protons with a stoichiometry of 1:1. The movement of TMG and protons was increased by SCN^- or valinomycin (see below).

(4) The chemiosmotic theory implies that artificially generated membrane potentials should be capable of driving active transport. Such electrochemical gradients (or proton-motive force) can be generated by passage, through the membrane of a readily permeant anion such as SCN^- or by inducing K^+ efflux by addition of the ionophore, valinomycin, the mode of action of which has been reviewed (Gale *et al.* 1972). Streptococci possess a high intracellular K^+ content. Addition of valinomycin to starved organisms in a K^+-free environment induces a K^+ efflux accompanied by uptake of β-galactoside or amino acids. Similarly, if *E. coli* vesicles are pre-loaded with K^+ followed by exposure to a low-K^+ environment containing valinomycin and proline, a membrane potential is established with concomitant transport of proline into the vesicles. This is a key experiment since neither N,N'-dicyclohexylcarbodiimide (DCCD), an inhibitor of ATPase, nor respiratory inhibitors prevented the uptake of proline, suggesting that the vesicles transported proline in response to the artificially generated membrane potential without the participation of vesicular metabolism (Simoni and Postma 1975).

(5) The theory also predicts that solutes should be transported in response to an artificially induced transmembranal pH gradient. Manipulation of external pH to create such a gradient promotes uptake of β-galactosides or amino acids in several bacterial species.

At physiological pH the amino acid lysine exists as a cation carrying a single positive charge, the anion glutamate carries a single negative charge and the neutral amino acids such as isoleucine are uncharged. Hamilton (1975) has predicted, and using intact *Staphylococcus aureus*, has verified, that the isoleucine carrier is a symporter, transport being driven, as with β-galactosides, by the total proton-motive force ($\Delta\psi - Z\Delta pH$). Transport of lysine is catalysed by a carrier acting as a uniporter and is driven by the membrane potential $\Delta\psi$ and, since a symporter carrier for glutamate and protons will carry no net charge, transport of glutamate is driven solely by the pH gradient ($-Z\Delta pH$).

Brief reference has already been made (see section 1.1) to the observations of Koch and his associates, namely that facilitated diffusion is, contrary to earlier belief, dependent on energy metabolism. Failure of stringently starved organisms to carry out facilitated diffusion is readily explained by the chemiosmotic theory. Under conditions of active uptake, solute and protons enter a bacterial cell via a symporter in response to a proton-motive force. Due to extrusion and lowering of internal proton concentration by the energy yielding reactions of electron transport or ATP hydrolysis, protons are released internally from the symporter and, since binding of solute by the symporter is dependent on concomitant binding of protons, solute is released internally. Consequently efflux of solute is prevented, thereby leading to accumulation. In the energy-depleted state, the mechanisms for proton extrusion are inoperative and symport of solute and proton leads to intracellular accumulation of protons, establishing a membrane potential (interior positive). This is the opposite situation to that normally obtaining and leads to cessation of solute accumulation by the symporter. Thus a facilitated diffusion process is converted to active transport, not by modification of the carrier, but by coupling to a steeper gradient of protons in the opposite direction.

The chemiosmotic theory also offers an explanation of the main weaknesses of Kaback's redox model, namely the effect of ionophores and especially the fact that uncouplers of oxidative phosphorylation inhibit active transport, even under anaerobic conditions. Uncouplers such as DNP or carbonylcyanide *m*-chlorophenylhyrazone (CCCP) dissolve in membranes and act as conductors of protons across the membrane. This collapse of the membrane potential accounts for inhibition of accumulation of solute by these compounds.

8.3 Energisation by ATP

Several workers have made a distinction between active transport systems involving periplasmic binding proteins and therefore sensitive to osmotic shock and those which are resistant to osmotic shock. At first sight such a distinction appears artificial and unnecessary since, as already discussed (section 4.2.3), the binding proteins may play only an ancillary role in translocation. Nevertheless the distinction may be real since shock-sensitive and shock-resistant systems in *E. coli* seem to be energised in different ways.

Whereas the transport of proline, serine, phenylalanine, glycine and cysteine (all shock-resistant) can be driven by either oxidative (electron transport) processes or hydrolysis of glycolytically produced ATP, transport systems for glutamine, arginine, diaminopimelic acid, leucine, isoleucine and ribose (all shock-sensitive) derive energy directly from ATP, as does the transport of galactose by the methyl-β-galactoside (Mgl) system. Evidence supporting this distinction includes the following observations:

(1) Arsenate, which causes a drastic depletion of intracellular ATP without preventing formation of the energised membrane state, strongly inhibits shock-sensitive systems, but not shock-resistant systems.

(2) Shock-sensitive systems are relatively resistant to uncoupling agents, whereas shock-resistant systems are strongly inhibited.

(3) Under anaerobic conditions, glycolytically-produced ATP can drive shock-sensitive systems, even in mutants lacking ATPase activity. On the contrary, although when driven by aerobic respiratory processes, shock-resistant systems are independent of ATPase activity, when driven by glycolytically-produced ATP, transport is dependent on a functional ATPase.

Henderson has demonstrated a similar correlation between binding protein-dependent systems and their apparent lack of coupling to proton movement. As already pointed out (section 6), galactose is transported by a variety of systems, including the so-called galactose permease (GalP) and also by the binding protein-dependent methyl-β-galactoside (Mgl) system. Similarly arabinose is transported by two systems; only one of which (specified at least in part by gene *araF*) is binding protein-dependent, the other, specified by *araE*, not being dependent on a binding protein for activity. When a pulse of oxygen is admitted to anaerobically maintained organisms of *E. coli* induced for the GalP and Mgl systems there is a rapid acidification of the medium (efflux of protons). Addition of galactose or 6-deoxygalactose

Table 12.3   Uptake of protons by proton—sugar symport
in *E. coli* mutants
(Data of P. J. F. Henderson, personal communication)

Transport substrate	*H*$^+$ *uptake (nmole)*		
	GalP$^+$ Mgl$^+$	GalP$^+$ Mgl$^-$	GalP$^-$Mgl$^+$
Galactose	47·7	30·6	0
Fucose	45·3	49·4	0
Methyl-β-galactoside	0·6	2·1	1·0

(fucose) results in alkalinisation of the medium (influx of protons) with a stoichiometry of solute/H$^+$ ratio of approximately 1:1, so lending strong support to the chemiosmotic theory. Methyl-β-galactoside does not produce this response. The influx of protons on addition of one of several solutes transported by one or both of these systems to organisms of different phenotypes shows that transport by the GalP (shock-resistant) system induces proton movement, whereas the binding protein-dependent Mgl system (shock-sensitive) does not (Table 12.3). Similarly, the shock-resistant arabinose transport system is a proton translocator, whereas the shock-sensitive system is not.

8.4 Summary

In summary, powering of active transport can be achieved by oxidation via the respiratory chain, ATP hydrolysis by Mg^{2+}-ATPase, by ion gradients or, in some bacteria, by photosynthesis. Each mode leads to establishment of an 'energised state' of the membrane. Both the permease model and Kaback's redox model envisage modification of the carrier during translocation. The latter model, in particular, is unacceptable since it fails to account for many of the experimental observations. The chemiosmotic theory, however, identifies the 'energised state' of the membrane as a proton-motive force generated by passage of protons across the membrane which, besides being coupled to transport, can be utilised for ATP synthesis and some other metabolic processes. Nevertheless, some transport systems derive their energy by other modes, in particular directly from ATP or PEP.

9 Concluding remarks

Although much has been learned about transport of solutes into microbial cells, we do not yet understand the molecular mechanism by which any single solute is actually translocated across the plasma membrane. Progress in this facet of transport will, of necessity, be closely linked with research currently in progress on the topography of membranes (see Ch. 11) and the study of carriers altered as a result of mutation. In this respect it is of interest that *E. coli* mutants have been isolated in which the β-galactoside carrier appears to be altered in such a way that β-galactoside transport is uncoupled from proton influx. The incorporation of a partially purified alanine carrier from a thermophilic bacterium into vesicles reconstituted from phospholipids extracted from the same organism and the finding that such vesicles will take up alanine is an exciting development which reveals a new facet for transport research in the future.

Knowledge of the uptake of nucleic acid bases and nucleosides is far from complete but studies so far have focused attention on two aspects, namely the important role of the periplasm of Gram-negative organisms in metabolism and the previously unsuspected complexity of the interrelationships between transport and metabolism. The link between transport and the first step in the metabolism of a solute is also apparent in group translocation of sugars and fatty acids.

Workers on biological transport were slow to accept Mitchell's far-reaching chemiosmotic theory. The relative simplicity of the concept providing, as it does, a uniform theory applicable to all organisms, makes it intellectually satisfying and it is now gaining universal acceptance, supported by an ever-increasing body of evidence.

References

BACHMAN, B. J., LOW, K. B. and TAYLOR, A. L. (1976) 'Recalibrated linkage map of *Escherichia coli* K-12', *Bacteriol. Rev.,* **40**, 116–67.

BOOS, W. (1974) 'Pro and contra carrier proteins; sugar transport via the periplasmic galactose-binding protein', *Curr. Topics Membr. Transp.,* **5**, 51–36.

CHRISTENSEN, H. N. (1975) *Biological Transport* (2nd edn.). Benjamin, Reading, Mass.

CORDARO, C. (1976) 'Genetics of the bacterial phosphoenolpyruvate: glucose phosphotransferase system', *Ann. Rev. Genetics,* **10**, 341–59.

DIETZ, G. W. (1976) 'The hexose phosphate transport system of *Escherichia coli*', *Adv. Enzymol.,* **44**, 237–59.

GALE, E. F., CUNDLIFFE, E., REYNOLDS, P. E., RICHMOND, M. H. and WARING, M. J. (1972) *The Molecular Basis of Antibiotic action.* Wiley, London.

HAMILTON, W. A. (1975) 'Energy coupling in microbial transport', *Adv. Microb. Physiol.,* **12**, 1–53.

HAROLD, F. M. (1972) 'Conservation and transformation of energy by bacterial membranes', *Bacteriol. Rev.,* **36**, 172–230.

HEPPEL, L. A. (1969) 'The effect of osmotic shock on release of bacterial proteins and on active transport', *J. Gen. Physiol.,* **54**, 95S–109S,

HOCHSTADT, J. (1974) 'The role of the membrane in the utilisation of nucleic acid precursors', *CRC Crit. Rev. Biochem.,* **2**, 259–310.

KABACK, H. R. (1970) 'The transport of sugars across isolated bacterial membranes', *Curr. Topics Membr. Transp.,* **1**, 36–99.

KENNEDY, E. P. (1970) 'The lactose permease system of *E. coli*', in *The Lactose Operon* (Eds. J. R. Beckwith and D. Zipser), pp. 49–92. Cold Spring Harb. Lab., New York.

KEPES, A. (1971) 'The β-galactoside permease of *Escherichia coli*', *J. Membrane Biol.,* **4**, 87–112.

KORNBERG, H. L. (1976) 'Genetics in the study of carbohydrate transport by bacteria', *J. Gen. Microbiol.,* **96**, 1–16.

KORNBERG, H. L. and JONES-MORTIMER, M. C. (1977) 'The phosphotransferase system as a site of cellular control', in *Microbial Energetics,* Soc. Gen. Microbiol. Symp. 27 (Eds. B. A. Haddock and W. A. Hamilton), pp. 217–240. Cambridge University Press, London.

MALONEY, P. C., KASHNET, E. R. and WILSON, T. H. (1975) 'Methods of studying transport in bacteria', *Methods in Membrane Biology,* **5**, 1–49.

MITCHELL, P. (1970) 'Membranes of cells and organelles: morphology, transport and metabolism', in *Organization and Control in Prokaryotic and Eukaryotic Cells,* Soc. Gen. Microbiol. Symp. 20 (Eds. H. P. Charles and B. C. J. G. Knight), pp. 121–66. Cambridge University Press, London.

MITCHELL, P. (1973) 'Performance and conservation of osmotic work by proton-coupled solute porter systems', *J. Bioenergetics,* **4**, 63–91.

OXENDER, D. L. (1972) 'Amino acid transport in microorganisms', in *Metabolic Transport* (Vol. 6 of *Metabolic Pathways*) (Ed. L. E. Hokin), pp. 133–85. Academic Press, New York.

PAYNE, J. W. and GILVARG, C. (1971) 'Peptide transport', *Adv. Enzymol.,* **35**, 187–244.

POSTMA, P. W. and ROSEMAN, S. (1976) 'The bacterial phosphoenolpyruvate:sugar phosphotransferase system', *Biochim. Biophys. Acta,* **457**, 213–57.

ROSEMAN, S. (1972) 'Carbohydrate transport in bacterial cells', in *Metabolic Transport* (vol. 6 of *Metabolic Pathways*) (Ed. L. E. Hokin), pp. 41–89. Academic Press, New York.

SIMONI, R. D. (1972) 'Macromolecular characterization of bacterial transport systems', in *Membrane Molecular Biology* (Eds. C. F. Fox and A. D. Keith), pp. 289–322. Sinauer Associates, Stamford, Conn.

SIMONI, R. D. and POSTMA, P. W. (1975) 'The energetics of bacterial active transport', *Ann. Rev. Biochem.,* **44**, 523–54.

SINGER, S. J. (1974) 'The molecular organization of membranes', *Ann. Rev. Biochem.,* **43**, 805–33.

SLAYMAN, C. W. (1973) 'The genetic control of membrane transport', *Curr. Topics Membr. Transp.,* **4**, 1–174.

13
Bacterial motility and chemotaxis

D. G. Smith

Department of Botany and Microbiology, University College London

1 Introduction

The mechanisms by which bacteria move actively through their environments have long intrigued microbiologists and exciting discoveries have been made in recent years. This has been coupled with renewed interest in bacterial chemotaxis, the movement towards or away from specific chemical stimuli, which can be regarded as a relatively simple model for the study of sensory reception and behaviour.

Active movement is by no means universal in bacteria and there are highly successful non-motile species, e.g. *Staphylococcus aureus*. Nevertheless, the ability to travel at speeds of up to about 18 cm h^{-1} and to respond to external stimuli could clearly be important for the success of motile species in the microenvironments in which they normally exist: Smith and Doetsch found that motile *Pseudomonas fluorescens* outgrew a non-motile mutant by 10:1 in 24 h under oxygen-limiting conditions. Furthermore, these capabilities require the possession of numerous genes for the synthesis of the organelles of motility and of the chemoreceptors for the detection of stimuli, indicative of a highly evolved system.

In this chapter three different mechanisms of active bacterial movement will be considered: flagellar, spirochaetal and gliding. This will be followed by a discussion of some of the recent developments in research on bacterial chemotaxis.

2 Flagellar movement

2.1 Arrangement of organelles

Flagella are arranged either at the ends of the bacterial cell (polar) or over the whole cell surface (peritrichous). The number per cell ranges from one to thousands and is a fairly stable characteristic of a species under standard conditions.

Individual bacterial flagella are too thin (20 nm) to be observed by light microscopy without staining, but where there are multiple flagella on a cell, bundles are produced which can be seen, in the living state, by phase contrast or dark field optics. Suitable staining increases the thickness of flagella rendering them visible: their number, arrangement, overall length, wavelength and amplitude can then be determined. Flagella are readily observed by electron microscopy after heavy metal shadowing or negative staining of the specimens (Fig. 13.1).

(a)   (b)

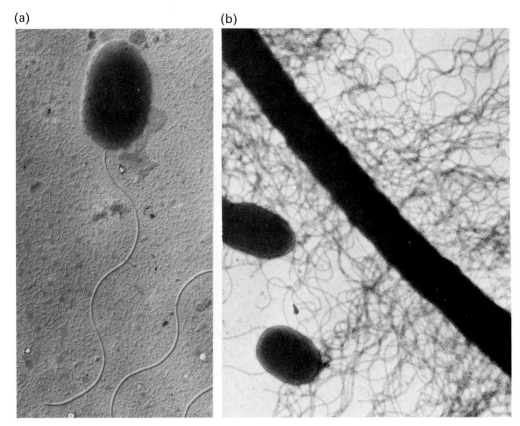

Fig. 13.1 Electron micrographs of flagellated bacteria: (a) *Pseudomonas fluorescens* with single polar flagellum, heavy metal shadowed preparation; (b) *Proteus mirabilis,* negatively stained, showing short cell with a few peritrichous flagella and part of a long swarming cell with numerous flagella.

2.2 Ultrastructure of flagella

Bacterial flagella have three main parts: a basal body inserted in the cell, a hook region and a shaft up to about 15 μm long (Fig. 13.2). The detailed ultrastructure of these components has been difficult to elucidate because of the limitations of specimen preparation and electron microscopy at the level of resolution required.

For many years there was controversy over whether the flagellum actually penetrated the cell wall and it was believed by some (notably Pijper and co-workers in the 1940s and 50s) that the flagella were extracellular twirls of material caused by but not causing bacterial motility. This debate was settled when it was shown that the flagella are essential for motility and that they originate from a basal body embedded in the cytoplasmic membrane. DePamphilis and Adler demonstrated the clear distinction between the basal bodies of Gram-positive and Gram-negative bacteria which correlates with the different organisation of their cell walls (Fig. 13.3).

The hook region provides a connection between the basal body and the flagellum shaft. In negatively stained preparations in the electron microscope the hook is seen as a structure up to about 100 nm long, slightly thicker than the shaft and apparently constructed of helically arranged subunits.

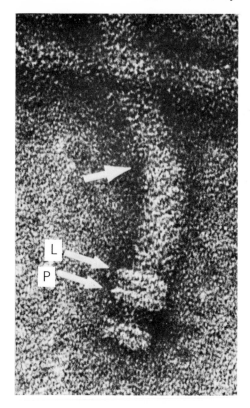

Fig. 13.2 Flagellar components in *E. coli*. In this electron micrograph of an isolated flagellum a portion of the shaft is seen joined to the hook (at arrow) which is attached to the basal body. (Reproduced, with permission, from DePamphilis and Adler (1971), *J. Bacteriol.*, **105**, 384.)

The flagellum shaft is a helix except in species where the cell itself is helical: like the hook it is constructed from subunits. Negatively stained specimens show an array of roughly globular particles about 5 nm in diameter but the symmetry of their packing is obscured by random 'noise' in the image and because both the topside and the underside of the shaft are seen in superimposition. This problem has been largely overcome by optical diffraction and filtering of electron micrographs to eradicate the noise and to show only one side of the structure (Fig. 13.4). The model that emerges from such studies consists of 11 nearly longitudinal rows of subunits (Fig. 13.5). Each subunit in the model has six adjacent subunits arranged hexagonally and the lattice can be considered as a series of helices with 1,5,6 and 11 starts but the actual operational helices present depend on the inter-subunit binding and little is known of this at present. This work has been done with mutants of *Escherichia coli* and *Salmonella typhimurium* having straight rather than helical flagella but it is assumed that only a slight change in subunit orientation would be necessary to give the helical form.

A tube formed from 11 rows of spherical subunits would have a large internal space. However, electron microscopy gave conflicting results about the existence of an internal passage and although it is now known that the flagellum does have a hollow centre it appears to be narrow. This suggests that the subunits are not spherical but elongated and a model based on this concept was suggested by Bode, Engel and Winklmair (Fig. 13.6). Externally, a structure of this sort would still appear to be composed of roughly spherical subunits.

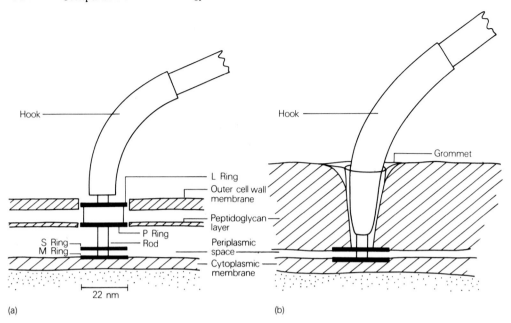

Fig. 13.3 Comparison of flagellum insertion in Gram-negative (a) and Gram-positive bacteria (b). In both cases the M (membrane) ring is embedded in the cytoplasmic membrane. The S (supermembrane) ring is closely associated with the M ring and in the Gram-positives may be attached to the inside of the thick peptidoglycan layer. In the Gram-negatives the additional P (peptidoglycan) and L (lipopolysaccharide) rings form a bearing for the rod to pass through the outer cell wall membrane. The hook of Gram-positives is longer than that of Gram-negatives but the rod is shorter, suggesting the arrangement shown in (b). A 'grommet' observed in *Clostridium sporogenes* may act as a bearing in Gram-positive bacteria.

The lattice structure of the bacterial flagellum model probably indicates the relative positions of the protein subunits of which it is composed but it does not necessarily represent the structure of the flagellum in the living state. Evidence from *Proteus mirabilis* and *Bacillus subtilis* suggests that, with accurate focusing and elimination of astigmatism in the electron microscope, flagella have an amorphous appearance.

Sheathed flagella are found in some species, e.g. *Vibrio metchnikovii* and *Bdellovibrio bacteriovorus*. The core is of similar structure to unsheathed flagella and the surrounding sheath is an extension of the cell wall giving an overall diameter of about 35 nm.

2.3 Composition of bacterial flagella

The composition of bacterial flagella is simple compared with that of eukaryotic flagella and cilia. The shaft of the flagellum consists of multiple units of a single protein known as flagellin and these are believed to correspond with the subunits observed by electron microscopy.

Being effective antigens, flagellins are useful in the classification and identification of bacteria, especially the Enterobacteria, e.g. the Kauffmann–White Scheme for *Salmonella* spp. and they provide a suitable system for investigating the relationship of molecular structure to antigenic specificity.

The only flagellin to be completely sequenced is that of *Bacillus subtilis* strain 168. This flagellin has a molecular weight of 32 600 and is a single polypeptide of 304 amino acid residues. The amino acid composition, shown in Table 13.1, is unusual in that there is no

(a)   (b)   (c)

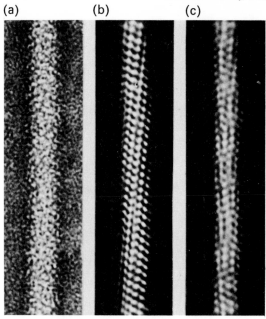

Fig. 13.4 Processing of an electron micrograph to show subunit arrangement in a straight flagellum mutant of *Salmonella typhimurium*: (a) original micrograph; (b) one-sided filtered image from (a); (c) two-sided filtered image. Magnification × 400 000. (Reproduced, with permission, from O'Brien and Bennett (1972), *J. Molec. Biol.*, **70**, 133.)

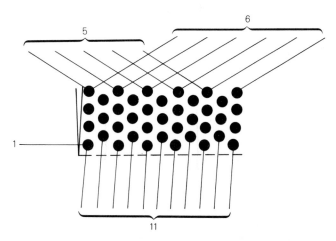

Fig. 13.5 Surface lattice of subunits in straight flagellum mutants. The directions of the 1,5,6 and 11 start helices are indicated. In the flagellum this lattice would be folded round the surface of a cylinder. In the wild type a small change in the geometry of the individual subunits forms a helical flagellum.

tryptophan or cysteine, only a single tyrosine and two proline residues. The total absence of cysteine from flagellin is possibly correlated with the need for a protein with a conformation not easily affected by changes in the redox potential of the environment. The distribution of hydrophobic amino acids throughout the sequence is random but there is asymmetry in the distribution of charged amino acids: the NH_2–terminal region is basic, the central region acidic

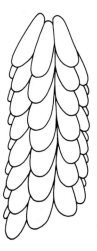

Fig. 13.6 Model of bacterial flagellum based on the evidence of Bode, Engel and Winklmair (1972) that the subunits are elongated rather than spherical.

and the COOH–terminal region weakly acidic. The tertiary structure of flagellin has not yet been elucidated so the conformations leading to monomer interactions and to the production of antigenic sites remain unknown. There is evidence from *P. mirabilis* flagellin, which contains five tyrosine residues per molecule, that two of the tyrosines are situated at the molecular surface and are necessary for the non-covalent bonding of one monomer to another.

Table 13.1   Amino acid composition of flagellin
from *Bacillus subtilis* strain 168*

Amino acid	*Number of residues per molecule*
Alanine	39
Leucine	29
Asparagine	27
Glutamine	24
Serine	24
Aspartic acid	22
Isoleucine	22
Glycine	19
Threonine	18
Glutamic acid	17
Lysine	15
Arginine	14
Valine	14
Methionine	8
Phenylalanine	5
Histidine	4
Proline	2
Tyrosine	1

*For the full sequence see the paper by DeLange
et al. (1976).

Although flagellins vary in detail from species to species the structure of a particular flagellin is critical for its proper functioning: a mutant of *B. subtilis* 168 with a single amino acid substitution (valine in place of alanine at position 233) has straight instead of helical flagella and is non-motile.

Flagellins from other bacterial species have only been partially analysed but there are general similarities to the flagellin of *B. subtilis*. The molecular weights are usually in the region of 30 000 to 60 000, tryptophan and cysteine are absent and phenylalanine, tyrosine, histidine and proline (the cyclic amino acids) are present in small amounts. There is also partial sequence homology between species as distantly related as *B. subtilis* (Gram-positive) and *S. typhimurium* (Gram-negative), strongly suggesting an evolutionary relationship. When more flagellin sequences become available they will provide valuable information on the phylogenetic relationships between flagellated bacteria.

Some serotypes of *Salmonella* have the unusual amino acid, ϵ-*N*-methyl lysine, in their flagellin. The methylation of the lysine residues occurs after synthesis of the flagellin protein. This amino acid is not found elsewhere in bacteria but it has been reported in histones from eukaryotic nuclei. However, methyl lysine is not essential and mutants lacking it function normally.

The hook region of the flagellum is antigenically distinct from the flagellin shaft. Kagawa *et al.* have isolated the hook protein from *Salmonella* spp. and compared its composition with that of flagellin from the same strain. The hook protein had a molecular weight of 43 000 compared to 56 000 for the flagellin; there was no methyl lysine in the hook but there were more cyclic amino acids. A similar hook composition has been reported in *E. coli* by Silverman and Simon.

2.4 Biogenesis of flagella

Salmonella typhimurium and *E. coli* have been the most intensively studied species for flagella biogenesis but work has also been done on *P. mirabilis* and *B. subtilis*.

There are about 20 genes concerned with the structure and operation of bacterial flagella. In *E. coli* they are all clustered around the 43 min. position on the linkage map (see Bachmann, Low and Taylor 1976). The primary sequence of the flagellin protein is determined by the *hag* (*H* antigen) gene. In addition, there are about ten other protein components of the flagellar apparatus (forming the hook and basal body) and these are the products of the *fla* (*fla*gella) genes. Other *fla* genes are concerned with regulation of flagella production and mutations lead to altered numbers of flagella per cell. There are also genes involved in the functioning of the flagella; these are designated *mot* (*mot*ility). The *mot* gene product in *E. coli* is a protein (M.W. 31 000) essential for the operation of the flagella: mutation leads to paralysis even though the protein is not a part of the flagellum itself but a component of the cytoplasmic membrane. Many of the genes concerned with motility in *E. coli* are grouped into multicistronic transcriptional units. Silverman and Simon have analysed these using bacteriophage mu-induced mutants selected by their resistance to the flagellotropic bacteriophage χ. Multicistronic transcriptional units reported for *E. coli* are: (*fla* R, *fla* Q, *fla* P, *fla* A), (*fla* E, *fla* O, *fla* C, *fla* B), (*fla* G, *fla* H) and (*fla* M, *fla* L, *fla* K). The other *fla* genes and the *hag* and *mot* genes are individually transcribed.

Salmonella is particularly interesting in that a given serotype can produce two different flagellins. This causes a change in antigenicity known as a phase change. Each flagellin is the product of a separate structural gene, i.e. *hag*1 and *hag*2 corresponding to flagellin of phase 1 and phase 2. Associated with the *hag*2 gene there is a regulatory gene *rh*1 (regulation of *hag*) which produces a repressor for the transcription of the *hag*1 gene. The *hag*2, *rh*1 operon oscillates between active and inactive states thus giving rise to the observed phase change of *Salmonella* spp.

The biosynthesis of flagellin and the other flagella components follows the normal process of protein synthesis. Evidence has been reported of unusually stable messenger RNA for flagellin but more recent reports give a half-life of about 7 min., which is not abnormal. The site of biosynthesis is probably at the base of each flagellum and a specialised ribosome system, the flagellosome, has been postulated by Iiono but not substantiated. Similarly, there is doubt about the existence of a pool of flagellin molecules in the cytoplasm.

In *E. coli* and probably in some other motile bacteria the biosynthesis of flagella is subject to catabolite repression: in the presence of glucose, flagella synthesis is decreased and when the glucose is exhausted flagella synthesis ensues. As in other cases of catabolite repression the effect on flagellation is mediated by cyclic AMP and the addition of exogenous cAMP to repressed cells stimulates flagella synthesis. According to Komeda *et al.* cAMP and the cAMP receptor protein modulate one of the *fla* genes (*fla* T) which regulates the *hag* gene in *E. coli*, thereby controlling flagellin synthesis. Catabolite-insensitive mutants have been obtained: these probably have an altered cAMP-dependent promoter region for the control of synthesis of the *fla* T product.

How the cell regulates exactly at which spot on the cell surface a flagellum is going to be produced is not known. Presumably the basal body is formed at the appropriate site on the membrane followed by development of the hook region and then the flagellum shaft.

Studies on the regeneration of flagella on mechanically deflagellated cells have shown that the flagellum shaft grows out from the cell at about $0.2\ \mu m\ min^{-1}$, the rate gradually slowing until the flagellum reaches its maximum length of about 15 μm. The flagellin molecules of which the shaft is constructed are self-assembling, i.e. they spontaneously aggregate *in vivo* into a filamentous structure without enzymic mediation. Under suitable pH and ionic conditions isolated flagellin assembles into flagella-like filaments *in vitro*: this provides a convenient system for studies on the properties of normal and mutant flagellins and for investigating how flagellin subunits interact to form flagella filaments. Flagellins from different *Salmonella* strains copolymerise *in vitro* to form hybrid filaments of intermediate properties. With mixtures of normal plus straight flagella mutant flagellins five distinct stable forms are produced depending on the ratio of the two components, but with normal or straight plus curly mutant (half normal wavelength) flagellin the change in wave form is continuous rather than stepwise.

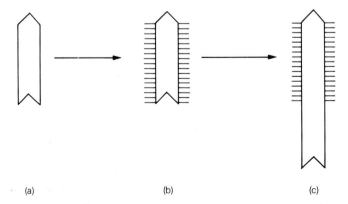

(a)   (b)   (c)

Fig. 13.7 Asymmetric growth of flagella fragments *in vitro* as demonstrated by Asakura, Eguchi and Iino (1968). The fragment (a) is labelled with specific antibody (b) which can be seen by electron microscopy. Removal of excess antibody followed by further polymerisation demonstrates the polarity of extension (c).

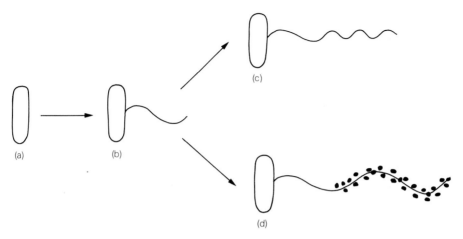

Fig. 13.8 Proof of terminal growth of bacterial flagella *in vivo*. Deflagellated cells (a) are allowed to partially regenerate flagella under normal conditions (b). Iino (1969) transferred cells of *Salmonella* to medium containing fluorophenylalanine the incorporation of which gives 'curly' flagella at the distal end (c). Emerson, Tokuyasu and Simon (1970) transferred *Bacillus* to medium with radioactive leucine: its site of incorporation was then detected by autoradiography. The location of the silver grains (d) confirms terminal growth of flagella.

Labelling of flagella fragments with antibody and then allowing further polymerisation in the absence of antibody has shown polarity in the growth of the filament, i.e. flagellin molecules are added to the structure at one end only (Fig. 13.7). The fragments have an asymmetric appearance: one end is pointed while the other end is notched. Monomers of flagellin are added only at the notched end.

It was natural to anticipate that in the living cell new flagellin molecules would be added to the growing flagellum at its base where it joins the hook, but surprisingly it was found by conclusive experiments with an amino acid analogue and a radioactive amino acid that flagella grow at their distal tips (Fig. 13.8). This means that flagellin molecules have to be transported to the tip of the flagellum before insertion into the growing structure. The probable route is via the internal passage in the flagellum since free flagellin is not detectable outside the cell.

2.5 Mechanism of flageilar movement

The flagella of eukaryotic organisms generate plane waves originating either from the base or the tip of the organelle. ATPase is present in these flagella and even when separated from the cell they can exhibit independent movement if ATP is added. The operation of bacterial flagella is totally different: the shaft is made up of a single non-enzymic protein and therefore there is no possibility of independent movement. The mechanism by which the bacterial flagellum operates has consequently been a particularly interesting problem stimulating many hypotheses and experiments.

Unlike the planar eukaryotic flagellum, the bacterial flagellum is normally a left-handed helical structure. It was therefore assumed that the flagellum is flexible and propagates helical waves from its base to its tip, the base being firmly fixed into the cell wall. Various mechanisms for generating the helical waves were suggested including: (*a*) formation of contractile fibres of flagellin; (*b*) wobbling of the basal body generating a wave passing down the flagellum; (*c*) dislocations in the arrangement of the flagellin molecules propagating a wave along the flagellum. However, none of these ideas explains why there is no damping out of the wave form as it passes down the flagellum, or how reversing of swimming direction occurs. In 1966, in a theoretical paper, Doetsch suggested that the bacterial flagellum could be a rigid helix and that motility would result from the rotation of this helix from its basal body. The idea seemed highly unlikely as there are no other known examples of rotary systems in biology. Nevertheless, recent research has supported the view that bacterial flagella operate in this way. Evidence for rotation includes the following: (*a*) bivalent antibody molecules cross-linking one flagellum to another stop motility but cells with a single flagellum are not affected; (*b*) polyhook mutants (*fla* E mutants with abnormally long hook regions) or straight flagella mutants both of which are non-motile show cell rotation when the hook or shaft is tethered to a surface with antibody; (*c*) polystyrene–latex beads coupled to a straight flagellum can be seen to rotate about the axis of the flagellum while the cell body rotates in the opposite direction.

The helical flagellum therefore acts like the propeller of a boat. For normal movement in polarly flagellate bacteria the flagellum rotates anticlockwise while the cell rotates clockwise, but more slowly due to viscous drag. A thrust is thereby generated and the cell moves forward with the flagellum trailing. Reversing the direction of the flagellum rotation causes the cells to back up and move flagellum first. In peritrichously flagellate bacteria the flagella still rotate anticlockwise but they tend to form a bundle, a feature made possible by the flexible nature of the hook region which acts like a universal joint. In this case reversing the flagella rotation causes the bundle to separate and smooth swimming is replaced by a characteristic random 'tumble'. Bacteria in the genus *Spirillum* have helical cells (a right-handed helix) but non-helical tufts of polar flagella. The flagella in this case produce cones of rotation at one or both ends of the cell causing rotation of the helical cell which then moves in the appropriate direction (Fig. 13.9).

If the flagellum rotates like a propeller then there must be a motor to cause the rotation. The complex basal body structure appears to fulfil this role and a model has been suggested by Berg. In this model the turning motion is generated between the S ring and the M ring of the basal body: the M ring is fixed to the flagellum while the S ring is fixed to the cell wall. In the

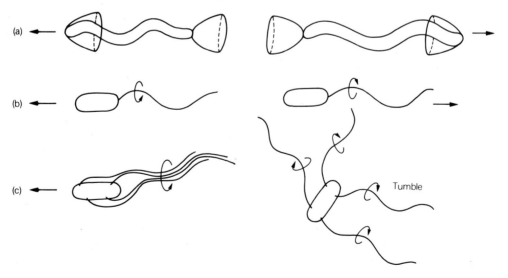

Fig. 13.9 Locomotion in flagellate bacteria. (a) The rigid helical cells of *Spirillum* reverse by the coordinated reversal of rotation of their flagellar tufts. (b) Bacteria with single polar flagella reverse by changing the direction of flagella rotation. (c) Anticlockwise rotation of flagella gives smooth swimming in peritrichously flagellate bacteria. Reversal of flagellar rotation causes tumbling.

Gram-negative cell the P and L rings act as bushes for the rotating rod. The molecular mechanism for the powering of the flagellum motor has not yet been determined, but it has been suggested that movement of ions between the M and S rings might be the cause of their relative movement. This is an energy-using process but, unlike muscle contraction and eukaryotic flagellar movement, ATP is not directly involved; instead the process appears to be powered by a high-energy intermediate.

3 Spirochaetal movement

3.1 Occurrence and description

Spirochaetes are flexible helical bacteria varying in size from 0·1 by 6–20 μm for *Leptospira* to 3·0 by 30–150 μm for *Cristispira*. Species of *Spirochaeta* can reach a length of 500 μm (0·5 mm) but are only about 0·5 μm in diameter. Some species of spirochaetes are saprophytes, others are harmless commensals or symbionts and some are pathogens causing diseases such as syphilis and relapsing fever.

All spirochaetes show active movement and since they are not easily stained for microscopic examination they are often best observed in the living state by dark-ground microscopy. Syphilis can be diagnosed in this way by taking material from a primary lesion and observing the characteristic swimming helical cells of *Treponema pallidum*.

Spirochaetal movement has been even more baffling than flagellar movement because the cells have no external organelles of motility and yet can perform a variety of movements such as flexing and spinning, free swimming through liquid and creeping on surfaces.

3.2 Structures associated with spirochaetal movement

The spirochaete often gives the impression of being wound around an axial filament which appears to run the length of the cell, and it has been suggested that the helical shape of the cell is maintained by the tension of the filament. However, this has been disproved by the demonstration that the isolated peptidoglycan from spirochaetes is itself helical.

After much confusion in the earlier literature the electron microscope has revealed that the axial filament is situated between the cytoplasmic membrane and the outer cell wall membrane. The filament is thus an internal (periplasmic) structure. Further investigations have shown that the axial filament is composed of 2–100 or more fibrils which resemble bacterial flagella both in structure and composition.

The fibrils are inserted into the cytoplasmic membrane near the pole of the cell with a structure identical to the basal body of flagella. Attached to this is a hook region and then a shaft which has the same appearance as the shaft of a flagellum. Fibrils originate from both ends of the cell and overlap in the central region. Analysis of the fibrils has shown them to consist of a protein, molecular weight 37 000, similar to flagellin in composition: aspartic and glutamic acids, alanine, leucine, glycine and serine are the most frequent amino acids, cysteine is absent and the cyclic amino acids present in low amounts. No sequence analyses have yet been reported.

Additional layers of material may ensheath the spirochaetal fibrils but it has become apparent that they are basically flagella which have become internalised in the course of evolution.

3.3 Possible mechanism of spirochaetal movement

The axial filament was assumed to be involved in spirochaetal movement: possibly contraction of the fibrils would cause a helical wave to pass along the cell. However the absence of a torque-resisting 'head' attached to the helix makes it impossible for forward movement to occur by this means. This difficulty would be overcome if the spirochaete is semi-rigid and rotates not only about its helix axis (like a corkscrew) but also spins around its local body axis. According to Wang and Jahn a helical wave of this sort, generated by the sequential contraction of axial fibrils, would allow translatory movement. This idea has been extended by Berg in the light of the knowledge that bacterial flagella rotate. He postulates that the axial fibrils of spirochaetes do not contract but rotate in the same way as flagella. Rotation of the fibrils in the periplasmic space would cause the protoplasmic cylinder to rotate in the opposite direction on its local body axis while the external cell wall layer would rotate in the opposite direction to the proto-plasmic cylinder (Fig. 13.10). This model presupposes that the protoplasmic cylinder is semi-rigid rather than rigid (like the cells of *Spirillum*) thus allowing the rotation on the local body axis. Free swimming is explained by the propagation of a helical wave and surface creeping by the rotation of the outer cell wall layer. Twisting of the protoplasmic cylinder in reaction to the rotating basal bodies will account for the flexing often observed in the more slender spiro-chaetes. There is, as yet, little direct evidence for this model but it would seem to provide a likely explanation and furthermore it suggests a unity in the mechanisms of bacterial swimming.

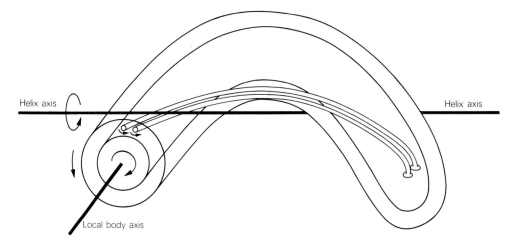

Helix axis

Helix axis

Local body axis

Fig. 13.10 Terminal coil of a spirochaete showing the arrangement of the axial fibrils in the periplasmic space and the possible rotary movements caused by the rotation of the fibrils against the cytoplasmic cylinder and the cell wall. The whole cell consists of up to about 30 of these coils.

4 Gliding movement

4.1 Occurrence and description

Gliding movement is distinct from the free swimming movements described above in that a solid surface is required for translocation of the cells. It is a property of numerous species of bacteria from widely different groups: (*a*) all species of the fruiting myxobacteria (Order Myxobacterales); (*b*) all species of the Order Cytophagales; (*c*) many unicellular and filamentous species of the Myxophyceae (blue–green algae/bacteria); and (*d*) some species of *Mycoplasma.* In addition a number of bacteria show a spasmodic type of surface translocation known as twitching.

The velocities achieved by gliding bacteria are low compared to free swimming, being of the order of the 3 μm s^{-1} at most.

4.2 Mechanism of gliding movement

It is likely that the mechanism of gliding differs in detail from one group of bacteria to another. Hypotheses for the mechanism have been grouped into four categories by Doetsch and Hageage: (*a*) movement due to a gradient of osmotic forces along the cell or filament of cells; (*b*) surface tension forces generated by localised secretion of surface-active material; (*c*) localised secretion of slime pushing the cells along; and (*d*) generation of contractile waves in the cell.

There are no external appendages or internal filaments visible in gliding bacteria but as traction occurs over a solid surface it is reasonable to conclude that the motive power is generated at the surface of the organism. Although slime is secreted by most gliding organisms its probable role is as an adhesive or lubricant between the cell and the substratum. The evidence available at present supports the contractile wave hypothesis although little is known about how these waves are propagated.

Halfen and Castenholz, working with the filamentous blue–green bacterium *Oscillatoria princeps*, reported that a parallel array of very fine fibrils was present in a layer near the cell surface. The fibrils were 5–8 nm thick and arranged helically around the cell with a pitch 30° from the longitudinal axis of the filament. They concluded that waves of bending passing along the fibrils would account for the rotation and forward movement observed in this organism.

Burchard has studied the unicellular myxobacterium *Myxococcus xanthus,* the cells of which aggregate to form fruiting bodies. Although no well-defined fibrillar structures have been seen in this organism he was able to find evidence that the basis of gliding motility resides near the cell surface. Gliding could be abolished by digesting the cells with the proteolytic enzyme pronase: the treated cells recovered their motility in about an hour but failed to recover in the presence of chloramphenicol, which inhibits protein synthesis. Similarly, release of surface and periplasmic proteins by EDTA treatment and osmotic shock stopped gliding and it could not be restored by adding back concentrated shock fluid. Attempts to identify a specific gliding protein were not successful but Burchard's results established that one or more surface or periplasmic proteins are required for gliding motility.

Unlike other gliding organisms, mycoplasmas lack a rigid cell wall and are therefore characteristically variable in morphology. Bredt *et al.* showed by microcinematography that reversible changes in shape occurred in *Mycoplasma hominis* on a glass surface and suggested that the cells have contractile material in their cytoplasm or membrane. The slow (maximum 0·75 μm s^{-1}) movement of mycoplasmas over a surface may perhaps be more accurately described as wriggling rather than gliding, although it is still possible that the underlying mechanism is similar to the gliding of the walled bacteria.

5 Bacterial chemotaxis

5.1 Directional movements in bacteria

Antony van Leeuwenhoek, who was the first to observe motility in bacteria three centuries ago, was aware of their ability to accumulate in favourable environments but the first detailed studies on bacterial chemotaxis were not made until 1881 when Engelmann investigated the chemotactic responses of bacteria to oxygen. However, elucidation of the underlying mechanism of chemotaxis was beyond the scope of available methodology and little further advance was made between the 1930s and the 1960s. Now, with the new techniques of genetic analysis and the vastly increased knowledge on structure provided by electron microscopy, progress towards understanding the phenomenon is proceeding rapidly.

5.2 Techniques for demonstrating chemotaxis

To demonstrate chemotaxis it is necessary to generate a concentration gradient of attractant or repellent. Motile bacteria move up or down the gradient and their populations are assessed at intervals in time and space.

Capillary tubes can be used to form gradients. When a capillary containing an attractant is dipped into a bacterial suspension diffusion from the mouth of the capillary creates a gradient up which the bacteria swim. The method is made quantitative by removing the tube after an arbitrary time and counting the number of bacteria that have accumulated inside the capillary. If a repellent is added to the suspension the bacteria tend to accumulate inside the capillary where there is a lower concentration of repellent.

Capillary or petri-dish techniques involving the formation of self-generating gradients of a metabolisable attractant have been developed. Motile bacteria are inoculated at one end of a tube or centrally in a petri dish containing soft agar medium through which bacteria can swim. The bacteria create a gradient by consuming the energy source in their vicinity and then migrate towards the area of higher concentration as a visible band. If more than one energy source is provided a separate band appears for each source. Colonies developing in the cleared areas between the inoculum and the chemotactic band are likely to consist of non-chemotactic mutants.

Several petri-dish methods have been devised for demonstrating negative chemotaxis. The simplest is to suspend bacteria in soft agar and to incorporate the repellent in a disc of firm agar placed in the dish. The bacteria swim away from the repellent leaving a cleared area around the repellent disc (Fig. 13.11).

The availability of large pore size (2 μm) polycarbonate membrane filters has made possible the development of chemotaxis chambers in which bacteria are situated on one side of the membrane and the attractant on the other. The attractant diffuses through the membrane and the bacteria are able to swim through the pores and accumulate on the attractant side of the membrane where they can be counted.

5.3 Effects of environmental conditions on chemotaxis

The degree of response by a bacterial population to an attractant or a repellent is related to the concentration of the chemical involved. At very low concentrations there is no migration but at a particular threshold concentration the chemical is detected, and as the concentration is increased, more bacteria move towards or away from the stimulus in a given time until a saturation point is reached. Above the saturation point there is no further increase in response or even a somewhat reduced response due to inhibition of motility or other cellular functions. The threshold for positive chemotaxis of *E. coli* to aspartate is about 3×10^{-8}M and the saturation is 10^{-1}M, whereas for galactose the values are 10^{-8}M and 3×10^{-6}M, respectively.

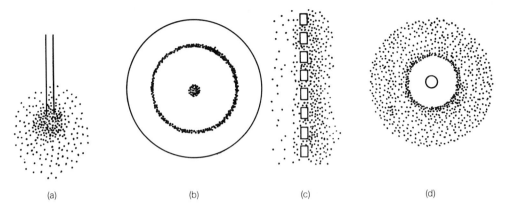

Fig. 13.11 Demonstration of bacterial chemotaxis. (a) A capillary containing attractant shows an accumulation of bacteria inside and around the aperture. (b) Bacteria centrally inoculated in a dish of soft agar containing a metabolisable attractant form a chemotactic band. (c) Membrane filters with 2 μm pore size allow passage of bacteria towards an attractant or away from a repellent. (d) A plug containing repellent causes a zone of clearing in suspension of bacteria in soft agar.

The thresholds for repellents like acetate are usually up to 10^4 times higher than the attractant thresholds. Since attractants are often usable substrates this means that if conflicting stimuli are present the bacteria are adapted to obtain the maximum from their environment. In the event of both an attractant and a repellent being present at above threshold levels the bacteria are able to process the opposing signals and respond to the stimulant which is present in the most effective concentration.

Chemotaxis is more sensitive to temperature than motility itself. Using his quantitative capillary technique Adler found that *E. coli* is motile but non-chemotactic below $15°C$. Raising the temperature from $20°C$ to $30°C$ causes an increase in motility by a factor of 2 to 3, as measured by cells finding their way into the capillary in the absence of a chemical gradient, while chemotactic accumulation increases 20-fold.

Other parameters which influence chemotaxis are any that affect the presence and activity of flagella, e.g. pH, heavy metal contamination and glucose repression of flagella synthesis. In addition some response mechanisms, e.g. for glucose, are constitutive while others, e.g. for galactose, are inducible and require the appropriate inducer in the growth medium.

5.4 Chemical signal reception in bacteria

The range of chemicals which elicit a chemotactic response varies with the bacterial species. Examples of positive and negative chemotactic agents for *E. coli* are shown in Table 13.2. Some other motile species, e.g. *P. mirabilis*, are relatively insensitive to their chemical environment (see Seymour and Doetsch 1973 for a survey of chemotactic behaviour in 10 motile species).

How such comparatively simple cells as bacteria detect specific chemicals has long been a matter for speculation. One popular idea was that the compound in question entered the cell, was metabolised and a metabolite or the ATP produced became the actual signal perceived by the cell. However, this hypothesis was disproved by Adler in an elegant series of experiments with *E. coli*. He showed that (1) a chemical may be metabolised without causing chemotaxis; (2) non-metabolisable chemicals may cause chemotaxis; (3) chemotaxis to a particular chemical is not blocked by the addition of a metabolisable compound; (4) chemotaxis is reduced by competition between structurally related compounds; and (5) mutants exist which cannot show

Table 13.2   Independent chemoreceptor systems
identified in *E. coli*

Receptors for attractants	Receptors for repellents
N-Acetyl glucosamine	Alcohols
Aspartate	Aromatic compounds
Fructose	Fatty acids
Galactose	Hydrogen ion
Glucose	Hydrophobic amino acids
Maltose	Hydroxyl ion
Mannitol	Indole
Mannose	Metal cations
Ribose	Salicylate
Serine	Sulphide
Sorbitol	
Trehalose	

chemotaxis to a compound even though they can metabolise it. These and other observations led to the conclusion that bacteria have specific chemoreceptor systems which detect certain compounds (or structurally related groups of compounds) and direct the cells in the appropriate direction. About 12 chemoreceptor systems for positive chemotaxis and 10 for negative chemotaxis have now been found in *E. coli* (Table 13.2) and genetic analysis has revealed that each has more than one component.

Hazelbauer and Adler discovered that in galactose chemotaxis the galactose is detected by the specific binding protein which is also responsible for the initial stage of transport of galactose into the cytoplasm (see Ch 12). Some structurally related compounds (glucose and fucose) are also detected by the galactose-binding protein. Mutation of the gene for the galactose-binding protein leads to loss of chemotaxis to galactose, but not to other unrelated groups of compounds, and to a simultaneous loss in galactose transport. Mutants which cannot transport galactose may nevertheless show chemotaxis towards it while other mutants can transport galactose but are not chemotactic. Thus the transport and chemotaxis systems are independent except for their joint utilisation of the binding protein. The situation with the galactose-binding protein has been found to be generally applicable to the detection of other chemical groups: in each case there is a binding protein involved in both transport and chemotaxis.

The cellular location of the galactose-, maltose- and ribose-binding proteins is the periplasmic space between the cytoplasmic membrane and the outer cell wall membrane. Osmotic shock treatment releases these receptors from the cell, resulting in simultaneous specific loss of chemotactic ability. The glucose and mannose receptors, however, are components of the phosphotransferase sugar transport system and these are integral parts of the cytoplasmic membrane.

5.5 Behaviour of single cells

An increased understanding of the chemotactic behaviour of bacteria has come from the direct observation of individual living bacterial cells with the light microscope. Berg has made detailed studies of individual cells by the development of a tracking microscope. This enables the observer to select an individual bacterium and the microscope stage then automatically moves to keep the cell in the same position in space (see Berg 1975 for a more detailed description of the apparatus). The movements made by the microscope thus correspond to the movements made by the chosen bacterium and can be monitored and recorded.

When a motile bacterium with peritrichous flagella is tracked in the absence of a gradient of chemotactic agent it moves in a characteristic way. After travelling in a roughly straight line for a few seconds (a run) it stops and briefly tumbles about in one spot, by reversing flagella rotation, before starting on a run in another direction. The cell thus moves about randomly in three dimensions without making net progress in any one direction. In a gradient of attractant the cell continues its three-dimensional movements but with a bias up the concentration gradient. This is achieved by the cell tumbling less frequently when it happens to be moving up the gradient but at normal frequency if moving down the gradient. In the case of a repellent the cell tumbles less frequently when it happens to be travelling down the gradient, as shown in Fig. 13.12.

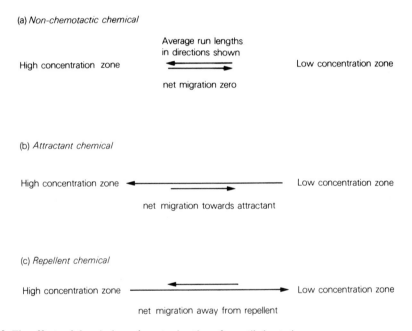

Fig. 13.12 The effects of chemicals on the net migration of a motile bacterium.

Thus bacteria do not orientate themselves in respect to an attractant or repellent (this would be topotaxis, which occurs in higher organisms) but show a net movement in the appropriate direction by biasing their random motion. As well as responding to concentration gradients in space, bacteria react to changes in concentration with time (temporal gradients): if cells are observed during the addition of an attractant the frequency of tumbling is reduced. A temporal decrease in concentration of a repellent has the same effect. The use of temporal gradients is often preferable for practical reasons but the distinction between spacial and temporal stimulation is rather artificial as bacteria travelling in a spatial gradient inevitably also experience concentration changes with time.

An alternative way of observing the behaviour of single cells is to tether them by their flagella to a microscope slide using flagella antibody. The tethered cell is seen to rotate since its flagellum cannot. A cell in this situation reverses its rotation at intervals corresponding to the tumbles of the untethered bacteria. The frequency of reversals can be controlled by changing the concentrations of attractant or repellent. The tumbling behaviour of larger numbers of individual cells has been observed by Spudich and Koshland using stroboscopic dark-ground microscopy. Rapid flashes of light during photographic exposure cause smooth swimming cells to appear as a series of illuminated dashes while tumbling cells show up as larger single spots.

Tumbling is clearly fundamental in the new understanding of bacterial chemotaxis and a 'tumble generator' has been postulated to account for the periodic reversal of the flagella. The tumble generator is modulated by the signals from the chemoreceptors. The nature of the signals and indeed the organisation of the tumble generator and the 'gear shift' system for the flagellar motor all remain to be elucidated. Methionine is known to be involved in tumble generation: its absence causes loss of tumbling and cells swim only smoothly and are consequently non-chemotactic. The methionine methylates a membrane protein probably via its derivative S-adenosylmethionine. The methylated protein possibly constitutes an activated component of the tumble generator and is therefore central to the mechanism of chemotaxis.

5.6 Chemotaxis mutants

Mutants incapable of chemotaxis are invaluable for investigations on the mechanism of the process. A chemotaxis mutant may be defective in (*a*) some part of the chemoreceptor; (*b*) the signal transmission system; (*c*) the tumble generator; (*d*) the flagellar motor; or (*e*) the flagellum itself. In a simplified form the flow of information with possible sites for mutation is shown in Fig. 13.13. The behaviour of a mutant in a gradient of attractant or repellent depends on which of these sites is affected.

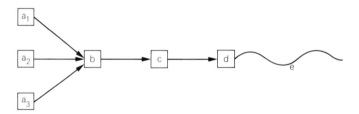

Fig. 13.13 Information flow in chemotaxis.

Chemoreceptor mutants may have an altered binding protein so that the compound specific for that receptor can no longer be detected. Chemotaxis which involves the other chemoreceptors is unaffected. Mutants of this type are usually also defective in the transport initiated by the binding protein.

The signal that a binding protein has bound an attractant or repellent molecule passes via a number of mutable steps to the tumble generator. The communication pathway is branched so that a mutation may affect some but not all responses, e.g. *trg* (*t*axis to *r*ibose and *g*alactose) mutants no longer detect ribose or galactose but are normal in other respects.

The stimuli from all the chemoreceptors converge on a common pathway; mutations here lead to a total loss of chemotactic ability. The genes involved are mainly concerned with components of the tumble generator, e.g. mutation of the *che A* (*che*motaxis) gene causes loss of tumbling whereas mutations of the *che B* gene cause either loss of tumbling or continual tumbling, indicating that the *che A* product is a part of the tumble generator and the *che B* product is a regulator of tumble frequency.

Mutants of the flagellar motor components (*fla* and *mot* genes) or of the flagellum itself (*hag* genes) also lead to a general loss of chemotaxis due to inability to assemble fully functional flagella.

6 Chemotaxis in non-flagellate bacteria

Bacteria with spirochaetal and gliding motility are also capable of chemotactic behaviour. Although these organisms have not yet been studied in any detail it is reasonable to make certain predictions about the mechanism of their response to external stimuli.

In the spirochaete changes in direction are probably caused by reversals in the rotation of the axial fibrils. Thus these cells are likely to have a chemoreceptor system which transmits signals to the fibril motors, thereby modulating the overall direction of cellular movement.

Similarly, with gliding bacteria, chemoreceptors must somehow change the direction of cell movement by reversing the direction of contractions in the cell surface. This appears to be an entirely different process from the flagellar and spirochaetal systems and is little understood at present. The aggregation of myxobacters to form fruiting bodies is a chemotactic process analogous to that of the slime moulds where the chemotaxis has been shown to be due to cAMP secreted by the cells themselves.

7 Summary and prospects

The movement of bacteria has provided many surprises. Flagella have turned out not to be contractile organelles or to transmit waves of motion but to be semi-rigid helices of protein which are rotated from basal bodies inserted in the cytoplasmic membrane. The axial filaments of spirochaetes are now thought to function in a similar way to flagella, i.e. rotation of the flagella-like fibrils in the periplasmic space generates motion of the helical cell. Gliding motion remains largely unexplained but there is evidence for contractile waves in the surface layers of gliding organisms. Some gliding bacteria resemble twitching bacteria in the possession of groups of polar pili which may provide an alternative explanation for surface translocation, especially as non-gliding mutants have been found to be free of pili.

Numerous details remain to be elucidated for all three types of movement. The unique rotary motor for driving the flagellum is a particularly intriguing structure and its operation, especially the ability to reverse, has yet to be fully explained.

Random movement is itself beneficial to a microbial cell because it increases the potential for finding nutrients but mobility is usually coupled with the additional ability to respond to certain stimuli from the environment. The positive or negative chemotactic response described in this chapter is only one aspect of the tactic abilities found among motile bacteria and other organisms. Photosynthetic species, for example, show phototaxis towards light of usable wavelengths. Other stimuli to which responses may occur are gravity, pressure, temperature, magnetism and electricity but these have not yet been extensively studied in bacteria.

The general outlines of the processes involved in the chemotactic response have become apparent but many details remain obscure. It is clear that attractants and repellents are detected by specific chemoreceptors but the nature of the signal to the flagella is still speculative: it might be conveyed by changes in membrane potential (analogous to neurotransmission), by changes in concentration of a chemical messenger or by direct protein—protein interactions in the cytoplasmic membrane. Whatever the signal is, it modulates the tumble generator probably through the mediation of S-adenosylmethionine and this represses the frequency of tumbling when the cell is moving in a favourable direction.

Bacterial behaviour provides a potentially valuable model system amenable to detailed genetic analysis and it may well provide guidelines for investigations on the more complex biochemical basis of behaviour in higher organisms.

References and further reading

ADLER, J. (1975) 'Chemotaxis in bacteria', *Ann. Rev. Biochem.*, **44**, 341–56.
An excellent review of recent developments.

ADLER, J. (1976) 'The sensing of chemicals by bacteria', *Scient. Am.*, **234**, No. 4, 40–7.

ASAKURA, S., EGUCHI, G. and IINO, T. (1968) 'Unidirectional growth of *Salmonella* flagella *in vitro*', *J. Molec. Biol.*, **35**, 227–36.

ASWAD, D. W. and KOSHLAND, D. E. (1975) 'Evidence for an *S*-adenosylmethionine requirement in the chemotactic behaviour of *Salmonella typhimurium*', *J. Molec. Biol.*, **97**, 207–23.

BACHMANN, B. J., LOW, K. B. and TAYLOR, A. L. (1976) 'Recalibrated linkage map of *Escherichia coli* K-12', *Bacteriol. Rev.*, **40**, 116–67.

BERG, H. C. (1975) 'Chemotaxis in bacteria', *Ann. Rev. Biophys. Bioeng.*, **4**, 119–36.
Relates the modern work to the early literature on chemotaxis.

BERG, H. C. (1975) 'How bacteria swim', *Scient. Am.*, **233**, No. 2, 36–44.

BERG, H. C. (1976) 'How spirochetes may swim', *J. Theor. Biol.*, **56**, 269–73.

BODE, W., ENGEL, J. and WINKLMAIR, D. (1972) 'A model of bacterial flagella based on small-angle X-ray scattering and hydrodynamic data which indicate an elongated shape of the flaggellin protomer', *Eur. J. Biochem.*, **26**, 313–27.

BURCHARD, R. P. (1974) 'Studies on gliding motility in *Myxococcus xanthus*', *Arch. Microbiol.*, **99**, 271–80.

CALLADINE, C. R. (1976) 'Design requirements for the construction of bacterial flagella', *J. Theor. Biol.*, **57**, 469–89.
An engineering approach to flagellar structure.

CARLILE, M. J. (1975) *Primitive Sensory and Communication Systems.* Academic Press, New York and London.
Covers behaviour of prokaryotes and eukaryotes.

CHET, I. and MITCHELL, R. (1976) 'Ecological aspects of microbial chemotactic behaviour', *Ann. Rev. Microbiol.*, **30**, 221–40.

CLAYTON, R. K. (1958) 'On the interplay of environmental factors affecting taxis and motility in *Rhodospirillum rubrum*', *Arch. Mikrobiol.*, **29**, 189–212.

CZAJKOWSKI, J., SOLTESZ, V. and WEIBULL, C. (1974) 'Absence of an electron microscopic substructure in intact flagella of *Proteus mirabilis* and *Bacillus subtilis*', *J. Ultrastruct. Res.*, **46**, 79–86.

DeLANGE, R. J., CHANG, J. Y., SHAPER, J. H. and GLAZER, A. N. (1976) 'Amino acid sequence of flagellin of *Bacillus subtilis* 168', *J. Biol. Chem.*, **251**, 705–11.

DePAMPHILIS, M. L. and ADLER, J. (1971) 'Fine structure and isolation of the hook–basal body complex of flagella from *Escherichia coli* and *Bacillus subtilis*', *J. Bact.*, **105**, 384–95.

EMERSON, S. U., TOKUYASU, K. and SIMON, M. I. (1970) 'Bacterial flagella: polarity of elongation', *Science.*, **169**, 190–2.

HILMEN, M., SILVERMAN, M. and SIMON, M. (1974) 'The regulation of flagellar formation and function', *J. Supramolec. Struct.*, **2**, 360–71.

IINO, T. (1969) 'Genetics and chemistry of bacterial flagella', *Bacteriol. Rev.*, **33**, 454–75.

IINO, T. (1969) 'Polarity of flagellar growth in *Salmonella*', *J. Gen. Microbiol.*, **56**, 227–39.

IINO, T. (1974) 'Assembly of *Salmonella* flagellin *in vitro* and *in vivo*', *J. Supramolec. Struct.*, **2**, 372–84.

IINO, T. and ENOMOTO, M. (1977) 'Motility', in *Methods in Microbiology* (Eds. J. R. Norris and D. W. Ribbons), Vol. 5A, pp. 145–63. Academic Press, New York and London.
Deals with practical aspects of the subject.

LAPIDUS, I. R. and SCHILLER, R. (1976) 'Model for the chemotactic response of a bacterial population', *Biophys. J.*, **16**, 779–89.
Mathematical treatment of chemotaxis.

O'BRIEN, E. J. and BENNETT, P. M. (1972) 'Structure of straight flagella from a mutant *Salmonella*', *J. Molec. Biol.*, **70**, 133–52.

PARKINSON, J. S. (1975) 'Genetics of chemotactic behavior in bacteria', *Cell*, **4**, 183–8.
A clear review of the subject.

PÉREZ-MIRAVETE, A. (1973) *Behaviour of Micro-organisms.* Plenum Press, London and New York.
A symposium covering prokaryotic and eukaryotic movement.

SEYMOUR, F. W. K. and DOETSCH, R. N. (1973) 'Chemotactic responses by motile bacteria', *J. Gen. Microbiol.*, **78**, 287–96.

SILVERMAN, M., MATSUMURA, P. and SIMON, M. (1976) 'The identification of the *mot* gene product with *Escherichia coli*-lambda hybrids', *Proc. Natl. Acad. Sci. USA*, **73**, 3126–30.

SMITH, R. W. and KOFFLER, H. (1971) 'Bacterial flagella', *Adv. Microb. Physiol.*, **6**, 219–339.
An extensive review covering much of the early literature.

TAYLOR, B. L. and KOSHLAND, D. E. (1976) 'Perturbation of the chemotactic tumbling of bacteria', *J. Supramolec. Struct.*, **4**, 343–53.

WEIBULL, C. (1960) 'Movement', in *The Bacteria* (Eds. I. C. Gunsalus and R. Y. Stanier), Vol. I. Academic Press, New York and London.
A wide-ranging review of the earlier literature.

14

Biological nitrogen fixation

J. R. Postgate
Unit of Nitrogen Fixation, University of Sussex

1 Introduction

The biological cycles of carbon, sulphur, nitrogen and other elements sustain the biosphere of this planet in a steady state with respect to the chemosphere. The importance of the nitrogen cycle to biological production, illustrated in Fig. 14.1, is well recognised: effectively, the rate at which the N atom is turned over determines biological productivity on most fertile (i.e. not arid

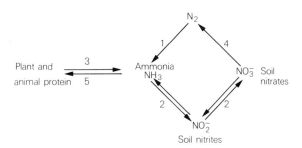

Fig. 14.1 A simple nitrogen cycle. The fixation of nitrogen is the name given to the conversion of atmospheric nitrogen to ammonia (1 above), a process which makes N available for the synthesis of plant and, hence, animal protein, 5 above). Other steps in the cycle include nitrification (the conversion of soil ammonia to nitrate via nitrites, 2 above) ammonification (the restoration of N to soil as ammonia in the decay of organic matter, 3 above) and dentrification (loss of N to the atmosphere by microbial reduction of nitrates, 4 above).

or frosted) agricultural areas of this planet. In all but the most advanced communities, where nitrogenous fertilizer was until recently cheap and plentiful, the rate-determining step in the cycle is the fixation of nitrogen. This process is conducted principally by microbes, usually in association with the roots of plants. Non-biological processes make a contribution; for example, nitrogen oxides can be formed in the atmosphere as a result of lightning flashes or combustion and become washed into the soil as nitrates. Man-made fertilizer also makes a significant contribution. Data on the relative importance of the various processes have been revised recently and may be revised again; Table 14.1, however, leaves little doubt that the biological process is pre-eminent on a global scale.

Table 14.1   Approximate scale of global fixation of nitrogen

The atmosphere contains about 3.9×10^{15} tonnes N as N_2.
The biosphere receives annually:
    175 million tonnes N by biological fixation.
    70 million tonnes N by spontaneous chemical fixation.
    30 million tonnes N as industrial fertilizer.
    The figure for nitrogen fixation is probably accurate to about 10 per cent, that for industrial fertilizer to about 1 per cent; that for spontaneous fixation is very approximate, based on total input of 200 million tonnes N of which about two-thirds is recycled NH_3 and nitrogen oxides in rain and dust.

From a human point of view, global protein production and distribution are inadequate to support the world's population, so protein deficiency and starvation were commonplace to some 40 million people in 1976. Provision of adequate and cheap fixed nitrogen for agriculture is therefore a matter of social urgency. Chemical fertilizer production consumes energy, now increasingly expensive, and requires sophisticated industry, which means high capital investment, maintenance costs and transport costs. Its use also entrains the run-off problem: relatively little of an ammonia fertilizer added to the soil is actually used by the plant, because more than half usually becomes oxidised, by the next steps of the cycle, to nitrates, which are less tightly bound by soil, and which tend to be washed away and to emerge in rivers, lakes and even drinking water. Eutrophication of waters may then occur, with attendant pollution problems; high nitrate levels in drinking water can actually be toxic. Biological nitrogen fixation is thus of great practical importance today, as well as for the future. The process also presents absorbing questions at a fundamental level: how, for example, do bacteria, in cold, damp, aerated soil, reduce nitrogen to ammonia when industry requires the high energy input and the anoxic, anhydrous conditions of the Haber–Bosch process?

The fact that free-living bacteria fix nitrogen and that some enter into a symbiotic relationship with plants was recognised even in the last century. Best known is a symbiosis in which bacteria of the genus *Rhizobium* invade the roots of leguminous plants (such as clover, lucerne, peas, beans and lupins) and fix nitrogen there. An infection thread is formed and rhizobia grow towards certain polyploid cells. These become engorged with rhizobia, which usually become morphologically altered and are then called 'bacteroids'. The bacteroids become surrounded by a gall-like nodule within which they fix nitrogen. An important technical advance was the introduction of heavy nitrogen, ^{15}N, in the 'forties by Burris and Wilson at the University of Wisconsin as an analytical tool. Fixed nitrogen as ammonia or nitrogen oxides is a universal contaminant of air and of allegedly pure chemical reagents, so direct analyses for N can give a false impression of N_2 fixation. Incorporation of ^{15}N from $^{15}N_2$ into the plant or microbe gave unequivocal evidence for fixation (though it did not compensate for bad microbiology – we now know that several putative nitrogen-fixing populations reported in the scientific literature must have been mixtures of fixing and non-fixing types). The next advance came in 1960, when Carnahan and colleagues from the laboratory of Dupont de Nemours (USA) extracted crude preparations of the nitrogen-fixing enzyme (called 'nitrogenase') from the anaerobe *Clostridium pasteurianum*. By that time, three important facts were known. Both molybdenum and iron had been shown to be essential elements for biological nitrogen fixation; nitrogenase was always associated with an enzyme called hydrogenase, capable of activating the hydrogen molecule (but there existed a number of microbes which possessed hydrogenase but which were unable to fix nitrogen); finally, the enzyme was not formed if the bacteria were grown with adequate fixed nitrogen as nitrates or ammonium salts.

The next historical landmark in the study of nitrogen fixation was the development of the acetylene test. In 1966, Burris, Schöllhorn and Dilworth showed that nitrogenase could reduce acetylene specifically and quantitatively to ethylene. This reaction occurs *in vivo*: living microbes, root nodules, soil, water and other natural samples will, if they contain functioning nitrogenase, form ethylene from acetylene. Ethylene can be detected and estimated rapidly and precisely by gas chromatography; the acetylene reduction reaction thus led to the development of a gas chromatographic test for nitrogen fixation which is much quicker than tests with $^{15}N_2$ or conventional N-analysis as well as being some 1 000 times more sensitive.

Nearly all the recent advances in biochemistry, physiology, ecology and genetics of nitrogen fixation have emerged from the use of the acetylene test as a research tool. In this chapter I shall discuss each of these four aspects in turn.

2 Biochemistry

In the original extraction of nitrogenase, the Dupont group grew *C. pasteurianum* in a nitrogen-free medium and, after disruption by drying, obtained an extract which, when supplied with sodium pyruvate, reduced $^{15}N_2$ to ^{15}N-ammonia. It behaved as a true protein solution in the ultra-centrifuge and the secret of their success was rigorous exclusion of oxygen: the nitrogenase was irreversibly inactivated if it became exposed to air. Table 14.2 illustrates

Table 14.2   Fixation of $^{15}N_2$ by an extract of *Clostridium pasteurianum* incubated with sodium pyruvate. (After Mortenson, Mower and Carnahan (1962). *Bact. Rev.*, **26**, 42).

Time (min.)	Pyruvate consumed (μmole)	^{15}N fixed (μg-atom)
5	400	0·5
10	490	3·5
15	580	5
20	640	6·5

the sort of result they obtained: large amounts of pyruvate were consumed and, though relatively little nitrogen was fixed, the process was reproducible. Pyruvate was the most effective substrate and only 2-oxobutyrate would replace it, rather ineffectively; in due course it transpired that pyruvate supplied not only reducing power but also ATP which proved to be necessary for nitrogenase function. ADP, formed from ATP in the reaction, also inhibited it. Similar extracts made from clostridia grown with ammonium salts did not fix nitrogen.

Comparable oxygen-sensitive nitrogenase extracts, some particulate rather than true solutions, were soon obtained from *Bacillus polymyxa* and *Klebsiella pneumoniae*, both of which can grow with fixed nitrogen in air but which fix N_2 only in the absence of air. The obligatory aerobic nitrogen-fixing bacterium *Azotobacter vinelandii* proved recalcitrant; pyruvate did not function with their extracts. When, in 1964, Bulen and his colleagues obtained active sub-

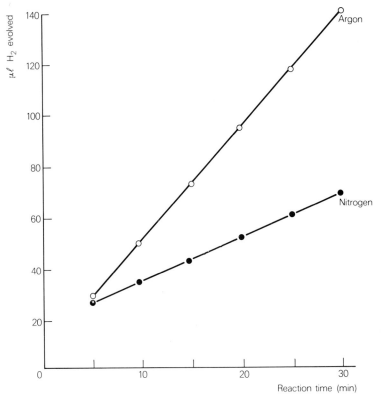

Fig. 14.2 Hydrogen evolution by cell-free nitrogenase. A crude particulate preparation from *Azotobacter chroococcum* was obtained by disruptuion in the French press and by centrifugation to remove gross debris. Conventional Warburg manometers contained Ar or N_2 over 2·5 ml of 25 mM 'tris' buffer, pH 7·4, extract equivalent to 7 mg protein, and an 'ATP-regenerating system'. This consists of $MgCl_2$ (5 μmole), creatine phosphate (10 mg) and creatine kinase (0·5 mg) and serves to supply ATP continually and remove inhibitory ADP as it is formed. The reaction was initiated by adding about 3 mg $Na_2S_2O_4$ and H_2 evolution was measured at 30°C.

cellular particles from *A. vinelandii,* two important new facts emerged. First, the particles were reasonably oxygen-tolerant: they could be sedimented over 2 h at 150 000 g in a centrifuge and could be handled in air without serious loss of activity. Secondly, though pyruvate was ineffective, sodium dithionite ($Na_2S_2O_4$) could be used with these particles and it promoted an unexpected reaction. Given some ATP and dithionite under an inert gas such as argon, the particles evolved hydrogen; under nitrogen, much less hydrogen was evolved and the difference was accountable as nitrogen reduced to ammonia. Figure 14.2 illustrates the hydrogen evolution reaction. Hydrogen evolution by hydrogenase is inhibited by carbon monoxide but the ATP-activated hydrogen evolution reaction of nitrogenase is not.

The need for ATP complicates the study of nitrogenase; ADP is formed and this inhibits the reaction, so laboratory tests on the enzyme generally include a system (e.g. creatine phosphate + kinase as in Fig. 14.2) to regenerate ATP from ADP. Too much ATP can also be inhibitory; a concentration of 5 μM is often used. Twelve to fifteen ATP molecules are hydrolysed for each N_2 molecule reduced to NH_3; in terms of ATP, nitrogenase is an extravagant enzyme.

Crude nitrogenases have now been obtained from about 25 organisms, including blue-green algae, photosynthetic bacteria and the bacteroids of legume root nodules. It has been purified

Table 14.3   Properties of some component proteins of nitrogenase

	Mo–Fe proteins			*Fe proteins*		
Organism	*C. pasteurianum*	*K. pneumoniae*	*A. chroococcum*	*C. pasteurianum*	*K. pneumoniae*	*A. chroococcum*
Molecular weight of protein	220 000	218 000	227 000	55 000	66 700	64 000
Subunits and molecular weight	50 700 59 500	51 300 59 600	$\left(\begin{array}{c}1 \text{ or } 2\\ \text{types}\end{array}\right)$	27 500	34 000	31 000
Mo atoms per molecule	2	2	1·9	0	0	0
Fe atoms per molecule	22–24	30–35	21–25	4	4	3·9
S^{2-} ions per molecule	22–24	(>18)	18–22	4	3·8	3·9
O_2-sensitivity (half-life in min. in air)	†	10	10	†	0·75	0·50

†Present, but not estimated.

to a high degree from five microbes and always consists of two distinct proteins, neither of which is active on its own. One of these is always a large tetrameric molecule, containing two atoms of molybdenum and many atoms of iron, the latter more or less matched by acid-labile sulphide (the Mo—Fe protein). The subunits of this protein are of two types, distinguishable by peptide mapping. The other is a smaller, dimeric protein, with no molybdenum but four atoms each of iron and labile sulphide (the Fe protein). Some properties of these proteins from these different bacteria are given in Table 14.3. An important point is that both proteins are damaged by oxygen, extremely rapidly in the case of the Fe proteins. New laboratory techniques have had to be developed in order to purify these proteins while preventing access of air, and it is a real achievement that the Mo—Fe protein of *A. vinelandii* has been crystallised.

The fact that the *Azotobacter* proteins are oxygen-sensitive raises a point which is important in considering the physiology of nitrogen fixation. The particulate preparation, which includes other proteins and membranous material, is oxygen-tolerant; so within the particle the oxygen-sensitive sites of the two proteins must be protected in some way from oxygen damage.

As Table 14.3 shows, the Mo—Fe or Fe proteins seem chemically similar no matter what their source. This similarity extends to function: one can often take a Mo—Fe protein from one source, add an Fe protein from another source and reconstitute an active but hybrid nitrogenase. For example, the Mo—Fe protein of *K. pneumoniae* forms fully active nitrogenase with the Fe proteins of *Azotobacter chroococcum* or of *B. polymyxa*, but only a weakly active one with that of *C. pasteurianum*. The Mo—Fe protein of *B. polymyxa,* however, forms only a partly active hybrid with the Fe protein of *K. pneumoniae* but a fully active one with that of *C. pasteurianum.*

Biochemical similarity extends beyond the proteins themselves. In all nitrogen-fixing systems so far examined the natural reducing agents are similar: they are flavodoxins or ferredoxins. They can be replaced by a viologen dye or $Na_2S_2O_4$ *in vitro*.

Nitrogenase requires the magnesium ions as well as ATP and a reductant such as ferredoxin or $Na_2S_2O_4$. With these accessories, nitrogenase reduces, in place of nitrogen, a number of small, triply-bonded molecules such as those given in Table 14.4. Investigation of the chemistry of methyl isocyanide reduction provided some of the first evidence that reducible substrates become bound to one of the transition metal atoms, iron or molybdenum, when nitrogenase

Table 14.4   Some substrates reduced by nitrogenase*

	Substrate†	*Products*	*Comments*
Dinitrogen,	$N \equiv N$	NH_3	
Acetylene,	$HC \equiv CH$	$H_2C = CH_2$	Rapid
Methyl acetylene,	$CH_3 - C \equiv CH$	$CH_3CH = CH_2$	Slow
Hydrogen cyanide,	$H - C \equiv N$	$CH_4 + NH_3$	Some CH_3NH_2 formed
Methyl cyanide,	$CH_3 - C \equiv N$	$C_2H_6 + NH_3$	Slow, some $C_2H_5NH_2$ formed
Methyl isocyanide,	$CH_3\overset{+}{N} \equiv \overset{-}{C}$	$CH_3NH_2 + CH_4$	Rapid, C_2H_4 and C_2H_6 formed
Hydrogen azide,	$H - \overset{-}{N} - \overset{+}{N} \equiv N$	$NH_3 + N_2$	Ratio of N_2 to NH_3 varies
Nitrous oxide,	$N \equiv \overset{+}{N} - \overset{-}{O}$	$NH_3 + H_2O$	—
Hydrogen ions,	H^+	H_2	Rapid

*All reactions require anaerobic conditions, Mg^{2+}, ATP and a reductant.
†Reduction of all substrates except H^+ inhibited by carbon monoxide, $\overset{-}{C} \equiv \overset{+}{O}$.

functions. In a test tube, a reducing agent such as sodium borohydride converts it to dimethylamine:

$$CH_3 - N \equiv C \rightarrow CH_3 - NH - CH_3$$

The enzyme splits the $N \equiv C$ triple bond to give methylamine and methane, together with small amounts of ethane and ethylene:

$$CH_3 - N \equiv C \rightarrow CH_3NH_2 + CH_4 (+ C_2H_4 + C_2H_6)$$

Parallel chemical studies showed that the enzymic products, even the C_2 by-products, were formed when an isocyanide complexed to a transition metal such as platinum was reduced:

$$Cl_2Pt(C \equiv N - C_6H_5)_2 \rightarrow C_6H_5NH_2 + CH_4 (+ C_2H_4, C_2H_6 \text{ etc.})$$

The metal atom which binds the substrate is still unknown, but investigation of the role of metals in nitrogenase has led to considerable understanding of the mode of interaction of the two proteins. So far, no effective biophysical probe has been developed to study the behaviour of molybdenum in the enzyme, but the iron atoms can be observed in three ways.

1 UV-visible spectroscopy. Though these brown proteins have rather featureless light spectra, small changes in absorption at 450 nm can be used to monitor the oxidation state of the iron in the Fe protein.

2 Mössbauer spectroscopy. This technique has been very informative. It is based on the ability of an iron isotope, ^{57}Fe to absorb γ-photons of energies differing over a small range, according to its chemical environment and valence state. The Mo-Fe protein purified from *K. pneumoniae* which had grown with ^{57}Fe proves to be able to exist with its iron atoms in three oxidation states: an oxidised form obtained by reaction with a dyestuff, an intermediate form (in which the protein is usually isolated) and a super-reduced form, detected only when the enzyme is functioning (i.e. when the Fe protein, ATP and Mg^{2+} are present in addition to sodium dithionite). Acetylene has no effect on the Mössbauer spectra and their changes.

3 Electron paramagnetic resonance (e.p.r.). This technique has also been informative, with both the Fe and the Mo–Fe proteins. The latter give a signal which is unique: it shows three clear resonances, all of which broaden if the ^{57}Fe protein is tested, so all are assignable to iron in the protein. The resonances disappear in both the oxidised and super-reduced form, but in the intermediate form their precise energy depends on the pH of the solution. The pK of this pH dependence (the pH for half-maximal shift) is altered by acetylene, indicating that a reducible substrate such as acetylene reacts with this protein species. The site of interaction is probably remote from that responsible for the e.p.r., or the effect of acetylene would be more dramatic.

The Fe protein too has an e.p.r. signal. This disappears if the protein is carefully oxidised (without loss of activity) using a dyestuff; it reappears when the protein is reduced again with sodium dithionite. Thus the Fe atoms in this protein can exist in two redox states. If ATP and Mg^{2+} are present, the shape of the signal changes, indicating that Mg^{2+} and ATP cause a structural change in the reduced form of the protein, altering the symmetry of its e.p.r. active site. Other properties of the protein change too: its sulphyldryl groups react more readily with group-specific reagents; the iron is more easily removed by chelating agents; the oxygen-sensitivity, already considerable, is augmented; the standard redox potential becomes more negative (from about -300 mV to -400 mV); the sedimentation properties in the ultra-centrifuge change. ATP apparently makes the reduced form of the Fe protein into a more powerful and reactive reducing agent than it was before.

Measurements of nitrogenase reaction rates with ATP and Mg^{2+} mixtures show that it is the species MgATP which reacts with the Fe protein, not free ATP or Mg_2ATP; studies involving equilibrium dialysis, as well as kinetic studies, suggest that each molecule of Fe protein has two sites at which MgATP can bind.

It seems probable that nitrogenase binds substrates such as acetylene or nitrogen itself at a metal atom within the protein. The question how a molecule such as N_2 binds to a metal can now be answered, in principle at least, in chemical terms. In the mid-1960s, a class of transition metal complexes called the dinitrogen complexes was discovered. These have triply bonded N_2 molecules ligated to metals, as in the example below.

$$[(NH_3)_5Ru \leftarrow N \equiv N]^{2+}$$

They can be binuclear, as in the compound:

$$[(NH_3)_5Ru \leftarrow N \equiv N \rightarrow Ru(NH_3)_5]^{4+}$$

The diruthenium compound above can, in alkaline solution, actually exchange its N_2 group with gaseous $^{15}N_2$, showing that N_2 can become bound to a transition metal in water. Such complexes can also carry two N_2 groups, and a tungsten complex of this kind exists which reacts with acid in a manner which results in reduction of one N_2 group almost quantitatively to ammonia (a small quantity of hydrazine is found as a by-product). During this process, the valency state of the tungsten atom changes from 0 to +6. A reaction scheme is shown in Fig. 14.3, in which most of the intermediates have been identified. An analogous complex with a molybdenum atom in place of tungsten gives about 30 per cent conversion of a N_2 group to ammonia.

14.3 Stages in the chemical conversion of a tungsten dinitrogen complex to ammonia. The *cis*-dinitrogen complex of tungsten and dimethylphenylphosphine (I) can be made using gaseous N_2. Protonation in methanol with sulphuric acid displaces one N_2 group but causes reduction of the other to ammonia in 90 per cent yield, with traces of hydrazine formed as by-product. A probable mechanism is illustrated. Intermediates such as II and III have been demonstrated using hydrochloric acid and/or a different phosphine ligand. Species IV is speculative but symbolises the fact that phosphine groups are removed in the reaction. During the reaction the formal valency of the tungsten atom changes from zero to +6 (see Chatt, Pearman and Richards (1975), *Nature*, **253**, 39–40).

The importance of this kind of reaction to biochemists is that it provides a chemical precedent for N_2 bound to a metal such as molybdenum becoming reduced to ammonia at the expense of a valency change in the metal, with no free intermediate being formed en route. Numerous attempts to detect intermediates at the hydrazine (N_2H_4) or di-imide (N_2H_2) level during nitrogenase action have failed and it is very tempting to assume that the molybdenum atoms of nitrogenase are the sites at which N_2 is bound and reduced. There is, however, no direct evidence for this view.

In general, however, there is now reasonable evidence that the Mo—Fe protein binds the reducible substrate (N_2, C_2H_2, H^+ etc.) at a metal atom which is probably not iron; the function of the Fe protein is to react with MgATP, so generating sufficient reducing power to form the super-reduced state of the Mo—Fe protein, which is part of the enzymically active species. A mechanism incorporating such a reaction is given in Fig. 14.4, though future research will doubtless reveal it to be an over-simplification.

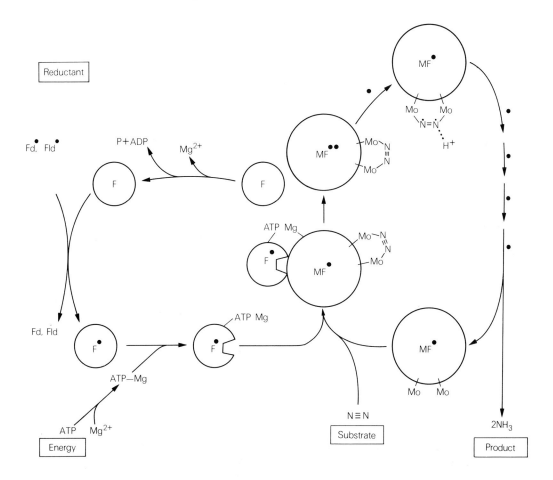

Fig. 14.4 A scheme for nitrogenase action. A ferredoxin (Fd) or flavodoxin (Fld) donates an electron to the iron atoms of an oxidised form of the Fe protein (F). Reduced F reacts with the monomagnesium salt of ATP undergoing a change in conformation. Meanwhile the Mo—Fe protein (MF) has combined with N_2 at a site which involves metal atoms but not iron (the role of molybdenum and the manner of binding N_2 in this scheme is purely speculative) and reacts with the F—Mg—ATP molecule to form a 1:1 complex. This complex is the nitrogenase enzyme; within it an electron is transferred from the iron atoms of F—Mg—ATP to the iron atoms of MF. As a result, F becomes oxidised and ATP hydrolyses to ADP + phosphate (P). Such an electron transfer event occurs six times, symbolised by the dotted arrows, before the MF protein releases its substrate as two ammonia molecules. Whether the same F molecule is involved throughout the reaction, being recharged with ATP—Mg once for each electron transferred, or whether it is replaced by a new F—Mg—ATP is not known.

3 Physiology

One outstanding property of nitrogenase is its sensitivity to oxidation, both proteins being irreversibly destroyed by exposure to air. Thus an important question arises which concerns the physiology of aerobic nitrogen-fixing bacteria. How can an aerobic nitrogen-fixing organism ever use nitrogenase in air, if the enzyme's components are so readily destroyed by oxygen? This question applies particularly to the blue-green algae, whose photosynthetic metabolism actually involves evolution of oxygen. In essence the answer is that evolution has provided aerobic microbes with a variety of physiological stratagems for partly or completely excluding oxygen from the nitrogen-site. Among these expedients are the following:

1 Respiration. The azotobacters can have the highest respiration rates among living things. Their respiration rate is closely regulated and increases rapidly if the oxygenation of the environment increases. This respiration maintains the dissolved oxygen concentration near the cell at a vanishingly low concentration and numerous lines of evidence are converging to support the view that respiratory protection of functioning nitrogenase is an essential part of aerobic fixation. The use of respiration in this way is called 'Respiratory Protection' and the process is quite sophisticated in, for example, *A. vinelandii.* At high oxygenation levels, this organism uses a different respiratory chain (cytochrome pathway) from that at low levels, one that is less effective at generating ATP. In consequence, highly aerated azotobacters expend a great deal of substrate when fixing nitrogen and they are, even then, more sensitive to inhibition by oxygen than if they are grown with ammonia as their nitrogen source.

Respiratory protection is less well developed in *Derxia gummosa* or *Mycobacterium flavum,* which consequently behave like microaerophiles when fixing nitrogen. Indeed, the majority of aerobic nitrogen-fixing bacteria behave in some degree like microaerophiles when fixing nitrogen. An extreme case is *K. pneumoniae* which is conventionally regarded as facultative: unable to fix nitrogen unless it is grown anaerobically, though it grows perfectly well aerobically when supplied with fixed nitrogen. This organism can be 'coaxed' into fixing at quite high oxygen tensions if the population is so dense that its respiratory activity maintains a local oxygen tension about zero.

2 Subcellular compartmentation. The subcellular nitrogenase particle which is extractable from *Azotobacter* spp. is oxygen-tolerant, though the enzymes it contains are oxygen-sensitive. It contains other proteins and there is some evidence that, when subjected to an oxygen stress with which respiration cannot cope, azotobacters can 'switch off' their enzyme rapidly and reversibly by combining it to a protein to make an oxygen-tolerant form. It then becomes non-functional.

The coccoid unicellular blue-green alga *Gloeocapsa* not only tolerates oxygen but evolves it during photosynthesis. It has a complex internal membrane structure which may represent an internal compartmentation of nitrogenase.

3 Cellular compartmentation. Many nitrogen-fixing blue-green algae possess large cells called heterocysts which are spaced at regular intervals along the algal filament. They do not appear when the organism is grown with ammonia. A feature of these heterocysts is their lack of the oxygen-evolving component of photosynthesis (photo-system 2). Much circumstantial evidence is now available supporting the view that nitrogen fixation is restricted to the heterocysts. They would seem to enable such microbes to fix at high tensions or illuminations because oxygen is at least partly excluded from them. There exist filamentous blue-green algae, such as *Plectonema,* which can fix without heterocysts and, consistent with this view, they only do so at low illumination levels or low oxygen tensions.

4 Clustering. The marine blue-green alga *Trichodesmium* is believed to have an even grosser form of compartmentation. In calm seas, the organism forms clusters of 100 or more trichomes, in the relatively anoxic centre of which photosynthesis declines, nitrogen is formed and fixation takes place. Rough seas inhibit fixation by dispersing the clusters.

5 An oxygen buffer. The legume nodule is an aerobic system. When it is effective, it always contains a high concentration of a haemoprotein called leghaemoglobin: a protein capable of being reversibly oxygenated, like the haemoglobin of mammalian blood. It has a very high affinity for oxygen and seems to serve as a specialized oxygen-transporting system: one that will pick up and deliver oxygen at very low concentrations, such that the bacteroids receive the oxygen they need, but insufficiently concentrated to damage nitrogenase. When rhizobia form bacteroids in the root nodule, their cytochrome pattern changes dramatically, notably the nature of the terminal cytochrome oxidase. The probability is that a component develops which can react specifically with oxyleghaemoglobin, but this is not established yet. The non-leguminous plant associations (see ecology) probably have similar oxygen buffering and transport in their nodules, but their physiology is largely unknown at the molecular level.

6 Gum and reserve materials. It is possible that the gum characteristic of nearly all aerobic nitrogen-fixing bacteria has some function in obstructing diffusion of oxygen to the cell. In *Azotobacter beijerinckii,* the storage material poly-β-hydroxybutyrate is laid down in oxygen-limited conditions and serves as a 'reserve' for respiratory protection in times of oxygen stress. Gum may also be mobilisable in this way and thus have a dual protective function.

Doubtless other oxygen-restricting systems exist in aerobic nitrogen-fixing bacteria; in principle, it is now evident that the organism has somehow to resolve an essential incompatibility between nitrogen fixation and the aerobic way of life.

A second physiological question arises from the enzyme's need for ATP. In the test tube, 12 to 15 molecules of ATP are converted to ADP for each N_2 reduced to $2NH_3$, corresponding to rather more than two molecules of ATP per electron transferred. This ratio seems to be independent of reducible substrate: it applies to the reduction of acetylene and the evolution of hydrogen, though the figure becomes larger with exotic substrates or with distorted ratios of Mo—Fe to Fe protein or if the Mo—Fe and Fe proteins come from different organisms. The question arises whether the enzyme in the living cell is also extravagant in its ATP economy. One can study this question crudely by continuous culture, using carbon-limited populations of organisms grown with ammonia and comparing yields of cells with comparable populations grown with N_2. The surprising result emerges that living *K. pneumoniae* and *C. pasteurianum* are in fact more inefficient that the isolated enzyme, consuming around 5 and 3·2 molecules of ATP per electron, respectively. In *A. chroococcum* the situation is complicated by the existence of respiratory protection, which can lead to excessive substrate consumption independent of nitrogen fixation, but chemostat experiments making allowance for this effect show unequivocally that *A. chroococcum* is more efficient than the other two organisms. The efficiency is necessarily more vague, but is closest to one ATP molecule consumed per electron transferred.

A third matter of physiological importance arises from the ability of the enzyme to evolve hydrogen. All nitrogen-fixing bacteria contain an enzyme called hydrogenase (it is also present in bacteria such as *Proteus* and *Escherichia,* which do not fix nitrogen) and the reason for its presence has for years been obscure. Hydrogenases fall into two classes: one type catalyses the reduction by hydrogen gas of dyes or ions such as ferricyanide; the second type can also catalyse such reactions but is reversible: given a strong reducing agent (such as sodium dithionite and the reduced form of the dye methyl viologen), it can catalyse the evolution of hydrogen. Reversible hydrogenases are often soluble iron–sulphur proteins; unidirectional hydrogenases are generally insoluble and of uncertain chemistry.

Nitrogenase, when evolving hydrogen, shows ATP-activated hydrogenase activity, but it is irreversible: it acts like a unidirectional hydrogenase but in the reverse direction. If A is an acceptor such as a viologen or ferredoxin, the three reactions can be illustrated as follows:

Unidirectional hydrogenase: $H_2 + A \rightarrow AH_2$
Reversible hydrogenase: $H_2 + A \rightleftharpoons AH_2$
Nitrogenase: $2H_2O + ATP \rightarrow H_2 + 2OH^- + ADP + P$

Nitrogenase evolves hydrogen even when nitrogen is available (see Fig. 14.2) and, when living bacteria fix nitrogen anaerobically, some hydrogen is evolved by the enzyme in a completely wasteful process as far as we know, since ATP is lost as well as reductant. A similar waste of biological resources occurs in the symbiotic systems: Professor H. J. Evans's group in Oregon have evidence that, in the field, some 30 per cent of the biological reserves available to rhizobia in symbiosis with soya beans are wasted as H_2 lost to the atmosphere. Azotobacter and other aerobic bacteria do not normally evolve hydrogen and they possess a unidirectional hydrogenase. By selective inhibition (with a mixture of acetylene and carbon monoxide) it is possible to inhibit the hydrogenase without interfering with nitrogenase function, in which circumstance, hydrogen is evolved by the bacteria (Table 14.5). Thus the hydrogenase seems to be involved, in the absence of inhibitor, in re-using any hydrogen produced by the nitrogenase. There is evidence that the hydrogen can be used to regenerate ATP from ADP, thus accounting for the improved biological efficiency with which aerobes such as Azotobacter fix nitrogen.

4 Ecology

The recognition of the oxygen-sensitivity of nitrogenase and of its ability to reduce acetylene, has led to important reassessments of the microbiology and ecology of the process. It has also revolutionised the microbiology, in the sense that the list of nitrogen-fixing microbes recognised in 1976 is very different from that before 1966. Table 14.6 lists the types of microbes now known to fix nitrogen; the most fundamental revision has been the disappearance of all eukaryotes from the list. Earlier reports of fixation by yeasts and other fungi have not been substantiated, and the property seems now to be exclusive to prokaryotes: bacteria, streptomycetes and blue-green algae. Table 14.6 includes many new groups of nitrogen-fixing

Table 14.5   Evolution of hydrogen provoked by acetylene and carbon monoxide in nitrogen-fixing *Azotobacter chroococcum* *

Gas	Partial pressure added to atmosphere (atm)	C_2H_2 reduced	H_2 produced
		(nmole h^{-1})	(nmole h^{-1})
	—	0	0
C_2H_2	0·1	780	20
CO	0·05	0	440
C_2H_2 CO	0·1 + CO 0·05	5	760

*Cells incubated under 0·05 atm O_2 at 30°C with N_2 plus other gases to 1 atm. Products assayed by gas chromatography.

Table 14.6   Some biological systems which fix nitrogen

Symbiotic systems:	Lichens	*Nostoc, Anabaena, Tolypothrix*
	Gunnera	*Nostoc*
	Azolla	*Anabaena azollae*
	Tropical grass	*Azotobacter paspal*
	Root nodules	Legumes + *Rhizobium*
		Non-legumes + *Frankia*
Free-living microbes:	Aerobic	*Azotobacter, Derxia, Berjerinkia*
		Mycobacterium flavum,
		Methane-oxidising bacteria
		Corynebacterium autotrophicum
		Spirillum lipoferum
		Thiobacillus ferrooxidans
	Aerobic, photosynthetic	*Nostoc, Anabaena, Plectonema*
	facultative	*Klebsiella* spp., *Bacillus*
		polymyxa, B. macerans
		Enterobacter spp.
	Facultative, photosynthetic	*Rhodospirillum,*
	anaerobic	*Rhodopseudomonas*
		Clostridium pasteurianum,
		C. butyricum,
		Methane-producing bacteria
		Desulfotomaculum,
		Desulfovibrio
	Anaerobic, photosynthetic	*Chromatium, Chlorobium*

bacteria, though some groups earlier thought to include nitrogen-fixing bacteria (*Pseudomonas, Azotomonas, Arthrobacter*) are excluded. The reason why *Azotomonas* and certain *Pseudomonas* species were once thought to fix nitrogen illustrates an important microbiological point. Representative species (*A. insolita, P. azotocolligans*) prove to be particularly adept at scavenging traces of fixed nitrogen from the atmosphere and from laboratory reagents; in all inhabited areas, traces of ammonia are present in the atmosphere, and wherever electrical machinery or internal combustion engines operate, oxides of nitrogen are formed. If an efficient scavenging organism is growing on a N-deficient agar medium, a diffusion gradient is set up in the direction of the colony and considerable growth may occur over several days; such diffusion gradients can be set up to liquid cultures in flasks or test tubes, and though diffusion is much hindered by a cotton wool plug, increments of the order of $6\,\mu g$ N ml^{-1} week^{-1} can be obtained which will correspond to 100 or more μg dry wt. of organism ml^{-1}. The unwary microbiologist may compare this with a boiled control, towards which there exists no diffusion gradient, and conclude that unequivocal evidence for nitrogen fixation has been obtained.

Scavenging accounts for some false reports of nitrogen fixation; others can be attributed to contamination. The acetylene test and ^{15}N incorporation are more rigid criteria of fixation, but the possibility of microbiological error must not be dismissed. Instances are known in which *Pseudomonas*-like isolates from soil gave positive acetylene reduction tests, and seemed pure when tested casually on agar plates in air, yet they harboured the anaerobe *Clostridium pasteurianum* within their colonies and these were the true nitrogen-fixing organisms.

The acetylene test is a very good check for nitrogen fixation, but with new strains it should always include a test for repression by ammonia (about 1 mg NH$_4$Cl per ml of culture) because, though no such example has yet arisen, the possibility exists that the system contains purely catalytic acetylene-reducing substances. There is one instance in which it does not work: acetylene is a good inhibitor of biological methane oxidation, so nitrogen fixation can be missed if methane is the sole metabolisable substrate. An example would be in tests involving the methane-oxidising bacteria; these reduce acetylene well with methanol as a substrate but not with methane.

Table 14.6 includes anaerobes, facultative organisms, aerobes and photosynthetic microbes. It is important to realise that not every strain of every species listed fixes nitrogen: in the genus *Klebsiella,* about 50 per cent of regular isolates fix nitrogen; among *B. polymyxa* and *B. macerans* the number is more like 90 per cent; in *Desulfovibrio* or *Chlorobium* the proportion of fixers is probably rather low but more data are needed. Facultative anaerobes such as *Klebsiella, Bacillus* or *Rhodospirillum* only fix nitrogen when grown in the absence of dissolved oxygen. This physiological fact is yet another reflection of the marked oxygen sensitivity of the nitrogenase proteins. Table 14.6 also shows the symbiotic systems, which are of great importance in terrestrial biological productivity and, in particular, in agricultural productivity. The symbioses between *Rhizobium* species and leguminous plants is well known and were thought, until 1975, to be obligate: scientists believed that rhizobia did not fix nitrogen in the absence of plant material. In fact this is not so: many species, notably the slow-growing 'cowpea' group, prove simply to be very sensitive microaerophiles, capable of fixing nitrogen in defined (and simple) laboratory media provided the dissolved oxygen tension is exceedingly low. Such rhizobia are now known to be able to form associations with a few non-legumes such as the subtropical shrub *Trema.*

A recent development of great importance has been the realisation of the enormous importance of non-leguminous symbioses in the terrestrial nitrogen balance. Alder, Ceanothus and Purshia are temperate trees or shrubs with a nitrogen-fixing symbiont which is not *Rhizobium.* Sub-tropical areas are widely colonized by Myrica and Casuarina, which have similar symbionts. Alder leaf fall alone can bring to the soil annually one-third of the fixed nitrogen that could be obtained by ploughing in a good crop of lucerne; the non-leguminous symbioses could be of enormous economic value in forest culture and specialised environments.

The nature of the symbionts in the non-leguminous symbioses is still not established, but cross-inoculation groups exist and the generic name *Frankia* has been given to the organisms. Morphological studies suggest that it is an actinomycete, but infective microbes have never successfully been cultivated away from the plant (though they seem able to persist in soils naturally).

An exciting recent development is the recognition of the associative symbioses, first described by Johanna Döbereiner of Brazil. An example is the sand-grass *Paspalum notatum,* widespread in tropical and sub-tropical areas, many cultivars of which have an azotobacter, *A. paspali,* specifically associated with their roots. *Azotobacter paspali* grows as a sheath round the root and then becomes almost dormant, though it continues to fix nitrogen. No nodule is formed, but the plant benefits dramatically; correspondingly, *A. paspali* survives poorly away from the plant. According to the acetylene test, the *P. notatum–A. paspali* association is capable of bringing as much nitrogen to the soil annually as a moderate legume crop, though whether it does so in natural conditions remains to be established. The immediate environment of the plant root (the rhizosphere) is almost anaerobic, and the symbiosis, though aerobic, is very sensitive to oxygen. To study it one must ensure that the plant–microbe system is not exposed to normal air during tests. Other associative symbioses include rice with *Beijerinckia,* and a grass called *Digitaria decumbens* with *Spirillum lipoferum.* Even some temperate weeds have nitrogen-fixing associates round their roots which enable them to grow well in poor soils.

As if to balance the new associative symbioses, two symbiotic systems once thought to be established have now to be rejected. One is the mycorrhiza of conifers, a fungus which grows in close association with the roots. It was once thought to fix nitrogen, largely because *Podocarpus,* a conifer with such mycorrhiza, fixed some $^{15}N_2$. More recent work has shown that the mycorrhiza is not needed for fixation: it is another case of associative symbiosis with free-living nitrogen-fixing bacteria, though these are as yet unidentified. The second is the leaf nodule system. Certain tropical plants such as *Psychotria* and *Ardesia* have leaf nodules which, when they are colonised by bacteria, benefit the plant enormously. Nitrogen-fixing bacteria can often, though not always, be isolated from such leaf nodules, and for about a decade the suggestion was that nitrogen was indeed fixed in the nodules. More recently, however, it has become clear that the nitrogen-fixing bacteria cease fixing when they colonise the leaf nodule; it is more probable that they benefit the plant by a hormone-like effect.

Free-living heterotrophic bacteria are probably relatively unimportant in the nitrogen cycle, except when they associate with a plant root, because they consume rather a lot of energy both to fix nitrogen and to exclude oxygen. Substrates to provide that energy are rarely plentiful in the biosphere. Photosynthetic microbes can sometimes divert solar energy to nitrogen fixation and, for this reason, the nitrogen-fixing blue-green algae are the most ecologically important free-living nitrogen-fixing microbes on this planet. They colonise arid and devastated zones; they are probably the primary nitrogen-fixing microbes in the sea and they are of immense importance in rice paddies, to which, in the growing season, they can bring nitrogen as rapidly as a good legume in a field. They form associations with fungi (as lichens and reindeer moss), with lower plants, ferns, cycads and even the angiosperm, *Gunnera*. Though their ability to fix nitrogen has been known for many years, the ecological importance of blue-green algae has been slow to dawn on the scientific community.

Reports appear occasionally of associations of nitrogen-fixing bacteria with animals including termites, ruminants (sheep, cows, deer), men (New Guinea natives) and guinea pigs. Bacteria of this kind are often present in the intestines of the animals studied but their importance to their nitrogen status is, where it has been studied, trivial. The probability is that the microbes are transient inhabitants of the gut: organisms which are ingested with food but which are of no serious physiological value to their host.

5 Genetics

Mutants of nitrogen-fixing bacteria such as *K. pneumoniae, C. pasteurianum* and *A. vinelandii* are known which are unable to fix nitrogen. Genetics really got started in 1971 when genetical techniques were used to restore to such mutants of *K. pneumoniae* the ability to fix nitrogen. Transduction of nitrogen fixation genes to the mutant by bacteriophage P1 was one procedure. Conjugation was the other procedure: the wild type was made fertile by introducing a drug resistance factor which had sex factor activity (a plasmid called R144drd3) and, in a low proportion of matings, portions of the *K. pneumoniae* chromosome were co-transferred with the R factor to the mutant. Geneticists' terminology for the transferrable nitrogen fixation genes is *nif*; both transduction and conjugation studies on transfer to *nif⁻* mutants carrying other markers showed that *nif* was very close to *his*, a determinant concerned with the biosynthesis of histidine. This fact was exploited for the inter-generic transfer of *nif*. A special strain of *K. pneumoniae* was selected in which the R factor was particularly prone to initiate transfer of *K. pneumoniae* chromosomal material at its *his* locus; this strain was mated with a 'restrictionless' *his⁻* strain of *E. coli* (a strain tolerant of foreign DNA) and a small number of His⁺ progeny were obtained. The majority of these proved to be able to fix nitrogen. The *nif* genes are now known to form a cluster which can be mutated at many loci. Among neighbouring determinants are *gnd*, a site specifying gluconate-6-phosphate dehydrogenase and *rfb*, a site determining lipopolysaccharide structure and hence sensitivity to some bacteriophages. In one of the *E. coli* hybrids, strain C-M7, genetical experiments and direct physicochemical examination of the DNA made it clear that a piece of *K. pneumoniae* DNA carrying *nif* had actually integrated itself into the *E. coli* chromosome. In other *E. coli* hybrids, the *K. pneumoniae* DNA formed plasmids. In fact, a confusing situation arose because several plasmids were formed and it was not always obvious which carried *nif*. However, the possibility arose of deliberately constructing a *nif* plasmid with its own sex factor activity and Dr R. Dixon did this successfully in 1974. The principle of the procedure was simple: the fertile strain of *K. pneumoniae* was mated with an Hfr strain of *E. coli* K-12, one known to initiate chromosome transfer near *his* and also having a *his*⁻ marker. The His⁺Nif⁺ progeny of the mating was then mated again to a *recA*⁻ mutant of *E. coli* which was also *his*⁻ (it carried other markers, too). The *recA* mutation prevented any recombination of introduced DNA, so any His⁺Nif⁺ progeny were obliged to conserve these genes extrachromosomally. Such progeny were obtained and indeed contained an F′ *his nif* plasmid; they had lost the original R factor, too. The F′ *nif* was somewhat unstable, but introduction of a different plasmid carrying resistance to the drug carbenicillin, followed by selection for His⁺ and carbenicillin resistance, yielded an F′ *his nif carb* plasmid (FN68) which was reasonably stable.

FN68 has been used to prepare various nitrogen-fixing strains of *E. coli,* as well as of *Salmonella typhimurium* and *Klebsiella aerogenes.* It has been isolated as a molecule of covalently closed circular DNA of molecular weight 1.36×10^8 and has been used in transformation experiments. Its relative instability and limited host range — the F sex factor is only self-transmissible among coliform bacteria — limits its usefulness and recently a somewhat more stable and versatile *nif* plasmid of the P class has been prepared. The P group of plasmids are self-transmissible among a variety of Gram-negative bacterial genera (*Pseudomonas, Agrobacterium, Rhizobium, Azotobacter,* the coliforms) and the carbenicillin-resistance determinant of FN68 originated from a plasmid of this class (see Ch. 4). Advantage was taken of this fact to transfer the *his nif* cluster of FN68 to RP4, a P plasmid carrying carbenicillin resistance among other markers, and a P prime plasmid was obtained, called RP41, which carries resistance to carbenicillin, kanamycin and tetracycline as well as *Klebsiella his nif.* Its molecular weight is about 5.9×10^7.

As an example of the use of such plasmids for the genetic analysis of *nif,* one can cite studies in the genetic regulation of these genes. Ammonium represses the expression of *nif*. Ammonium also regulates the expression of several other genes involved in nitrogen metabolism, including genes responsible for the utilisation of histidine (*hut*), for urease formation, for nitrate reduction (*nar*) and for proline oxidation (*put*). Ammonium also regulates its own assimilation pathway. When plenty of ammonia is available, the microbe makes glutamate dehydrogenase (GDH) and uses this to assimilate ammonia:

$$\text{NAD(P)H} + \text{NH}_3 + \text{2-oxo-glutarate} \xrightarrow{\text{GDH}} \text{NAD(P)} + \text{glutamate}$$

This enzyme has a high K_m for ammonium and becomes ineffective if the ammonia concentration is low. In such circumstances a different pathway is used. First, some existing glutamate is converted to glutamine by glutamine synthetase (GS) which requires ATP:

$$\text{glutamate} + \text{NH}_3 + \text{ATP} \xrightarrow{\text{GS}} \text{glutamine}$$

Secondly, this glutamine reacts with 2-oxo-glutarate via another enzyme, glutamate synthase (called GOGAT, from the initials of its systematic name):

$$\text{glutamine} + \text{2-oxoglutarate} + \text{NADPH} \xrightarrow{\text{GOGAT}} 2 \text{ glutamate} + \text{NADP}.$$

The GS + GOGAT sequence is biologically expensive, because it consumes ATP. It proves to be characteristic of nitrogen fixation: 'fixed' nitrogen is assimilated by the GS + GOGAT pathway. The enzyme GS proves to be a very remarkable molecule; it not only exists in enzymically active or inactive forms, but also activates the synthesis of itself and of the other enzymes of nitrogen metabolism. It can activate *hut* and *put* genes and it can repress GDH synthesis. What exactly it does in any particular physiological state of the cell depends on the extent to which it has reacted with ATP, ammonium ion, or both. This remarkable regulator of nitrogen metabolism also proves to activate expression of *nif*: there exist mutants of *K. aerogenes* which are constitutively rich in GS and, when the F′*nif* plasmid was introduced into such a mutant, or if the constitutive GS gene was transduced into nitrogen-fixing *K. pneumoniae*, then *nif* was constitutively expressed: it ceased to be subject to repression by ammonia.

The study of regulation is only one aspect of the many research uses to which one can put plasmids carrying *Klebsiella nif* genes. The conditions for expression can be studied in experiments such as one in which RP41 was transferred to mutants of *A. vinelandii* lacking either the molybdoprotein or the iron protein. Ability to fix nitrogen was restored, implying not only that *A. vinelandii* was using Klebsiella protein for at least part of its nitrogen-fixing activity but also that the regulator in *A. vinelandii* could initiate use of Klebsiella genes. In addition, though *K. pneumoniae* only fixes nitrogen in the absence of air, the 'corrected' *A. vinelandii* mutants could fix in air, so the *Klebsiella* proteins shared in the oxygen exclusion processes of *Azotobacter*. Not all transfers of *Klebsiella nif* genes yield nitrogen-fixing bacteria; RP41 in *Agrobacterium tumefaciens* or *Rhizobium meliloti* does not enable them to fix nitrogen, though in both cases there is immunological evidence that nitrogenase proteins are formed. The reasons for the failure of *nif* to be fully expressed in these hosts could be very informative.

Most experiments on transfer of nitrogen fixation genes have been performed with *Klebsiella nif,* but transfer of *nif* within *Azotobacter* mutants and from *Rhizobium trifolii* to *K. aerogenes* has been described and preliminary reports of *nif* gene transfer among photosynthetic bacteria have appeared.

An important aspect of the genetics of *nif* is the analysis of the gene cluster. How many cistrons are there? What do they specify? Transductional crosses between mutants and complementation studies with mutations on RP41 enable one to assign *nif*⁻ mutants to genetic groups, and biochemical examination of the members of the groups enable one to assign functions to the group. Two loci have been defined which determine the synthesis of the two polypeptides of the Mo-Fe protein, at least one for the Fe protein, one and possibly two regulatory sites, a locus involving some aspect of molybdenum incorporation and one involving transfer of electrons to nitrogenase. There are other genetic classes which are less well defined, including one which may not be part of the transferable *nif* cluster. Deletion mutants exist, lacking several cistrons, and temperature-sensitive ones have been obtained which make heat-sensitive, and therefore modified, enzyme protein. A few suppressible mutants exist and show evidence of polarity, but no constitutive types have yet been obtained apart from the GS constitutive type mentioned earlier, which is not mutated in the *nif* gene cluster. The genetic analysis of *nif* is, however, developing rapidly at the time of writing (1977) and for recent developments the primary literature should be consulted.

6 Some future prospects

The importance of nitrogen fixation in world food supply was emphasised at the beginning of this chapter. Escalating costs of production and transport of fertilizer, both in monetary and energy terms, make exploitation of biological processes increasingly attractive and probably obligatory in the next century. Means of making fertilizers must be cheapened and crops of legumes may be improved, but such expedients give the world a breathing space rather than solve its problems. In the long term more imaginative possibilities must be evaluated.

The value of establishing stable associations of cereal crops with organisms such as *A. paspali* or *S. lipoferum* must be obvious; even the extension of this type of association to useful forage grasses would be a great advance. The facts that some *Rhizobium* species fix nitrogen away from a plant host, and that nitrogen-fixing associations can be obtained between such *Rhizobium* and non-leguminous plant tissue cultures, open the possibility of generating rhizobial associations with unusual and useful plants. Hybridisation of legumes or other nitrogen-fixing plants with non-fixing flowers and vegetables is being considered by plant breeders. Transfer of *nif* genes to bacteria which naturally associate with plants could provide new, deliberately contrived associations; it might even be possible to devise an organelle, derived like the chloroplast from an endosymbiotic microbe, which would fix nitrogen in the plant cell itself. Ultimately, *nif* genes could probably be transferred directly to plants, but it must be obvious that their expression there would be a most complex problem: supporting genes would be needed to handle oxygen-sensitivity, regulation, ATP generation, direction of Mo to the enzyme, ferredoxin formation and doubtless functions yet to be discovered. If the way were easy to plants which fix nitrogen without benefit of bacteria, they would exist already, for selective pressure in favour of the evolution of such plants must have existed for 5×10^8 years if not longer. At present there seems to be some biological obstacle to the expression of *nif* in eukaryotes, though prokaryotes have overcome all the physiological and environmental obstacles that we are at present aware of. Perhaps there is in fact no real obstacle to nitrogen-fixing eukaryotes; some have argued that nitrogen fixation is a biochemical property of relatively recent origin ($< 10^8$ years) and that environmental considerations (principally the Earth's oxygen-rich atmosphere) have so far restricted its spread to prokaryotes.

Further reading

The subject of biological nitrogen fixation is advancing rapidly and any attempt at an up-to-date reference list would be hopeless. The following is a list of recent publications which contain background information relevant to this chapter.

Books

STEWART, W. D. P. (1966) *Nitrogen Fixation in Plants.* Athlone Press, London.
MISHUSTIN, E. N. and SHIL'NIKOVA, V. K. (1968) *Biological Fixation of Atmospheric Nitrogen* (Eng. trans., 1971). MacMillan, London.
POSTGATE, J. R. (Ed.) (1971) *The Chemistry and Biochemistry of Nitrogen Fixation.* Plenum Press, New York.
QUISPEL, A. (Ed.) (1974) *The Biology of Nitrogen Fixation.* North Holland, Amsterdam.

Symposia

NEWTON, W. E. and RYMAN, C. J. (Ed.) (1976) *Proceedings of the 1st International Symposium on Nitrogen Fixation,* Vol. 1 & 2. Washington State University Press, Pullman, Wash.

STEWART, W. D. P. (Ed.) (1976) *Nitrogen Fixation by Free-Living Organisms,* International Biological Programme 6. Cambridge University Press, London.

NUTMAN, P. S. (Ed.) (1976) *Symbiotic Nitrogen Fixation in Plants,* International Biological Programme 7. Cambridge University Press, London.

Reviews

YATES, M. G. and JONES, C. W. (1974) 'Respiration and nitrogen fixation in *Azotobacter*', *Adv. Microbial Physiol.*, **11**, 97–136.

DILWORTH, M. J. (1974) 'Dinitrogen fixation', *Ann. Rev. Plant Physiol.,* **25**, 81–114.

BRILL, W. J. (1975) 'Regulation and genetics of bacterial nitrogen fixation', *Ann. Rev. Microbiol.,* **29**, 109–30.

ZUMFT, W. G. and MORTENSON, L. E. (1975) 'The nitrogen-fixing complex of bacteria', *Biochim. Biophys. Acta,* **416**, 1–52.

WINTER, H. C. and BURRIS, R. H. (1976) 'Nitrogenase', *Ann. Rev. Biochem.,* **45**, 409–26.

15

Bioenergetics of chemolithotrophic bacteria

D. P. Kelly
Department of Environmental Sciences, University of Warwick

1 Introduction

Chemolithotrophic bacteria are those specialised organisms capable of obtaining all the energy necessary for growth from the oxidation of *inorganic* substances. In contrast to the photolithotrophic bacteria, which use light energy and oxidise inorganic sulphur compounds or hydrogen only to provide electrons, chemolithotrophs can meet all their needs for energy in the form of NADH and ATP by the oxidation of substances such as ammonia, nitrite, hydrogen, sulphur compounds or iron in the absence of light and have no mechanism for making use of light energy (Fig. 15.1).

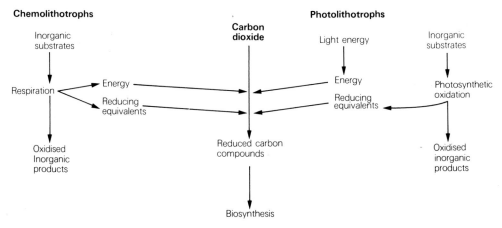

Fig. 15.1 The basic processes of lithotrophic autotrophy.

Virtually all chemolithotrophs are also autotrophs: that is they are able to synthesise all of their organic constituents from carbon dioxide as a sole source of carbon, using energy from the inorganic oxidations they catalyse (Fig. 15.1) and are thus non-photosynthetic primary producers of fixed carbon in the biosphere. Chemolithotrophic autotrophs are characterised by their ability to develop in a wholly inorganic aqueous environment when provided with their oxidisable substrate, carbon dioxide, and essential mineral nutrients. They can also usually assimilate organic nutrients (often at the expense of chemolithotrophic energy supplies) and may in nature actually exhibit a 'mixotrophic' metabolism in which they obtain energy from inorganic oxidations and carbon from both CO_2 and organic compounds (Whittenbury and Kelly 1977). A few 'chemolithotrophic heterotrophs' are known which are deficient in the ability to use carbon dioxide and use chemolithotrophic energy-yielding oxidations only to assimilate organic nutrients (Rittenberg 1969). Some chemolithotrophic autotrophs are also able to grow on organic media and then exhibit typically heterotrophic metabolism; others are obligate chemolithotrophs and can grow only if supplied with their inorganic energy substrate (Kelly 1971, Rittenberg 1969, Whittenbury and Kelly 1977).

In this chapter I am concerned only with the energy metabolism of chemolithotrophs growing on inorganic substrates and fixing carbon dioxide as the main carbon source.

2 The organisms and their substrates

The first thing to be clearly understood is that different groups of chemolithotrophs use different oxidisable inorganic substrates. Thus the nitrifying bacteria which convert ammonia to nitrate actually comprise two distinct groups of organisms, one type oxidising ammonia to nitrite, the other oxidising nitrite to nitrate; neither group can oxidise the substrate specific to the other.

2.1 Ammonia and hydroxylamine oxidisers

$$NH_4^+ + 1{\cdot}5\,O_2 = NO_2^- + H_2O + 2H^+$$
$$NH_2OH + O_2 \;\; = NO_2^- + H_2O + H^+$$

The best known ammonia oxidiser is *Nitrosomonas,* but others are also well established and are doubtless important in various natural environments. These include *Nitrosospira, Nitrosococcus, Nitrosolobus* and *Nitrosovibrio* (Watson 1971, Watson and Waterbury 1971, Harms *et al.* 1976). These all show similar guanine + cytosine content (mole % GC) in their DNA, ranging from 50–55 per cent, but show a remarkable diversity of ultrastructural morphology. All are autotrophic.

2.2 Nitrite oxidation

$$NO_2^- + 0{\cdot}5\,O_2 \;\; = NO_3^-$$

Nitrobacter is the best known example of this group but others exist, including *Nitrospina* and *Nitrococcus,* all having GC 57·7–61·2 per cent but differing considerably in ultrastructural morphology (Watson 1971, Watson and Waterbury 1971).

2.3 Sulphur oxidisers

$$S_8 + 12O_2 + 8H_2O = 8H_2SO_4$$
$$H_2S + 2O_2 = H_2SO_4$$
$$Na_2S_2O_3 + 2O_2 + H_2O = Na_2SO_4 + H_2SO_4$$
$$Na_2S_4O_6 + 3{\cdot}5\,O_2 + 3H_2O = Na_2SO_4 + 3H_2SO_4$$

$$Na_2S_3O_6 + 2O_2 + 2H_2O = Na_2SO_4 + 2H_2SO_4$$
$$NaSCN + 4O_2 + 4H_2O = Na_2SO_4 + H_2SO_4 + 2CO_2 + 2NH_3$$

The most studied genus of sulphur-oxidisers is *Thiobacillus*, which contains a variety of species varying considerably in physiological characteristics (Vishniac and Santer 1957, Trudinger 1967, Hutchinson, Johnstone and White 1969). Their GC percentage ranges from 51 for *T. thiooxidans* (which grows at pH 1–5), 57 for *T. neapolitanus* (pH range 3–8) or *T. ferrooxidans* (growing best at pH 2·5 on tetrathionate) up to 70 for *T. thioparus* (pH range 5–9) or *Thiobacillus* A2 (pH range 7–10) and *T. novellus*, the latter two of which are facultative heterotrophs. All species oxidise thiosulphate and sulphide using molecular oxygen but only one species, *T. denitrificans*, can grow anaerobically on sulphur or thiosulphate, when it reduces nitrate instead of oxygen (Baalsrud and Baalsrud 1954, Taylor and Hoare 1971, Timmer-ten-Hoor 1976).

$$5Na_2S_2O_3 + 8NaNO_3 + H_2O = 9Na_2SO_4 + H_2SO_4 + 4N_2$$

Other sulphur and sulphide oxidisers include the filamentous *Beggiatoa* and the thermophilic *Sulpholobus* (Brock *et al.* 1972) although little is yet known about their lithotrophic energy metabolism. Others exist whose physiology is probably similar to *Thiobacillus*, such as *Thiobacterium* and *Thiomicrospira.*

2.4 Iron oxidisers

$$4FeSO_4 + O_2 + 2H_2SO_4 = 2Fe_2(SO_4)_3 + H_2O$$
$$2Fe^{2+} + 0·5O_2 + 2H^+ = 2Fe^{3+} + H_2O$$

Filamentous bacteria depositing hydrated ferric salts around themselves were among the earliest iron bacteria identified and were used along with sulphur and nitrifying bacteria by Winogradsky in formulating the concepts of what we now know as chemolithotrophic autotrophy (Kelly 1971, Schlegel 1975, Whittenbury and Kelly 1977).

The only organism in which iron oxidation has been studied in detail is *Thiobacillus ferrooxidans*, which can grow using energy from the oxidation of iron as well as sulphur compounds and can also oxidise minerals like pyrite (FeS_2) and chalcopyrite ($CuFeS_2$) in which both iron and sulphur become oxidised, and the sulphides of copper, lead, cadmium, nickel, zinc and cobalt. Iron is also oxidised by organisms like *Sulpholobus, Leptospirillum ferrooxidans* and a partially characterised thermophilic thiobacillus.

2.5 Hydrogen oxidation

$$H_2 + 0·5O_2 = H_2O$$

A number of facultatively heterotrophic bacteria use this oxidation to support autotrophic growth (Davis *et al.* 1969, Schlegel 1975) under aerobic or in some cases anaerobic conditions. The unifying feature of these organisms is their ability to couple energy from hydrogen oxidation to the fixation of CO_2 as the sole carbon source. Their energy metabolism is, however, essentially conventional as hydrogen is oxidised in much the same way through the electron transport chain as would be hydrogen released from the oxidation of organic substrates. No particular energetic problems are exhibited by the hydrogen-oxidising system (Gunderson 1968, Schlegel 1975) and these organisms are not considered further in this chapter.

2.6 Hydrogen oxidation to methane

$$4H_2 + CO_2 = CH_4 + 2H_2O$$

The anaerobic methanogenic bacteria (e.g. *Methanobacterium*) couple the oxidation of hydrogen to CO_2 reduction to produce methane and use the energy obtained to assimilate CO_2-carbon to cell material. They are thus chemolithotrophic autotrophs but little more can be said of the biochemistry of their energy-coupling mechanisms (Wolfe 1972).

2.7 Other inorganic oxidations

It is theoretically feasible for any chemical oxidation that releases available free energy to be a source of energy for growth for an organism evolved to couple that oxidation to metabolic energy generation. Among oxidations for which evidence has been presented of organisms capable of using them for autotrophic growth are the conversion of cuprous to cupric copper, selenide to selenium, and oxidations of antimony, selenium and manganese.

It has also been suggested that uranium oxidation could be an energy-yielding reaction used by bacteria:

$$UO_2 + 0.5O_2 = UO_3$$

No substantive biochemical study has been made of these possible metabolic processes and they will not be discussed further.

3 Energy requirements for autotrophic growth and the nature of chemolithotrophic energy generation

3.1 Carbon dioxide fixation

It is characteristic of the chemolithotrophic autotrophs that the greater part of their carbon is obtained by fixation of carbon dioxide using the Calvin reductive pentose phosphate cycle (Kelly 1971, Schlegel 1975, Whittenbury and Kelly 1977). The fixation of CO_2 and its reduction to the level of carbohydrate is a very energy-expensive biochemical process (Fig. 15.2). Energy (as ATP) is consumed in the conversion of ribulose monophosphate to diphosphate and more energy (as ATP and NADH) is consumed in converting the primary product of CO_2-fixation (3-phosphoglyceric acid) to triose sugars. Thus to convert 6 CO_2 molecules to a molecule of hexose sugar (fructose) by the Calvin cycle, 18 ATP and 12 NADH molecules are needed:

$$6CO_2 + 18ATP + 12NAD(P)H + 12H^+ = \text{Fructose-6-P} + 18ADP + 17P_i + 12NAD(P)^+$$

Supplying the Calvin cycle with energy is thus the main energy need to be met by the chemolithotrophs and may account for at least 80 per cent of their total energy requirement for cell biosynthesis (Forrest and Walker 1971). Thus, in considering energy conservation and use in chemolithotrophs we are concerned, as with all other organisms, with the generation of ATP and reduced NAD(P).

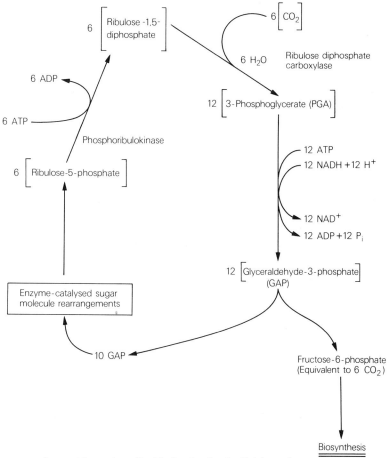

Fig. 15.2 Energy requirement for carbon dioxide fixation by the Calvin cycle.

3.2 The principle of energy generation

The work of the past 20 years has established that the principle of the unity of biochemical processes (Kluyver and Donker 1926) extends also to the chemolithotrophs, in contrast to the earlier view that they might be very different from other organisms (Stephenson 1931). All the chemolithotrophs using sulphur or nitrogen compounds or iron contain cytochromes and seem to possess respiratory electron transport chains not unlike those of other bacteria and basically like that of the mitochondrion. Consequently one can regard them as organisms that use conventional mechanisms for electron transport and ATP synthesis by oxidative phosphorylation but are specialised in being able to feed electrons into their electron transport systems from their specific inorganic oxidations. The basic electron transport chain as found in organotrophs is illustrated (Fig. 15.3) and the approximate electrode potentials for each couple in the chain is indicated. In the case of the hydrogen-oxidising bacteria, hydrogen oxidation (by hydrogenase) couples directly to the reduction of NAD^+ and thence into a complete electron flow to oxygen, with all three coupling sites for energy conservation by ATP synthesis likely to be able to function.

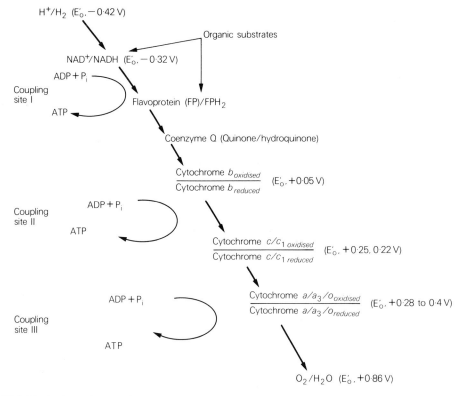

Fig. 15.3 The basic respiratory electron transport chain showing typical electrode potentials for the components (E_o' V).

4 The energetic problems facing the chemolithotrophs

A superficial examination of the chemolithotrophs would suggest they had solved their energetic requirements in a conventional manner, since they would seem to have evolved a metabolism fulfilling the criteria for successful respiration based on an oxidisable substrate. However, when the nature of the oxidations on which they depend is looked at more closely it can be seen that certain problems exist.

For an oxidation reaction to provide metabolically-useful energy, several criteria need to be met:

a The oxidation should be exergonic, meaning that free energy is made available as a result of the reaction.

b The amount of energy released needs to be sufficient at any one stage to support the synthesis of ATP so that useful conservation of the energy is possible.

c The hydrogen equivalents released from the oxidation need to be used to reduce NAD^+ so that metabolic reduction reactions (such as CO_2 fixation, ammonia conversion to amino acids) can be supported.

It is soon found that some chemolithotrophic oxidations (e.g. of ammonia, nitrite, iron) release relatively little energy: perhaps no more per mole oxidised than 1–10 per cent of the energy available to a heterotroph from the oxidation of one mole of glucose. In some cases with growing organisms, this energy is apparently not coupled with great efficiency to growth, so reducing the energy apparently available. Consequently, chemolithotrophic bacteria either grow slowly or have developed oxidation mechanisms of great activity, allowing very rapid turnover of their energy substrates so that an energy supply supporting relatively rapid growth becomes possible. Thus *Thiobacillus ferrooxidans* may grow on ferrous iron with a doubling time of 5 or 6 h in batch culture (or faster in the chemostat) when it oxidises 30 times its own weight of iron per hour!

More serious is the finding that (except for hydrogen) none of the inorganic substrates is capable of being used for the direct reduction of NAD^+, so a further energetic impediment is placed on the chemolithotrophs. The following sections consider these problems in detail, then demonstrate how they are overcome.

4.1 The energy available from inorganic oxidations

As a means of estimating available energy it can be useful to use thermodynamic calculation of free energy change ΔF° during a reaction and to estimate the energy available by calculation from the free energy of formation ΔF_f° or heats of formation ΔH_f° of the components of a reaction.

This kind of calculation was first used extensively for autotrophs by Baas Becking and Parks (1927). They rightly pointed out that measuring heat of reaction $(-\Delta H)$ gives the total amount of energy available from the process, but that this must be corrected for the temperature (degrees K) at which the process occurs and the increase in entropy to give the decrease in *free energy* $(-\Delta F)$ which is the maximum useful work available from the reaction (i.e. $-\Delta H + T\Delta S$). The value of ΔF must be negative for a process to be exergonic and to occur spontaneously. Clearly the larger the value of $-\Delta F$ the greater is the potentially available energy for metabolism. In practice ΔH and ΔF are often very similar (particularly in physiological conditions of temperature and concentrations; Fromageot and Senez 1960) and for our purposes when this is so, either quantity is a guide to the potential metabolic usefulness of lithotrophic oxidations. The energy made available can be compared with that needed to form ATP from ADP and orthophosphate and so some idea of the biochemical value of the oxidation can be obtained. In the past, ΔF and ΔH values have been expressed as kcal mole^{-1} for the reactions supporting growth and ATP formation. The data I have used below are derived from various literature sources and are all expressed in kilojoules per mole (using a conversion factor of 4·187 kJ kcal^{-1}).

Formation of ATP requires between 37 and 59 kJ mole^{-1} (Krebs and Kornberg 1957, Burton 1957):

$$ADP + H_3PO_4 \rightleftharpoons ATP + H_2O \qquad -\Delta F_{298}^\circ = 30\cdot6 \text{ kJ (Lehninger 1975)}$$

The fixation of carbon dioxide was calculated by Bass Becking and Parks (using atmospheric concentration of CO_2 and O_2) to require about 495 kJ mole^{-1}:

$$6CO_2(0\cdot0003 \text{ atm}) + 6H_2O = C_6H_{12}O_6 + 6O_2(0\cdot2 \text{ atm})$$
$$\Delta F_{298} = 2\,968 \text{ kJ} = 495 \text{ kJ mole}^{-1}$$

The ΔH° for this process is about 470 kJ mole^{-1} (Eyring *et al.* 1960). Most calculations of ΔF and ΔH refer to standard conditions with high concentrations of reactants and gases at atmospheric pressure, a situation less likely to prevail in nature. When the reactants are highly diluted or are gases at very low partial pressures, then ΔF tends towards ΔH, so in most biological systems these values will be similar.

For some of the important chemolithotrophic oxidations the following amounts of energy can be available from the *overall* oxidations catalysed by the bacteria (these data are taken or calculated from Baas Becking and Parks, 1927, Latimer 1952, Fromageot and Senez 1960, Kiesow 1967, Kelly, unpublished data):

$$Na_2S_2O_3(aq) + 2O_2(g) + H_2O \qquad = Na_2SO_4(aq) + H_2SO_4(aq) \qquad (1)$$
$$\Delta H°(kJ) -1\ 118 \quad -16 \quad -286 \quad -1\ 388 \quad -910$$

For the overall reaction:

$$\Delta H = [(-1\ 388) + (-910)] - [(-1\ 118) + 2(-16) + (-286)]$$
$$= -862 \text{ kJ mole}^{-1} \text{ thiosulphate oxidised}$$

$$S_2O_3^{2-} + 2O_2 + H_2O \qquad = 2SO_4^{2-} + 2H^+ \qquad (2)$$
$$\Delta H = -858 \text{ kJ}; \quad \Delta F = -936 \text{ kJ}$$

$$S_2O_3^{2-} + 2\cdot5O_2 \qquad = 2SO_4^{2-} \qquad (3)$$
$$\Delta F_{298} = -995 \text{ kJ}$$

$$5Na_2S_2O_3 + 8KNO_3 + 2NaHCO_3 = 6Na_2SO_4 + 4K_2SO_4 + 4N_2 + 2CO_2 + H_2O \qquad (4)$$
$$\Delta F_{298} = -3\ 740 \text{ kJ} = -748 \text{ kJ mole}^{-1} \text{ thiosulphate}$$

$$5S_2O_3^{2-} + 8NO_3^- + H_2O \qquad = 10SO_4^{2-} + 2H^+ + 4N_2 \qquad (5)$$
$$\Delta F° = -3\ 707 \text{ kJ } (741 \text{ kJ mole}^{-1} \text{ thiosulphate});$$
$$\Delta H° = -4\ 092 \text{ kJ } (818 \text{ kJ mole}^{-1} \text{ thiosulphate})$$

$$2S_2O_3^{2-} + 0\cdot5O_2 + H_2O \qquad = S_4O_6^{2-} + 2OH^- \qquad (6)$$
$$\Delta H° = -44 \text{ kJ}$$

$$S_4O_6^{2-} + 3\cdot5O_2 + 3H_2O \qquad = 4SO_4^{2-} + 6H^+ \qquad (7)$$
$$\Delta H° = -1\ 577 \text{ kJ}; \quad \Delta F° = -1\ 654 \text{ kJ}$$

$$H_2S + 0\cdot5O_2 \qquad = S° + H_2O \qquad (8)$$
$$\Delta F° = -210 \text{ kJ}$$

$$S° + 1\cdot5O_2 + H_2O \qquad = H_2SO_4 \qquad (9)$$
$$\Delta F_{298} = -496 \text{ kJ}$$

$$5S° + 6KNO_3 + 2CaCO_3 \qquad = 3K_2SO_4 + 2CaSO_4 + 2CO_2 + 2N_2 \qquad (10)$$
$$\Delta F_{298} = -2\ 763 \text{ kJ} = -553 \text{ kJ mole}^{-1} \text{ sulphur}$$

$$HS^- + 2O_2 \qquad = SO_4^{2-} + H^+ \qquad (11)$$
$$\Delta F_{298} = -995 \text{ kJ}$$

$$NH_3 + 0\cdot5O_2 \qquad = NH_2OH \qquad (12)$$
$$\Delta F° = +15\cdot5 \text{ kJ}$$

$$NH_4^+ + 1\cdot5O_2 \qquad = NO_2^- + H_2O + 2H^+ \qquad (13)$$
$$\Delta F_{298} = -272 \text{ kJ}$$

$$NH_2OH + O_2 \qquad = HNO_2 + H_2O \qquad (14)$$
$$\Delta F° = -288 \text{ kJ}$$

$$NO_2^- + 0\cdot5O_2 \qquad = NO_3^- \qquad (15)$$
$$\Delta F_{298} = -73 \text{ kJ}$$

$$4FeSO_4 + O_2 + 2H_2SO_4 \qquad = 2Fe_2(SO_4)_3 + 2H_2O \qquad (16)$$
$$\Delta H° = -100 \text{ kJ mole}^{-1}FeSO_4; \quad \Delta F° = -47 \text{ kJ mole}^{-1}FeSO_4$$
For physiological conditions, $\Delta F = -25 \text{ kJ mole}^{-1}$ (pH 3)
$$= -30 \text{ kJ mole}^{-1} \text{ (pH 2)}$$
$$= -33 \text{ kJ mole}^{-1} \text{ (pH 1·5)}$$

Thermodynamic data of this kind are useful in two ways.

a They enable an estimation of the total *maximum* ATP synthesis that might accompany substrate oxidation if the reaction were thermodynamically 100 per cent efficient; this is more useful and reliable the simpler the reaction. Thus nitrite oxidation (liberating 73 kJ mole^{-1} under physiological conditions) and ferrous iron oxidation (30 kJ mole^{-1} at pH 2) could not be expected to support the synthesis of more than one ATP per mole nitrite or per two moles iron oxidised. The data also indicate that ammonia oxidation to hydroxylamine cannot be energy-yielding and is in fact slightly energy-consuming. One can therefore conclude from the thermodynamic data alone that all energy conservation during ammonia oxidation must accompany the oxidation of hydroxylamine to nitrite. With the oxidation of sulphur and its compounds, where large amounts of energy (500–1 700 kJ mole^{-1}) are available from the overall reactions, the arithmetic deduction of potential for ATP synthesis is less reliable since the bacterial oxidations may be broken down into a number of steps, each giving some energy release, but only some of these (for example, those supporting electron transport processes) may be coupled to energy-conservation through ATP synthesis (see sections 4.6 and 5).

b Thermodynamic data enable an estimation of the total overall efficiency of conversion of energy from the oxidation reactions to metabolic energy as measured by production of new bacterial growth. Thus if the amount of growth produced (or CO_2 fixed) during the oxidation of a known amount of inorganic substrate is determined, the 'growth efficiency' can be estimated since the energy required for converting CO_2 to the level of $C_6H_{12}O_6$ (which approximates the reduction level in the cell, and represents the initial reduction level of all CO_2 fixed by the Calvin cycle: i.e. 70–90 per cent depending on the organism) is known to be 495 kJ mole^{-1} CO_2. The heat of combustion of the bacteria could, of course, also be determined directly for the purpose of such calculations. Using the data of Meyerhof (1916), Baass Becking and Parks (1927) calculated that about 5·9 per cent of the energy from ammonia oxidation and 7·9 per cent of that from nitrite oxidation was coupled to CO_2 fixation and growth. Thus for nitrite oxidation, one mole of CO_2 ($\equiv$ 495 kJ) was fixed for the oxidation of 86 moles nitrite ($\equiv$ 6 278 kJ) giving an efficiency of 495/6 298 $\times$ 100 = 7·9 per cent. Similar calculations with *Thiobacillus ferrooxidans* growing on iron are as follows: growth in batch culture at pH 1·5 gives about 0·35 g organism per mole $FeSO_4$ oxidised (Tuovinen and Kelly 1973). The bacteria contain 48 per cent carbon, so fix about 0·014 mole CO_2 per Fe^{2+}, or have to use about 71·4 moles Fe^{2+} (2 356 kJ) to fix one mole CO_2 (495 kJ), giving an overall efficiency of 21 per cent. For this thiobacillus oxidising thiosulphate, 5·9 g organism were obtained per mole oxidised, representing one CO_2 fixed per four moles $Na_2S_2O_3$. Using the $-\Delta F$ value of 936 kJ mole^{-1} this gives an observed efficiency of 13 per cent conservation of the energy apparently available. Older calculations with aerobic and anaerobic sulphur bacteria gave efficiencies of 6–9 per cent, although efficiencies as high as 25–50 per cent have been reported in short-term experiments with thiobacilli (Baas Becking and Parks 1927, Fromageot and Senez 1960).

The relatively low energy efficiencies reported for growing cultures of chemolithotrophs (excluding hydrogen bacteria) indicate three things: (*a*) that in common with other processes governed by the laws of thermodynamics, chemolithotrophic energy conversion systems (e.g. the mitochondrion) overall efficiency of energy conversion cannot be 100 per cent efficient; (*b*) in common with other bioenergetic conversion is unlikely to be better than 50 per cent; (*c*) the low efficiencies (<20 per cent) reported for thiobacilli oxidising sulphur compounds may well indicate that some of the intermediate oxidation steps, while being exergonic, are not coupled to energy-conserving processes, and consequently lower the maximum efficiency possible. Moreover, the calculations do not consider the *reducing equivalents* that are directly transferred from the substrate oxidation reaction to reduce CO_2 and thus cannot pass into the electron transport chain to generate ATP.

4.2 The reducing potential available from inorganic oxidations

The most important function of biological oxidations in respiring organisms like chemolithotrophs is the production of reducing equivalents for NAD^+ reduction and for the electron
transport chain. As in heterotrophic respiration the *dehydrogenation* of the substrate is thus the
primary phase of energy conservation.

   To understand the major problem that had to be solved in the evolution of chemolithotrophic energy-generating metabolism, one must recall that chemical oxidations and reductions
all have characteristic oxidation-reduction potentials. Thus one oxidation reaction may generate
reducing equivalents that are sufficiently electronegative to allow the reduction of a substance
of more positive potential than that of the oxidation substrate (Fig. 15.4). This is, of course, a

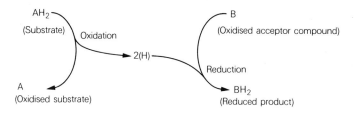

Fig.15.4

familiar concept and is the one on which the functioning of the electron transport chain
depends. The point to be stressed is that under standard conditions AH_2 will always tend to
reduce B, but BH_2 would not be expected spontaneously to reduce A. Thus all coupled
oxidation-reduction reactions are a 'downhill' flow of reducing potential from more reducing
(more electronegative) to less reducing (or more oxidising; i.e. more electropositive, less
electronegative) reactions. The reducing capacity of a particular oxidation can be defined by
means of the standard reduction potential (E_o') of the oxidation reduction reaction (see

Table 15.1   Standard reduction potentials
for NAD^+ reduction and inorganic oxidations

System	E_o' (V)
H_2/H^+	-0.420
$NAD^+/NADH$	-0.320
H_2S/S^0	-0.280
SO_3^{2-}/SO_4^{2-}	-0.250
S^0/SO_3^{2-}	$+0.050$
$2S_2O_3^{2-}/S_4O_6^{2-}$	$+0.240$
$S_2O_3^{2-}/2SO_4^{2-}$	$+0.484$
NH_2OH/NO_2^-	$+0.066$
NH_4^+/NO_2^-	$+0.344$
NH_4OH/NH_2OH	$+0.562$
NH_4^+/NH_2OH	$+0.899$
Fe^{2+}/Fe^{3+}	$+0.770$
$O_2/2H_2O$	$+0.860$

Lehninger 1975). Table 15.1 lists these potentials for the important inorganic oxidations and for NAD^+ reduction. It is immediately apparent that none of the oxidations (apart from H_2) can be used directly for the reduction of NAD^+ since they are all energetically 'downhill' from the $NAD^+/NADH$ couple. Consequently, an *indirect* method for NAD^+ reduction has to exist in the chemolithotrophs.

Caution needs to be exercised in applying the values for electrode potentials available from the literature, since these generally relate to standard conditions and factors such as concentration, pH and the complexing of reactants with other compounds could modify true potential prevailing in an organism. For example, the potential for iron oxidation is extremely electropositive but the potentials for different cytochromes may range from -0.2 to $+0.4$ V, and these depend primarily on the oxidation-reduction of the *iron* complexed in the cytochrome molecules. However, the potentials do serve as a guide in most cases to the probable point of entry of electrons from chemolithotrophic oxidations into the electron transport chain.

In the case of the sulphur compounds, recalling that these potentials are only a guide and absolute values depend on absolute concentrations and ratios of the reactants, some values come close to being sufficiently negative to couple directly to NAD^+. Taking published chemical data for *overall* oxidative hydrolysis of sulphur compounds, some E'_o values are very electronegative:

$$S_2O_3^{2-} + 3H_2O = 2H_2SO_3 + 2H^+ + 4e^-; \quad E'_o = -0.4 \text{ V}$$
$$S_3O_6^{2-} + 3H_2O = 3H_2SO_3 + 2e^- \quad ; \quad E'_o = -0.3 \text{ V}$$
$$S_4O_6^{2-} + 6H_2O = 4H_2SO_3 + 4H^+ + 6e^-; \quad E'_o = -0.51 \text{ V}$$

These reactions are not, however, known to occur as single-step processes in thiobacilli, and thus reducing equivalents of such low potential are unlikely to be available to bacteria that metabolise inorganic sulphur compounds.

4.3 Electron transport in chemolithotrophs

It has been established by much work over the past 20 years that all the chemolithotrophs contain electron transport components (flavoproteins, quinones, cytochromes) for the operation of a complete respiratory chain, and NADH oxidation mediated by the cytochrome chain can occur in experiments with cell-free extracts. The points at which electrons from the various chemolithotrophic oxidations enter the electron transport chain are shown in Fig. 15.5. Cytochrome *c* is the level for entry for electrons for oxidations of sulphur, ammonia, hydroxylamine and possibly nitrite. There is some evidence (from spectrophotometric observa-

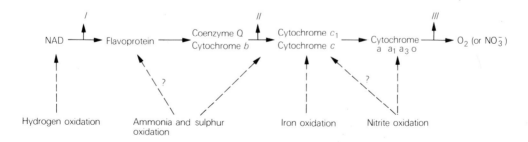

Fig. 15.5 Points of entry into the electron transport chain of reducing equivalents from inorganic oxidations (I, II and III indicate probable coupling sites for oxidative phosphorylation).

tions and the effect of specific inhibitors of the chain) that hydroxylamine and sulphur-compound oxidation might initially reduce a cytochrome *b* or a quinone and possibly involve flavin. A ferrous iron oxidase has been demonstrated that catalyses cytochrome *c* reduction in *Thiobacillus ferrooxidans.* This is apparently possible because of the lowered E'_o of the Fe^{2+} by complexing during the oxidation reaction. The oxidation of nitrite is unlikely on thermodynamic grounds to be able to reduce cytochrome *c,* and cytochrome a_1 is believed to be the initial acceptor. Evidence for this is the inhibition by dibromophenol of cytochrome *c* reduction, which does not inhibit either nitrite oxidation or cytochrome a_1 reduction. Details of the electron transport systems are discussed further elsewhere (Aleem 1970, Tuovinen and Kelly 1972, Suzuki 1974, Schlegel 1975).

As with the conventional electron transport chain the energy released by electron flow from cytochrome *c* to oxygen is coupled in the chemolithotrophs to ATP synthesis by oxidative phosphorylation (see section 4.6).

4.4 NAD$^+$ reduction in chemolithotrophs

Since the oxidation reactions they catalyse cannot directly reduce NAD$^+$ the chemolithotrophs have evolved alternative mechanisms for generating NADH, elucidated largely by the work of Aleem and his collaborators in the 1960s (Aleem 1969 and 1972, Aleem and Sewell 1969, Kiesow 1967, Peck 1968, Schegel 1975).

The problem may be envisaged as an energetic one from the equations describing theoretical NAD$^+$ reduction using hydroxylamine or nitrite:

$$NH_2OH + 2NAD^+ + H_2O = NO_2^- + 2NADH + 3H^+; \quad \Delta F = +149 \text{ kJ}$$
$$NO_2^- + NAD^+ + H_2O \quad = NO_3^- + NADH + 2H^+ \; ; \quad \Delta F = +159 \text{ kJ}$$

Taking the ΔF for ATP hydrolysis at 30 kJ mole^{-1}, this means that up to 5 ATP molecules or their energy equivalent would be *consumed* in these reactions to produce NADH.

It is now clear that in the case of all chemolithotrophs (except H_2 users), some of the electrons transferred to cytochrome *c* (or *a*) from NH_4^+, NO_2^-, Fe^{2+} or sulphur compounds are driven 'uphill' through the electron transport chain to NAD$^+$ at the expense of energy conserved during the *'downhill'* transfer of electron from cytochrome *c* (or *a*) to oxygen (or nitrate in the case of *T. denitrificans*) (Fig. 15.6).

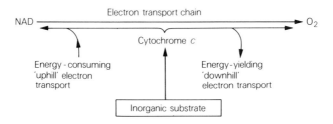

Fig. 15.6

The mechanism for 'uphill' electron flow is essentially a complete reversal of the normal 'downhill' flow reactions, involving a reversal of the energy-conserving reactions so that ATP or high-energy intermediate stages of oxidative phosphorylation are consumed and reducing equivalents are transferred via cytochrome *b,* quinone and flavins to NAD$^+$. For thiobacilli oxidising thiosulphate the sequence would be as shown in Fig. 15.7.

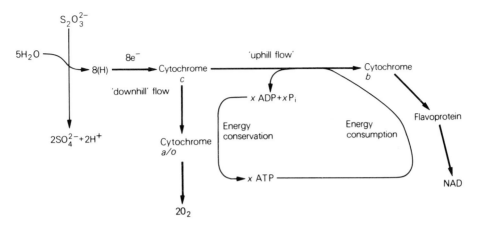

Fig. 15.7

Evidence for this process in cell-free extracts and whole cells of chemolithotrophs comes from the following observations:

a NAD$^+$ reduction is dependent on ATP (or on energy from 'downhill' oxidation of cytochrome c).

b NAD$^+$ reduction is paralleled by reduced cytochrome c oxidation (ratio: one NAD: two cytochrome c).

c ATP-dependent NAD$^+$ reduction is inhibited by oligomycin and uncouplers of oxidative phosphorylation such as dinitrophenol and dibromophenol, which prevent the use of energy to drive electron flow 'uphill'.

d Energy-linked reverse electron flow from nitrite to NAD$^+$ is blocked by inhibitors of electron transport, such as mepacrine, rotenone, amytal or antimycin A (all of which block between NAD and cytochrome c), malonate and cyanide.

e By blocking electron transport between cytochrome b and c with antimycin A it can be shown (Fig. 15.8) that (as in mitochondria):

(i) ATP-dependent NAD$^+$ reduction at the expense of succinate occurs in *Thiobacillus novellus,* indicating energy consumption at coupling site I in the chain.

(ii) Similarly, anaerobically with substrate levels of cytochrome c as electron acceptor, energy can be generated by succinate oxidation at coupling site II and coupled to site I to drive NAD$^+$ reduction.

(iii) Aerobically, using reduced cytochrome c as substrate, energy can be generated at coupling site III and consumed at sites I and II to effect electron flow from c to NAD$^+$.

f Direct observation of intact thiobacillus organisms demonstrated NAD$^+$ and NADP$^+$ reduction during thiosulphate or sulphide oxidation (Roth *et al.* 1973). Uncoupling agents and amytal inhibited NAD(P)$^+$ reduction but did not inhibit ATP synthesis. This indicated that even sulphide oxidation (E'_o = −0·28 V) could not couple directly to NAD$^+$ reduction.

g In *Nitrobacter*, dibromophenol inhibits cytochrome c reduction without affecting cytochrome a reduction or nitrite oxidation.

h The uncoupling agent dinitrophenol inhibits thiosulphate-dependent CO$_2$ fixation (requiring both ATP and NADH) by *Thiobacillus neapolitanus* at concentrations that have little effect on the intracellular ATP content during thiosulphate oxidation (Kelly and Syrett 1963 and 1966).

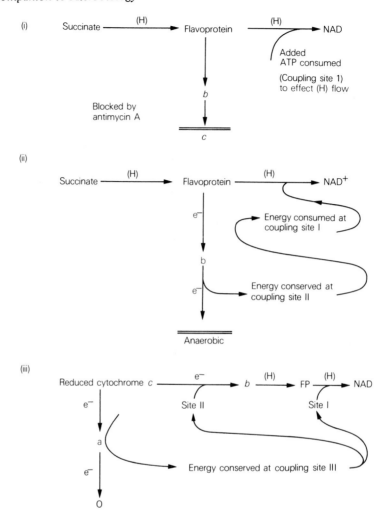

Fig. 15.8

This indicated that dinitrophenol-insensitive ATP production could occur in this thiobacillus, but that inhibition of DNP-sensitive NAD^+ reduction caused the suppression of CO_2 fixation.

4.5 The energy requirement for NAD^+ reduction

Various estimates have been made with cell-free extracts of different chemolithotrophs of the amount of ATP needed to effect the reduction of NAD^+. For *Nitrobacter* oxidising nitrite, up to 5 moles ATP are required for each mole NAD^+ reduced. With thiobacilli oxidising thiosulphate, up to 2 or 3 ATP per NAD^+ were needed, consistent with an 'energy gap' of 100 kJ between *Thiobacillus* cytochrome c and NAD^+, although lower values of one or 1·3 ATP per NAD^+ have been claimed (see Hempfling and Vishniac 1967, Ross *et al.* 1968, Aleem 1970, Cole and Aleem 1970, Tuovinen and Kelly 1972).

4.6 ATP generation by oxidative phosphorylation

Experimentally, P/O ratios up to 1·0 have been shown for cell-free extracts of *Nitrobacter* oxidising nitrite and thiobacilli using thiosulphate, indicating that each pair of electrons transferred to oxygen via cytochrome *c* or *a* is coupled to dinitrophenol-sensitive energy conservation by ATP synthesis at site III. Recent studies on the respiration-driven proton translocation in *T. neapolitanus* oxidising thiosulphate, sulphite or sulphide indicated maximal P/O ratios of 0·8–0·9 on each substrate (Drozd 1977). This is interesting since sulphide oxidation, being strongly reducing and liberating considerable energy, might be expected to support a P/O ratio of 2·0 if flavins and cytochrome *b* are involved in electron transport from sulphide and both coupling sites II and III were involved in energy conservation during sulphide oxidation (Aleem 1975). Similarly in *Nitrosomonas,* if electrons do enter the chain at the flavin level, a P/O ratio of 2·0 is feasible. So far, experimental P/O ratios are very low and this point is unresolved. For iron oxidation, two Fe^{2+} ions need to be oxidised to produce an electron pair for electron transport and a P/O ratio of 1·0 is likely (i.e. 0·5 ATP per Fe^{2+} oxidised) assuming electrons enter the respiratory chain at cytochrome *c* (Tuovinen and Kelly 1972). *In vivo* evidence for oxidative phosphorylation in intact thiobacilli came from the observation that ATP synthesis and carbon dioxide fixation was inhibited by various uncoupling agents, including dinitrophenol, during oxidation of sulphide (Fig. 15.9), sulphur, thiosulphate and polythionates (Kelly and Syrett 1964 and 1966).

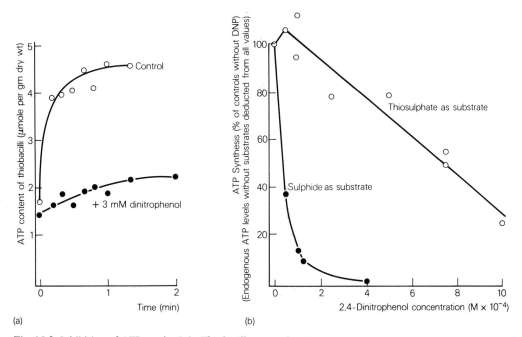

Fig. 15.9 Inhibition of ATP synthesis in *Thiobacillus neapolitanus* by 2,4-dinitrophenol provides evidence of oxidative and substrate level phosphorylation during sulphide and thiosulphate oxidation. (a) 0·6 mM Na_2S added at zero time. (b) ATP content of bacteria oxidising thiosulphate or sulphide for 1 to 6 min. with various amounts of 2,4-dinitrophenol.

4.7 Calculating an energy balance for the chemolithotrophic growth of an autotroph

We can now see that the inorganic oxidation has to provide three necessary components for growth on carbon dioxide: (*a*) energy as ATP, probably mainly by oxidative phosphorylation; (*b*) NADH generated by energy-dependent electron transport from cytochromes to NAD^+; (*c*) reducing equivalents for the reduction of NAD^+ and then CO_2. These must also be derived directly from the inorganic oxidation.

We can use nitrite oxidation by *Nitrobacter* as a model for these processes in operation. To fix CO_2 by the Calvin cycle requires 3 ATP and 2 NADH per mole. Consequently with a production of one ATP per nitrite oxidized (P/O = 1), 13 NO_2^- must be oxidised to provide enough energy to form 2 NADH (5 ATP per mole for reduction) and 3 ATP. In addition, 2 NO_2^- must be oxidised to provide the reducing equivalents (H) for NAD reduction, making a total of 15 NO_2^- for the production of 3 ATP + 2 NADH + $2H^+$. Aleem *et al.* (1965) demonstrated that water was the source of the (H) for NAD^+ reduction by showing that tritium (^3H) from water was transferred to produce NAD^3H during 'uphill' reduction of NAD^+. The process for CO_2 fixation in *Nitrobacter* is summarised in Fig. 15.10. Aleem (1970) has further calculated that the NADH and ATP requirements for other metabolic processes such as glutamate formation from ammonia, amino acid activation and incorporation into protein probably lead to an overall requirement of at least 21 moles NO^- for growth on each CO_2 molecule fixed. The figure (section 4.1) of 1 CO_2 fixed for 86 NO_2^- oxidised thus becomes a *biochemical efficiency* of 21/86 × 100 or 24·4 per cent rather than a thermodynamic efficiency of 7·9 per cent.

Calculation of this kind can be performed with less certainty for ammonia and sulphur compound oxidations since the oxidation pathways and sites and numbers of dehydrogenation reactions are not so well defined.

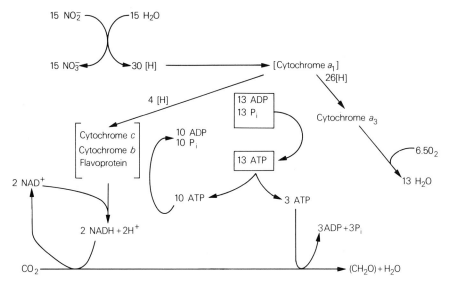

Fig. 15.10 Nitrite oxidation to support fixation of one molecule of carbon dioxide by *Nitrobacter*. [H] represents $H^+ + e^-$.

5 Pathways of oxidations in chemolithotrophs

For iron and nitrite oxidation little more need be said, since no free intermediates between Fe^{2+} and Fe^{3+} or NO_2^- and NO_3^- are likely, although the initial complexing and states in which these ions are actually metabolised are not fully understood. Basically they are simply oxidised to produce electrons that supply the electron transport chain. For ammonia and sulphur the mechanisms are more complex.

5.1 Ammonia oxidation

A probable mechanism (Aleem 1970, Suzuki 1974, Schlegel 1975) is as follows:

(*a*) $NH_3 + O_2 + AH_2$ (reduced cytochrome P-460?) $\xrightarrow{\text{'ammonia hydroxylase'}}$
    $NH_2OH + H_2O + A$ (oxidised cytochrome P-460?)

(*b*) NH_2OH + cytochrome c = (NOH) + reduced cytochrome c + $2H^+$
    nitroxyl

(*c*) $(NOH) + H_2O$ $= NO_2^- + AH_2 + H^+$

This sequence produces a $2e^-$ stepwise change in the oxidation state of the nitrogen atom from -3 (ammonia) to -1 (hydroxylamine) to $+1$ (nitroxyl) to $+3$ (nitrite) and explains ammonia conversion to hydroxylamine as a hydroxylation (oxygenation) reaction. Purified hydroxylamine dehydrogenase converts NH_2OH to nitrite and the fate of the postulated AH_2 is not then known although it could presumably be oxidised through the respiratory chain. If this is so, hydroxylamine oxidation should be more energy-yielding than overall ammonia oxidation as in the former both reactions (*b*) and (*c*) generate reducing equivalents that can be used in energy conservation; while in ammonia oxidation only reaction (*b*) could be used to support electron transport-linked phosphorylation. Aleem (1970) has postulated that nitrohydroxylamine (NO_2NH_2OH) might be an intermediate between (NOH) and NO_2^-.

5.2 Sulphur compound oxidation by thiobacilli and the probability of substrate-level phosphorylation producing ATP during sulphite oxidation

Although studied now for nearly 80 years the pathway of sulphur compound oxidation is still not unequivocally established in thiobacilli. A feasible mechanism for aerobic oxidation of sulphur and a number of its compounds is shown in Fig. 15.11. Polythionates ($S_4O_6^{2-}$, $S_3O_6^{2-}$) would enter this pathway as sulphite, sulphur or thiosulphate after reductive or hydrolytic cleavage. Detailed discussion of these reactions is given elsewhere (Trudinger 1967 and 1969, Kelly 1968, Peck 1968, Roy and Trudinger 1970, Suzuki 1974, Schlegel 1975).
    There is considerable uncertainty about the fate of the sulphane atom of thiosulphate and the mechanism of sulphur and sulphide oxidation. It is clear that sulphite is produced from these and there is enzymatic and ^{18}O evidence that the sulphur-oxidising enzyme may be an oxygenase. If this is the case for sulphide and sulphane—sulphur oxidation, some of the oxygen consumed in these oxidations will be incorporated into sulphite and not used in the oxidation of the electron transport chain, thus reducing the available energy yield of the oxidations. The presence or absence of an oxygenase reaction in sulphur oxidation is thus of great significance in assessing the bioenergetic capabilities of thiobacilli. The presence of the oxygenase reaction as a central one in the scheme is not proved, and its likely role is weakened by the fact that the anaerobic *T. denitrificans* successfully oxidises sulphur and sulphide anaerobically, when it cannot be using an oxygenase reaction.
    During aerobic thiosulphate oxidation, two moles of oxygen are consumed per mole of thiosulphate. If an oxygenase functions, only one oxygen could be used for terminal oxidation of the respiratory chain. With a P/O = 1, only two ATP could thus be formed by oxidative

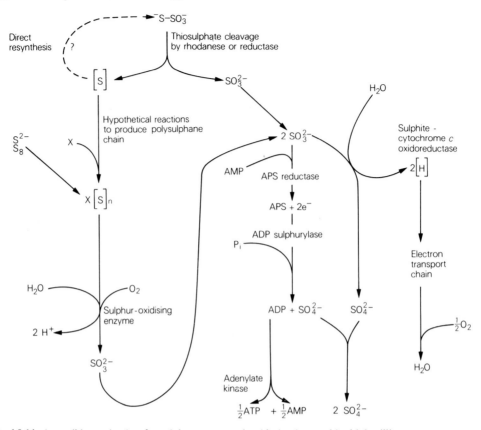

Fig. 15.11 A possible mechanism for sulphur compound oxidation by aerobic thiobacilli.

phosphorylation for each thiosulphate oxidised. If no oxygenase functions, *four* ATP could be formed during the consumption of two moles of oxygen. If, however, sulphide (or sulphane-sulphur) oxidation allows a P/O = 2 (Aleem 1975), one could predict that six ATP could be formed during thiosulphate oxidation by the scheme shown in Fig. 15.12.

An additional ATP might also be available from a substrate level phosphorylation during sulphite oxidation. Two distinct pathways for sulphite oxidation have been demonstrated (Fig. 15.11). One is the cytochrome *c*-linked sulphite oxidase that is commonly found in thiobacilli. The other sequence allows the synthesis of ADP (and ATP) by the 'APS pathway' involving the intermediate formation of adenosine phosphosulphate (adenylyl sulphate; APS). This synthesis of ATP is unaffected by dinitrophenol and has been indicated to occur in intact *T. neapolitanus* by showing that ATP formation was not inhibited during thiosulphate oxidation by concentrations of DNP that severely depressed ATP formation during sulphide oxidation (Fig. 15.9). The explanation given was that oxidative phosphorylation was more important during sulphide oxidation (Kelly and Syrett 1963). The APS pathway may be absent from *Thiobacillus* A2, but seems to be significant in *T. thioparus* and *T. denitrificans,* in which APS reductase may be 3–5 per cent of the total protein (Suzuki 1975). Both mechanisms for sulphite oxidation may function simultaneously in some thiobacilli.

This discussion demonstrates the uncertainty over the maximum amount of biochemically available energy from sulphur compound oxidation. The possibility exists that fundamental differences exist among the thiobacilli: thus in some species sulphane–sulphur oxidation might

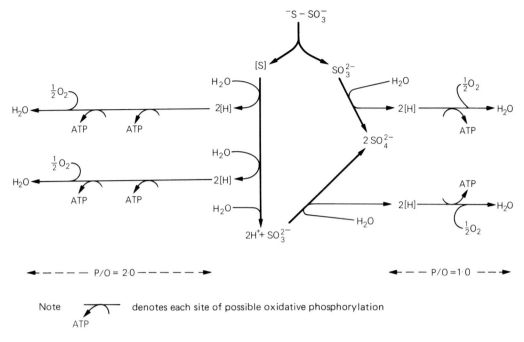

Fig. 15.12 Hypothetical scheme for energy coupling during thiosulphate oxidation by thiobacilli.

be energy-yielding; in others, using an oxygenase, it might not. A new approach to the problem of assessing bioenergetic capacity is given in the following section.

6 Continuous flow culture as a tool in elucidating chemolithotrophic energy-coupling mechanisms

When heterotrophic bacteria are grown in continuous culture on a growth-limiting energy or carbon substrate, the growth yield (= g biomass produced or g-mole carbon assimilated per mole substrate oxidised) generally increases as the specific growth rate (= dilution rate) is increased (Bull 1974, Pirt 1975) and it is possible to calculate the maximum theoretical growth yield attainable on a particular substrate under 'ideal' conditions. The same has been found for thiobacilli (Hempfling and Vishniac 1967, Justin and Kelly 1976, Timmer-ten-Hoor 1976, Eccleston and Kelly 1977). For thiosulphate-limited cultures, aerobic *T. neapolitanus* gave a theoretical true growth yield (Y_G) of 13·9 g dry wt per g-mole thiosulphate, while *T. denitrificans* gave values up to 18·5 g mole^{-1}. Elemental analysis of *T. denitrificans* showed it to contain 44–48 per cent carbon and to have an approximate elementary composition of $C_1H_{1.82}O_{0.49}N_{0.24}$. Two kinds of calculation can be derived using these data, which enable an assessment of the biochemical energy (ATP and NADH) generated during thiosulphate oxidation.

6.1 Energy-coupling in *Thiobacillus denitrificans* calculated from growth-linked CO_2-fixation

From the Y_G of 18·5, assuming the organism to be 46 per cent carbon, it can be calculated that 8·51 g carbon or 0·71 mole CO_2 are fixed per mole thiosulphate oxidised. Assuming all this carbon to be fixed by the Calvin cycle, the initial fixation to the level of fructose (CH_2O) requires 3 ATP and 2 NADH per mole CO_2. Hence 0·71 CO_2 requires 2·13 ATP and 1·42 NADH. Consequently at least 2·13 ATP + 1·42 NADH are formed during oxidation of one thiosulphate molecule. As the formation of 1·42 NADH also requires 2·84 of the 8·0 reducing equivalents (i.e. electrons) available from thiosulphate oxidation, only 5·16 (or 64·5 per cent) of the reducing equivalents from the oxidation of one molecule of thiosulphate can actually be used for electron transport-dependent phosphorylation. We therefore revise the growth equation to state that the electron transport-linked oxidation of 0·645 mole thiosulphate supports the fixation of 0·71 mole CO_2 by producing 2·13 ATP + 1·42 NADH. The production of each NADH will require one, two or three ATP molecules. If the requirement is one ATP mole^{-1} NAD$^+$ reduced, the total ATP available from oxidising 0·645 mole thiosulphate will be 2·13 + 1·42 or 3·55, indicating that 5·5 ATP are produced per mole thiosulphate. If 2 ATP per mole NAD$^+$ are required, the ATP yield could be 7·7 per mole thiosulphate. If substrate-level phosphorylation (Fig. 15.7) occurs, then one ATP for fixing 0·71 mole CO_2 would come from the APS pathway, leaving 1·13 ATP and 1·42 NADH to be generated by oxidative phosphorylation from 0·645 mole thiosulphate. Then (calculated as above) 4 or 6 ATP would be formed per mole thiosulphate by oxidative phosphorylation (cf. Fig. 15.12).

6.2 Alternative energy-yield estimations based on carbon assimilation data

The calculations given above would be modified if allowance for alternative CO_2 fixing mechanisms were made. Up to 10 per cent of CO_2 fixation in *T. denitrificans* occurs by carboxylation of phosphoenolpyruvate (PEP) (derived from phosphoglyceric acid (PGA) produced by the Calvin cycle). Consequently, some carbon is fixed by a less energy-expensive process since the conversion of 3 CO_2 to PGA effectively requires 9 ATP and 6 NADH (Fig. 15.2), but the fixation of a further molecule by PEP carboxylation requires no more energy. Since this fact is likely to affect the calculations of section 6.1 by no more than 10 per cent, it is not a serious source of error.

Timmer-ten-Hoor (1976) used the gross cell composition to calculate the proportions of thiosulphate used for carbon reduction and for energy generation. Her calculations showed a requirement for 4·3 reducing equivalents per mole carbon, consequently introducing an *underestimate* in the calculations of section 6.1 of up to 10 per cent.

We can conclude that the estimate of one ATP by substrate level phosphorylation and 4 or 6 by oxidative phosphorylation per mole thiosulphate oxidised by *T. denitrificans* to support CO_2 fixation is reasonable. Y_G calculations with other aerobic thiobacilli indicate that less ATP might be available per mole thiosulphate or tetrathionate (Justin and Kelly 1976, Eccleston and Kelly 1977).

Further ATP is required to effect cell biosynthesis from PGA (section 3.1), but this may be no more than 14–20 per cent of the total ATP requirement (Stouthamer 1973), again not greatly altering the calculation in section 6.1.

6.3 Energetics of *Thiobacillus ferrooxidans* on iron

The true growth yield of this organism is 1·33 g mole^{-1} iron oxidised, equivalent to 0·053 CO_2 fixed or 18·8 moles Fe^{2+} oxidised per CO_2 fixed. Since four (H) are required for CO_2 reduction, 14·8 Fe^{2+} support the energy requirement, and with a P/O of 1·0 could produce 7·4 ATP. The requirement for CO_2 fixation is therefore indicated to be 3 ATP plus 2 or 4 ATP per 2 moles NADH used by the Calvin cycle (Jones and Kelly 1977).

6.4 The relation of biochemical and thermodynamic energy efficiency calculations

It can be concluded from this discussion that biochemically the chemolithotrophs have developed their metabolism to a level of bioenergetic efficiency equivalent to that of hetero-trophic organisms and actual growth yields in the chemostat reach at least 75 per cent of the calculated Y_G values, which are themselves consistent with efficient energy conservation. Using the data of section 4.1, thermodynamic efficiency based on Y_G values can be calculated. Thus *T. denitrificans* (section 4.1, equation 2) fixes one mole CO_2 (495 kJ) at the expense of 1·4 mole thiosulphate (*including* the requirement for reducing equivalents) which gives 1 310 kJ; efficiency = 38 per cent. *Thiobacillus ferrooxidans* (section 4.1, equation 16) fixes one CO_2 per 18·8 Fe^{2+} (620 kJ); efficiency = 80 per cent. Corrected for the (H) requirement, CO_2 fixation is supported by 14·8 Fe^{2+} with a thermodynamic efficiency of 100 per cent probably indicating underestimation of $-\Delta F$ for iron oxidation at pH 1·5.

7 Concluding remarks

The chemolithotrophic bacteria are organisms that have evolved the ability to couple energy-yielding inorganic oxidations to essentially conventional electron transport and energy con-servation mechanisms. They have solved the problems of handling substrates of relatively low molar energy yield and poor reducing capacity with great success, as reflected by their high energy to growth conversion efficiency under suitable conditions. Much work is still needed on the biochemistry and culture of some groups, but the foundation of understanding their energetic processes is firmly laid.

8 Bibliography

The majority of the references listed are expert review articles, to which reference is made in the sections of this chapter. Reference to research papers has been minimised and those included in the following list are either fundamentally important as the first or only contribu-tion in the field or contain useful basic discussion of principles.

References

ALEEM, M. I. H. (1969) 'Generation of reducing power in chemosynthesis', *Antonie van Leeuwenhoek. J. Microbiol. Serol.*, **35**, 379–91.

ALEEM, M. I. H. (1970) 'Oxidation of inorganic nitrogen compounds', *Ann. Rev. Plant Physiol.*, **21**, 67–90.

ALEEM, M. I. H. (1972) 'Generation of reducing power in chemosynthesis', *Arch. Microbiol.*, **84**, 317–26.

ALEEM, M. I. H. (1975) 'Biochemical reaction mechanisms in sulphur oxidation by chemosynthetic bacteria', *Plant and Soil*, **43**, 587–607.

ALEEM, M. I. H., HOCH, G. E. and VARNER, J. E. (1965) 'Water as the source of oxidant and reductant in bacterial chemosynthesis', *Proc. Natl. Acad. Sci. USA*, **54**, 869–73.

ALEEM, M. I. H. and SEWELL, D. L. (1969) 'Generation of reducing power in chemosynthesis', *Biochim. Biophys. Acta.*, **172**, 467–75.

BAALSRUD, K. and BAALSRUD, K. S. (1954) 'Studies on *Thiobacillus denitrificans*', *Arch. Mikrobiol.,* **20**, 34–61.

BAAS BECKING, L. G. M. and PARKS, G. S. (1927) 'Energy relations in the metabolism of autotrophic bacteria', *Physiol. Rev.,* 7, 85–106.

BROCK, T. D., BROCK, K. M., BELLY, R. T. and WEISS, R. L. (1972) '*Sulfolobus:* a new genus of sulfur-oxidizing bacteria living at low pH and high temperature', *Arch. Mikrobiol.,* **84**, 54–68.

BULL, A. T. (1974) 'Microbial growth', in *Companion to Biochemistry* (Eds. A. T. Bull, J. R. Lagnado, J. O. Thomas and K. F. Tipton), pp. 415–42. Longman, London.

BURTON, K. (1957) 'Free energy data of biological interest', *Ergebn. Physiol.* **49**, 275–85.

COLE, J. S. and ALEEM, M. I. H. (1970) 'Oxidative phosphorylation in *Thiobacillus novellus*', *Biochem. Biophys. Res. Comm.,* **38**, 736–43.

DAVIS, O. H., DOUDOROFF, M. and STANIER, R. Y. (1969) 'Proposal to reject the genus *Hydrogenomonas*', *Int. J. System Bact.,* **19**, 375–90.

DROZD, J. W. (1977) 'Energy conservation in *Thiobacillus neapolitanus:* sulphide and sulphite oxidation', *J. Gen. Microbiol.,* **98**, 309–12.

ECCLESTON, M. and KELLY, D. P. (1977) 'Kinetics of *Thiobacillus ferrooxidans* on tetrathionate in the chemostat', *Proc. Soc. Gen. Microbiol.,* **4**, 72–3.

EYRING, H., BOYCE, R. P. and SPIKES, J. D. (1960) 'Thermodynamics of living systems', in *Comparative Biochemistry* (Eds. M. Florkin and H. S. Mason), Vol. I, pp. 15–73. Academic Press, New York.

FORREST, W. W. and WALKER, D. J. (1971) 'The generation and utilization of energy during growth', *Adv. Microbial Physiol.,* **5**, 213–74.

FROMAGEOT, C. and SENEZ, J. C. (1960) 'Aerobic and anaerobic reactions of inorganic substances', in *Comparative Biochemistry* (Eds. M. Florkin and H. S. Mason), Vol. I, pp. 347–409. Academic Press, New York.

GUNDERSON, K. (1968) 'The formation and utilization of reducing power in aerobic chemoautotrophic bacteria', *Z. allg. Mikrobiol.,* **8**, 445–57.

HARMS, H., KOOPS, H.-P. and WEHRMANN, H. (1976) 'An ammonia-oxidizing bacterium, *Nitrosovibrio tenuis*, nov.gen.nov.sp.', *Arch. Microbiol.,* **108**, 105–11.

HEMPFLING, W. P. and VISHNIAC, W. (1967) 'Yield coefficients of *Thiobacillus neapolitanus* in continuous culture', *J. Bacteriol.,* **93**, 874–8.

HUTCHINSON, M., JOHNSTONE, K. I. and WHITE, D. (1969) 'Taxonomy of the genus *Thiobacillus:* the outcome of numerical taxonomy applied to the group as a whole', *J. Gen. Microbiol.,,* **57**, 397–410.

JONES, C. A. and KELLY, D. P. (1977) 'Energetics of *Thiobacillus ferrooxidans* grown on ferrous iron in the chemostat', *Proc. Soc. Gen. Microbiol.,* **4**, 73.

JUSTIN, P. and KELLY, D. P. (1976) 'Growth of *Thiobacillus denitrificans* in continuous-flow culture', *Proc. Soc. Gen. Microbiol.,* **4**, 25.

KELLY, D. P. (1968) 'Biochemistry of oxidation of inorganic sulphur compounds by microorganisms', *Aust. J. Sci.,* **31**, 165–73.

KELLY, D. P. (1971) 'Autotrophy: concepts of lithotrophic bacteria and their organic metabolism', *Ann. Rev. Microbiol.,* **25**, 177–210.

KELLY, D. P. and SYRETT, P. J. (1963) 'Inhibition of formation of adenosine triphosphate in *Thiobacillus thioparus* by 2:4 -dinitrophenol', *Nature,* **202**, 597–8.

KELLY, D. P. and SYRETT, P. J. (1964) 'The effect of uncoupling agents on carbon dioxide fixation by a *Thiobacillus*', *J. Gen. Microbiol.,* **34**, 307–17.

KELLY, D. P. and SYRETT, P. J. (1966) 'Energy coupling during sulphur compound oxidation by *Thiobacillus* sp. strain C', *J. Gen. Microbiol.*, **43**, 109–18.

KIESOW, L. A. (1967) 'Energy-linked reactions in chemoautotrophic organisms', *Current Topics in Bioenergetics*, **2**, 195–233.

KLUYVER, A. J. and DONKER, H. J. L. (1926) 'Die Einheit in der Biochemie', *Chem. Zelle u. Gewebe*, **13**, 134–90.

KREBS, H. A. and KORNBERG, H. L. (1957) 'Energy transformation in living matter', *Ergebn. Physiol.*, **49**, 212–98.

LATIMER, W. M. (1952) *The Oxidation States of the Elements and Their Potentials in Aqueous Solutions* (2nd edn.). Prentice-Hall, Englewood Cliffs, N.J.

LEHNINGER, A. L. (1975) *Biochemistry* (2nd edn.). Worth, New York.

MEYERHOF, O. (1916) 'Untersuchungen über den Atmungsvorgang nitrifizierender Bakterien', *Pflugers Arch. ges. Physiol.*, **164**, 353–427.

PECK, H. D. (1968) 'Energy-coupling mechanisms in chemolithotrophic bacteria', *Ann. Rev. Microbiol.*, **21**, 489–518.

PIRT, S. J. (1975) *Principles of Microbe and Cell Cultivation.* Blackwell, Oxford.

RITTENBERG, S. C. (1969) 'The roles of exogenous organic matter in the physiology of chemolithotrophic bacteria', *Adv. Microbial Physiol.*, **3**, 159–96.

ROTH, C. W., HEMPFLING, W. P., CONNERS, J. N. and VISHNIAC, W. (1973) 'Thiosulphate- and sulfide-dependent pyridine nucleotide reduction and gluconeogenesis in intact *Thiobacillus neapolitanus*', *J. Bacteriol.*, **114**, 592–9.

ROSS, A. J., SCHOENHOFF, R. L. and ALEEM, M. I. H. (1968) 'Electron transport and coupled phosphorylation in the chemoautotroph *Thiobacillus neapolitanus*', *Biochem. Biophys. Res. Commun.*, **32**, 301–6.

ROY, A. B. and TRUDINGER, P. A. (1970) *The Biochemistry of Inorganic Compounds of Sulphur.* Cambridge University Press, London.

SCHLEGEL, H. G. (1975) 'Mechanisms of chemoautotrophy', in *Marine Ecology* (Ed. O. Kinne), Vol. II, pp. 9–60. Wiley, London.

STEPHENSON, M. (1931) *Bacterial Metabolism.* Longman, London.

STOUTHAMER, A. H. (1973) 'A theoretical study on the amount of ATP required for synthesis of microbial cell material', *Antonie van Leeuwenhoek*, **39**, 545–65.

SUZUKI, I. (1975) 'Mechanisms of inorganic oxidation and energy coupling', *Ann. Rev. Microbiol.*, **28**, 85–101.

TAYLOR, B. F. and HOARE, D. S. (1971) '*Thiobacillus denitrificans* as an obligate chemolithotroph', *Arch. Mikrobiol.*, **78**, 193–204; 262–76.

TIMMER-TEN-HOOR, A. (1976) 'Energetic aspects of the metabolism of reduced sulphur compounds in *Thiobacillus denitrificans*', *Antonie van Leeuwenhoek, J. Microbiol. Serol.*, **42**, 483–92.

TRUDINGER, P. A. (1967) 'The metabolism of inorganic sulphur compounds by thiobacilli', *Rev. Pure Appl. Chem.*, **17**, 1–24.

TRUDINGER, P. A. (1969) 'Assimilatory and dissimilatory metabolism of inorganic sulphur compounds by microorganisms', *Adv. Microbial. Physiol.*, **3**, 111–58.

TUOVINEN, O. H. and KELLY, D. P. (1972) 'Biology of *Thiobacillus ferrooxidans* in relation to the microbiological leaching of sulphide ores', *Z. Allg. Mikrobiol.*, **12**, 311–46.

TUOVINEN, O. H. and KELLY, D. P. (1973) 'Studies on the growth of *Thiobacillus ferrooxidans*', *Arch. Mikrobiol.*, **88**, 285–98.

VISHNIAC, W. and SANTER, M. (1957) 'The thiobacilli', *Bacteriol. Rev.*, **21**, 195–213.

WATSON, S. W. (1971) 'Taxonomic considerations of the family Nitrobacteriaceae Buchanan', *Int. J. System. Bact.,* **21**, 254–70.

WATSON, S. W. and WATERBURY, J. B. (1971) 'Characteristics of two marine nitrite oxidizing bacteria, *Nitrospina gracilis* nov.gen. nov.sp. and *Nitrococcus mobilis* nov.gen. nov.sp.', *Arch Mikrobiol.,* **77**, 203–30.

WHITTENBURY, R. and KELLY, D. P. (1977) 'Autotrophy: a conceptual phoenix', *Soc. Gen. Microbiol. Symp.,* **27**, 121–49.

WOLFE, R. S. (1972) 'Microbial formation of methane', *Adv. Microbial Physiol.,* **6**, 107–46.

16

Photosynthetic endosymbionts of invertebrates

D. C. Smith
Department of Botany, University of Bristol

1 Introduction

1.1 Concepts and meanings of the term 'symbiosis'

A variety of aquatic lower invertebrates contain photosynthetic endosymbionts, either uni-cellular algae or algal chloroplasts. The relationship of these photosynthetic units to their animal host is usually described as 'symbiotic'. Unfortunately, the meaning of this term has become confused. It was originally devised by de Bary in 1876 to describe any intimate association between two dissimilar organisms, including parasitism. Subsequently, some biologists continued to use this broad definition but others restricted the use of 'symbiosis' to those associations believed to be mutualistic, i.e. where each partner gained a distinct advantage or 'benefit' from the alliance.

The interactions between two organisms in intimate association are usually very complex and it may be difficult to assess whether the sum total of these interactions represents a net 'benefit' to both partners. For example, an organism may receive certain essential nutrients from its host, but at the same time its growth rate and capacity for sexual reproduction may be severely curtailed. Whether or not the organism is 'benefiting' compared to free-living forms becomes a matter of human opinion. 'Benefit' is often not a simple, precise and easily measured characteristic, nor is it a concept which generates fruitful avenues of experimental investigation. While there is an obvious practical convenience in classifying associations into 'parasitic' or 'mutualistic', too rigid an adherence to this division may obscure the fact that certain types of interaction, e.g. mechanism of recognising host, may be the same in both cases. For these reasons, the term 'symbiosis' will be used here in its original broad sense.

Until recently, the study of symbiosis involved investigations of an assortment of distinct associations with little attempt to establish any common or consistent pattern between them. Photosynthetic endosymbionts represent a particularly convenient class in which to attempt such comparisons: they are a distinct ecological group, they are easy to study by light and electron microscopy; photosynthetic fixation of ^{14}C is a very convenient marker for studying endosymbiont metabolism within the host; and in some cases symbiont and host can be grown separately as well as together so that consequences of the association can be analysed.

Table 16.1   Survey of Alga – invertebrate associations

	Dinoflagellates (all marine)	Chlorella (all freshwater)	Cyanellae – blue – green algae (marine and freshwater)	Chloroplasts (marine)	Platymonas (marine)	Miscellaneous (all marine)
Protozoa:	Various	Various	Various reports (e.g. *Cyanophora paradoxa, Paulinella* etc.) but poorly studied	*Mesodinium rubrum* (with ?cryptomonad chloroplasts + other ciliates ? some foraminifera	—	(a) *Chlamydomonas hedleyi* in foraminifer *Archaias angulatus* (b) Unidentified green symbionts in *Urceolaria viridis* and *Noctiluca*
Porifera:	—	Some reports but very poorly studied	Present in some marine sponges – *Verongia* and *Ircina*	—	—	—
Coelenterates:	Numerous, including all hermatypic corals	Green *Hydra* spp.	—	—	—	Unidentified green symbionts in *Myrionema amboinense*, and some strains of *Anthopleura elegantissima*
Platyhelminths Turbellaria:	*Amphiscolops langerhansi*	*Phaenocora* spp. and reports of others	—	—	Some *Convoluta* spp.	?Diatom reported from *Convoluta convoluta*
Molluscs Gastropoda:	Various	e.g. *Limnaea peregra*	—	Many Elysioid sacoglossans (with chloroplasts from siphonaceous seaweeds)	—	Unidentified green symbiont in Pacific cockle *Clinocardium nuttallii*
Molluscs Bivalvia:	All giant clams	*Anodonta cygnea Unio pictorum*	—	—	—	Unidentified green unicellular symbionts in marine *Placopecten magellanicus*

1.2 Types of association between photosynthetic endosymbionts and invertebrates

Table 16.1 summarises the different kinds of association reported between invertebrates and photosynthetic endosymbionts. Originally, endosymbionts were simply classified on the basis of their colour into three main groups: zooxanthellae (yellow—brown), zoochlorellae (green) and cyanellae (blue—green). However, zooxanthellae and zoochlorellae contain diverse types of algae and this old classification should be abandoned in favour of that given in the table.

Some animal groups have only been superficially investigated, so there is little doubt that further associations remain to be discovered. Nevertheless, it seems clear that photosynthetic endosymbionts are probably absent from higher invertebrates, while, in the molluscs and platyhelminths, they are only found in certain groups. The commonest kind of endosymbiont in marine habitats are dinoflagellates, and in freshwater, *Chlorella* species. This chapter will concentrate on those associations which have been investigated experimentally.

2 Symbiotic dinoflagellates

2.1 Introduction

Species of dinoflagellate are the commonest photosynthetic endosymbiont of marine invertebrates. Ecologically they are particularly important because of their abundant and universal presence in the tissues of reef-building corals, which cover an estimated $2 \times 10^8 \, km^2$ of the earth's surface. In tropical waters they also occur in virtually all other coelenterates except non-reef-building corals. Giant clams — the most prominent bivalve in the Red Sea, Indian Ocean and Western Pacific — contain a rich dinoflagellate flora in their mantle tissue. Symbiotic dinoflagellates also occur in temperate invertebrates, and on some rocky shores symbiotic sea-anemones may be the dominant invertebrate.

The dinoflagellates are a large and widespread group of algae. They are mostly motile unicells, each with two unequal flagella and a wall or theca giving a characteristic shape; of particular note, the chromosomes remain condensed and coiled during interphase. Within host cells, symbiotic forms usually do not produce flagella and there is no rigid wall so that the alga loses its characteristic shape, becoming approximately spherical and bounded by a complex of several membranes. However, the highly characteristic dinoflagellate nucleus remains. In pure culture, most isolates rapidly form flagella and wall, thereby regaining all the characteristic features of typical dinoflagellates.

2.2 Taxonomic identity of symbionts and relationship to free-living forms

Because the algae change in morphology when they enter symbiosis, reliable taxonomic identification can be made only with forms isolated into pure culture. However, this has been attempted by only a minority of research workers and very few have compared isolates from an adequate range of hosts. Bearing in mind the consequently fragmentary nature of the evidence, the following conclusions can be drawn.

All endosymbionts so far isolated from benthic hosts have been assigned to the single species *Gymnodinium microadriaticum.* There are obvious doubts that a single species should be the symbiont of such a large number of different hosts in tropical and temperate waters throughout the world. Apart from a suggestion that the symbiont of giant clams should be called *Gymnodinium tridacnorum,* no sound taxonomic arguments to support these doubts have been produced, but there is clear evidence that at least different races of *G. micro-adriaticum* can occur.

Gymnodinium microadriaticum has never been reported living free in nature. Exhaustive and comprehensive searches for it have never been made but, nevertheless, its apparent absence is perplexing, especially since it is densely abundant within coral tissues, and since motile forms can be extruded from zoanthids and anemones. Furthermore, some invertebrate hosts need to be reinfected at each generation (section 2.5). These taxonomic and ecological problems will remain until there is a full and thorough study of this important algal symbiont.

In pelagic hosts, the identity of the symbionts is even more sketchy. It is believed that *Amphidinium (Endodinium) chattonii* is the commonest symbiont, although rigorous evidence is available only for the chondrophores *Velella velella* and *Porpita porpita*. The symbiont of some foraminiferans and radiolarians has been ascribed to *Endodinium nutricola*, but on slender evidence. However, the symbiont of the primitive flatworm *Amphiscolops (Convoluta) langerhansii* is quite distinct and belongs to *Amphidinium klebsii*. This endosymbiont retains its wall and flagella and closely resembles free-living forms.

2.3 Morphology

In metazoan hosts, the endosymbionts are restricted to particular tissue regions. In coelenterates, they always occur in the endodermis, except for a few species where they are found in the mesoglea, and for some eggs and a few coral planulae where they are in the ectoderm. In almost all cases they are intracellular — even in the mesoglea they are within amoebocytes.

In giant clams the symbionts occur mainly in blood sinuses throughout the dorsal part of the mantle, becoming so abundant in the distal part as to completely fill the spaces in the margins. They also occur in the visceral mass and adjacent tissues in adults. There is controversy as to whether the algae occur free or within blood cells in the sinuses. A recent explanation for the conflicting reports is that the algae occur adjacent to but never completely enclosed by amoebocytes. They are intracellular in the visceral mass.

In both molluscs and coelenterates, ultrastructural studies give no positive evidence that the symbionts are enclosed in host vacuolar membranes, though these might be difficult to discern since the symbiont is already bounded by a complex of what appears to be its own membranes. All symbiotic dinoflagellates possess a structure of uncertain function, the 'accumulation body' — found in free-living forms only in the non-motile stage. The symbiont *A. klebsii* is unusual because it is the only dinoflagellate to have an accumulation body and at least the potential for motility. This raises the question of whether formation of an accumulation body is promoted by the consequences of symbiotic interactions.

There have been few estimates of the proportion of alga to animal tissue, although this may be important in considering the quantitative role of symbionts in host nutrition. In temperate anemones, plant-to-animal dry weight ratios are of the order of 1 to 300, but in corals, algae may account for one half of the protein nitrogen in the tissue.

2.4 Metabolic interactions

2.4.1 Photosynthate transfer to host

Virtually all experiments on the transfer of photosynthate from alga to animal concern *Gymnodinium,* the hosts being either coelenterates (especially sea-anemones, zoanthids and corals) or giant clams. The starting point of investigations is to incubate hosts in light and dark in sea-water containing ^{14}C-carbonate. The transfer of photosynthetically fixed ^{14}C to animal tissues is then studied by one or more of five techniques.

1 Autoradiography of sections of host. This is particularly valuable in showing which organs or animal tissues accumulate the fixed ^{14}C. The main disadvantages are that it is difficult to make accurate quantitative estimates of transfer, and that unless special techniques are used, water- or ethanol-soluble metabolites are lost during preparation.

2 Separation of algae-containing from algae-free regions of the animal. This permits the unequivocal demonstration of movement of ^{14}C into purely animal tissue and, by appropriate analysis of tissue extracts, will indicate how the animal metabolises ^{14}C-compounds. The main disadvantage is that a significant amount of ^{14}C released from the alga remains in the alga-containing tissue, and no estimate of this can be made.

3 Separation of algae by centrifugation of tissue homogenates. Tissues are gently homogenised so that animals' cells, it is hoped, are disrupted without damaging the algae, which are separated by centrifugation. All fixed ^{14}C not in the algae is assumed to be in the animal tissue. This is the main technique for estimating the total transfer of ^{14}C and studying the initial metabolism of ^{14}C compounds, but there are two potential sources of error. First, since the algae have no rigid wall in symbiosis, they are fragile and may become ruptured during homogenisation. Secondly, the continued release of ^{14}C from algae during homogenisation, centrifugation and washing may again give inflated estimates of transfer.

4 Release of fixed ^{14}C from isolated algae. If isolated algae are resuspended in the light in a homogenate of animal tissue containing ^{14}C-carbonate, they will photosynthesise and release fixed ^{14}C at an appreciable rate. However, it is not known whether conditions in the homogenate are the same as those in the host cell and consequently there are obvious drawbacks to using this technique alone.

5 Isotope trapping. If non-radioactive glycerol, glucose or alanine is added to the sea-water media of marine hosts during or shortly after ^{14}C fixation, then the radioactive forms of these compounds are released to media. Apparently, external and internal molecules exchange by 'counterflow' or 'exchange diffusion' phenomena. The effect is very specific. Isotope trapping (sometimes inappropriately named 'inhibition technique') is a simple and sensitive method of measuring the rapidity with which photosynthetically fixed carbon enters animal tissues and of identifying the mobile compounds containing this carbon. It has the obvious disadvantage that host metabolism may become distorted by the external compounds. If compounds entering the animal tissue are respired, then the $^{12}CO_2$ released may reduce the specific activity of the $^{14}CO_2$ fixed by the algae, again complicating interpretation of experimental data.

   None of these techniques on its own can give a complete and accurate picture of photosynthate transfer. However, appropriate combinations in a variety of associations present a reasonably consistent picture.

   The amount transferred to the host is usually in excess of 40 per cent total net fixation and values up to 80 per cent have been found in corals. The movement is rapid and fixed ^{14}C can be detected in animal tissues within a few minutes. Freshly isolated algae release glycerol and alanine, suggesting that these are the principal compounds transferred. Although algae may well behave differently inside host cells, it is significant that in animal tissues ^{14}C is found principally in lipids (in the glycerol moiety) and also in protein. There are reports of nucleotide phosphates being involved in transfer, but these come from experiments in which breakage of algal cells may have occurred.

Almost all animal hosts retain their conventional holozoic feeding mechanisms, so the question arises of the importance of the nutrient supply from symbiotic algae. Absolute rates of photosynthesis in symbiotic animals have only been made for the 'upside-down' jellyfish, *Cassiopea,* where a rate of fixation of $900 \, \mu g \, C \, cm^{-2} (12h)^{-1}$ sunlight was found; respiratory losses were estimated at $50 \, \mu g \, C \, cm^{-2} (24h)^{-1}$, implying a substantial excess of carbon available to the association. This result should not be extrapolated to other animals since the ratio of alga to animal varies widely between different associations.

Most interest in the nutritional importance of symbionts centres upon coral reefs. These are among the most productive ecosystems, yet they occur in oceans which are very poor in animal nutrients. Various estimates of the amount of zooplankton available to corals have been attempted but, because of the size and complexity of reef ecosystems, these estimates are at best highly approximate; however, they suggest that conventional feeding mechanisms are unlikely to supply more than 10 per cent of the energy requirements of the reef. Photosynthesis by symbiotic algae is therefore probably the major factor in the high productivity of the ecosystem.

Autoradiographical studies of fixed carbon in animal tissues show that it has a variety of morphological destinations. In giant clams ^{14}C appeared within 10 min. of fixation in mucous glands, the byssal gland and style sac — all organs which are likely to have a high consumption or turnover of carbon. Such investigations are less informative in structurally simple animals such as coelenterates.

2.4.2 Recycling of nitrogen and phosphorus between alga and host

Tropical marine waters are generally low in dissolved and particulate nutrients, with nitrogen being in especially low supply — indeed, it is the major limiting nutrient for phytoplankton production in the tropical Pacific. Set against this, the high productivity of tropical reef ecosystems seems paradoxical: reef communities fix $4-12 \, g \, C \, m^{-2} \, day^{-1}$, whereas in the adjacent open ocean fixation is only of the order of $0.06-0.50 \, g \, C \, m^{-2} \, day^{-1}$.

Two key factors probably cause the productivity of coral reefs to be substantially less limited by nitrogen than the surrounding ocean: (1) nitrogen fixation by blue–green algae on the reef; and (2) recycling of nitrogenous waste within animal tissues by symbiotic dinoflagellates.

The discovery of nitrogen fixation by blue–green algae on reefs is relatively recent and data are so far available for only one community, Enewetak Atoll. Nevertheless, the results are striking and may apply to many other reefs where blue–green algae are common. On Enewetak, fixation was primarily due to *Calothrix crustacea* which covered large areas of the reef flat as a thin film. Fixation per unit area was comparable to the highest found in terrestrial ecosystems, and comparison of sea-water before and after flowing over the reef showed a net export of nitrogen.

The role of symbionts in recycling nitrogen is suggested by two observations. First, corals with symbiotic dinoflagellates release ammonia in the dark but not in the light, while those without symbionts release in both light and dark. Secondly, if ammonium chloride is added to the sea-water medium of symbiotic corals, ammonia is taken up, with a marked increase in the amount of photosynthetically produced alanine in animal tissues (i.e. as if the algae convert ammonia to amino acids and release them back to animal cells). Again, experiments directly demonstrating that symbiotic dinoflagellates recycle nitrogen are few, but the results are striking.

Tight nutrient cycling may be even more important for phosphorus because there is no massive phosphorus input to the reef ecosystem comparable to that provided by nitrogen fixation. Experimental addition of phosphate as well as nitrogen fertilisers to reefs enhances primary productivity. Phosphate release by corals with symbiotic algae is several orders of magnitude less per unit body weight than marine invertebrates without symbiotic algae. Some experiments with isolated dinoflagellates indicate they may release nucleoside polyphosphate, but whether this occurs within symbiotic animals has yet to be investigated experimentally.

Virtually all the observations on the role of symbionts in nutrient recycling have come from corals. These observations are still few in number, but this must not obscure the possibility that it is a role of major importance. Invertebrates with symbiotic dinoflagellates are very much more common in nutrient-poor tropical waters than in nutrient-rich temperate waters. Other kinds of symbiotic association, such as mycorrhizas and lichens, show similar increased frequency in nutrient-poor habitats.

2.4.3 'Factors' in host stimulating release of nutrients from alga

Animal tissues contain factor(s) which substantially increase photosynthate release from isolated symbionts. This is consistently shown by many experiments in which a homogenate or aqueous extract of animal tissue is added to freshly isolated algae. Increase by a factor of 2 to 12 in release of photosynthetically fixed ^{14}C invariably results. It is particularly striking that only the release of specific compounds is stimulated, especially glycerol and alanine. The effect on glucose release — the predominant intracellular carbohydrate — is only slight. 'Factors' from different animal hosts have similar effects on symbiotic dinoflagellates, even when they originate from such taxonomically diverse types as giant clams and corals.

The effect of tissue homogenates on net photosynthetic ^{14}C fixation is variable; in some experiments it is stimulated, in others reduced. This is perhaps to be expected in view of the variety of compounds present in homogenates (including, for example, nematocyst toxins in some coelenterates). In the sea-anemone *Anthopleura elegantissima,* common on the West Coast of North America, it can be shown that the release-stimulating factors are produced in response to the presence of symbionts. Aposymbiotic races of this anemone occur in certain habitats such as deep, dark crevices. The tissues of such anemones do not stimulate release. Aposymbiotic anemones can become reinfected over a period of weeks if symbiotic algae are added to the food they eat, and after reinfection, tissue homogenates develop stimulatory properties. It is believed that 'factors' operate within the intact association to regulate release.

The nature of the factors remains unknown. If homogenates of host tissue are boiled, stimulatory properties are lost, suggesting that the factor may be proteinaceous.

2.4.4 Stimulation of calcium carbonate deposition on coral reefs

In reef-building corals the rate of calcium carbonate deposition in the light is approximately 10 times that in the dark. This light-stimulated acceleration of deposition is inhibited by DCMU, showing that it is linked to photosynthesis rather than any other kind of photobiological phenomenon. However, since some calcification can occur in the dark, and since non-reef-building corals (which lack symbiotic algae) also deposit calcium carbonate, symbiotic algae are not essential for deposition, but their presence substantially enhances the rate.

The problem of how photosynthesis by algae in the endodermis enhances calcium carbonate deposition by the ectodermis has not been resolved. The most widely accepted theory postulates the following reactions:

(1) Ca^{2+} (from sea-water) + HCO_3^- (animal cell metabolism) $\longrightarrow Ca(HCO_3)_2$

This occurs on some kind of matrix external to the ectodermal cells.

(2) $Ca(HCO_3)_2 \longrightarrow CaCO_3\downarrow + H_2CO_3$

Tropical marine surface waters are supersaturated with respect to calcium carbonate, so it precipitates.

(3) $H_2CO_3 \longrightarrow CO_2 + H_2O$

This reaction is catalysed by carbonic anhydrase, abundantly present in corals.

(4) $CO_2 \longrightarrow$ dinoflagellate photosynthesis

The continual removal of CO_2 by photosynthesis in reaction (4) is presumed to accelerate carbonate deposition in reaction (2).

This theory has several difficulties. First, it is not easy to envisage how carbonate can be simultaneously supplied for calcification and removed for photosynthesis. Secondly, in the aptly-named Stag's Horn coral, *Acropora cervicornis*, the highest rates of calcification occur at the tips, yet these are the regions with fewest symbiotic algae. In some other corals, cells of the superior endodermis contain 10 times as many algae as the inferior endodermis, so that the main site of photosynthesis is not even adjacent to the cells depositing calcium carbonate. Thirdly, if corals are supplied in the dark with the excreted products of dinoflagellate photosynthesis, glycerol and alanine, calcification is depressed rather than enhanced.

Several other theories for explaining calcification have been advanced. One suggests that deposition is regulated by the organic matrix present in the skeleton and that the effect of photosynthesis is to increase the rate of matrix synthesis. Another postulates that, since crystal growth of calcium salts is inhibited by phosphate, the light-enhancement of calcification is due to light-enhanced phosphate uptake by the algae. There is little experimental evidence to support either of these theories.

2.4.5 Selective advantage to host of presence of symbionts

Although various metabolic interactions have been described in the preceding sections, direct experiments demonstrating selective advantages are few. The simplest are those concerning the nutritional value of photosynthesis; for example, when starved in the light, symbiotic specimens of the sea-anemone *Anthopleura elegantissima* lost weight at half the rate of aposymbiotic specimens. However, the rarity of aposymbiotic hosts in nature is strong indirect evidence of the selective advantage of symbiosis. The relative importance of the various interactions can change with different associations. For example, there is distinct variation between different species of coral in the degree to which they feed holozoically, and these different species occupy different niches in the reef ecosystem. In tropical oceans, nutrient conservation may be more important than in temperate.

The possibility that oxygen released in photosynthesis has a beneficial effect on animal tissues has not been investigated.

2.5 Regulation of symbiont population by host

The ability of an animal host to regulate the population size of its symbionts is obviously vital to the long-term success and stability of the association. Digestion of surplus algae is one potential method, but unequivocal evidence for this is completely lacking except for the possible case of digestion of senescent algae in the giant clam.

Ejection of algae by coelenterate hosts has been observed on various occasions. These observations are of two kinds. (*a*) Mass expulsion of algae triggered by environmental stress or shock. Such expulsion from corals has occurred after hurricanes, where the specific stimulus was probably lowered salinity caused by heavy rain. Under laboratory conditions, expulsion has been caused by factors such as high or low temperatures. (*b*) Natural extrusion of surplus algae by healthy animals, usually in the form of mucus strings, packets or pellets released from the mouth. This appears to be a normal regulatory mechanism in a variety of coelenterates, and becomes particularly pronounced in conditions favouring high algal growth rates such as high light intensities.

In the sea-anemone *Aiptaisia,* various stages in the life cycle of the alga are present in the extruded packets, including healthy dividing cells, sometimes even accompanied by motile stages. This is of particular importance in considering reinfection under natural conditions.

It is not known if animal cells also have specific mechanisms for regulating algal cell division, or whether there is precise control over the number of algae per cell. The mechanism of extrusion has not been studied.

2.6 Transmission of symbionts

In most symbiotic coelenterates — scleractinian corals, zoanthids and some scyphozoans — the eggs are commonly infected with algae so that transmission of symbionts from one generation to the next is ensured through sexual reproduction as well as asexual budding. In a few symbiotic coelenterates, notably in all gorgonians (octocorallia) so far investigated, the eggs are definitely uninfected and larvae have to be reinfected at each generation. It is not clear why this group of coelenterates should have diverged from the others. It has been suggested that since different strains of *Gymnodinium* occur, it may be advantageous to the animal to acquire the strain most suitable to the region where the larva settles.

Giant clams also produce uninfected eggs and therefore must be reinfected at each generation.

2.7 'Recognition' of symbionts by host, and infection mechanisms

Many species of dinoflagellate occur in the sea, but only a few can become symbiotic, implying that hosts have efficient mechanisms for recognising potential symbionts. No experiments have been carried out on 'recognition' beyond the observation that when uninfected gorgonian larvae were exposed to a range of different dinoflagellate strains, they became reinfected only by *Gymnodinium.* The genus *Amphidinium,* a symbiont of pelagic invertebrates, failed to reinfect.

Mechanisms of infection likewise have received little study beyond the general observation that the initial route of entry seems to be through normal digestive processes. In coelenterates it is not known how, after acquisition by digestive cells lining the body cavity, symbionts appear in the tentacles. It is also not known if algae become sequestered in specific regions of the cell, as in Hydra. In those animals which require reinfection at each generation, symbionts might enter directly or, since extruded algae have been occasionally observed to be ingested by ciliates, reinfection might occur through digestion of such zooplankton.

3 Symbiotic *Chlorella*

3.1 Introduction

Chlorella species are the commonest photosynthetic endosymbionts of freshwater invertebrates, although there are no major habitats where they assume an ecological importance comparable to symbiotic dinoflagellates in tropical marine environments.

On the other hand, two associations, *Hydra viridis* and *Paramecium bursaria,* are easily grown in mass culture under laboratory conditions and are much easier subjects for experimental investigation than associations involving dinoflagellates. Various methods exist for producing aposymbiotic strains of *Hydra* and *Paramecium.* Some of the symbiotic *Chlorella* spp. can be grown in pure culture and techniques have been developed for reinfecting aposymbionts. Thus, many aspects of the interaction between symbiont and host can be investigated.

Chlorella is a genus of unicellular green algae, each unicell containing a single parietal chloroplast. The life cycle is simple and reproduction occurs exclusively by autospores; motile stages are never formed. The cells are typically small, usually being below 10 μm in diameter. A pyrenoid is present in some species, including symbiotic forms. The genus is widespread, cosmopolitan and encountered in virtually all water habitats.

Some *Chlorella* spp. contain the highly indestructible compound sporopollenin in their cell walls. The symbiont of *Paramecium bursaria* has this compound, but its occurrence in other symbiotic forms has not been studied.

3.2 Taxonomic identity and relationship to free-living forms

All symbiotic *Chlorella* spp. so far studied resemble the ubiquitous *Chlorella vulgaris.* However, strains of symbiotic *Chlorella* can differ from each other in ultrastructure, cultural characteristics and ability to reinfect aposymbiotic hosts.

The turbellarian *Phaenocora typhlops* has a loose, facultative association with free-living *Chlorella vulgaris* var. *vulgaris* but, apart from this, no other evidence exists as to whether symbiotic *Chlorella* occur free-living and whether they are transferred in nature between different host species. Unlike symbiotic dinoflagellates, they undergo little morphological modification upon entry to symbiosis, although detailed ultrastructural investigations of this problem still remain to be carried out.

The symbionts of *Paramecium* are easily brought into pure culture but until very recently it was found impossible to culture symbionts from *Hydra.*

3.3 Morphology

The distribution of symbionts within host animals has been studied only for *Hydra viridis* and *Paramecium bursaria.*

In green *Hydra,* the algae are restricted to the endodermis. If portions of the endodermis are excised and allowed to differentiate into new *Hydra*, cells which de-differentiate to become ectodermal cells expel their algae. Within the endodermis, the algae occur in digestive cells, the precise number per cell varying with the part of the body, the strain of *Hydra* and the environmental conditions. Each algal cell is within a vacuole, and on division daughter algal cells segregate into their own vacuole. The algae, typically about 5 μm diameter, are grouped together at the end of the digestive cell distal to the body cavity — presumably leaving the proximal end free for digestive functions.

In *Paramecium bursaria,* the number of algae per animal may vary widely between one and over 1 000. As in *Hydra,* each alga is enclosed inside a vacuole. Within the animal, the algae are most numerous in the peripheral cytoplasm, but may also be found in interior positions closely adjacent to the buccal apparatus and macronucleus. So far as can be ascertained, the algae are relatively stationary in the cytoplasm.

3.4 Metabolic interactions

3.4.1 Photosynthate transfer to host

Most experiments have been carried out with green *Hydra*. The commonest method of study is to incubate animals for a period in the light in media containing $NaH^{14}CO_3$, homogenise, separate the algae by centrifugation, and measure fixed ^{14}C in alga and animal fractions. Because the *Chlorella* symbionts retain their cell walls and are robust, the technique is more reliable than in associations involving the more easily ruptured symbiotic dinoflagellates (cf. section 2.4.1). Transfer is rapid, with about 40 per cent of net fixed ^{14}C moving to the animal tissue. If, instead of homogenising animals, the ectoderm is separated from the endoderm, fixed ^{14}C is found to move rapidly from endoderm to ectoderm, but at a constant rate of about 12 per cent of net fixation. Thus, the greater proportion of the released photosynthate is retained within endodermal cells.

Although there is qualitative evidence for transfer of fixed ^{14}C from algae to animal in *Paramecium,* no quantitative measurements have been attempted.

Chlorella from both *Hydra* and *Paramecium,* when freshly isolated, release substantial quantities of maltose. The maltose is synthesised at or near the cell surface from an internal pool of hexose phosphates: no maltose can be detected within algal cells even though internal pool of hexose phosphates: no maltose can be detected within algal cells even though large quantities are released. Synthesis is very sensitive to the pH of the medium presumably because it occurs at the cell surface. Maltose is not known as an excretion product from any other kind of alga, whether symbiotic or not. Indeed, it is the only known example amongst green plants where free maltose is produced by synthetic rather than degradative mechanisms. The only other labelled compounds released during ^{14}C fixation by freshly isolated symbionts are small amounts of labelled glycolic acid and glucose. In culture, *Hydra* symbionts do not release maltose, although some *Paramecium* symbionts continue to do so.

The animal tissue of *Hydra* contains maltase, so that it can convert maltose to free glucose. After ^{14}C fixation, various products in the animal tissues become labelled, especially glycogen, but also lipids and proteins, indicating that products of photosynthesis are utilised by the animal.

3.4.2 Movement of nitrogen and other compounds from alga to host

Given the involvement of other kinds of symbiotic algae in nutrient conservation and recycling waste nitrogen, it would be surprising if this did not occur in associations containing symbiotic Chlorellae. However this possibility has not been investigated experimentally.

3.4.3 'Factors' stimulating release from algae

Homogenates of animal tissue have no stimulatory effect on release of fixed ^{14}C by *Chlorella* from *Hydra.* There is no experimental evidence that the host uses the marked sensitivity to pH of maltose release as a regulating mechanism. Animals kept in continuous darkness for 7 days have higher rates of photosynthesis on return to the light than those kept continuously illuminated, but the proportion of fixed carbon transferred is the same. The animal might therefore be able to exert some measure of control over photosynthetic rates rather than release.

3.4.4 Movement of compounds from animal to alga

Because the algae are intracellular, all nutrients entering them are derived from animal cells. The culture medium used for green *Hydra* contains only Ca^{2+}, Mg^{2+}, Na^+, K^+, Cl^- and HCO_3^-. All other nutrients essential for algal growth must be derived from the food given to the animals. When this food contains ^{14}C or ^{35}S, label can be recovered from algae.

There have been no published studies of the effect of adding inorganic nutrients such as nitrate, phosphate or sulphate to culture media of *Hydra*. Such nutrients are traditionally added to media of *Paramecium* in experiments where sterile, bacteria-free suspensions of the animal are required, and algae are able to grow in the host.

3.4.5 Selective advantage to host of presence of symbionts

With excess of food there is little difference in growth rate between symbiotic and aposymbiotic *Paramecium* and *Hydra*. With limited food symbiotic forms grow much more quickly in the light than aposymbiotic. Since, in nature, food supply is normally limiting, this illustrates the selective advantage of the presence of symbionts. Aposymbiotic *Paramecium* and *Hydra* have not been found in nature.

The results of experiments in continuous darkness, but limited food supply, are less clear-cut. Some indicate that symbiotic *Paramecia* grow better than aposymbiotic, but, unfortunately, it was not known if algae were being digested. In continuous darkness *Paramecium* will outgrow its algae and become aposymbiotic. Some strains of green *Hydra* do not lose their algae even after 6 months in the dark, so that food presumably has to sustain algal as well as animal growth, and symbiotic forms might not grow as well in the dark. The problem requires investigation.

3.5 Regulation of symbiont population by host

In green *Hydra*, the population of algae per cell of laboratory strains grown under constant conditions of feeding and illumination remains very constant, with mean values lying in the range 9 to 22 algae per cell. This value varies according to the following factors. (1) *Region of animal*. In general there may be up to 30 per cent more algae per cell in the stomach and budding region than in the parts of the body above and below this. (2) *Strain of animal*. In the stomach and budding region, the number of algae per cell may be as low as 12 in some laboratory strains and as high as 18 to 22 in others. In wild material values may be much more variable. (3) *Feeding and illumination*. If animals are maintained in constant darkness the number of algae per cell declines, the rate of decline being accelerated if animals are also fed. However, even after 6 months' maintenance in the dark a few algae remain showing that animals do not completely outgrow their symbionts, and can indeed maintain a low population of algae entirely from their own resources. Animals maintained in DCMU likewise never completely lose their algae.

If animals maintained in darkness or DCMU are returned to photosynthesing conditions the algae rapidly start dividing until the previous population size is reached. Evidently, there are factors which set an upper limit to the algal population size, though it is not known what these are. Apart from the production of cell division regulators, other possibilities are: (*a*) a physical limitation of the space available in the digestive cell (unfortunately relative sizes of digestive cells with different populations have not been compared); or (*b*) limitation by the host cell of nutrients essential for algal growth. Digestion of algal cells has not been reported under normal conditions, while the possibility that surplus algae are ejected — as occurs from dinoflagellate hosts — has never been satisfactorily investigated.

Complete expulsion of algae from *Hydra* has been reported as a consequence of incubation in 0·5 per cent glycerol for periods of up to 8 days. More recently, many laboratory strains have ceased to respond to glycerol treatment, although some wild strains still respond.

Most strains of *Hydra* can be induced to expel their algae by exposure to high light intensities in the range of $600-1\,900\,W\,m^{-2}$. The length of exposure required for expulsion varies with the strain, the most recalcitrant requiring 5 days and with addition of DCMU to the medium needed to increase the effect. It is not known why high light has this effect. Possibly the algae are destroyed and expulsion of the moribund algae is a normal response (certainly if heat-inactivated algae are taken into digestive cells, then they become expelled again).

In *Paramecium* the population of *Chlorella* shows wider variations than in *Hydra*. The differences between various strains are more marked and the effects of combinations of feeding, and constant light or dark much more pronounced. Under optimum conditions, some strains may contain about 1 000 Chlorellae per animal, but in constant darkness with abundant food, the algal population declines until, after about 4 to 6 weeks, animals become aposymbiotic.

In general, host control over algal symbionts seems more lax in *Paramecium* than in *Hydra*. An even looser relationship exists between the turbellarian *Phaenocora typhlops* and its symbiont for, under natural conditions, there was wide seasonal variation in numbers of chlorellae per animal, numbers falling from 77 000 per animal to less than half this value between June and July.

3.6 Transmission of symbionts

In *Hydra,* buds are invariably infected with algae, as are a high proportion of eggs. *Paramecium* has been less well studied, although it is known that symbionts remain throughout conjugation. Transfer of algae during reciprocal nuclear exchange at conjugation seems relatively rare. In *Phaenocora,* infection has to occur anew each generation. Transmission of *Chlorella* symbionts in other hosts has not been studied.

3.7 Reinfection of aposymbiotic animals and 'recognition' of symbionts

3.7.1 Reinfection of *Hydra*

Reinfection of *Hydra* is studied by injecting a dense suspension of algae into the coelenteron of aposymbionts, and following their fate by light and electron microscopy. Five phases can be recognised:

(1) Contact. Algae become attached to the digestive cell membrane, which shows increased electron opacity at points of contact; this phase is relatively non-specific.

(2) Engulfment. This resembles normal phagocytosis. Microvilli develop in the region of the algal cell and begin to enclose it. Algae then become engulfed and sequestered into individual vacuoles at the tip of the cell. Up to about 20 algae can be taken into each cell. Engulfment is fairly rapid with over half the cells being taken in within a few minutes, and the remainder over the next hour.

(3) Recognition. Soon after engulfment a marked difference develops between the fate of the

eventual algal symbionts and that of unacceptable algae, carmine particles, latex spheres, etc. The latter are collected into large apical phagosomes and expelled from the cell by exocytosis; symbiotic algae remain sequestered in individual vacuoles. The mechanism of recognition remains unstudied.

(4) Intracellular migration. Symbiotic algae migrate to the distal end of the cell during the next 2 to 5 h. Migration is inhibited by reagents such as colchicine and vinblastine, so that a microtubular system is implicated.

(5) Regulation. Algae divide until the normal cell population size is achieved: this may take between 6 and 28 days, depending upon the strain of *Hydra* and the experimental conditions.

3.7.2 Reinfection of *Paramecium*

Symbiotic algae enter the cell by the same digestive route as food particles and after phago-cytosis lie in the cytoplasm in vacuoles or phagosomes. In the case of food particles, non-symbiotic algae and non-living particles, digestive enzymes are secreted into the vacuoles. In the case of symbiotic algae, some digestive enzyme activity (notably alkaline phophatase) is detectable in the early stages of infection, but later none can be detected in the perialgal vacuoles. It is assumed that symbiotic algae suppress secretion of digestive enzymes but it is not known whether this is by secretion of inhibitors or by inducing a change in the phagosome membrane so that it can no longer fuse with enzyme-containing lysosomes.

The 'recognition' phase is less well studied with *Paramecium* than with *Hydra,* and the extent to which non-symbiotic algae may become expelled rather than digested is not clear.

3.7.3 Reinfection of other associations

The turbellarian *Phaenocora typhlops* becomes infected through algal contaminants of its food. In the early stages of infection algae occur in the pharyngeal area, but later spread throughout the rest of the animal.

Casual observation of other hosts, such as *Stentor* and *Dalyellia,* also suggests that infection occurs from algal contaminants of food, followed by some presumed resistance to host digestion.

3.8 Specificity and 'recognition' of algal symbionts by *Hydra* and *Paramecium*

When injected with a range of symbiotic *Chlorella,* aposymbiotic *Hydra* immediately re-establishes a completely successful association with algae freshly isolated from *Hydra.* Some cultured strains from *Paramecium bursaria* fail to infect, and no infection has been achieved so far by any strain of free-living *Chlorella* isolated from nature. *Paramecium bursaria* establishes stable relationships with a wider range of symbiotic *Chlorella,* and the symbiont from *Hydra* promotes host growth as successfully as its own.

These results suggest that the 'recognition' mechanism in *Hydra* may be more specific than in *Paramecium.* The ecological advantage to *Hydra* of a higher level of specificity is not easy to understand. *Hydra* lives in habitats where a range of other symbiotic invertebrates occur, and one might have assumed it would be advantageous if it could be infected by Chlorellae from other hosts. Relatively few laboratory experiments on reinfection have been carried out, so it is still possible that misleading results arise because techniques have not been fully perfected. It is also possible that the laboratory strains of *Hydra* and *Paramecium* come from widely different habitats.

4 *Convoluta roscoffensis*

4.1 Introduction

Convoluta roscoffensis is a small acoelous platyhelminth occurring in the intertidal zone of certain sandy beaches in North Brittany and the Channel Islands. Colonies, each consisting of millions of worms, appear as dark green glistening patches on the sand, particularly in tiny drainage rivulets or margins of shallow pools around rocks. The worms are motile, and entire patches can rapidly disappear under the sand in response to mechanical disturbance.

The life cycle is simple. Eggs laid by adult worms hatch to produce white, uninfected larvae, and these feed voraciously. If, during feeding, they acquire algal symbionts and become infected, they develop into mature green worms. If they are uninfected, they do not develop to maturity.

The particular interest of this association is that, as soon as the infected animals reach maturity, they cease to ingest particulate matter. They thus pass the whole of their adult life entirely dependent upon their symbionts. Indeed, the association has been grown in the laboratory through several generations in a bacteria-free sea-water medium containing only nitrate, phosphate, vitamins and minerals. *C. roscoffensis* is therefore a good subject for studying the nutritional interrelationships between symbionts, since there are no complications due to the occurrence of holozoic nutrition.

4.2 Taxonomic identity of algal symbiont and relationship to free-living forms

The algal symbiont is *Platymonas convolutae,* a member of the Prasinophyceae. In the free-living state it possesses four anterior flagella, a rough-surfaced theca, a cup-shaped chloroplast and a pyrenoid of a distinctive structure. In natural conditions it has been isolated free-living from sea-water samples taken from the neighbourhood of *Convoluta* colonies.

4.3 Morphology

Convoluta belongs to the simplest group of ciliated platyhelminths, the acoelous turbellarians. It lacks an intestine and the gut is formed from a syncytium of endodermal cells. The symbionts are not intracellular, and occur between the host cells in the epidermal and subepidermal tissues.

After entry into symbiosis, the alga shows the following structural changes. (1) The theca is lost, so that the alga becomes a seemingly naked protoplast. The shape of the symbiont becomes irregular, with finger-like projections containing lobes of the chloroplast penetrating between host cells. This produces a large area of contact between the symbionts. The anterior ends of the algae point inwards, with the ramifications of the chloroplast facing the outside of the animal. (2) Structures associated with locomotion and phototaxis – the flagella, flagellar pit and eyespot – are lost, although flagellar roots remain. (3) Golgi bodies are smaller and appear quiescent – in the free-living state they are very active and packed with particles released to form the extracellular theca. (4) Mitochondria have fewer and smaller cristae, and are less electron-dense. Adjacent animal mitochondria have many cristae.

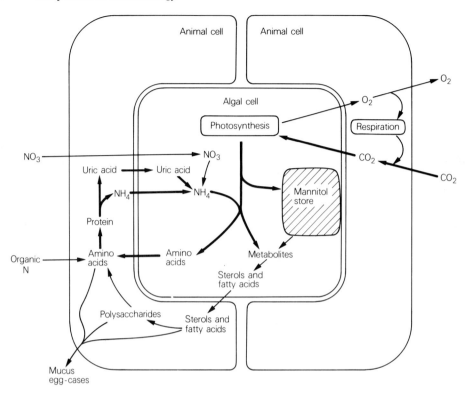

Fig. 16.1 Diagram summarising biochemical interactions between alga and host animal in *Convoluta roscoffensis.*

4.4 Biochemical interactions

4.4.1 Nutrient interchanges

The symbionts of *C. roscoffensis,* being virtually naked protoplasts, are very fragile and difficult to isolate intact from animal homogenates. However, worms given a short exposure of CO_2 will eject most of their algae intact; the reason for this is not known. A detailed picture of nutrient relationships has been compiled from experiments with uninfected larvae, adult animals, freshly isolated and cultured algae (Fig. 16.1). The main experimental observations used to construct this scheme are as follows:

a Freshly ejected algae release fixed carbon to the medium mainly as neutral amino acids, especially glutamine.

b Cultured algae are able to grow on uric acid as a sole nitrogen source. Symbiotic algae have uricase.

c Platyhelminths reportedly lack uricase and excrete uric acid: acoels such as *Convoluta* have no flame excretory system and so accumulate insoluble urate. Ammonia is also formed as a waste product.

d If photosynthesis is inhibited in adult worms by DCMU, uric acid accumulates. There is also some accumulation of lactic acid, presumably because photosynthetic oxygen production is stopped.

e The main product of photosynthesis in both cultured and symbiotic algae is mannitol. Mannitol cannot be utilised by aposymbiotic animals and is not released by cultured or ejected algae. The mannitol pool in symbiotic algae turns over very slowly, but as soon as they are ejected, it turns over rapidly. Thus mannitol appears to be a convenient carbohydrate reserve for the alga which the animal cannot use. It may also function to maintain an adequate osmotic pressure within the alga.

f Aposymbiotic animals cannot incorporate ^{14}C-acetate into sterols, or synthesise fatty acids *de novo*. The sterols of young and adult animals are primarily phytosterols identical to those in the alga. Animal cells are able to carry out chain elongation and desaturation of algal fatty acids, but in the dark, incorporation of ^{14}C-acetate into fatty acids is much reduced, indicating a basic dependence of fatty acid metabolism on the algal cells.

g Photosynthetically fixed ^{14}C can be found in both mucus and eggs released from animals, where it is prominent in amino acid and carbohydrate moieties. Estimates of transfer have been made by exposing animals to a pulse of $NaH^{14}CO_3$ in the light, and then subsequently determining the proportion of fixed ^{14}C which is recoverable in ejected algae. This gave values of 37 to 58 per cent for the amount of fixed ^{14}C transferred to the animal.

h Rates of photosynthesis per unit chlorophyll in intact worms are comparable to free-living algae.

i Analysis of drainage water on a sandy beach before and after passage over *Convoluta* colonies showed that there is an appreciable net uptake of both nitrate and organic nitrogen by the worms.

4.4.2 'Factors' stimulating release from symbionts

Freshly ejected algae release much less fixed ^{14}C during photosynthesis than algae suspended in animal homogenates. It is inferred from this that animal tissues contain 'factors' stimulating release from symbionts, but direct studies of 'factors' are lacking.

4.4.3 Selective advantage to host

The selective advantage to the host is clear from their ability to abandon the ingestion of particulate material for their adult lives, and from the absence of uninfected adults.

4.5 Regulation of symbiont population

It is not known how or whether the host regulates the size of its symbiont population. Since CO_2 induces algal ejection, the animal can presumably squeeze surplus algae out between epidermal cells if necessary. In senescent worms in the field, and in some laboratory cultures, digestion of algae has been inferred from electron micrographs.

4.6 Transmission

Motile, free-living forms of *Platymonas convolutae* are attracted by and rapidly settle on egg-cases, so enhancing the chances of aposymbiotic larvae ingesting the symbiotic algae. However, the response is not species-specific, as certain other algae, such as the related *Platymonas tetrathele*, are also attracted.

4.7 Specificity of association and 'recognition' by host of symbiont

In laboratory conditions, aposymbiotic larvae can become infected by species of *Prasinocladus*, a genus related to *Platymonas*. Worms infected with these other algae grow more slowly and take longer to become fully green. When worms infected with the 'unnatural' symbiont *Prasinocladus marinus* are presented with *Platymonas convolutae,* the former are ejected and replaced by the latter within a few days. Even amongst the unnatural symbionts, there is a 'pecking' order in the ability to displace each other.

It is not known how recognition operates. It could be a specific mechanism involving recognition of cell surfaces, or it could be a more general phenomenon in which the 'natural' symbiont is able to grow more quickly, and sheer physical pressure causes expulsion of the other algae.

5 Chloroplast symbiosis

5.1 Introduction

The phenomenon of chloroplast symbiosis was discovered when green bodies regularly present in cells lining the digestive tract of some marine sacoglossan molluscs were conclusively identified as chloroplasts. The chloroplasts could still carry out photosynthesis even after several weeks, and products of photosynthesis were detected in various regions of the animal.

The chloroplasts are derived from the seaweeds upon which animals feed. The mere presence of chloroplasts in cells lining the digestive tract is not in itself proof of symbiosis. It is necessary to show that the chloroplasts continue to function for an extended period of at least a week, that photosynthetic activity of animals is not due simply to chloroplasts passing through the gut, and that animals utilise products of photosynthesis. Some molluscs have unusually high rates of heterotrophic carbon dioxide fixation, with the products of such fixation more diverse than in green plants (including, e.g. glucose and glycogen). Careful experiments are therefore needed to demonstrate that CO_2 fixation is genuinely due to photosynthesis.

Chloroplast symbiosis has been unequivocally demonstrated only for elysioid sacoglossans feeding on seaweeds of the order Siphonales, such as *Codium, Caulerpa* and *Bryopsis*. Some closely related groups of sacoglossan molluscs also feed on these as well as other types of seaweed, and typically appear green; however, some are definitely not symbiotic, and for others the critical evidence is still lacking.

Since higher plant chloroplasts are so fragile, it might seem remarkable that such organelles can withstand the shock of transfer from one type of cell to another. However, chloroplasts from siphonaceous algae are remarkably robust. They are difficult to rupture, and even continue photosynthesis for at least 7 days after isolation into simple mineral media in the laboratory.

Chloroplast symbiosis will probably be discovered in other lower invertebrates. Chloroplast-like structures have been reported from the marine ciliates, *Prorodon* and *Stombidium,* while the foraminiferan *Metariotaliella* apparently contains chloroplasts from its food diatom. The marine ciliate *Mesodinium rubrum* was originally reported to contain a photosynthetic structure comprising a chloroplast, stalked pyrenoid, several mitochondria, ribosomes and some membrane systems all enclosed within an outer double membrane, but lacking a nucleus. More recent studies show that a nucleus is present and that the symbiont is a substantially modified cryptophycean alga. It can fix $^{14}CO_2$ and evolve oxygen photosynthetically.

The remainder of this section will be limited to elysioid sacoglossans.

5.2 Taxonomic identity of the symbionts

Seaweeds of the order *Siphonales* contain siphonein and siphonoxanthin, and the presence of these unique pigments in animals is conclusive proof that they have been feeding upon siphonaceous seaweeds – important for those animals such as *Tridachia crispata* which have not been observed feeding in the field.

Chloroplasts of red algae occur in some sacoglossans, and in the laboratory, the symbiotic mollusc *Elysia viridis* (normally found on *Codium* sp.) will feed on red seaweeds. Although proof is still lacking, there is no reason why chloroplasts of red algae should not exist symbiotically.

5.3 Morphology

Arising from the stomach of elysioid sacoglossans is a highly branched tubular system, the digestive diverticulum. The cells limiting the diverticulum are of two types, lime cells of unknown function and phagocytic digestive cells. The latter sequester the chloroplasts, and in healthy animals the cells become quite densely packed with plastids. Some are enclosed by a membrane, but many appear to lie free in the cytoplasm.

These molluscs have a radula comprising a single row of dagger-shaped teeth. The radula punctures the wall of the alga, so that the contents can be sucked out. There was presumably an original advantage in feeding on siphonaceous seaweeds since their coenocytic habit and absence of cross walls in the filaments enabled more food material to be sucked from a single puncture than with non-siphonaceous seaweeds.

Within the digestive cells the chloroplasts retain their morphological integrity; the plastid envelope remains intact and the thylakoid membranes, starch grains, etc. all appear identical to the plastids in seaweeds from which they were derived.

5.4 Metabolic interactions

5.4.1 Rates of photosynthesis and transfer of photosynthate from plastid to animal tissues

The rate of photosynthetic ^{14}C fixation per unit chlorophyll in the mollusc *Elysia viridis* is approximately 60 per cent of that of the *Codium* upon which it feeds. Since *Elysia* and *Codium* contain similar amounts of chlorophyll measured on a fresh weight basis, amounts of carbon fixed by the two organisms are of a comparable order.

Several different techniques applied to a variety of molluscs show that the animals utilise products of photosynthesis. Within one hour of fixation, ^{14}C could be found by autoradiography in *Tridachia crispata* and *Tridachiella diomedea* in organs such as the renopericardium, cephalic ganglia, intestine and pedal gland. All animals produce copious mucus, and in *T. crispata,* 30 per cent of a pulse of fixed ^{14}C could be found in various carbohydrate moieties of mucus released 5 h after exposure to the isotope. Simple metabolites may become labelled more rapidly, and in *Elysia viridis,* ^{14}C-alanine can be released from animals by isotope trapping within 20 min. of exposure.

During photosynthetic fixation of ^{14}C in *E. viridis,* free glucose is the first sugar to become labelled, soon accompanied within the first few minutes by small amounts of galactose. After about 20 min., a rapid rise in labelling of the galactose begins and it soon becomes a major product of fixation. Neither isolated chloroplasts nor intact *Codium* can synthesis free galactose and it is evidently an animal product. If animal tissues are destroyed by homogenising, incorporation of label into galactose ceases, though photosynthesis by intact chloroplasts in the homogenate continues unabated. When homogenates are centrifuged, about 40 per cent of the fixed ^{14}C remains in the supernatant, mostly as glucose but accompanied by small amounts of glycolic acid. This suggests that glucose is the principal product released from chloroplasts, though plastids in the homogenate may well be covered by a surface layer of contaminating animal cytoplasm which converts released products into free glucose.

The amount of photosynthate moving from chloroplast to animal has been estimated as approximately 36 per cent by the amount of a pulse of fixed ^{14}C passing through the galactose pool.

5.4.2 Cycling of nutrients

Chloroplasts do not seem to be involved in nutrient cycling. They may release some alanine (see section 5.4.3) but there is no evidence that animal waste products enter them, or that they have nitrate reductase.

5.4.3 'Factors' in host stimulating release from chloroplasts

Chloroplasts isolated from *Codium* release approximately 10 per cent of their fixed ^{14}C to the medium, primarily as glycolic acid. If an aqueous extract of *Elysia* is added to them there is a substantial increase in the release of fixed ^{14}C, mostly as glucose, but also with some amino acids such as alanine. Other labelled compounds are retained within the chloroplasts, showing that release is not due to lysis. An aqueous extract of the symbiotic anemone, *Anemonia sulcata,* has the same effect on the chloroplasts, just as the extract from *Elysia* stimulates release from the dinoflagellate symbiont of the anemone. The 'factor', which is thermolabile, is present only in the chloroplast-containing tissues and is entirely absent from related, non-symbiotic sacoglossans.

5.4.4 Advantages to host of chloroplasts

Photosynthesis makes an important quantitative contribution to the nutrition of the animals, since specimens of *E. viridis* starved in the light lose weight much less rapidly than those starved in the dark (Fig. 16.2). The nutritional importance of photosynthesis may be even greater than is implied by Fig. 16.2, since feeding animals are continuously acquiring fresh chloroplasts highly active in photosynthesis, while chloroplasts in starving animals are not being turned over and their photosynthetic capacity is continually declining (section 5.5).

It is not known if photosynthetic oxygen production is of major benefit to the animal.

5.5 Autonomy of symbiotic chloroplasts and regulation of their population by host animals

Chloroplasts in starving *Elysia* are still capable of some photosynthesis after 2 months, and in extreme cases after 3 months. Throughout the period of starvation, rates of photosynthesis progressively decline (Fig. 16.3), but the fact that chloroplast activity persists for such remarkably long periods raises the question of whether they have an unusual degree of autonomy for an organelle. Many chloroplast functions are controlled by the plant cell nucleus, but this is not transferred to the animal cell.

In *E. viridis,* chloroplasts have some capacity for protein synthesis, but they cannot synthesise chlorophyll, glycolipids or DNA. Neither in this animal nor in *T. crispata* can plastids synthesise ribulose diphosphate carboxylase, the primary CO_2-fixing enzyme in photosynthesis. Chloroplast division has not been observed in animals, and no increase in chlorophyll or

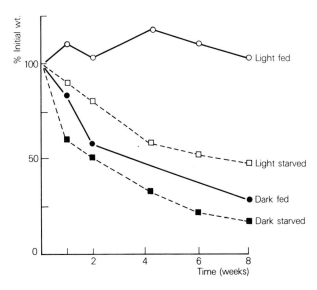

Fig. 16.2 Changes in weight of *Elysia viridis* kept for 2 months in light or dark, with or without its food plant, *Codium fragile.*

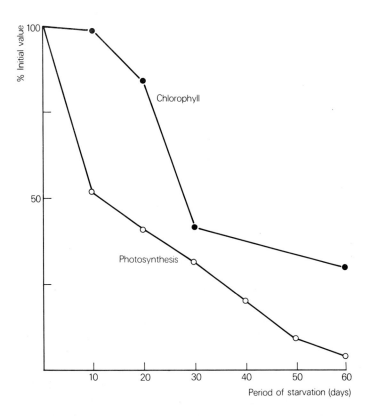

Fig. 16.3 Changes in chlorophyll content and photosynthesis (carbon fixation) of *Elysia viridis* during starvation in the light.

chloroplast number occurs during starvation. There is, therefore, no evidence that symbiotic chloroplasts have any greater autonomy than other types.

Since healthy, freshly collected animals always show high rates of photosynthesis, they are presumably always acquiring fresh chloroplasts when feeding. Since chloroplast digestion has not been convincingly demonstrated in healthy animals, old plastids are presumably expelled, though the process has not been observed. In some starved animals, degenerating plastids can be seen, and it is possible they are being digested. In *T. crispata,* acid phosphatases cannot be detected in digestive cells containing intact plastids, but after 12 weeks' starvation they could be detected in digestive cells containing defunct plastids.

5.6 Transmission

Eggs do not contain chloroplasts, nor do newly hatched veligers. The stage at which chloroplasts begin to be acquired has not been observed.

5.7 Recognition

Clearly, animals must have some mechanism for recognising their host seaweed, but this has never been studied.

In laboratory experiments, *E. viridis* preferentially feeds on the tip rather than the middle or base of *Codium fragile.* Chloroplasts from the tip have much higher rates of photosynthesis than those from other regions, but it is not known if animals prefer tips because of the active plastids, or because of other factors like softer cell walls.

If animals feed on *Codium* previously labelled with ^{14}C, they acquire plastids with radioactive chlorophyll. Since, within animals, plastids cannot synthesise chlorophyll, the rate of acquisition and loss of ^{14}C-chlorophyll can be used to estimate plastid turnover in animals. Such experiments show that, during feeding, plastids are being turned over, but during starvation they are not. This implies that a digestive cell can 'recognise' fresh plastids, and that uptake will then trigger elimination or digestion of 'old' plastids.

Elysia cauzescops feeds on *Caulerpa sertularioides,* which has both photosynthetic and amylogenic plastids. During feeding, only photosynthetic plastids are acquired, again showing some ability of the animal cells to 'recognise' symbionts.

6 Cyanellae

Blue—green algal symbionts occur only very sparsely among lower invertebrates, and they have been inadequately studied. The testate amoeba *Paulinella chromatophora* contains two cyanellae (believed to be *Synechococcus* spp.) whose division is closely linked to that of the host. When the amoeba divides, one cyanelle goes to each daughter and then divides to give the daughter the full complement of two. Another non-testate amoeba has been described as having about 100 cyanellae, but in general, the few protozoans with cyanellae have 1 to 6, with cyanelle division not being as tightly linked to host division as in *Paulinella.*

Cyanellae of Protozoans have not yet been cultured; in symbiosis their cell wall is much reduced and may be absent. Amongst flagellates it is often difficult to determine the affinity of the host, as if the symbiosis had been in existence for a long time. *Cyanophora paradoxa* with 1 to 6 cyanellae is believed to be a cryptomonad, as is *Cryptella cyanophora* with one cyanella. *Peliaina cyanea,* also with 1 to 6 cyanellae, has no clear-cut affinity at all.

The function of the cyanellae has not been investigated, beyond the solitary observation that in *Peliaina,* no starch is formed unless cyanellae are present. In *Cyanophora,* there is a close physical association of the cyanellae with a host organelle, as if the latter were some kind of primitive peroxisome.

Some marine sponges, such as *Ircinia variabilis, Petrosia ficiformis, Pellina semitubulosa, Verongia aerophoba* and *Chondrilla nucula* also contain unicellular blue–green symbionts, usually assigned to the genus *Aphanocapsa. Ircinia* has two species of *Aphanocapsa* as dense populations in its cortical layer: *A. feldmanii* which may be extracellular or intracellular in thesocytes and pinacocytes, and the larger *A. raspaigellae* which is only extracellular. In both situations and for both species, disintegrating algal cells occur alongside normal ones. It is not known if disintegration results from active digestion or from senescence.

Again no experimental investigations have been carried out into the function of the symbionts. There is an ecological correlation in that, in shaded environments, the sponge may occur without algae, but in sunny situations they are always present. This does not, of course, necessarily prove transfer of fixed carbon from symbiont to host.

In the few other associations with invertebrates the blue–green symbionts are usually extracellular, as with certain Echiuroid worms found in corals, some ascidians and some freshwater bryozoans. Whether these associations are mutualistic or commensal is not known.

7 Discussion

7.1 Introduction

The advantage of establishing a successful association with an autotrophic endosymbiont is clear: hosts receive an abundant supply of products of photosynthesis, typically 40 to 60 per cent of all the carbon fixed by the symbiont. Other nutritional advantages often accrue, and some algae — especially symbiotic dinoflagellates — play a crucial role in the nitrogen economy of the association.

If these associations are advantageous to the host, why are they not more widespread? Intracellular symbionts are only common in protozoa and coelenterates. Apart from a few sponges and some specific types of sacoglossan and bivalve mollusc, they scarcely occur in any other group — although intracellular parasites are found in all animal groups. Intercellular symbionts are much less common: they occur sporadically through the Turbellaria (Platyhelminths) and in only a few examples in certain other groups.

Even more striking is the paucity of algal species involved in symbiosis: *Gymnodinium microadriaticum, Chlorella vulgaris* and siphonaceous chloroplasts probably account for over 90 per cent of all associations described. Although it is generally believed that chloroplasts evolved from symbiotic blue–green algae, it is notable how very few modern examples exist of intracellular symbiotic blue–green algae, and none are ecologically prominent.

As a habitat for other organisms, the living cell can be viewed ecologically as an extreme environment. As with other extreme environments the diversity of species which can inhabit a cell is low, and mixed infections are almost unknown. Although a cell contains all the nutrients required for an invading symbiont, they are not necessarily easily available. The host cell is presumably a very efficient competitor for its own metabolic pools and it can display a variety of reactions to destroy or eliminate invading cells.

Establishment of a successful symbiosis requires a series of obstacles to be overcome:

a Symbiont must gain entry to the host.
b Symbiont must resist digestion.
c Symbiont must be 'recognised', and appropriate responses triggered in host.
d Symbiont must receive nutrients and be able to multiply in host.
e Host must be able to limit symbiont population.
f Host must induce nutrient-release without destroying symbiont.
g Mechanisms for the re-establishment or transmission of the association from generation to generation must be developed.

7.2 Entry of symbiont to host

In all cases symbionts enter aposymbiotic hosts by normal digestive process and entry into host cells is by phagocytosis. Since cells do not phagocytose all objects with which they come into contact, the symbiont has to have the necessary surface or other property to induce the mechanism. Usually the symbiont is either closely similar to the food of the host, or is first eaten by ciliates, etc. which are then eaten by the host. Entry is consequently not very specific and it is not the stage at which a host specifically 'recognises' or selects its symbiont.

7.3 Digestion resistance

Since symbionts normally enter by digestive processes, the ability to resist or avoid digestion is essential; the few observations available suggest two possibilities:

(1) Symbionts may be able to repress the formation of digestive enzymes and this has been demonstrated for alkaline phosphatase in *Paramecium bursaria*. It is also implicit for symbiotic chloroplasts since this enzyme is only detectable in the region of senescent, not healthy chloroplasts. The mechanism of enzyme repression is less clear. It could be by secretion of inhibitors, catabolite repression, or − as in the case of *Toxoplasma* parasitic in mouse macrophages − by preventing fusion of lysosomes with the symbiont-containing vacuole.

(2) Symbionts may be resistant to digestive enzymes. There is some evidence that symbiotic *Chlorella* are among those members of the genus with the highly indestructible compound sporopollenin in their cell walls. The outer envelope of siphonaceous chloroplasts is also remarkably resistant to various treatments, and this could be important in the establishment of symbiosis, especially as they mostly lie free in host cytoplasm.

Unfortunately there are no relevant observations concerning the largest group, symbiotic dinoflagellates. They are surrounded by a complex of membranes, but it is not known if they are digestion-resistant.

7.4 'Recognition' of symbiont

Soon after entering the host, the fate of the symbiont diverges from food or other objects. In intracellular symbiosis, the first visible signs of 'recognition' therefore occur within host digestive cells, and not at the point of first external contact. The characteristics of the algal cell 'recognised' by the host are not known. In *Hydra*, digestive cells take in both living and heat-killed *Chlorella*, but the latter are subsequently expelled, implying that 'recognition' depends upon a property of the living symbiont.

The characteristic response triggered by recognition is the movement of the symbiont away from the digestive region. In *Hydra*, a microtubular system is involved in the movement of

Chlorella to the distal end of vegetative cells; in *Convoluta,* the symbiotic algae migrate towards the epidermal regions; in *Paramecium,* algae also move towards the surface.

7.5 Multiplication of symbiont in host

The rates of growth of symbionts in their hosts have been little studied, but they generally seem to be less than in culture. During the period of multiplication, the host cell presumably permits access to essential nutrients (though again, flow of nutrient from host to symbiont has been very little studied). It would be interesting to determine whether the host exercises a positive regulation of nutrient flow to symbionts, or whether there is simply a finite and limited pool made available to them.

There is generally some degree of synchrony between division of host and symbiont. The extreme situation is represented by *Paulinella* and its two cyanelles, but in most other associations division is less precisely coordinated. Indeed, in most cases it is not known whether division of symbionts precedes or follows host cell division. The formation of motile or sexual stages in symbionts is extremely rare, even though they may be produced abundantly in culture.

Where symbionts are confined within individual host vacuoles, as in most *Chlorella* symbioses, the products of algal cell division invariably become sequestered within their own vacuole and are not allowed to aggregate with a common vacuole. The significance of this is not clear.

7.6 Host limitation of symbiont population

For obvious reasons there have to be mechanisms which prevent the symbiont from outgrowing and bursting the host. Four possibilities exist.

a Expulsion of surplus algae. This appears to be an important mechanism in sea anemones. Although most other hosts can be induced to expel algae under laboratory conditions, the importance of expulsion as a natural regulatory mechanism is not known.

b Digestion. Moribund and disintegrating symbionts are occasionally observed in some hosts, but not with sufficient frequency to suggest that digestion is a normal control mechanism, with the possible exception of giant clams where the evidence is a little stronger. Symbiont digestion occurs at the end of the life of *C. roscoffensis* and sometimes during adult life under laboratory conditions. Likewise, under abnormal conditions such as prolonged darkness, some digestion of chloroplasts occurs in *E. viridis*. In general, however, hosts expel rather than digest symbionts under conditions of stress.

c Limitations of physical space. The tight packing of chloroplasts in the digestive cells in *E. viridis* suggest that physical space may limit the number of plastids which can be taken up. In *C. roscoffensis,* the algae likewise seem to occupy all the available space between the subepidermal cells. In most other associations, space seems a less important limiting feature. For example, in *Hydra* and *Paramecium*, the symbiont population of a cell varies with environmental conditions and strain of alga, indicating other limitations. There are unfortunately no relevant observations for hosts of symbiotic dinoflagellates.

d Regulation of symbiont cell division. There is no direct evidence that this attractive possibility exists. The constancy of the symbiont population in *Hydra* and *Paramecium* under a given set of conditions certainly suggests that cell-division rates may be regulated by the host, but unfortunately the occurrence of expulsion as a control mechanism has not yet been properly investigated.

7.7 Induction of nutrient release from symbiont

Symbionts release to their hosts a substantial amount of fixed carbon, typically in the form of only one or a few types of molecule (i.e. there is no general, non-specific relaxation of membrane permeability barriers). The latter fact presumably offers the symbiont a measure of control if carbon loss becomes excessive; it also means that key metabolites such as coenzymes are not at risk of being released.

Except for some strains of *Chlorella*, symbionts rapidly cease to release massive amounts of photosynthate if they are isolated, and the compounds exported in symbiosis become very much less prominent. There is, therefore, some aspect of existence in symbiosis which induces carbon loss.

In all of the marine associations so far investigated the animal tissue contains 'factors' stimulating release of photosynthate from isolated symbionts. The 'factor' is thermolabile and cannot be detected in aposymbiotic animals or symbiont-free tissues. The 'factor' may be a compound normally present in cells at low levels and involved in some aspect of membrane regulation, whose synthesis is stimulated by presence of symbionts.

The situation in symbiotic *Chlorella* is confusing. All strains release maltose in symbiosis. After isolation into culture, some strains (including one derived from *Hydra*) lose this characteristic, but others retain it for several years before it is eventually lost — it is not known if loss is due to selection of non-releasing cells. No free-living *Chlorella* synthesises maltose, so that release of this compound is a consequence of some aspect of residence in the host, but the problem has not been investigated. So far, 'factors' have not been reported.

7.8 Transmission

The most effective method of transmitting symbionts from one generation to the next is the movement of symbionts directly from parent to offspring. In protozoans and coelenterates, it is normal for products of asexual reproduction — by cell fission or budding — to contain symbionts. Passage of symbionts to products of sexual reproduction seems not quite so common and has been little studied in protozoans. In coelenterates, Gorgonians produce uninfected eggs, while other hosts produce eggs which are usually, but not universally, infected.

Gorgonians, together with all invertebrate phyla above coelenterates, do not produce infected eggs. Since the symbiotic associations are mostly ecologically obligate, mechanisms for promoting reinfection of the next generation have evolved. The egg-cases of *C. roscoffensis* seem to attract motile unicellular algae. Elysioid sacoglossans normally feed on siphonaceous algae, though it is not known how they recognise their hosts. Giant clams and gorgonians live in coral reefs, where symbionts *presumably* abound if corals control their algal population by continual expulsion.

Bibliography

General

JENNINGS, D. H. and LEE, D. L. (1975) *Symbiosis,* 29th Symp. Soc. Exp. Biol. Cambridge University Press, London.
This contains review articles on almost all of the main kinds of association discussed here.

MUSCATINE, L., POOL, R. R. and TRENCH, R. K. (1975) 'Symbiosis of algae and invertebrates: aspects of the symbiont surface and the host—symbiont interface,' *Trans. Amer. Micros. Soc.* **94,** 450-69.
A valuable comparative review article.

SMITH, D. C. (1974) 'Transport from symbiotic algae and symbiotic chloroplasts to host cells', *Symbiosis,* 28th Symp. Soc. Exp. Biol., pp. 437—508.
A comparative survey of photosynthate transport.

Specific aspects

This section deals either with papers subsequent to reviews in Jennings and Lee (1975) or with papers on topics not covered by that book, or with papers forming the basis of figures.

BOYLE, J. E. and SMITH, D. C. (1975) 'Biochemical interactions between the symbionts of *Convoluta roscoffensis', Proc. R. Soc. Lond.,* **B. 189,** 121—35.
A description of experimental studies leading to many of the interactions shown in Fig. 16.1.

GEITLER, L. (1959) 'Syncyanosen', *Encyclopedia of Plant Physiology,* **XI,** 530—45 (Ed. W. Ruhland), Springer, Berlin.
The most recent review on cyanellae in invertebrates.

HIBBERD, D. J. (1977) 'Obervations on the ultrastructure of the cryptomonad endosymbiont of the red-water ciliate *Mesodinium rubrum', J. Mar. Biol. Ass. U.K.,* **57,** 45—61.
A recent thorough description which contradicts earlier observations that the symbiont chloroplast was unaccompanied by a nucleus.

HINDE, R. and SMITH, D. C. (1975) 'The role of photosynthesis in the nutrition of the mollusc *Elysia viridis', Biol. J. Linn. Soc.,* **7,** 161—71.
Describes experiments which produced Figs. 16.2 and 16.3.

MUSCATINE, L., POOL, R. R. and CERNICHIARI, E. (1972) 'Some factors influencing selective release of soluble organic material by zooxanthellae from reef corals', *Marine Biology,* **13,** 298—308.

PARDY, R L. (1976) 'The production of aposymbiotic green *Hydra* by the photodestruction of green *Hydra* zoochlorella', *Biol. Bull.,* **151,** 225—35.

STEELE, R. D. (1976) 'Light intensity as a factor in the regulation of the density of symbiotic zooxanthellae in *Aiptasia tagetes* (Coelenterata, Anthozoa)', *J. Zool. Lond.,* **179,** 387—405.
Describes the most recent work on extrusion of symbiotic dinoflagellates.

VANDERMEULEN, J. H. and MUSCATINE, L. (1974) 'Influence of symbiotic algae on calcification in reef corals: critique and progress report', in *Symbiosis in the Sea* (Ed. Winona B. Wernberg), pp. 1—19. University of South Carolina Press, Columbia.
A useful summary of the present situation.

WEBB, K. L., DUPAUL, W. D., WIEBE, W., SOTTILE, W. and JOHANNES, R. E. (1975) 'Enewetak (Eniwetok) Atoll: aspects of the nitrogen cycle on a coral reef', *Limnol. Oceanogr.,* **20,** 198—210.

WIEBE, W. J., JOHANNES, R. E. and WEBB, K. L. (1975) 'Nitrogen fixation in a coral reef community', *Science,* **188**, 257–59.

17

Current problems and developments in medical mycology

D. Kerridge
Department of Biochemistry, University of Cambridge

1 Introduction

The fungi comprise a complex group of eukaryotic protista characterised by either a mycelial or yeast-like morphology. There are probably in excess of 100 000 species and their economic importance reflects their ubiquity and metabolic diversity. Not only are they of benefit to man in that certain fungi are used in the fermentation industry for production of food and antibiotics, but also other species are responsible for food spoilage and a number of plant and animal diseases (Wolstenholme and Porter 1968, Emmons, Binford and Utz 1970).

The three major types of mycotic disease in humans are (1) *Mycoses* which result from the invasion of living tissue by a fungus, (2) *Allergies* which result from hypersensitivity to fungal antigens, (3) *Toxicoses* which result from ingestion of either toxic fungal metabolites in contaminated food (*mycotoxicoses*) or toxic fungal fruiting bodies (*mycetisms*) (Austwick 1972). The allergic and toxic reactions are important in agriculture and other industries where fungal contamination is common, but the problem is prevention rather than cure and outside the scope of this chapter. Numerically and economically fungal infections are not as important as those caused by bacteria, viruses and protozoa, yet it has been estimated that there are probably 2 000 000 new cases of systemic mycoses each year, which result in about 2 000 deaths. The frequency of dermatophytic infections is much greater and most people are infected at some stage in their life. The hundred or so fungal species pathogenic to man form a heterogenous group with the one common feature that they will grow at 37°C. The principal pathogenic fungi are shown in Table 17.1. The first group includes all airborne infections where the primary lesion is in the respiratory tract. In severe infections, the fungus may spread to other parts of the body. This group includes the two most important pathogens *Histoplasma capsulatum* and *Coccidioides immitis*. The second group includes those diseases where infection occurs by contamination of a wound and the primary lesion is in the subcutaneous or subepithelial tissue, although here again the pathogen may spread to other tissues. The final group comprises the commensal and dermatophytic fungi. These organisms are normally found associated with man (or his domestic animals) and these infections are contagious.

The fungi in the first two groups have a normal saprobic existence in soil and their pathogenic activities are incidental to their normal life cycle. They are, in fact, 'facultative pathogens' and can be further sub-divided into those that can infect a healthy host and others, such as *Aspergillus fumigatus,* which normally do not infect healthy humans but are a serious

Table 17.1   Principle mycotic infections

Fungus	Normal habitat	Disease	Morphological form Saprobic state	Morphological form Pathogenic state
Fungi causing pulmonary infections				
Histoplasma capsulatum	Soil enriched with bird or	Histoplasmosis	Mycelium	Yeast
Histoplasma dubosii	bat droppings			
Coccidioides immitis	Endemic in desert areas of America	Coccidioidomycoses	Mycelium	Spherule
Blastomyces dermatitidis	Soil, N. America	Blastomycoses	Mycelium	Yeast
Paracoccidioides brasiliensis	Soil, S. America	Blastomycoses	Mycelium	Yeast
Cryptococcus neoformans	Worldwide in pigeon droppings	Cryptococcosis	Yeast	Encapsulated yeast
*Aspergillus fumigatus**	Soil	Aspergillosis	Mycelium	Mycelium
Fungi infecting the skin and subcutaneous tissue through a wound				
Phialophora spp.	Tropics and subtropics	Chromomycoses	Mycelium	Mycelium
Cladosporium carrionnii				
Sporothrix schenkii	Tropics and subtropics	Sporotrichosis	Mycelium	Yeast
Various (16 spp. identified)	Soil	Mycetomas	Mycelium	Mycelium
Commensals and dermatophytic fungi				
*Candida albicans**	Commensal	Candidiasis	Both mycelium and yeast present	
Microsporum spp.				
Trichophyton spp.	Obligate parasites	Dermatophytosis		Mycelium
Epidermatophyton spp.				
Pityrosporum furfur	Obligate parasite	Pityriasis versicolor	Yeasts predominate in culture	Mycelium

*Opportunistic pathogens.

problem for patients whose normal defense mechanisms are impaired. Fungi in this latter group are often referred to as 'opportunistic pathogens'. The major predisposing factors to infection by such fungi are shown in Table 17.2. Essentially anything which impairs the immune response or alters the microenvironment in which the fungus can grow, will affect man's susceptibility to infection. The frequency of such infections has increased recently with the widespread use of antibacterial antibiotics, immunosuppressants and cytostatic drugs. The clinical importance of fungal infections stems largely from the absence of any really satisfactory antifungal drug.

Table 17.2   Predisposing factors to opportunistic fungal infections

Diabetes	Organ transplants	Cytostatics
Hemoblastoses	Heart surgery	Immunosuppressants
Malignant disease	Venous catheters	Contraceptive steroids
Tuberculosis	Prolonged antibiotic therapy	Pregnancy
Congenital immune defects	Corticosteroids	Nutritional disorders
Drug abuse		

In one chapter it is impossible to provide a comprehensive account of the current status of medical mycology and I have selected four topics for discussion:

1 Detection. Fungi are ubiquitous in the environment and the presence of a fungus in a clinical specimen is not always proof of an infection. In patients at risk, early detection is important since it is not feasible to give antifungal drugs systemically as a prophylactic measure.

2 The role of exocellular fungal products in infection. The fungi are renowned for the diversity of their exocellular products, ranging from the carcinogenic aflatoxins to proteolytic enzymes; these compounds may be important in the establishment and spread of infection.

3 Structural dimorphism. In a number of fungal pathogens the saprobic form has a normal mycelial morphology and the parasitic form is unicellular. This is of obvious importance in understanding the host—parasite relationship and as a model for studying morphogenesis.

4 Chemotherapy. Unfortunately since both host and parasite are eukaryotes, treatment of systemic mycoses lags behind that of bacterial infections. The molecular basis of action of the clinically important antifungal drugs will be discussed, emphasising the problem associated with drug resistance.

Three of these features are common to all pathogenic fungi and the fourth — structural dimorphism — to many fungi responsible for systemic infections. It is unfortunate that biochemical investigations have so far concentrated on a limited number of fungi and the discussion will of necessity be restricted to these.

2 Detection

The only reliable diagnosis is detection of the fungus in tissues and pathological specimens by conventional microscopic, histological and cultural techniques. Unfortunately this is not always possible and in some cases fungi may be isolated from pathological specimens and incriminated erroneously as aetiological agents of disease. In these instances serological tests provide a useful adjunct, and for histoplasmosis, coccidioidomycoses, blastomycoses and cryptococcosis are entirely satisfactory (Wolstenholme and Porter 1968, Prier and Friedman 1974). There is, however, a class of patients whose normal immune response is impaired either as a result of disease or the treatment they are receiving and these people are susceptible to bacterial and fungal infections. Treatment with antibacterial antibiotics may encourage the development of saprobic fungi in the gut and respiratory tract and severe infections may develop. These infections are commonly caused by *Candida albicans, Aspergillus fumigatus* and *Cryptococcus neoformans* and are difficult to diagnose. The organism may be isolated easily from certain sites, e.g. urine and faeces, but from others, such as the brain, it is difficult, and if detected in respiratory excretions, the results are difficult to interpret. A reliable serological method for diagnosis would be invaluable for such infections. There are problems however when the fungus is normally associated with man as in the case of *C. albicans.* The presence of agglutinating antibodies to *C. albicans* in humans is widespread and in fact their absence from anyone over 10 years old has been considered to support a diagnosis of an immunological deficiency. Even a rising antibody titre may not necessarily indicate a clinically important infection. Circulating antibodies to *C. albicans* rise in patients after open heart surgery and this correlates with an increase in the yeast flora of the gut occurring shortly after the operation (Evans and Forster 1976). These patients, although clearly at risk, do not have a clinical infection requiring immediate therapy.

In immunologically deficient patients, the techniques for precipitin detection may be unreliable and in a fulminating infection too late to help the patient. It is possible that here detection of circulating fungal antigens would be a better diagnostic test. In cryptococcal infections this is certainly so; production of capsular antigen is such that although antibody is produced, it is neutralised by the excess antigen and free antigen can be detected in the blood stream. Here the latex agglutination test is invaluable both in diagnosis of the disease and in assessing response of the patient to antibiotic therapy. There is already evidence that such a method might prove satisfactory for other fungal infections. Circulating aspergillus antigens were detected in immunosuppressed mice with aspergillosis (Richardson and White 1976) and circulating candida antigens in a patient with candidiasis (Axelson and Kirkpatrick 1973). The success of such a method depends upon the presence of common antigenic determinants in all strains of *C. albicans* and *A. fumigatus.* Many antigenic preparations used at present are crude cell extracts and there is a need for a detailed antigenic analysis of these organisms not only as an aid to clinical diagnosis but also as an aid to analysis of the fungal cell wall.

3 The role of exocellular products in fungal infections

Pathogenic fungi produce a variety of toxic metabolites and exocellular enzymes and these have been implicated in fungal infections. However, with the exception of dermatophytic fungi where the role of exocellular enzymes is well established, the evidence for the involvement of specific exocellular products in other infections is equivocal.

Dermatophytes occupy a specific ecological niche in a tissue which normally provides a physical barrier to infection. Colonisation of this region is aided by production of exocellular proteolytic enzymes capable of hydrolysing keratin and by the ability of the fungus to grow on the products. Keratin is normally resistant to proteolysis but keratinases are produced by both insects and fungi. However keratinases alone are unable to degrade keratin completely since disulphide bridges impede proteolytic action. The breakdown of the disulphide links in keratin by *Keratinomyces ajelloi* results from a reaction of the type:

$$R-S-S-R + HSO_3^- \longrightarrow R-SH + R-S-SO_3^-$$

The mechanism for sulphite production by the fungus is unknown and both proteolysis and sulphitolysis probably occur simultaneously during keratin breakdown (Ruffin *et al.* 1976). It is not known how widespread this mechanism of keratin degradation is among the dermatophytic fungi. Exocellular proteolytic enzymes may also be responsible for pathological changes in the deeper layers of the epidermis and dermis in situations where the fungus is confined to the stratum corneum.

The role of exocellular products in systemic infections is hard to establish. Even where toxic metabolites are produced in culture as for example the aflatoxins by *Aspergillus flavus*, gliotoxin, helvolic acid and fumagillin by *A. fumigatus* and phenethylalcohol and tryptophol by *C. albicans* there is no evidence for their involvement in the host–parasite interaction, or for that matter of their production *in vivo*.

Exocellular proteolytic enzymes produced by *C. albicans* have been implicated in the aetiology of denture stomatitis but no distinction was made between secreted proteins and those released by cell lysis, a factor of some importance in assessing the significance of this finding. Candida-infected chick embryo chorioallantoic membrane is a good model for muco-cutaneous candidiasis since both hyphal invasion and epithelial hyperplasia simulate the human response. The infecting cells adhere strongly to each other and to the host tissue and although individual cells invade the epithelial tissue a minimum-sized aggregate is apparently required to induce epithelial hyperplasia. One explanation for this is that hyperplasia is induced by exocellular enzymes and that cell aggregates are required to achieve a critical concentration of exocellular enzymes. This difference between the tissue response to individual cells and cell aggregates could explain the distinction between the asymptomatic carrier state and active mucocutaneous candidiasis (Wain *et al.* 1975).

What proteins are secreted by *C. albicans*? An exocellular peptidase is present when *C. albicans* is grown in a serum albumin medium; the pH optimum of this enzyme is 3·2 so its effect on infected tissues may not be significant. Chattaway, Odds and Barlow (1971) were unable to detect any enzyme activity in the culture medium after growth of either the yeast or mycelial form, but soluble cell extracts had peptidase and acid phosphatase activity. In *Saccharomyces fragilis*, enzymes normally present in the periplasmic space, i.e. those involved in cell wall synthesis and utilisation of nutrients unable to penetrate the plasma membrane, are also present in the culture medium and a similar situation occurs in *C. albicans* (Davies and Wayman 1975 and personal communication). It is not known if these enzymes are released by *C. albicans* during an infection or if additional ones are produced. The importance of these exocellular proteins in development of infection cannot be assessed until mutant strains lacking them are isolated and the virulence of these strains is tested.

4 Structural dimorphism

Probably the simplest structural change associated with the development of a fungal infection is the synthesis of a complex polysaccharide capsule by *C. neoformans*. Electron micrographs taken of organisms *in situ* in their normal environment in pigeon droppings show quite clearly little or no capsule, but once the infection is established the yeasts are encapsulated. The encapsulated cells are resistant to the host's normal defence mechanisms. An acapsular variant of *C. neoformans* is avirulent for mice and obviously capsule synthesis is an important factor in the establishment of infection.

Structural dimorphism although first observed in fungal infections can be reproduced *in vitro* and is not limited to the pathogenic fungi (Smith and Berry 1974). Usually the saprobic form is mycelial and the parasitic form is unicellular, a factor of obvious importance in disseminating the fungus through the body. An exception is *C. albicans* where both yeast and mycelial forms are found in lesions but the mycelial form is considered more important in the establishment of infection (Emmons, Binford and Utz 1970). There are four aspects of structural dimorphism to consider: (1) morphology; (2) chemical composition of the cell envelope; (3) environmental factors influencing the change; and (4) the biochemical basis for this dimorphism.

4.1 Morphology

The simplest structural dimorphism occurs in *C. albicans* where under suitable conditions the yeast cells (blastospores) produce elongated germ tubes, which eventually give rise to long, occasionally branched, septate hyphae. Pseudomycelia may be produced when elongated yeast cells are joined end to end. In the reverse transformation yeasts arise from both the hyphal tips and sites along the hyphae.

The fungi *Histoplasma capsulatum*, *Blastomyces dermatitidis*, *Paracoccidioides brasiliensis* and *Sporothrix schenkii* are all mycelial in their saprobic form at 20°C and yeast-like at 37°C in their parasitic form. Infection occurs by inhalation of airborne spores which germinate to produce short hyphae from which yeasts arise. The yeasts may also arise directly from spores. The reverse transformation from yeast to mycelium can be readily demonstrated by lowering the temperature and in *B. dermatitidis* and *H. capsulatum* the yeasts divide giving rise to transitional cells from which hyphae ultimately develop. During the transition the parent yeasts and the transitional cells contain an increased number of mitochondria and within 48 h the newly formed hyphae have the intracytoplasmic membranes normally seen in established mycelia. *Coccidioides immitis* is an exception to this general pattern. Infection results from inhalation of arthrospores derived from the fungal mycelium. As a result of synchronous nuclear division the arthrospores develop into multinucleate giant cells. After about three days cytoplasmic cleavage occurs and the mature endosporulating spherules are formed. Bursting of the spherules releases mature uninucleate endospores which may be dispersed to other parts of the body where they develop into endosporulating spherules and so repeat the cycle.

4.2 Composition of the cell envelope

These morphological changes are associated with changes in the composition of cell envelope. Although fungal cell walls contain polysaccharides, lipids and proteins, attention has been concentrated on the polysaccharide constituents. There are changes in the relative proportions of the polysaccharide constituents and, although data are limited, a consistent pattern seems to be present. In *H. capsulatum, P. brasiliensis* and *B. dermatitidis* the yeast cell walls contain more chitin than do those of the mycelium (Domer, Hamilton and Harkin 1967, Kanetsuna *et al.* 1969). The converse applied in *C. albicans* where the mycelium contains three times more chitin than the yeasts (Chattaway, Holmes and Barlow 1968). There are also changes in the relative proportions of α- and β-linked glucans in the cell walls of *B. dermatitidis* and *P. brasiliensis* with the mycelial walls possessing more β-linked glucans (Kanetsuna and Carbonell 1970). It is not surprising that the different fungi have certain features in common, since the composition and ultrastructure of the cell wall determines cell shape. Unfortunately, there are, as yet, insufficient data to correlate structure and composition in fungal walls.

Plasma membranes have only been isolated and characterised from *C. albicans* (Marriott 1975). There are significant differences between the two forms with the mycelial membrane having a much higher carbohydrate content than membranes from exponentially growing yeasts. Caution must be observed in interpreting these data, since membranes from stationary phase yeasts also had a high carbohydrate content. There are differences in the phospholipid and neutral lipid compositions of membranes from the different forms. The significance of these differences cannot be assessed without further information on the influence of the environment on fungal membranes under conditions where morphological changes do not occur but since the synthesis of the cell wall involves cell membranes, this clearly deserves further study.

4.3 Environmental factors affecting structural dimorphism

The morphological changes observed in these fungi are phenotypic and although a number of environmental factors mediate this transformation, the dimorphic fungi can be grouped into three classes. In the first the transformation is controlled by temperature, in the second by temperature and nutrition, and in the third by nutrition only. With the exception of *C. albicans* which normally grows at 37°C, the most important factor is temperature with the mycelial form growing at lower temperatures and the yeasts at higher temperatures when additional nutritional factors may be required.

Candida albicans is the most studied of the pathogenic dimorphic fungi and there are numerous methods of producing germ tubes and hyphae. The environmental factors that have been used include: pH value of the growth medium; temperature; nitrogen source; carbon source; inorganic phosphate concentration; and zinc or biotin deficiency. There are difficulties in providing a common explanation for these results since many of the methods are strain-specific (Shepherd and Sullivan 1976; Land *et al.* 1975). In some cases the results are completely conflicting: Nickerson and Falcone (1956) found the presence of cysteine maintained *C. albicans* as the yeast even under conditions which would normally support mycelial growth, whereas Wain, Price and Cawson (1975) did not. The situation is further complicated since the physiological state of the cells is important in germ tube formation. (Chaffin and Sogin 1976). Yeasts harvested during the exponential phase of growth are apparently unable to produce germ tubes in the amino acid medium of Lee, Buckley and Campbell (1975) whereas those harvested from stationary phase cultures can. There is also evidence that once the yeasts are committed to either budding or germ tube formation the process cannot be reversed by an environmental change (Evans, Odds and Holland 1975).

4.4 Biochemical basis for structural dimorphism

Two mechanisms can be invoked to explain structural dimorphism in fungi: (1) The environmental change may result in quantitative differences in flow of intermediary metabolites through pre-existing pathways and so affect the composition and structure of the cell envelope. (2) New enzymes may be synthesised or old ones repressed so that qualitative as well as quantitative differences occur in the pattern of intermediary metabolism.

Phenotypic variation is common among the protista and the relevance of any metabolic change to structural dimorphism must be carefully assessed. If we consider structural dimorphism in *C. albicans* and *H. capsulatum* can we assign them to one or other of the hypotheses?

Macromolecular synthesis during the yeast mycelial transformation in *C. albicans* has been studied but most investigations have emphasised the patterns of metabolism and their influence on cellular structure. One early proposal was that dimorphism results from a changed balance between cell growth and division and that a protein disulphide reductase is involved. In the mycelial form it was thought that the disulphide bridges between the cell wall mannoproteins were not reduced and in the absence of plastic deformation of the wall, bud formation could not occur and germ tubes were produced. This was supported by the findings that mycelium formation did not occur in a cysteine-containing medium and that a mycelial mutant lacked the protein disulphide reductase (Nickerson 1963). Changes in cell wall morphology are associated with alterations in the polysaccharide composition and Chattaway *et al.* (1973) found the different patterns of glucose metabolism in the yeast and mycelial culture were consistent with the concept that NADPH has a role in cell division and that the balance between mannan and chitin synthesis may in part be controlled by modulation of phosphofructokinase. Additional support for a changed pattern of metabolism rather than differential gene expression in *C. albicans* comes from studies on germ tube formation by starved cells (Land *et al.* 1975 and 1976). The ability of amino acids to induce germ tube formation is limited to those whose catabolism is via α-oxoglutarate, succinyl CoA and acetoacetyl CoA. The transition from the yeast to mycelial form was associated with a shift from aerobic to anaerobic metabolism, again consistent with the hypothesis correlating germ tube formation with a changed pattern of carbohydrate metabolism and an interruption of electron flow within the cell. Somewhat different results were obtained by Shepherd and Sullivan (1976) in an examination of the growth characteristics of *C. albicans* in continuous culture. At comparable dilution rates there were no significant differences in the rate of oxygen uptake or in the respiratory quotient of the two forms.

The ability of yeasts to produce germ tubes in serum is a diagnostic feature of *C. albicans* and Chattaway *et al.* (1976) isolated a peptide fraction from seminal plasma which also induces germ tube formation. This is not specific and an acid hydrolysate of the peptide fraction or a mixture of the amino acids in the same relative proportions will also induce germ tube formation. Another amino acid mixture has been used as a nitrogen source for *C. albicans* in a temperature—mediated transformation (Lee, Buckley and Campbell 1975). Their choice of amino acids was based on the ability of cells to hydrolyse α-naphthylamide derivatives of the amino acids, the rationale for which is obscure. The diversity of results reflects both strain differences within the species and the wide variety of environmental conditions used to induce germ tube formation. The metabolic changes responsible for structural dimorphism in *C. albicans* have yet to be defined but the morphological changes in this organism can be explained by quantitative changes in intermediary metabolism rather than by qualitative ones.

Studies on the more complex transformation observed in *H. capsulatum* have emphasised the changing pattern of nucleic acid synthesis rather than the synthesis of the fungal cell wall, and it is implicit that differential gene expression is involved. There is direct evidence for this since the yeast form at 37°C has an absolute growth requirement for cysteine and the mycelial form at 23°C has not. This results from the loss of ability to synthesise the enzyme sulphite reductase (Boguslawski, Akagi and Ward 1976). Germ tube formation in *C. albicans* occurs within 4 h once the yeasts have been transferred to a suitable environment, whereas in *H. capsulatum* transferring the mycelium from 23°C to 37°C results in the production of yeast cells after about 4 days. Initially the total RNA content of the mycelium falls but after 24 h increases at a time when buds appear at the hyphal tips. Throughout this period, incorporation of radioactive precursors into RNA occurs although the rate is only 20 per cent of that into mycelial RNA at 23°C. Absolute comparisons are difficult since the specific activities of the

RNA precursors were not determined (Cheung *et al.* 1974). There are differences also in the RNA polymerase activities of the two forms. Three polymerases occur in the yeast soluble fraction, one of which comprises some 80 per cent of the total activity and is sensitive to α-amanitin, whereas the others are not. The specific polymerase activity of the mycelium is much lower and only one polymerase was detected in the soluble extract. An inhibitor of RNA polymerase is present in the mycelial extract. This compound, histin, is a heat-stable acidic protein with a molecular weight of 24 000 and appears to inhibit those polymerases that are insensitive to α-amanitin (Boguslawski *et al.* 1974 and 1975).

While much of the evidence points to an effect on gene expression being responsible for the mycelium-to-yeast transition there are insufficient data to elaborate this statement. There have been few investigations on the phenotypic changes occurring in the pathways of metabolism in this organism, but a recent paper by Maresca *et al.* (1977) has provided evidence for the involvement of cysteine-mediated control of cyclic AMP levels in the yeast mycelial trans-formation and it is probable that the structural dimorphism of *H. capsulatum* involves both quantitative and qualitative changes in intermediary metabolism.

5 Chemotherapy of fungal infections

The basic requirement for treatment of a systemic mycotic infection is at least one difference between man and the invading fungus to provide a target for antibiotic action. Since both are eukaryotes the difference will result from a variation on the basic organisational theme. The major differences between the pathogenic fungi and their host reside in the cell envelope. Fungi have a complex cell wall containing polysaccharides as the major structural components together with proteins and lipids, whereas mammalian cells lack a complex wall. There are also differences in the composition of the plasma membranes with fungi having ergosterol as their major sterol, whereas cholesterol is present in mammalian membranes. Not one of the clinically important antifungal agents acts by interfering with cell wall synthesis, a situation in direct contrast to bacteria where two important groups of antibiotics, the penicillins and cephalo-sporins inhibit the synthesis of cell wall mucopeptide. Polyoxin D, an analogue of UDP-*N*-acetylglucosamine is a competitive inhibitor of chitin synthetase, yet it cannot be used as an effective antifungal agent since the concentrations required to inhibit growth are frequently much greater than those required to inhibit chitin synthetase *in vitro* and reflect the presence of a permeability barrier to the antibiotic (Endo, Kakiki and Misato 1970).

There are numerous antifungal compounds but the number that can be administered systemically is limited by the toxicity of many of these compounds. The principle antifungal agents and the organisms against which they are effective are shown in Table 17.3. Antifungal antibiotics have been comprehensively reviewed by Cartwright (1975) and the problems of selectivity and drug resistance will be discussed in this section. Many of the clinically important antibacterial antibiotics inhibit the synthesis of cellular macromolecules, whereas the antifungal agents interact with cellular structures causing loss of function or are incorporated into cellular macromolecules, also resulting in loss of function.

5.1 Mode of action of the clinically important antifungal agents

The polyene antibiotics, amphotericin B, nystatin and candicidin interact with sterol-containing membranes causing an impairment of function, leakage of cellular constituents and ultimately cell death. The extent of the interaction depends on both the polyene and the membrane sterol.

Table 17.3   Clinically important antifungal agents

Agent	Target organism	Target site
Polyene antibiotics		
Amphotericin B	*Candida albicans*	Cell membrane
	Aspergillus fumigatus	
	Cryptococcus neoformans	
	Histoplasma capsulatum	
	Blastomyces dermatitidis	
	Coccidioides immitis	
Nystatin	*Candida albicans*	Cell membrane
Candicidin	(superficial infections)	
Imidazole derivatives		
Clotrimazole	*Candida albicans*	Cell membrane
Miconazole	*Aspergillus fumigatus*	
Econazole		
5-Fluorocytosine	*Candida albicans*	RNA metabolism
	Cryptococcus neoformans	
Griseofulvin	Dermatophytic fungi	Microtubules

In general the smaller polyenes cause more extensive membrane damage than the larger ones and the differences in the affinities of the polyenes for ergosterol-containing fungal membranes and cholesterol-containing human membranes make it possible to use certain of these compounds for the treatment of mycotic infections (Hamilton Miller 1974, Kerridge and Russell 1975). The presence of sterol in the membrane may not be essential for its interaction with the antibiotic but it is the fluidity of the membrane that determines polyene sensitivity (Hsu-Chen and Feingold 1974, Archer 1976). In natural membranes the fatty acid composition is such that the sterol affects lipid mobility and as a result polyenes can interact with them. The incorporation of either amphotericin B or nystatin into sterol-containing membranes is thought to result in the production of aqueous pores. For amphotericin B, physical studies provided the basis for a model consisting of an annulus of amphotericin B and sterol in which the hydrophilic region of the antibiotic faces the interior of the pore. The length of the annulus is such that two half pores span the lipid bilayer (De Krujff and Damel 1974). The interaction of the polyene with the plasma membrane of mammalian cells and fungal protoplasts is reversible (Cass and Dalmark 1973, Kerridge, Koh and Johnson 1976) and death of the cells results from loss of low molecular weight constituents rather than the interaction of the antibiotic with the cell membrane.

The antifungal drugs, clotrimazole, econazole and miconazole also interact with membranes of fungi and erythrocytes causing impairment of function but the molecular basis of the interaction and the reasons for the selectivity of these compounds are not known (see Cartwright 1975).

Griseofulvin, unlike polyene antibiotics, is absorbed from the gut and is given orally in the treatment of certain dermatophytic infections, since by this route higher concentrations of this poorly soluble antibiotic are achieved at the site of infection. Griseofulvin inhibits growth of sensitive fungi and tissue culture cells by interfering with microtubular function in both the nuclei and cytoplasm (Weber, Wehlande and Herzog 1976, Gull and Trinci 1974). This results from the binding of the antibiotic to microtubule-associated protein rather than tubulin (Roobol, Gull and Pogson 1977). The molecular basis for its selectivity and chemotherapeutic efficiency is not known but may be related to its solubility properties and accumulation in keratinised tissues. After oral administration it is incorporated into keratinised tissues and as a result these tissues are less susceptible to proteolytic digestion by dermatophytic fungi. At least part of the efficacy of this compound in treating fungal infections may result from a reduction in available sources of nutrient and elimination of the fungus by renewal of the infected tissue (Yu and Blank 1973).

The other major antifungal agent used in the treatment of systemic mycoses is 5-fluoro-cytosine. This compound was originally developed as a potential antileukaemic drug but has proved efficacious in treating systemic candidiasis and cryptococcosis. The mode of action is well understood (Polak and Scholer 1975). The analogue is taken up into sensitive cells by a cytosine permease where it is deaminated by a cytosine deaminase to 5-fluorouracil which is subsequently incorporated in place of uracil into cellular RNA. Growth inhibition results from the production of malfunctioning RNA. Selectivity is good since mammalian tissues lack both cytosine permease which is responsible for uptake of 5-fluorocytosine and also cytosine deaminase which converts it into the toxic 5-fluorouracil. 5-Fluorocytosine is not itself incorporated into cellular RNA.

5.2 Antibiotic resistance

In treating systemic mycoses we are often relying on quantitative rather than qualitative differences between host and parasite. Many antifungal drugs are also toxic to man with unpleasant side effects on prolonged administration, and this affects the maximum serum levels that can be achieved during therapy. Hamilton Miller (1973) determined the minimum growth inhibitory concentrations of the commonly used antifungal agents for a number of clinical isolates of *C. albicans* and compared these values with the antibiotic concentration in serum during therapeutic treatment. An appreciable number of strains were insensitive to the anti-mycotic at the serum levels and for many others the inhibitory concentrations were comparable to the serum concentrations. Clearly minor changes in the sensitivity of the fungus to the therapeutic agent can affect the efficacy of the treatment. Changes in antibiotic sensitivity arise both phenotypically and genotypically and examples of both are found among the fungi.

5.2.1 Genotypic drug resistance

Clinically this is only of importance with 5-fluorocytosine and here it is essential to monitor the resistance of the fungus during treatment to ensure that the therapeutic measures used do not result in the selection of resistant strains. Resistance to 5-fluorocytosine can arise as a result of mutation at any one of five genetic loci (Polak and Scholer 1975). These mutations may cause: (1) loss of cytosine permease; (2) loss of cytosine deaminase; (3) a deficiency in UMP pyrophosphorylase; (4) an increase in *de novo* synthesis of pyrimidines due to loss of feedback regulation of aspartate transcarbamylase; and (5) stimulation of orotidylic acid pyrophos-phorylase and orotidylic acid decarboxylase. If resistance results from a loss of the cytosine permease, a synergistic mixture of 5-fluorocytosine and amphotericin B may be satisfactory since interaction of the polyene with the plasma membrane will allow entry of the other drug. The selection of drug-resistant fungi during treatment with polyene antibiotics, griseofulvin and the imidazole derivatives is not a problem.

5.2.2 Phenotypic drug resistance

There is considerable variation in the sensitivity of *C. albicans* to amphotericin B methyl ester (AME) associated with growth in batch culture. When grown in either a complex medium or a simple salt medium there is little change in polyene sensitivity during the exponential phase of growth. However, during the stationary phase there is a marked increase in resistance to the antibiotic (Gale 1974). This phenotypic variation is not restricted to the polyenes and occurs also with miconazole and clotrimazole. An analogous situation occurs in *A. fumigatus* where the conidia are completely resistant to polyene antibiotics but, once germination commences, rapidly become sensitive (Russell, Kerridge and Gale 1975). This variation in antibiotic sensitivity may be of clinical importance since in any infection there will be considerable heterogeneity with both growing and non-growing cells present and could explain the persistence of fungal infections and the need for prolonged therapy. Changes in the cell wall are responsible for the increase in antibiotic resistance since protoplasts prepared from resistant cultures of *C. albicans* are sensitive. There are small changes in the chemical composition of cell walls of *C. albicans* harvested at different growth phases but it is unlikely that these are directly responsible for antibiotic resistance. Variation in antibiotic sensitivity of *C. albicans* can also be induced by chemical modifications of the cell wall (Gale *et al.* 1975).

Amphotericin B is a large molecule with a molecular weight of 938 and interacts with many components of the cell wall, so clearly penetration of the antibiotic to the plasma membrane is a complicated process and the molecular organisation of the macromolecular constituents of the cell wall will be an important factor in determining the inhibitory effects of the antibiotic. The phenotypic variation in polyene sensitivity could result from either *de novo* synthesis of new components for insertion into the wall or from *in situ* modification of pre-existing structures. Unfortunately there is insufficient evidence to allow this distinction to be made. Extension of the yeast wall during budding occurs at specific sites and in these regions there is a balance between hydrolysis of existing polymers and synthesis and insertion of new constituents. This continual metabolic activity could result in regions of wall growth providing a ready access for antibiotic penetration to the plasma membrane. If this is so then once cell wall synthesis ceases antibiotic resistance might be expected to increase. Chemical modification would affect the entire cell surface. This explanation of phenotypic variation does not exclude the antibiotic binding to other sites on the cell wall or even penetrating there, provided penetration is more rapid at the growing point (Kerridge *et al.* 1976).

6 Prospects

What of the future? There is obviously a great need for improving both diagnosis and therapy of fungal infections. Developments in diagnosis will no doubt come from analyses of the antigenic determinants in pathogenic fungi and the application of immunological methods to the detection of fungal antigens and antibodies in the infected host. Improvements in antifungal therapy present more serious problems. We are dealing with a eukaryotic pathogen and the differences between the pathogen and its host reside largely in the composition and synthesis of the cell envelope. At present no clinically important antifungal agent inhibits cell wall synthesis but polyoxin D might provide a model for the development of future antifungal agents. Amphotericin B is still the most effective compound for the treatment of systemic infections but its toxic side effects are such that it is only administered when absolutely essential. Methylation of this compound improves its solubility properties and lowers its toxicity but since the molecule is complex and unstable the scope for chemical modification is limited. It is possible that selection of suitable biosynthetic mutants of the producing streptomycete in conjunction with a more detailed knowledge of the interaction of the polyenes with the cell envelopes of the sensitive organisms might result in improvements in this class of antibiotics. If we look further ahead then inhibitors of the dimorphic transition from the saprobic mycelial form to the pathogenic yeast form might provide the next generation of antibiotics. But before that can be achieved a much greater knowledge of morphogenesis in the pathogenic fungi is required.

References

ARCHER, D. B. (1976) 'Effect of the lipid composition of *Mycoplasma mycoides* subspecies *capri* and phosphatidylcholine vesicles upon the action of polyene antibiotics', *Biochim. Biophys. Acta,* **436**, 68–76.

†AUSTWICK, P. K. C. (1972) 'The pathogenicity of fungi', *Symp. Soc. Gen. Microbiol.,* **22**, 251–68.

AXELSON, N. H. and KIRKPATRICK, C. H. (1973) 'Simultaneous characterisation of free Candida antigens and Candida precipitins in a patient's serum by means of cross immuno-electrophoresis with intermediate gel', *J. Immunol. Methods,* **2**, 245–9.

BOGUSLAWSKI, G., AKAGI, J. M. and WARD, L. G. (1976) 'Possible role for cysteine biosynthesis on conversion from mycelial to yeast form of *Histoplasma capsulatum*', *Nature,* **261**, 336–8.

BOGUSLAWSKI, G., KOBAYASHI, G. S., SCHLESSINGER, D. and MEDOFF, G. (1975) 'Characterisation of an inhibitor of ribonucleic acid polymerase from the mycelial phase of *Histoplasma capsulatum*', *J. Bacteriol.,* **122**, 532–7.

BOGUSLAWSKI, G., SCHLESSINGER, D., MEDOFF, G. and KOBAYASHI, G. S. (1974) 'Ribonucleic acid polymerases in the yeast phase of *Histoplasma capsulatum*', *J. Bacteriol.,* **118**, 480–5.

†CARTWRIGHT, R. Y. (1975) 'Antifungal drugs', *J. Antimicrobiol. Chemotherap.,* **1**, 141–62.

CASS, A. and DALMARK, M. (1973) 'Equilibrium dialysis of ions in nystatin-treated red cells', *Nature,* **244**, 47–9.

CHAFFIN, W. L. and SOGIN, S. J. (1976) 'Germ tube formation from zonal rotor fractions of *Candida albicans*', *J. Bacteriol.,* **126**, 771–6.

CHATTAWAY, F. W., HOLMES, M. R. and BARLOW, A. J. E. (1968) 'Cell wall composition of mycelial and blastospore forms of *Candida albicans*', *J. Gen. Microbiol.,* **51**, 367–76.

CHATTAWAY, F. W., BISHOP, R., HOLMES, M. R., ODDS, F. C. and BARLOW, A. J. E. (1973) 'Enzyme activities associated with carbohydrate synthesis and breakdown in the yeast and mycelial forms of *Candida albicans*', *J. Gen. Microbiol.,* **75**, 97–109.

CHATTAWAY, F. W., O'REILLY, J., BARLOW, A. J. E. and ALDERSLEY, T. (1976) 'Induction of the mycelial form of *Candida albicans* by hydrolysates of peptides from seminal plasma', *J. Gen. Microbiol.,* **97**, 317–22.

CHATTAWAY, F. W., ODDS, F. C. and BARLOW, A. J. E. (1971) 'An examination of the production of hydrolytic enzymes and toxins by pathogenic strains of *Candida albicans*', *J. Gen. Microbiol.,* **67**, 255–63.

CHEUNG, S. S. C., KOBAYASHI, G. S., SCHLESSINGER, D. and MEDOFF, G. (1974) 'RNA metabolism during morphogenesis in *Histoplasma capsulatum*', *J. Gen. Microbiol.,* **82**, 301–7.

DAVIES, R. and WAYMAN, F. J. (1975) 'The effects of thiols on *Saccharomyces fragilis*', *Antonie van Leeuwenhoek J. Microbiol. Serol.* **41**, 33–58.

DE KRUIJFF, B. and DEMEL, R. A. (1974) 'Polyene antibiotic–sterol interactions in membranes of *Acholeplasma laidlawii* cells and lecithin liposomes III Molecular structure of the polyene antibiotic–cholesterol complexes', *Biochim. Biophys. Acta,* **339**, 57–70.

DOMER, J. E., HAMILTON, J. G. and HARKIN, J. C. (1967) 'Comparative study of the cell walls of the yeast-like and mycelial phases of *Histoplasma capsulatum'*, *J. Bacteriol.,* **94**, 466–74.

*EMMONS, C. W., BINFORD, C. H. and UTZ, J. P. (1970) *Medical Mycology* (2nd edn.). Lea and Febiger, Philadelphia.

EVANS, E. G. V. and FORSTER, R. A. (1976) 'Antibodies to Candida after operations on the heart', *J. Med. Microbiol.,* **9**, 303–8.

EVANS, E. G. V., ODDS, F. C. and HOLLAND, K. T. (1975) 'Resistance of the *Candida albicans* filamentous cycle to environmental change. *Sabouraudia'*, **13**, 231–8.

ENDO, A., KAKIKI, K. and MISATO, T. (1970) 'Mechanism of action of the antifungal agent Polyoxin D', *J. Bacteriol.,* **104**, 189–96.

GULL, K. and TRINCI, A. P. J. (1974) 'Effects of griseofulvin on the mitotic cycle of the fungus *Basidiobolus ranarum'*, *Arch. Mikrobiol.,* **95**, 57–65.

GALE, E. F. (1974) 'The release of potassium ions from *Candida albicans* in the presence of Polyene antibiotics', *J. Gen. Microbiol.,* **80**, 451–65.

GALE, E. F., JOHNSON, A. M., KERRIDGE, D. and KOH, T. Y. (1975) 'Factors affecting the changes in amphotericin sensitivity of *Candida albicans* during growth', *J. Gen. Microbiol.,* **87**, 20–36.

†HAMILTON-MILLER, J. M. T. (1973) 'Chemistry and biology of the polyene macrolide antibiotics', *Bacteriol., Rev.,* **37**, 166–96.

HSU-CHEN, C. C. and FEINGOLD, D. S. (1973) 'Polyene antibiotic action on lecithin liposomes: Effect of cholesterol and fatty acyl chains', *Biochem. Biophys. Res. Commun.,* **51**, 972–8.

KANETSUNA, F. and CARBONELL, L. M. (1970) 'Cell wall glucans of the yeast and mycelial forms of *Paracoccidioides brasiliensis'*, *J. Bacteriol.,* **101**, 675–80.

KANETSUNA, F., CARBONELL, L. M., MORENO, R. E. and RODRIQUEZ, S. (1969) 'Cell wall composition of the yeast and mycelial forms of *Paracoccidioides brasiliensis'*, *J. Bacteriol.,* **97**, 1036–41.

†KERRIDGE, D., KOH, T. Y., MARRIOTT, M. S. and GALE, E. F. (1976) 'The production and properties of protoplasts from the dimorphic yeast *Candida albicans'*, in *Microbial and Plant Protoplasts* (Eds. Peberdy, J. F., Rose, A. H., Rogers, H. J. and Cocking, E. C.). Academic Press, London and New York.

KERRIDGE, D., KOH, T. Y. and JOHNSON, A. M. (1976) 'The interaction of amphotericin B methyl ester with protoplasts of *Candida albicans'*, *J. Gen. Microbiol.,* **96**, 117–23.

†KERRIDGE, D. and RUSSELL, N. J. (1975) 'Polyenes; actions and prospects', in *Chemotherapy Progress,* Vol. 6, Proceedings of the Sixth International Congress of Chemotherapy, pp. 111–16. Plenum, New York and London.

LAND, G. A., McDONALD, W. C., STJERNHOLM, R. L. and FRIEDMAN, L. (1975) 'Factors affecting filamentation in *Candida albicans*: Relationship of the uptake and distribution of proline to morphogenesis', *Infect. Immun.,* **11**, 1014–23.

LAND, G. A., McDONALD, W. C., STJERNHOLM, R. L. and FRIEDMAN, L. (1976) 'Factors affecting filamentation in *Candida albicans*: Changes in respiratory activity of *Candida albicans* during filamentation', *Infect. Immun.,* **12**, 119–27.

LEE, K. L., BUCKLEY, H. R. and CAMPBELL, C. (1975) 'An amino acid liquid synthetic medium for development of mycelial and yeast forms of *Candida albicans'*, *Sabouraudia,* **13**, 148–53.

MARESCA, M., MEDOFF, G., SCHLESSINGER, D. and KOBAYASHI, G. S. (1977) 'Regulation of dimorphism in the pathogenic fungus *Histoplasma capsulatum.*' *Nature, London,* **266**, 447–8.

MARRIOTT, M. S. (1975) 'Isolation and chemical characterisation of plasma membranes from the yeast and mycelial forms of *Candida albicans*', *J. Gen. Microbiol.,* **86**, 115–22.

‡NICKERSON, W. J. (1963) 'Molecular bases of form in yeasts', *Bacteriol. Rev.,* **27**, 305–24.

NICKERSON, W. J. and FALCONE, G. (1956) 'Identification of protein disulphide reductase as a cellular division enzyme in yeasts', *Science,* **124**, 722–3.

POLAK, A. and SCHOLER, H. J. (1975) 'Mode of action of 5-fluorocytosine and mechanism of resistance', *Chemotherapy,* **21**, 113–30.

†PRIER, F. E. and FRIEDMAN, H. (Eds.) (1974) *Opportunistic Pathogens.* Macmillan, London.

RICHARDSON, M. D. and WHITE, L. O. (1976) 'The detection of circulating *Aspergillus fumigatus* antigens in immunosuppressed mice with systemic aspergillosis', *Proc. Soc. Gen. Microbiol.,* **3**, 188–9.

ROOBOL, A., GULL, K. and POGSON, C. I. (1977) 'Evidence that griseofulvin binds to a microtubule associated protein', *FEBS Letters,* **75**, 149–53.

RUFFIN, P., ANDRIEU, S., BISERTE, G. and BIGUET, J. (1976) 'Sulphitolysis in keratinolysis: biochemical proof', *Sabouraudia,* **14**, 181–4.

RUSSELL, N. J., KERRIDGE, D. and GALE, E. F. (1975) 'Polyene sensitivity during germination of conidia of *Aspergillus fumigatus*', *J. Gen. Microbiol.,* **87**, 351–8.

*SMITH, J. E. and BERRY, D. R. (1974) *An Introduction to Biochemistry of Fungal Development.* Academic Press, New York and London.

SHEPHERD, M. G. and SULLIVAN, P. A. (1976) 'The production and growth characteristics of yeast and mycelial forms of *Candida albicans* in continuous culture', *J. Gen. Microbiol.,* **93**, 361–70.

WAIN, W. H., PRICE, M. F. and CAWSON, R. A. (1975) 'A re-evaluation of the effect of cysteine on *Candida albicans*', *Sabouraudia,* **13**, 74–82.

WAIN, W. H., PRICE, M. F. and CAWSON, R. A. (1976) 'Factors affecting plaque formation by *Candida albicans* infecting the chick chorio-allantoic membrane', *Sabouraudia,* **14**, 149–54.

WEBER, K., WEHLAND, J. and HERZOG, W. (1976) 'Griseofulvin interacts with microtubules both *in vivo* and *in vitro*', *J. Molec. Biol.,* **102**, 817–29.

†WOLSTENHOLME, G. E. W. and PORTER, R. (Eds.) (1968) *Systemic Mycoses.* Churchill, London.

YU, R. J. and BLANK, F. (1973) 'On the mechanism of action of griseofulvin in dermatophytosis', *Sabouraudia,* **11**, 274–8.

*Text books. †Symposia. ‡Reviews.

18

Epidemiology and control of protozoal pathogens of man

J. R. Baker

M.R.C. Biochemical Parasitology Unit, Molteno Institute, University of Cambridge

1 Introduction

Ten genera of protozoa contain one or more species commonly pathogenic to man, four others are pathogenic but only rarely or accidentally infect man, and six parasitize man without producing any harmful effect. This rather bold statement is over-simplified. The distinction between pathogenic and not pathogenic is not always sharp; many people are infected with *Entamoeba histolytica,* for example, apparently without being harmed by it, while others may suffer severe dysentery and even fatal complications. Some of these organisms parasitize man (and woman) only at one particular stage of their life cycle, while others may also inhabit mammals of other genera, so-called 'reservoir hosts'. Some are transmitted from man to man by vector hosts — usually insects — in which they undergo a part of their life cycle different from that which occurs in the human host. Some parasites do not undergo any development in their vector, utilising the latter merely as an animated hypodermic syringe, but such non-cyclical or mechanical transmission is rare among the parasites of man, it probably occurs to a significant degree only during epidemic spread of *Trypanosoma brucei rhodesiense* (section 4.2.1). Other parasites again have no vector, relying for transmission on their passive acquisition by the human (or other mammalian) host. The term 'parasite' itself is not sharply defined. Derived from the Greek words *para* (beside) and *sitos* (food), it need include no idea of harm to the host beyond the diversion of some of the latter's food. I shall use the word in this chapter in just this sense — 'an association between two animals of such a kind that one lives and feeds, temporarily or permanently, either in or on the body of the other' (Baker 1973), though others restrict it to more or less harmful associations. All the parasitic protozoa of man live in, not on, his body, i.e. they are endoparasites.

Detailed information on the morphology and life cycles of the parasites discussed below is given in various standard texts, including those by Levine (1973) for all groups, Garnham (1966) for malaria parasites, Hoare (1972) for trypanosomes and Mulligan (1970) for African trypanosomiasis only. An elementary introduction has been attempted by Baker (1973). A more clinical treatment is provided by Faust *et al.* (1970) and Wilcocks and Manson-Bahr (1972); the latter deal especially well with pathology and treatment, but some of their information on morphology and life cycles (Appendix 1) is rather unreliable. Information on diagnostic techniques and treatment must be sought in these or other books.

Table 18.1   Outline classification of pathogenic protozoan parasites of man*

Phylum Protozoa
   Subphylum Sarcomastigophora
      Superclass Mastigophora ('flagellates')
         Class Zoomastigophorea
            Order Kinetoplastida: **Leishmania, Trypanosoma**
            Order Retortamonadida: *Retortamonas, Chilomastix, Enteromonas*?
            Order Diplomonadida: **Giardia**
            Order Trichomondida: **Trichomonas,** *Dientamoeba*
      Superclass Sarcodina ('amoebae')
         Class Rhizopodea
            Subclass Lobosia
               Order Amoebida: **Entamoeba,** *Iodamoeba, Endolimax,* [**Hartmannella,**
                  Naegleria]
   Subphylum Sporozoa ('sporozoa')
      Class Coccidia
         Order Eucoccida
            Suborder Eimeriina ('coccidia'): **Isospora, Toxoplasma, Sarcocystis**
            Suborder Haemosporina ('malaria parasites'): **Plasmodium**
      Class Piroplasmea: [**Babesia**]
   Subphylum Microspora
      Class Microsporidea
         Order Microsporida: [**Encephalitozoon**]
   Subphylum Ciliophora ('ciliates')
      Class Ciliatea
         Subclass Holotrichia
            Order Trichostomatida: **Balantidium**

---

*Genera containing species pathogenic to man are shown in bold type, those in square brackets causing very rare, probably accidental infections only; other genera with species parasitizing man are in italics.

2 Classification

Detailed discussion of this topic is beyond the scope of this chapter, but in order to provide 'pegs' on which to hang the parasites concerned, and to indicate their interrelationships, I give in Table 18.1 an outline classification of the protozoa including only those which can parasitize man. Diagnostic characters and definitions are excluded, having been given elsewhere (e.g. Baker 1973, Baker 1977, Levine 1973). The possible evolution of the group has also been discussed elsewhere (Baker 1974, 1975).

3 Parasites acquired orally

The common pathogenic protozoa of man acquired by this route belong to the amoebae, flagellates, ciliates and coccidia; each will be briefly described and then their epidemiology and control will be discussed together. The microsporidan genus *Encephalitozoon* (sometimes synonymized with *Nosema*), a very rare and accidental parasite of man (only three definite, well-established cases being known; see Canning 1975) will not be considered further.

3.1 Life cycles and pathology

3.1.1 Amoebae

Entamoeba. Four species infect man, *E. histolytica, E. hartmanni* and *E. coli* in the large intestine and *E. gingivalis* in the mouth; only *E. histolytica* is sometimes pathogenic, causing amoebic dysentery (amoebiasis).

Entamoeba histolytica infection is probably of worldwide distribution, though the disease amoebiasis is commoner in tropical and subtropical areas. The picture is not clear-cut because of confusion between this parasite and its non-pathogenic, smaller but otherwise very similar relative, *E. hartmanni*. The infective stage is a spherical cyst (Fig. 18.1) containing four nuclei and, sometimes, cylindrical blunt-ended inclusions (chromatoid bodies) which are regular arrays of ribosomes. The cysts measure 9·5—15·5 μm in diameter. They resist digestion in the stomach but in the intestine the quadrinucleate amoeba emerges; its nuclei divide once and cytoplasmic cleavage produces eight uninucleate trophozoites which enter the colon and grow to the full size of about 10—20 μm (Fig. 18.2). The trophozoites multiply by binary fission, no sexual process occurring, and feed by phagocytosis on bacteria and small food particles present in the gut lumen. Some secrete a thin but tough cyst wall to form the transmissive cyst which is passed out in the faeces. The wall increases the cyst's resistance to drying (though it cannot withstand complete desiccation), so that it can survive for some time outside the host. As well as man, *E. histolytica* can infect other primates, dogs, cats, pigs and rodents, which may sometimes serve as reservoirs of human infection.

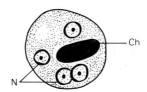

Fig. 18.1 *Entamoeba histolytica* cyst, × 1800. Ch, chromatoid body; N, nucleus. (From Baker, 1973.)

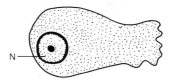

Fig. 18.2 *E. histolytica* trophozoite, × 1800. N, nucleus. (From Baker, 1973.)

Often the trophozoites live only in the lumen of the large intestine and do no significant harm. Occasionally, however, for unknown reasons, they penetrate the mucosa and multiply in the submucosa, forming a flask-shaped ulcer in which they may grow to about twice the size of those in the gut lumen. The spread of the ulcer results in much bleeding, and characteristic dysentery or 'bloody flux' results. Sometimes the ulcer perforates the gut wall and causes peritonitis; more commonly the amoebae are carried by the blood vessels to other viscera, particularly the liver, where they multiply and produce amoebic abscesses. Cysts are not formed in any tissue. Death may result from loss of blood and body fluid and from peritonitis.

Hartmannella and Naegleria. Within the last 15 years or so it has been recognised that these 'soil amoebae', normally free-living in soil or muddy water, may cause symptomless or mild nasopharyngeal infections of children and occasionally adults. Also, fortunately rarely, they may cause fatal meningoencephalitis as a result of infection via the nasal mucosa of persons swimming in infected water and subsequent migration up the olfactory tract to the brain; most if not all of the fatalities seem to be due to *Naegleria fowleri* (Carter 1972).

3.1.2 Flagellates

Only one species of orally-acquired flagellate is pathogenic to man, namely *Giardia lamblia* (= *G. intestinalis*). This parasite is distributed throughout the world; it is commoner in children and in them, especially (though in adults too), it may cause a diarrhoeic disease (with no blood in the faeces) called giardiasis or lambliasis. Infection occurs, as with *E. histolytica*, by swallowing a cyst of characteristic morphology (Fig. 18.3); it is oval, about $8-14 \times 6-10 \, \mu m$, and contains four nuclei and assorted other structures including flagellar remains. The cyst hatches in the duodenum and the emergent parasite divides into two binucleate trophozoites. These also have a very characteristic appearance, often likened to a small face (Fig. 18.4). The trophozoites are shaped like a pear cut in half longitudinally, with a sucking disc at the broad (anterior) end of the flat (ventral) surface. They are bilaterally symmetrical, having two nuclei, four pairs of flagella and, usually, two median bodies of unknown function; they are about $10-20 \, \mu m$ long, $5-10 \, \mu m$ broad and $2-4 \, \mu m$ thick. Reproduction is by asexual binary fission, no sexual process being known. Encystment occurs as with *E. histolytica.*

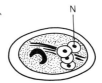

Fig. 18.3 *Giardia lamblia* cyst, × 1800. N, nucleus. (From Baker, 1973.)

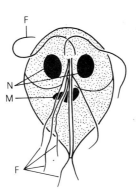

Fig. 18.4 *Giardia lamblia* trophozoite, × 1800. F, flagella; M, median body; N, nucleus. (From Baker, 1973).

As well as man, *G. lamblia* can infect monkeys and pigs. It lives in the lumen of the duodenum and upper ileum, attached to the mucosal surface by the anterior sucker. In large number the trophozoites are sufficiently irritant to produce acute diarrhoea, and also possibly interfere with absorption of fat and fat-soluble vitamins. Some workers believe that *G. lamblia* can invade cells, though the pathogenic significance of this — if true — is not yet clear.

3.1.3 Ciliates

The only ciliate parasitizing man is *Balantidium coli,* a parasite of worldwide distribution but rather rarely prevalent in man, which behaves in many ways like *E. histolytica.* It lives in the large intestine (caecum and colon), often harmlessly, but sometimes produces the disease known as balantidiosis or balantidial dystentery. Infection results from the swallowing of the resistant cyst (50–60 µm in diameter) (Fig. 18.5), which hatches in the large intestine to release a single oval trophozoite measuring 60–70 × 40–60 µm (Fig. 18.6). It is easily recognisable by its size and the fact that it is covered with cilia – short contractile hair-like locomotory processes (Latin *cilium* = eyelash). Like almost all ciliates, *B. coli* has two nuclei – a large polyploid macronucleus and a small micronucleus which functions only in sexual reproduction (races or strains lacking micronuclei occur and seem quite viable, though doubtless deprived of sex). *Balantidium coli* also has specialised feeding structures (a deep anterior groove or vestibule, lined with longer cilia, leading to a mouth or cytostome), a permanent excretory pore or 'anus' (cytopyge) and two contractile vacuoles to pump out excess water entering the parasite when it engulfs its food (mainly bacteria and particles of debris from the host's food).

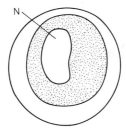

Fig. 18.5 *Balantidium coli* cyst (unstained × 600). N, nucleus. (From Baker, 1973.)

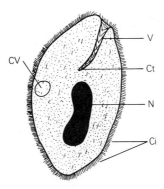

Fig. 18.6 *Balantidium coli* trophozoite (× 600). Ci, cilia; Ct, cytostome; CV, contractile vacuole; N, nucleus; V, vestibule. (From Baker, 1973.)

Reproduction is either asexual – by transverse binary fission – or sexual – by conjugation. The latter is a specialised process characteristic of ciliates which involves exchange and subsequent fusion of 'daughter' haploid micronuclei and macronuclear degeneration and reformation from one of the products of postzygotic micronuclear division: for more detail, see Mackinnon and Hawes (1961). Resistant cysts are formed and passed out in the faeces to ensure the onward transmission of the parasite. As well as man, *B. coli* parasitizes other primates and pigs. The pig is its major host, in which its prevalence ranges from 21 to 100 per cent; in man it is less than 1 per cent. It has been recorded rarely from dogs, rats, sheep and cattle. *Balantidium coli* is probably sometimes pathogenic to man and other primates only. Like *E. histolytica*, it usually lives harmlessly in the gut but sometimes, for reasons unknown, invades the submucosal tissues to form ulcers; this results in a bloody dysentery. Spread of the parasites to other viscera has not been recorded.

3.1.4 Intestinal coccidia

Commonly called 'coccidia' in spite of the taxonomic use of the word for higher-ranking groups of sporozoa (see Table 18.1), these protozoa are represented in man by three genera – *Isospora, Toxoplasma* and *Sarcocystis*. Recognition of the last two as members of this suborder occurred only within the last ten years or so (see references in Markus *et al.* 1974); before that their taxonomic position was uncertain. All three genera have a basically similar life cycle, which will be described first for *Isospora*, but *Toxoplasma* and *Sarcocystis* have some specialised additions to it which are of considerable epidemiological importance.

Isospora belli. This is the only species of *Isospora* which infects man. It has been recorded only rarely, but probably occurs throughout the world, producing a mild condition called iso-sporiasis or human coccidiosis. Infection follows the swallowing of the oocyst, a resistant body measuring on average 30 × 12 μm and containing two sporocysts each enclosing four sporo-zoites (Fig. 18.7). The oocyst hatches, presumably in the small intestine; studies on other species indicate the involvement of bile and trypsin in stimulating hatching. The emerged motile sporozoites enter gut epithelial cells and commence multiple asexual division (schizogony). The resultant merozoites re-enter other epithelial cells and either repeat the process or differentiate into sexual cells (gametocytes) and subsequently gametes. Fertilization results in an intra-cellular zygote which encysts to form the resistant oocyst; this ruptures its host cell and matures as it passes down the gut and out with the faeces and, when mature, constitutes the infective stage. Oocysts are very resistant to drying and chemical attack and can survive for a considerable time outside the host, particularly in cool moist micro-climates (up to 18 months; Davies *et al.* 1963). This complex life cycle is summarised in Fig. 18.8.

As far as is known *I. belli* has no host other than man. It is mildly harmful to its host, the main pathogenic effect probably being mechanical damage to the gut epithelium resulting from

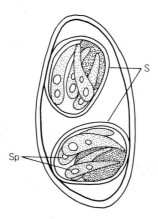

Fig. 18.7 *Isospora belli* oocyst, × 1800. S, sporocysts; Sp, sporozoites. (From Baker, 1973.)

the rupture of parasitized cells. Some abdominal pain and diarrhoea occur, but the disease in man is self-limiting after a few weeks and treatment is unnecessary.

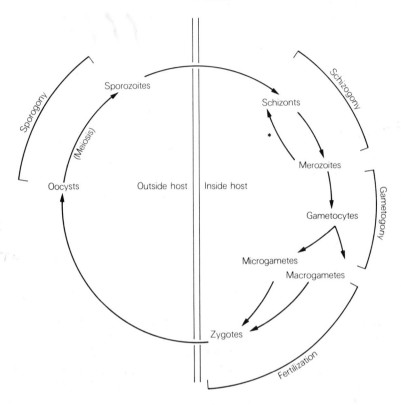

Fig. 18.8 Diagram of life cycle of intestinal coccidia. (From Baker, 1973.)

Toxoplasma gondii. The exciting and relatively recent recognition of a phase in this parasite's life cycle closely similar to that undergone by *Isospora* has been briefly reviewed in the *British Medical Journal* (Anon. 1970). *Toxoplasma gondii* is an ubiquitous parasite, producing a clinical condition ranging from mild to fatal known as toxoplasmosis. Inapparent infections must also occur, since one-quarter to one-third of the adult population of Britain and the USA (and even higher proportions elsewhere – e.g. 94 per cent in Guatemala) have antibodies to *Toxoplasma* in their serum.

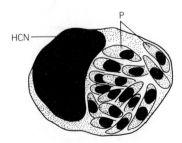

Fig. 18.9 *Toxoplasma gondii* pseudocyst, × 1800. HCN, host cell nucleus; P, parasites (endozoites). (From Baker, 1973.)

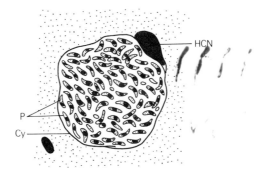

Fig. 18.10 *Toxoplasma gondii* cyst, × 1000. Cy, cyst wall; HNC, host cell nucleus; P, parasites (cystozoites). (From Baker, 1973.)

In addition to a typical isosporan life cycle in the intestinal epithelium as described above, *T. gondii* may spread throughout the body of man and a variety of other hosts; it then multiplies repeatedly in tissue cells (usually macrophages) to form intracellular aggregates of small (5 × 1–2 μm) organisms (endozoites) called pseudocysts (Fig. 18.9). In addition, true cysts bounded by a thin but tough cyst wall develop in some tissues, notably the brain (Fig. 18.10) and musculature; they contain similar small organisms called cystozoites. The isosporan phase results in the production in the faeces of cats of oocysts (Fig. 18.11) measuring about 12·5 × 10·5 μm; they may in the past have been confused with those of *Isospora bigemina*.

Fig. 18.11 *Toxoplasma gondii* oocyst. (Reproduced by permission from Dubey, J. P. *et al., Journal of Experimental Medicine* (1970), **132**, p. 641.)

Toxoplasma gondii has an unusually wide host range; it can accomplish its extra-intestinal development (the toxoplasmic phase) in most if not all warm-blooded vertebrates. The isosporan phase is apparently much more restricted, seemingly occurring only in some Felidae, including domestic cats – in which it has been most studied. The occurrence of cysts in the meat of a wide range of mammals, including rabbits, sheep and cattle, means that these may serve as reservoir hosts for carnivorous men; cats undoubtedly act as reservoirs of infection for both vegetarians and carnivores. The endozoites can cross the human placenta, so that the foetus of a mother with an acute infection may itself become infected; this produces the severest clinical form of toxoplasmosis, such foetuses often dying or surviving only with massive brain damage resulting from unchecked multiplication of the endozoites. Apart from this congenital condition the endozoites and pseudocysts may produce a generalised febrile illness with enlargement of the lymph glands, which, as stated, may range from inapparent to fatal. Chronic infection, presumably suppressed by the host's antibodies, with only cystozoites surviving, is symptomless and may last for months or years.

Sarcocystis. There have been fewer than a score of records of large cysts (200–300 μm long or more) in human muscle, which were named *Sarcocystis lindemanni*; as the infection is at most mildly pathogenic, it may however be commoner than this. Recently it has been shown that *Sarcocystis* spp. of various ungulates (sheep, cattle, pigs) are, like *Toxoplasma*, extra-intestinal stages of isosporan coccidia, and this is presumably true of *S. lindemanni*. It is also known that what used to be regarded as the second human species of *Isospora, I. hominis,* is the intestinal phase of bovine and porcine *Sarcocystis*. It has been suggested that the muscle cysts of *S. lindemanni* are a stage in the life cycle of one of these species which normally undergoes its sexual development in the human intestine, as a result perhaps of the rare development in man of oocysts ingested from faeces of the same, or another, person (such oocysts normally infecting the ungulate host). The *Sarcocystis* life cycle is summarised in Fig. 18.12, in which man would normally be cast in the role of 'predator' but, in the case of *S. lindemanni,* may also play the part of 'prey' (see review by Markus *et al.* 1974).

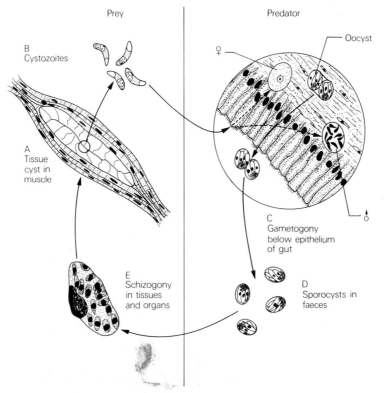

Fig. 18.12 Life cycle of *Sarcocystis fusiformis.* (Reproduced by permission from Markus, M. B. *et al., Journal of Tropical Medicine and Hygiene* (1974), 77, 251. Not drawn to scale.)

3.2 Epidemiology

The intestinal amoebae and flagellates have a simple epidemiology. Reservoir hosts, though probably occasionally a source of human infection with *E. histolytica,* are generally unimportant and the main source of infection is undoubtedly the ingestion of human faeces containing cysts. This usually results from contamination of food or water, by the use of human faeces as a fertiliser of crops, by uncontrolled defaecation near food plants or water supplies, or by unclean habits of food handlers who are themselves chronically infected. With *G. lamblia* infection in children, especially, faecal ingestion may result more directly from unclean habits during play, and a similar situation may occur in mental hospitals, orphanages, etc. in which there are often unusually high prevalences of intestinal parasites. The pig is an important source of human infection with *B. coli*; most human cases probably originate from it and hence the condition is commoner in rural populations. Professor F. E. G. Cox of London University recounts a report of balantidiosis in a small child whose habits included sitting in the backyard eating pig faeces from a wooden spoon — an example of unusually close reservoir-to-man contact!

The situation is more complex with the coccidia. Reservoir hosts are important — except, as far as is known, with *I. belli* — and, particularly with *Toxoplasma,* routes of infection other than oocyst ingestion may be significant. Probably the major source of human toxoplasmosis is ingestion of oocysts in cat faeces, either on uncooked vegetables or on food contaminated after cooking directly or *via* the unwashed hand of a cat-loving food handler. Fondling the domestic pet may also result in more direct contamination, and children's sand-pits are probably a fruitful source of infection. The cystozoites, protected by their cyst from the action of gastric juice, are also infective after ingestion and transmission can occur by eating infected raw or undercooked meat (cystozoites are thermolabile). Congenital transmission of *T. gondii* will be considered later (sections 5.2. and 5.3). The epidemiology of *Sarcocystis* is less well understood but may be basically similar, with the probable exception of congenital transmission (see section 3.1.4 and Fig. 18.12).

The 'soil amoebae' *Hartmannella* and *Naegleria* are rather distinct epidemiologically owing to their unusual mode of infection. Symptomless pharyngeal infections presumably follow oral introduction of cysts in soil, dust or water and possibly trophozoites in water too. The pathogenic infections however all seem to result from nasal introduction of, presumably, cysts or trophozoites in contaminated water, followed by their local multiplication and subsequent migration up the olfactory tract to the forebrain. Amoebae of these and other genera are common in mud and dirty water and have also been found in swimming pools.

3.3 Control

It should be clear from the previous section that prevention and control of the intestinal protozoal infections discussed is basically a matter of hygiene. If faecal contamination of food and water is avoided and if foodhandlers are not allowed to remain chronically infected and also follow elementary rules of hygiene in the preparation of food, then the incidence of infection would be markedly reduced. Adequate cooking of food destroys the infected stages of all the organisms concerned but 'rare' steak is not adequately cooked from this point of view. Thorough washing of raw vegetables can help to remove contaminating cysts but of itself is probably inadequate; the use of hypochlorite disinfectants at adequate concentration is effective, but the time-honoured tropical custom of washing salads with potassium permanganate is not, at least for *E. histolytica.* Filtration, chlorination or boiling of water of course prevents infection, though contamination can occur after treatment, a classical example being an epidemic of amoebiasis during the Chicago World Fair in 1933, which was traced to an hotel in which a careless plumber had mistakenly connected a sewer to a water pipe.

There is no doubt that pet cats constitute a risk of *Toxoplasma* infection, though since it seems that infected animals cease to pass oocysts in their faeces fairly soon as a result of developing immunity, this may be less serious than had been feared. The infection is rarely serious anyway, except when acquired congenitally (see section 5.3).

No effective prophylactic treatment is known for any of these intestinal protozoa, though all are fairly readily treated by a variety of drugs (see Wilcocks and Manson-Bahr 1972).

Prevention of infection with *Naegleria* and *Hartmannella* is relatively simple: avoidance of swimming in dirty fresh water and careful maintenance of swimming pools with adequate filtration and chlorination of the water. No treatment for the infection is known.

4 Parasites acquired via the blood

This group includes species of *Trypanosoma, Leishmania, Plasmodium* and *Babesia*. The last genus is normally restricted to rodents and ungulates (and to its tick vectors) but occasionally infects man — usually a person who has earlier been splenectomised or, perhaps, one whose immune system is impaired for other reasons. Only nine human infections are known and, because of its rarity, the conditions will not be discussed further; a good brief review has appeared in the *Lancet* (Anon. 1976).

4.1 Life cycles and pathology

4.1.1 *Trypanosoma brucei*

The vectors of this species are tsetse flies (*Glossina* spp.), large pupiparous insects restricted to tropical Africa, both sexes of which feed exclusively on blood. There are two main groups — those which breed and live extensively over vast areas of fairly dry 'bush' country, feeding mainly on wild ungulates, the so-called 'game tsetse', typified by *G. morsitans,* and others which inhabit only moister areas such as forested river banks, the 'riverine tsetse', exemplified by *G. palpalis.* Man often forms a larger though not exclusive part of the diet of the latter group (see Nash 1969 and references therein).

Tsetse flies transmit trypanosomes of several subgenera, of which only one (*Trypanozoon*) contains a species, *T. brucei*, which can infect man. *Trypanosoma brucei* consists of three morphologically indistinguishable subspecies, *T. b. brucei, T. b. rhodesiense* and *T. b. gambiense*; only the latter two infect man. *Trypanosoma brucei*, like its vectors, is restricted to tropical Africa, between about 20°N and 20°S. *T. b. rhodesiense* occurs mainly on the eastern side of the continent, *T. b. gambiense* mainly in western and central areas. The disease they produce in man is called African human trypanosomiasis or sleeping sickness.

When an infected fly bites a man or other suitable mammal, it injects into the blood or tissue fluid anti-coagulatory saliva containing trypomastigotes, about 14–18 µm long and 2 µm broad, called 'metacyclic' trypomastigotes as they occur at the end of the developmental cycle in the vector (Fig. 18.13). These elongate and divide, often initially restricted to the area around the bite where they may produce a swelling or chancre. After a few days they spread throughout the blood and tissue fluids. The first forms to develop in mammalian hosts are long, thin trypomastigotes (about 30 × 1·5 µm) which are essentially multiplication forms (Figs. 18.14 and 18.15). Their respiration does not involve the Krebs cycle or cytochrome chain, hence

Fig. 18.13 *Trypanosoma brucei brucei* metacyclic trypomastigote, × 2000. (From Baker, 1973.)

Fig. 18.14   *T. b. brucei* slender haematozoic trypomastigote, × 2000. (From Baker, 1973.)

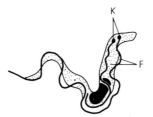

Fig. 18.15   *T. b. brucei* dividing slender haematozoic trypomastigote, × 2000. F, replicating flagella; K, replicating kinetoplasts. (From Baker, 1973.)

Fig. 18.16   *T. b. brucei* stumpy haematozoic trypomastigote, × 2000. (From Baker, 1973.)

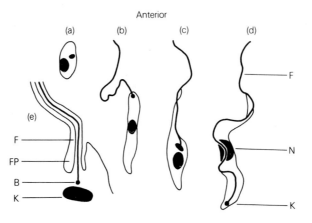

Fig. 18.17   Diagram of morphae and ultrastructure of Trypanosomatidae, (a)–(d) × 2000: (a) amastigote; (b) promastigote; (c) epimastigote; (d) trypomastigote; (e) ultrastructure of region of flagellar origin. B, basal body; F, flagellum; FP, flagellar pocket ('reservoir'); K, kinetoplast; N, nucleus. (From Baker, 1973.)

their mitochondrion is non-functional and structurally reduced; they consume large amounts of oxygen and glucose, which is oxidised in extramitochondrial terminal respiratory bodies to pyruvic acid and excreted. After a few days the long forms differentiate into short, stumpy trypomastigotes, about $18 \times 3 \cdot 5\ \mu m$ (Fig. 18.16); these do not divide in the mammal and continue to develop only if ingested by a tsetse. They have a partly functional mitochondrion. After ingestion by feeding tsetse, the stumpy forms may change into elongate procyclic trypomastigotes in the fly's midgut and enter the space between the peritrophic membrane (which lines the midgut like the inner tube of a tyre) and the gut wall, where they are safe from expulsion as the fly digests and excretes its meal. They move forwards to the fly's proventriculus and, eventually, salivary glands, where they change into epimastigotes (see Fig. 18.17 for illustrations of the meaning of this and related terms) and continue to divide. Finally they differentiate into the non-multiplicative metacyclic trypomastigotes, whose function is to continue development in the mammalian host; these already have a reduced and at least partially inactive mitochondrion, whereas all the other stages in the vector have a functional mitochondrial enzyme system and fully oxidise glucose to carbon dioxide and water (see Vickerman, in Fallis 1971). Only a small proportion of tsetse is susceptible to trypanosome infection; the parasites may fail to overcome two 'barriers' – one being the establishment outside the peritrophic membrane and the second being migration to the salivary glands. The latter, which is essential for re-infection of a mammal, occurs in no more than about 10 per cent of flies even under optimal experimental conditions and in fewer than 1 per cent of natural populations.

Characteristically, *T. brucei* subspecies have cyclically fluctuating populations in their mammalian hosts (Fig. 18.18). After a few days of infection, the host produces IgM antibody against the parasites and the majority are killed. Some, however, in a way not yet understood, change the structure of their surface antigen (to which the host responds) and thus survive. The host then produces more antibody against this second antigen; and again some parasites change their surface antigen. Two dozen or more variants have been recorded. Eventually the host, its IgM level enormously raised, usually succumbs to the unequal struggle – though some wild ungulates can support infections for years with little apparent harm. The relevant antigens are glycoproteins forming a cell coat or glycocalyx outside the parasite's limiting unit membrane. They occur on all forms in the mammal and on the metacyclic trypomastigotes but are missing from the other stages in the vector (see Vickerman, in Porter and Knight 1974).

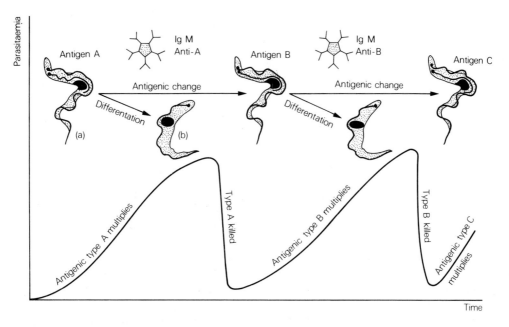

Fig. 18.18 Diagram illustrating antigenic variation of haematozoic trypomastigotes of *T. brucei* spp. Dividing slender trypomastigotes (a) either differentiate into stumpy forms (b) which are infective to the vector, or change their surface antigen, or are killed by host antibody.

Experimentally, *T. b. rhodesiense* readily infects most primates (but not baboons), rodents, carnivores and ungulates. It has been isolated from naturally infected cattle, *Tragelaphus scriptus* (bushbuck) and *Alcelaphus buselaphus* (hartebeest).

T. b. rhodesiense produces a much more acute disease than does *T. b. gambiense,* but the general clinical picture is similar in both diseases (Apted, in Mulligan 1970). Initially it is a generalised febrile illness. Especially with *T. b. gambiense* infection, the lymph glands, particularly those in the neck, become enlarged — 'Winterbottom's sign'. Within a few weeks, with *T. b. rhodesiense,* or several months, with *T. b. gambiense,* the parasites enter the cerebrospinal fluid and signs of central nervous system involvement occur. There is an immunosuppressive effect and death of untreated patients is often due to intercurrent infection such as pneumonia; it ensues within 6 to 9 months in acute infections or after several years in the chronic type. The two types of disease tend to intergrade. Occasionally apparent 'healthy carriers' may be seen.

4.1.2 *Trypanosoma cruzi*

The vectors of this species are insects of the family Reduviidae (cone-nosed or kissing bugs), a largely but not exclusively American family of Hemiptera, suborder Heteroptera. Both sexes and all instars (larval stages) feed on the blood of man and a wide range of animals including chickens, mainly at night. The so-called domestic species, which inhabit cracks, nooks and crevices in low-quality housing, are the most important vectors of trypanosomes to man. Species of *Rhodnius, Triatoma* and *Panstrongylus* transmit the pathogenic *Trypanosoma (Schizotrypanum) cruzi* and also the less common *T. (Herpetosoma) rangeli*; the latter, being non-pathogenic to its human hosts, will be ignored here (see Hoare 1972 for further information). *Trypanosoma cruzi* occurs in man only in South and Central America (40°S to 30°N); in mammals other than man it probably extends into the southern United States to about 42°N. It causes South American human trypanosomiasis or Chagas's disease (see Elliot *et al.* 1974).

Metacyclic trypomastigotes (17–22 μm long) develop in the hind gut of the bugs (Fig. 18.19). While feeding, or soon after, the bugs defaecate; if the bug is infected, metacyclic trypomastigotes are ejected with the faeces and enter the vertebrate's blood either via the wound made by the bug's proboscis or as a result of the host's scratching the bite. The metacyclic trypomastigotes enter macrophages, neuroglia and muscle fibres and transform into amastigotes (about 4 μm in diameter), which undergo repeated binary fission. After about a week they transform into trypomastigotes, about 20 μm long (Fig. 18.20), and re-enter the host's blood. They then either invade other tissue cells and recommence multiplication or — if ingested by a suitable feeding bug — continue their development in it. It is not known whether *T. cruzi* undergoes antigenic variation, nor whether, like *T. brucei* (section 4.1.1) it 'switches off' its mitochondrial respiration at certain phases of the life cycle (see review by Brener 1973). In the vector, trypomastigotes transform into epimastigotes and divide extracellularly in the midgut. After about 1–2 weeks, metacyclic trypomastigotes develop in the hind gut. Bugs, once infected, probably remain so for life; there is no evidence that they suffer from the infection, and probably all individuals and all instars of a suitable species are susceptible. Initially there is an acute febrile illness which may be fatal, particularly in young children. If it is not, after 3–4 weeks a lengthy chronic phase may supervene, with a few or no signs or symptoms apart perhaps from general malaise which is often overlooked in a generally undernourished population suffering from a plethora of other parasitic and infectious diseases. When, as often happens, cardiac muscle is infected, electrocardiography may be abnormal (Anselmi and Moleiro, in Elliott *et al.* 1974). If the neuromuscular control of oesophagus or colon is upset, as occurs commonly in certain regions (e.g. Brazil) but less often in others (e.g. Venezuela), gross dilation of these organs results in megaoesophagus and megacolon. Probably

Fig. 18.19 *Trypanosoma cruzi* metacyclic trypomastigote, × 2000. (From Baker, 1973.)

Fig. 18.20 *Trypanosome cruzi* haematozoic trypomastigote, × 2000. (From Baker, 1973.)

the commonest cause of death of chronic sufferers from Chagas's disease is cardiopathy (cardiomegaly and apical aneurism); 90 per cent of autopsies of persons infected with *T. cruzi* in Brazil showed this condition, 20 per cent had megacolon and 18 per cent had megaoesophagus (Köberle, in Elliott *et al.* 1974). These two latter conditions can themselves result in death from peritonitis or starvation, respectively. Although many infections may be more or less symptomless, there is no proven record of self-cure. The woman in whom, at the age of two, the disease was first recorded by Chagas in 1909, was alive and well, but still infected, some 50 years later.

The essential mechanism underlying the pathogenic effects of chronic *T. cruzi* infection, whether manifested in the heart, hollow viscera or elsewhere, appears to be selective destruction of ganglion cells (Köberle, in Elliott *et al.* 1974). This results in destruction of the muscle fibres in the organs concerned and hence interference with their vital functions. The precise mechanism of this denervation is not understood; it may result from toxic secretions of the parasites or toxic products liberated from ruptured host cells, or, it may be due to to allergic or autoimmune phenomena. Santos-Buch and Texeira (1974) showed that lymphocytes sensitised to *T. cruzi* antigen killed rabbit heart muscle cells *in vitro*, whether or not the cells were infected with *T. cruzi*; there seemed to be common antigenic determinants in both parasites and cells. Also, a circulating IgM which reacts with endocardium and other heart tissue constituents has been demonstrated in the sera of persons infected with *T. cruzi* (Cossio *et al.* 1974).

4.1.3 *Leishmania*

Several species of this genus infect man and all are transmitted by 'sandflies' of the genus *Phlebotomus* or its near relatives. These small insects breed in damp grass, refuse, soil, etc. often in burrows made by rodents or other animals and, in Africa, in termite hills; only the females feed on blood, so only they serve as vectors of the parasite. The species of *Leishmania* infective to man are distributed throughout all the tropical and subtropical parts of the world, excepting Australia, including the southern part of Europe. Collectively the conditions they cause are called leishmaniasis, of which there are two major types — cutaneous and visceral leishmaniasis; the former is known by a variety of descriptive local vernacular names, 'oriental sore' in the Mediterranean, Middle and Far Eastern regions, while the latter is often referred to by its Indian name 'kala-azar'.

When a sandfly feeds on blood, it injects saliva through its proboscis and with it, if the fly is infected with *Leishmania,* are carried promastigotes (Fig. 18.21) from the fly's buccal cavity

Fig. 18.21 *Leishmania* sp. promastigote, × 2000. (From Baker, 1973.)

and pharynx. These flagellates are about 10–20 μm long, but once in the vertebrate host they quickly enter macrophages, dedifferentiate their flagella and become rounded amastigotes, 2–4 μm in diameter (Fig. 18.22). The host cells, for reasons not yet fully understood, do not kill the invaders as they would normally destroy engulfed foreign organisms, but within them the parasites multiply by binary fission. A subsequently feeding sandfly may ingest parasitized macrophages from the skin or blood; the amastigotes emerge within its midgut, differentiate into promastigotes, multiply asexually by binary fission and migrate forwards to the insect's pharynx and buccal cavity, thus rendering it infective to subsequent mammals on which it feeds.

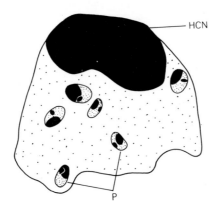

Fig. 18.22 *Leishmania* sp. amastigotes within host macrophage *in vivo,* × 2000. HCN, host cell nucleus; P, parasites (amastigotes). (From Baker, 1973.)

The pathogenesis of the disease obviously differs markedly between the two forms, visceral and cutaneous. In the former, infected macrophages collect, as the name implies, in the viscera – particularly spleen and liver – and in the bone marrow. Their proliferation, and that of the contained parasites, interferes with the vital functions of these organs and leads to a chronic debilitating disease with intermittent fever which, if untreated, ends fatally. In the cutaneous forms of the disease the parasitized macrophages are confined to the skin, apparently because the amastigotes cannot tolerate temperatures above about 35°C, and usually the infected cells are contained within one or a few discrete cutaneous lesions which are often self-limiting. However, in one form of the disease, 'espundia', caused by *L. braziliensis braziliensis* in Brazil and neighbouring areas, lesions may develop in the nasopharyngeal mucosa, with destructive results. Occasionally, in persons with an impaired immune system, other forms of the disease (notably those caused by *L. aethiopica* in Ethiopia and *L. braziliensis guyanensis* in Central and

Fig. 18.23 *Plasmodium* sp. sporozoites, × 2000. N, nucleus. (From Baker, 1973.)

Fig. 18.24 *Plasmodium falciparum* young trophozoite ('ring') within erythrocyte, × 2000. (From Baker, 1973.)

northern South America) the parasites may spread throughout all or most of the unfortunate patient's skin to produce a condition resembling leprosy (Bryceson 1970).

4.1.4 *Plasmodium*

Four species of *Plasmodium* infect man; all are transmitted by mosquitoes of the genus *Anopheles*. Mosquitoes breed in usually stagnant water and, like sandflies, only the females are vectors since only they feed on blood. *Plasmodium* species which infect man are distributed throughout the tropical and subtropical regions and extend into cooler areas even as far north as Holland and, in earlier times, England (see Table 18.2). All species cause the disease known as malaria, and all have a similar life cycle.

When an infected mosquito bites, it injects saliva containing small parasites called sporozoites (Fig. 18.23), about 15 × 1 μm, which penetrate liver parenchyma cells and commence asexual multiple fission (schizogony). After a week or so (Table 18.2), these exoerythrocytic schizonts mature and liberate thousands of small parasites (merozoites), some of which enter erythrocytes (Fig. 18.24). It used to be thought that, in all species of mammalian malaria except *P. falciparum,* other merozoites re-entered liver cells to maintain the infection cyclically and thus explain the long-term relapses characteristic of this disease. However, although persistent exoerythrocytic schizonts are a source of relapses, there is now doubt as to whether they represent the progeny of continuing cycles of schizogony or whether they are dormant parasites reactivated in some way (Garnham 1967). Of the merozoites entering erythrocytes, some continue asexual multiple fission (erythrocytic schizogony) while others differentiate into male or female sexual cells (gametocytes). The latter continue development only after ingestion by a mosquito of the susceptible species. They emerge from their host erythrocytes and the male produces eight motile microgametes while the female matures into a single macrogamete. The ensuing motile zygote, the ookinete, penetrates the mosquito's gut wall and encysts to form an oocyst. The contents undergo multiple fission to produce, one to two weeks later, large numbers of uninucleate sporozoites which, after rupture of the mature oocyst, congregate in the mosquito's salivary glands ready to infect the next susceptible mammal on which the insect

Table 18.2   Species of *Plasmodium* which infect man

Species	Geographical distribution	Erythrocytic schizogony		'Ring' form
		Duration (h)	No. of merozoites	
P. vivax	Worldwide, in tropical, subtropical and warmer temperate regions	48	12−24	At least one-third diameter of erythrocyte
P. malariae	Worldwide but scattered, mainly tropical and subtropical	72	6−12	At least one-third diameter of erythrocyte
P. ovale	Tropical Africa; also occasionally in other parts of tropics and subtropics (possibly of exogenous origin)	48	6−12 (12−24 in relapses)	At least one-third diameter of erythrocyte
P. falciparum	Worldwide, in tropics, subtropics and warmer temperate regions	48	8−24	Very small at first; 2 nuclei commonly; some apparently on edge of cell (accolé forms)

*These alterations do not develop until some growth of the parasite has occurred.

feeds. Meiosis occurs during the first post-zygotic nuclear division in the oocyst, the parasites being therefore haploid for most of their life cycle.

Species of *Plasmodium* generally have a restricted range of hosts; of those infecting man, only one, *P. malariae,* can naturally infect another species — the chimpanzee, an animal sufficiently rare to be of no importance as a reservoir of human infection. However, experimental manipulation, usually splenectomy, can reduce the resistance of other primates to the malaria parasites of man.

Only the intraerythrocytic stages of *Plasmodium* are pathogenic. The rupture of the red blood cells in which schizogony occurs may produce anaemia, but only *P. falciparum* can, on its own, kill an infected person. The fatal effect results mainly from damage to brain capillaries resulting from their mechanical blockage by parasitized erythrocytes and from damage caused to their endothelium by antigen—antibody reactions and the consequent release of kinins, histamine, etc. Fatal haemorrhages into the brain may result (cerebral malaria); more detail is given by Garnham (1966).

4.2 Epidemiology

The epidemiology of the parasites in this group depends on the differing habits of their various vectors, so it is not possible (as it was in section 3.2) to consider them all together; however the several epidemiologies will be compared wherever possible.

4.2.1 *Trypanosoma brucei*

Wild and domestic ungulates act as reservoir hosts of *T. b. rhodesiense* and this, coupled with the widespread distribution of the main vectors (*G. morsitans* group), means that infection with this subspecies is usually acquired away from villages, particularly by men such as hunters and

Main morphological characteristics of erythrocytic forms			
Trophozoite	Schizont diameter (μm)	Gametocytes	Host erythrocyte*
Amoeboid	10	Round or ovoid: male, 9 μm; female, 10–11 μm	Enlarged, stippled ('Schüffner's dots')
Compact, often bandlike	7	Round or ovoid, 7 μm	Not enlarged; very faint ('Ziemann's') stippling after prolonged staining only
Compact	7 (larger in relapses)	Round or ovoid, 9 μm	Slightly enlarged, stippled ('Schüffner's dots'); may be distorted and elongated
Compact; rarely seen in peripheral blood	5 (rarely seen in peripheral blood)	Crescentic: male, 9–11 μm long; female, 12–14 μm	Not enlarged; often with fewer larger dots ('Maurer's clefts')

honey-gatherers whose occupation takes them into the bush. Tourists in the game parks are another group at risk. Man–tsetse–man transmission undoubtedly occurs during epidemics, sometimes perhaps by non-cyclical transmission (see section 1). The disease in man is so acute that patients seldom remain ambulant and in contact with vectors long enough for this type of transmission to maintain the disease. It is, therefore, usually a zoonosis – a disease acquired by man from a wild animal reservoir (Fig. 18.25). Ungulates have a much more chronic infection.

T. b. gambiense, on the contrary, is much less infective to non-human mammals and rarely produces in them parasitaemia adequate to infect tsetse regularly. Thus it cannot be maintained zoonotically, and has essentially a man–tsetse–man cycle (Fig. 18.25). Hence its association with riverine tsetse, in the diet of which man plays a much larger part, and the fact that infection is usually acquired peridomestically – in or near villages, especially at watering places and fords where tsetse are abundant. Consequently, men, women and children are equally at risk, whereas *T. b. rhodesiense* is more commonly acquired by men. There is no good evidence of a mammalian reservoir host for *T. b. gambiense,* other than man. Although these broad statements are generally true, there are of course occasional exceptions (see Baker, in Elliott *et al.* 1974).

4.2.2 *Trypanosoma cruzi*

The human infection is essentially a disease of poverty; as mentioned above, the important vectors inhabit dilapidated and low-standard housing. Probably at least 7 million people in South America are infected (Hoare 1972). The parasite infects a wide range of mammals — opossums, armadilloes, rodents including *Rattus rattus*, carnivores including dogs and cats, and monkeys. Obviously domestic animals such as rats, dogs and cats may be important reservoirs of human infection, as are other infected persons. When humans, dogs, cats and bugs all live in the same one- or two-roomed house, as they often do, an ideal situation for focal infection is

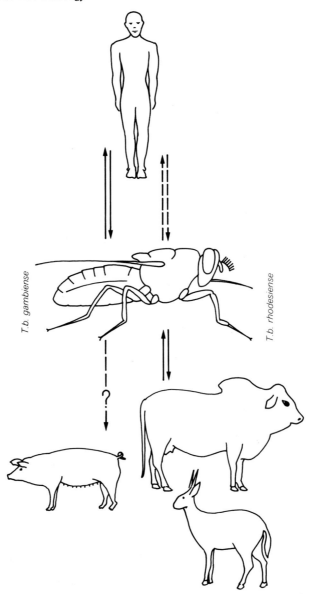

Fig. 18.25 Diagram illustrating epidemiology of *Trypanosoma brucei gambiense* (*left*) and *T. b. rhodesiense* (*right*). (Not to scale.)

set up, since Reduviidae are as catholic in their taste as are the trypanosomes – in fact more so, since they feed on chickens which are insusceptible to *T. cruzi*. The discovery of presumed *T. cruzi* in wild mammals and bugs in areas where human infection is virtually unknown (e.g. the southern United States) indicates that sylvatic cycles of transmission must be commonly proceeding in which man is unnecessary and from which he is normally excluded. Such cycles do however pose a threatening potential hazard. Study of the epizootiology of *T. cruzi* infection is hampered by the absence of any conclusive criterion for specific identification of the parasite other than its infectivity to man. There are known to be species of the subgenus (e.g. *T. vespertilionis* and *T. dionisii* of bats) which do not infect other mammals, including man, but which are morphologically indistinguishable from *T. cruzi*; there may be more which, although perhaps infective to laboratory rodents and hence at present identified as *T. cruzi* or '*T. cruzi*-like', are not in fact able to infect man (cf. *T. brucei brucei*, section 4.1.1). Further investigation is urgently needed.

The domesticity of the infection means that all age groups are at risk. In endemic areas infection is often acquired in early childhood, and provided the acute attack is survived (see section 4.1.2), persists into adulthood for the life of the patient. It is not uncommon to find an entire family, except perhaps the babe-in-arms, and their domestic 'pets', chronically infected with *T. cruzi*.

4.2.3 *Leishmania*

The epidemiology of the leishmaniases is largely determined by the habits of the reservoir host, if any, and vector. Basically both cutaneous and visceral leishmaniasis can exist in 'urban' or 'rural' forms. The urban epidemiology results from the presence of sandfly vectors in or near houses and either a domesticated animal reservoir or none. This pattern is exhibited by *L. infantum* around the Mediterranean, *L. donovani* in parts of China and central Asia, *L. tropica* in India and the Middle East (Iran etc.) and *L. peruviana* high in the Peruvian Andes; in all these situations the main reservoir host is the domestic dog and the vectors too are 'domestic' in their breeding habits. This close contact between man and vector, in areas of dense population, increases the likelihood of direct man–sandfly–man transmission. In India *L. donovani* is apparently maintained in this way, no reservoir other than man having been identified. The rural epidemiology is exemplified by *L. major* in parts of Soviet Asia, where the major reservoirs are gerbils (*Rhombomys* and others) and the vectors breed, not near houses, but often in the burrows of the gerbils themselves; man can thus be regarded as an accidental host, not essential to the cyclical maintenance of the parasite. Similar epidemiologies pertain to *L. aethiopica* in Ethiopia, with hyraxes as reservoirs and vectors often breeding in their burrows, and to the South and central American cutaneous leishmaniases caused by subspecies of *L. braziliensis* and *L. mexicana*. The reservoirs of the latter two are usually small forest-dwelling rodents, and the vectors also breed in the forest, so the disease is seldom acquired within towns or villages. South American visceral leishmaniasis (*L. chagasi*) has a dual epidemiology, both domestic dogs and wild foxes having been incriminated as reservoirs. Little is known of the reservoir hosts of *Leishmania* in tropical Africa (apart from Ethiopia): *L. donovani* and *L. tropica* are mainly rural in distribution, the latter probably having small wild rodents (*Arvicanthis*) as reservoirs as does the former in Sudan (*Rattus, Acomys, Arvicanthis*). In East Africa there is doubt as to whether *L. donovani* does or does not have a reservoir but, if it does, rodents are likely candidates. The sandfly vectors often breed in termite hills, in which various species of rodent may also dwell; such termitaria in or near villages may form local foci of infection and thus the disease can sometimes be urban in distribution. This subject is fully discussed by Lainson and Shaw (1972), who suggest that visceral leishmaniasis has evolved from a primitive rodent–sandfly–man cycle, through the dog–sandfly–man association, to a purely human infection (man–sandfly–man), as in India; see also Bray (in Elliot *et al.* 1974).

4.2.4 *Plasmodium*

With no significant reservoir host (see section 4.1.4), the epidemiology of *Plasmodium* species is relatively simple. Infection is transmitted from man to man by the female anopheline vectors, most of which breed in standing water in or near houses, compounds, etc. – sometimes even in water contained in old tin cans or discarded car tyres. Thus contact between man and vector may be very close. *Anopheles* species usually feed mainly in the evening, often inside houses, so that infection is often truly domestic and may occur as readily in large towns as it does in villages. The absence of reservoir hosts other than man means that the disease is unlikely to be acquired away from human habitations. In holoendemic areas, where transmission is maximal, infants are often infected very early in life, and unless they die from cerebral malaria due to *P. falciparum* (see section 4.1.4), may develop considerable resistance by the time they are adult. Antigenic variation of the parasites (cf. section 4.1.1) combines with their persistence in the liver (section 4.1.4) to prevent complete recovery, but chronically infected individuals may restrict the infection to such an extent that it seems to have little or no harmful effect; the immunological basis of this is discussed by Brown (in Porter and Knight 1974). Thus whole populations may have infection rates approaching 100 per cent, with constant reinfection providing the antigenic stimulus needed to maintain their resistance. If such people leave an endemic area, the absence of this stimulus may lead to a lowering of their immunological defences resulting in a subsequent acute attack of malaria.

4.3 Control

As in the previous section, each parasite in this group will be considered separately, since the diversity of their epidemiologies leads also to a diversity of potential control measures.

4.3.1 *Trypanosoma brucei*

Prevention of sleeping sickness involves avoidance of infective tsetse bites. Casual visitors to endemic or enzootic areas can use repellants (e.g. dimethylphthalate) or/and wear protective clothing, e.g. long-sleeved bush jackets, long trousers; tsetse can however easily pierce thin cloth. For those living or working in risky areas it is less easy. A better approach is to prevent man–fly contact by removing the tsetse, usually by bush clearing or insecticides (see Nash 1969). Although the low infection rate in tsetse makes any one bite statistically unlikely to be infective, repeated or constant exposure increases the chance of infection. Prophylactic drug treatment is seldom if ever used against *T. b. rhodesiense,* lest it merely suppress clinical signs until involvement of the central nervous system has made the disease much more difficult to treat. Salts of pentamidine [4,4′-(pentamethylenedioxy) dibenzamidine] , however, give fairly effective short-term protection (ca. 2 months). Pentamidine has been used much more in large-scale campaigns against *T. b. gambiense,* and was reasonably successful in interrupting transmission, although the same risk of suppression was present at the personal level. The dosage recommended is 4 mg kg^{-1} of the base (max. 300 mg) given intramuscularly, and protection lasts for about 6 months (see Williamson, and Waddy, in Mulligan 1970).

4.3.2 *Trypanosoma cruzi*

Prevention of infection involves avoidance of the faeces of infected insects. No prophylactic drug is available. Mere avoidance of the bite of bugs may not be enough; in heavily infested houses infective bug faeces may rain down upon the unsuspecting sleeper even if he is protected by a mosquito net. Improvement of rural housing would be the single most important preventive step which could be taken at present, coupled with improved standards of domestic hygiene to discourage bugs from living and breeding under beds, etc. Spraying of houses with insecticides is effective, if expensive, provided it is repeated sufficiently often to prevent reinfestation and provided insecticide resistance does not become a problem. Removal of infected domestic mammals and rats is also important. Since the bugs feed on chickens, which are insusceptible to *T. cruzi*, it has been suggested that the keeping of fowls in or near houses should be encouraged as a means of reducing both the number of potentially infective blood meals available to the vector and also the frequency with which the latter, once infected, feeds on man.

4.3.3 *Leishmania*

Prevention of infection with leishmaniasis means avoidance of the bites of infected sandflies; no prophylactic drug is available. At a personal level this can best be done by using insect repellants and insecticidal sprays if visits to areas of infection are unavoidable. In towns, attendance to 'municipal hygiene', augmented by insecticidal spraying of houses, can interrupt transmission, but such measures are not applicable to the control of rurally-acquired infection. Where dogs are important reservoirs, some control may be achieved by identifying and killing infected animals but this is costly and difficult. Where man is the major reservoir, mass diagnosis and treatment should reduce the prevalence of infection as well as being of value to the individual sufferers. In parts of Asia, people have for many years practised crude immunisation against cutaneous leishmaniasis by inoculating children, on a suitably concealed part of the body, with material from a lesion on some other person. As the infection does not spread and is self-limiting, and solid immunity follows recovery, this prevents subsequent disfiguring infection of a more cosmetically important area such as the face. 7

Treatment of the various types of leishmaniasis is usually based on drugs containing pentavalent antimony; it is uncertain and toxic. Single lesions caused by *L. tropica, L. major, L. aethiopica* or *L. mexicana* usually spontaneously regress within a year or so, and may not need treatment (see Wilcocks and Manson-Bahr 1972).

4.3.4 *Plasmodium*

Control of malaria means essentially control of the vector. Though treatment of infected persons can also reduce the reservoir of infection, it may not always be individually beneficial in endemic areas because, unless transmission is occurring at a very high rate, the treated individual may be deprived of antigenic stimulation long enough to lose his partial immunological resistance to the parasites (see section 4.2.4) with the consequent risk of an ensuing acute attack. Vector control can be achieved by insecticidal spraying of houses to kill adult mosquitoes, or, by draining breeding sites or treating them with suitable substances to kill larvae and pupae.

Personal protection involves the use of screening over windows, mosquito nets over beds, and drug prophylaxis. The last measure is generally most effective, in spite of occasional reports of drug-resistant strains of parasites. The most commonly used prophylactic drugs are proguanil, amodiaquine or chloroquine. These compounds, together with pyrimethamine, primaquine and certain sulphonamides — as well as, sometimes, the long-established natural alkaloid quinine — are also used in treatment (Peters 1970 and Wilcocks and Manson-Bahr 1972 give details).

5 Parasites acquired congenitally

There are rare records of presumed congenital infection with African trypanosomes, malaria parasites and leishmanias, but — although this may occur — some at least may represent infection of the baby during delivery *via* a superficial wound contaminated with maternal blood. The only two parasitic protozoa of man to which the placenta does not usually present an uncrossable obstacle are *Toxoplasma gondii* and *Trypanosoma cruzi*; transplacental infection is particularly common with the former.

5.1 Life cycles and pathology

These topics have been discussed in sections 3.1.4 and 4.1.2. Only special features related to congenital transmission will be mentioned here.

5.1.1 *Toxoplasma gondii*

Transplacental infection occurs only if the mother has an early acute infection during pregnancy; presumably the placenta is crossed by actively motile endozoites liberated from ruptured macrophages in the endometrium and maternal placental tissue. In the immunologically incompetent foetus, multiplication of the parasite apparently continues unchecked with disastrous results. If the foetus is infected early in pregnancy, severe brain damage may occur. Hydrocephaly is a common sequel, and many such foetuses abort or are still-born; or the infant may die soon after birth. If the foetus is infected later, it may be born with a severe generalised infection accompanied by jaundice. Sometimes, however, the infection may be chronic and inapparent in the new-born child, though examination with an ophthalmoscope may reveal a characteristic choroidoretinitis.

5.1.2 *Trypanosoma cruzi*

Little is known of the mechanism of transplacental infection. Haematozoic and metacyclic trypomastigotes of *T. cruzi* are suspected of being able to penetrate undamaged mucous membranes, and it may be that the former can also cross intact placentae. However, it is possible that minor damage is needed, and this could result from endometritis produced by the mother's infection. There seems to be no special behaviour of the parasite once it enters the foetus; the chronicity of the infection means that foetuses may be born infected with no obvious sign of ill-health. Abortion may result, however — perhaps more as a result of the concomitant endometritis than the foetal infection itself (see Zeledón, in Elliott *et al.* 1974).

5.2 Epidemiology

As in section 5.1, only aspects related to transplacental infection will be discussed here; the general topics have been dealt with in sections 3.2 and 4.2.2.

5.2.1 *Toxoplasma gondii*

A detailed study of the epidemiology of congenital toxoplasmosis was made by Desmonts *et al.* (1965), who surveyed 14 828 pregnant Parisiennes, of whom 2 424 (16 per cent) were initially negative, i.e. had presumably never been infected. Subsequent tests on this group allowed the authors to calculate an incidence of infection of about 0·8 per cent during the nine months of pregnancy. Twenty-four offspring of mothers who became infected during pregnancy were studied: 2 died neonatally but were not autopsied, so the cause of death was unproven but may well have been toxoplasmosis; 10 were definitely diagnosed serologically but only one showed any other clinical sign — choroidoretinitis; the remaining 12 were uninfected. No baby born to a mother initially seropositive, i.e. chronically infected, showed any sign of infection. Thus, though the chances of a woman becoming infected during pregnancy were below 1 per cent, if she did so the chance of her foetus becoming infected was about 50 per cent.

5.2.2 *Trypanosoma cruzi*

Estimates of the frequency of transplacental transmission of *T. cruzi* are usually about 1–2 per cent of unselected pregnancies in endemic areas; among mothers who are themselves infected, the incidence rises to about 10 per cent. Even if one child is born infected, subsequent ones may not be (Zeledón, in Elliot *et al.* 1974).

5.3 Control

There is little that can be said under this heading. With *T. gondii,* the risk being only to foetuses of mothers who acquire the infection during pregnancy, it would seem reasonable for a serological test to be done as soon after conception as possible and, if the mother was then seronegative, to advise her to take special care in handling and cooking meat, and, as far as possible, to avoid exposure to cats. The faeces of any pet cat in the household could also be examined for oocysts.

   As there is no treatment for *T. cruzi,* nothing can be done to avoid the risk of foetal infection other than the general precautions mentioned in section 4.3.2.

6 Parasite acquired venereally — *Trichomonas vaginalis*

Trichomonas vaginalis is the only venereally transmitted protozoan parasite of man (and woman); it occurs in all parts of the world, with infection rates up to 40 per cent.

6.1 Life cycle and pathology

Trichomonas vaginalis has a simple life cycle. It inhabits the male urethra and female vagina and is directly transmitted during sexual intercourse. There is no cyst and no sexual reproduction of the parasite; it merely multiplies by asexual binary fission. The organism measures up to $30 \times 15\,\mu m$, but is usually about $15 \times 10\,\mu m$ (Fig. 18.26). Infection in males is usually symptomless but in females the irritant effect of large numbers may, but not always, produce severe vaginitis, often associated with a raising of vaginal pH from about 4·5 to 5·5.

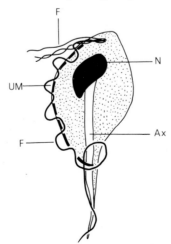

Fig. 18.26 *Trichomonas* sp. trophozoite, × 2000. Ax, axostyle; F, flagella; N, nucleus; UM, undulating membrane. (From Baker, 1973.)

6.2 Epidemiology

This parasite's epidemiology is equally simple and obvious. Infection is commoner among sexually promiscuous people.

6.3 Control

This too is obvious. It is useless to treat only one member of a stable sexual partnership, since both will almost certainly be infected and the untreated person will reinfect the other. No prophylactic drug is known, but the infection usually responds to metronidazole.

7 Conclusions

Enough, perhaps too much, has been said to indicate the variety of epidemiologies associated with pathogenic protozoal infections of man. Many aspects of the subject have been inadequately studied. Generally, the complexity of the epidemiology increases with the complexity of the parasites' life cycles and the number of hosts involved. This is not surprising since, with a simple cycle and only one species of host, e.g. *T. vaginalis* (section 6.2), the variables are limited. As soon as a vector is introduced, its behaviour too becomes important and the possibilities increase, e.g. *Plasmodium* (section 4.2.4). When a third, 'reservoir' host appears the situation becomes more complex still since all four variables — parasite, human host, vector and reservoir — interact, e.g. *Leishmania, Trypanosoma* (sections 4.2.1 to 4.2.3). Such 'webs of causation' (MacMahon and Pugh 1970) are often inadequately understood.

The elucidation of epidemiologies is important in the designing of adequate control methods. Drug prophylaxis, even if possible, is unsatisfactory as it may produce known or unknown side-effects; parasites may, and do, become resistant to the compounds used (see Peters 1970), and it requires constant vigilance. The only certain method to control parasitic protozoa is to interrupt the life cycle at some point, and it is the job of the epidemiologist to identify the most sensitive and efficacious point at which this can be done.

References

ANONYMOUS (1970) 'Transmission of toxoplasmosis', *Brit. Med. J.,* **1**, 126–7.
ANONYMOUS (1976) 'Human babesiosis', *Lancet,* **1**, 1001–2.
BAKER, J. R. (1965) 'The evolution of parasitic Protozoa', *Symp. Brit. Soc. Parasitol.,* **3**, 1–27.
BAKER, J. R. (1973) *Parasitic Protozoa* (2nd edn.). Hutchinson, London.
BAKER, J. R. (1974) 'The evolutionary origin and speciation of the genus *Trypanosoma'*, *Symp. Soc. Gen. Microbiol.,* **24**, 243–66.
BAKER, J. R. (1977) 'Systematics of Parasitic Protozoa', in *Parasitic Protozoa* (Ed. Kreier, J. P.), Vol. 1, pp. 35–56. Academic Press, New York.
BRAY, R. A., ASHFORD, R. W. and BRAY, M. A. (1973) 'The parasite causing cutaneous leishmaniasis in Ethiopia', *Trans. Roy. Soc. Trop. Med. Hyg.,* **67**, 345–8.
BRENER, Z. (1973) 'Biology of *Trypanosoma cruzi'*, *Ann. Rev. Microbiol.,* **27**, 347–82.
BRYCESON, A. D. M. (1970) 'Diffuse cutaneous leishmaniasis in Ethiopia, III, IV', *Trans. Roy. Soc. Trop. Med. Hyg.,* **64**, 380–93.

CANNING, E. U. (1974) 'The medical importance of Microsporida', *Folia Parasitol.* (Prague), **22**, 10.

CARTER, R. F. (1972) 'Primary amoebic meningoencephalitis', *Trans. Roy. Soc. Trop. Med. Hyg.,* **66**, 193–208.

DAVIES, S. F. M., JOYNER, L. P. and KENDALL, S. B. (1963) *Coccidiosis.* Oliver and Boyd, Edinburgh.

DESMONTS, G., COUVREUR, J. and BEN RACHID, M.-S. (1965) 'Le toxoplasme, la mère et l'enfant', *Arch. Fr. Pédiat.,* **22**, 1183–200.

ELLIOTT, K., O'CONNOR, M. and WOLSTENHOLME, G. E. W. (Eds.) (1974) *Trypanosomiasis and Leishmaniasis with Special Reference to Chagas' Disease.* Ciba Foundation Symposium 20, Elsevier-Excerpta Medica-North Holland Associated Scientific Publishers, Amsterdam, London and New York.

FALLIS, A. M. (Ed.) (1971) *Ecology and Physiology of Parasites.* University of Toronto Press, Toronto.

FAUST, E. C., RUSSELL, P. F. and JUNG, R. C. (1970) *Craig and Faust's Clinical Parasitology* (8th edn.). Lea and Febiger, Philadelphia.

GARNHAM, P. C. C. (1966) *Malaria Parasites and other Haemosporidia.* Blackwell Scientific Publications, Oxford.

GARNHAM, P. C. C. (1967) 'Relapses and latency in malaria', *Protozoology (J. Helminthol.* supplement), **2**, 55–64.

HOARE, C. A. (1972) *The Trypanosomes of Mammals.* Blackwell Scientific Publications, Oxford.

LAINSON, R. and SHAW, J. J. (1972) 'Leishmaniasis of the New World: taxonomic problems', *Brit. Med. Bull.,* **28**, 44–8.

LEVINE, N. D. (1973) *Protozoan Parasites of Domestic Animals and of Man.* (2nd edn.). Burgess Publishing Co., Minneapolis.

MACKINNON, D. L. and HAWES, R. S. J. (1961) *An Introduction to the Study of Protozoa.* Oxford University Press, Oxford.

MACMAHON, B. and PUGH, T. F. (1970) *Epidemiology: Principles and Methods,* Little Brown, Boston. (Cited by Bray, in Elliott *et al.,* 1974, pp. 87–100.)

MARKUS, M. B., KILLICK-KENDRICK, R. and GARNHAM, P. C. C. (1974) 'The coccidial nature and life-cycle of *Sarcocystis', J. Trop. Med. Hyg.,* **77**, 248–59.

MULLIGAN, H. A. (Ed.) (1970) *The African Trypanosomiases.* George Allen and Unwin, London.

NASH, T. A. M. (1969) *Africa's Bane: the Tsetse Fly.* Collins, London.

PETERS, W. (1970) *Chemotherapy and Drug Resistance in Malaria.* Academic Press, London and New York.

PORTER, R. and KNIGHT, J. (Eds.) (1974) *Parasites in the Immunized Host: Mechanisms of Survival.* Ciba Foundation Symposium 24, Elsevier-Excerpta Medica-North Holland Associated Scientific Publishers; Amsterdam, London and New York.

SANTOS-BUCH, C. A. and TEIXEIRA, A. R. L. (1974) 'The immunology of experimental Chagas' disease. III. Rejection of allogeneic heart cells *in vitro', J. Exp. Med.,* **140**, 38–53.

WILCOCKS, C. and MANSON-BAHR, P. E. C. (1972) *Manson's Tropical Diseases* (17th edn.). Baillière Tindall, London.

19

Microbial control of pest insects

J. R. Norris
Agricultural Research Council, Meat Research Institute, Bristol

1 Introduction

The idea of using one living species to control another is by no means a recent one, but it is only during the last 20 years that our understanding of the factors influencing the production and use of microbiological control agents has enabled microbial insecticides to emerge as significant pest control agents. We know today that insects are susceptible to a wide variety of infectious agents. Several hundred insect pathogens including bacteria, viruses, fungi, protozoa, rickettsia and nematodes have been described and many of them have been tested for their ability to control field infestations. Concern over the possible toxicological side effects of many of the chemical insecticides currently used has led to a rapid development of interest in microbial control agents in recent years and many laboratories around the world are at present engaged in field-testing a wide variety of agents. It is not my intention in this chapter to attempt a survey of the whole field; rather, I shall concentrate on the small number of agents that have achieved success or near success as control agents in a commercial sense. My object is to extract principles from the work which is now available and to attempt to illustrate the factors which I believe are important for the future development of effective microbiological control agents. For an excellent treatment of the whole subject I would refer the reader to the comprehensive book by Burges and Hussey (1971).

One of the problems to be overcome when using a disease-causing organism to control a field pest is the production of the infective agent in a form in which it will remain viable during storage and following application in the field for long enough to ensure its effective distribution and use. It is not surprising therefore that most success has so far attended the use of microorganisms which have resistant, or dormant, phases in their life cycles; bacterial or fungal spores or viruses enclosed in protective proteinaceous polyhedra. These at present constitute the most successful microbiological control agents and the majority of what follows will concern two species of aerobic spore-forming bacteria, *Bacillus popilliae* and *Bacillus thuringiensis* and the polyhedrosis viruses.

1.1 A definition

Biological control is usually defined in terms of the use of one living species to control another. Microbiological control would therefore be concerned with the control of a pest (macro) organism by an infective microorganism but such a definition is oversimple since it fails to distinguish between the multiplication of a microorganism as such in the body of a susceptible host and the use of toxins elaborated by a growing microorganism for control purposes. *Bacillus popilliae* and viruses control their hosts by virtue of their ability to multiply extensively in their infected bodies but the efficacy of *B. thuringiensis* is concerned very largely with a toxin synthesised during the growth of the bacterium, the multiplication of the microorganism in the body of the target insect playing a secondary, and often minor, role. With *B. popilliae* and the viruses the object of field use is to set up a spreading infection in a pest population and in some cases infection may persist from season to season and even from year to year. The dissemination of *B. thuringiensis* in the field, however, is normally not followed by a persisting infection. The mixture of spores and toxins of this bacterium, which constitutes the insecticidal preparation, is used in the field in exactly the same way that a chemical insecticide would be used and may call for repeated application during the growing season in order to protect a crop.

Clearly it would be nonsense to consider as a biological control agent any toxic product which is produced by a living organism; such an approach would imply that an antibiotic was a biological control agent. For our present purposes, therefore, we must fall back on a definition of convenience and I shall consider as a microbiological control agent any living microorganism used either alone or together with its toxic products to control another species where the introduction of living material into the host population is an essential feature of the control process. This definition involves no consideration of the spread or persistence of infection in a natural community and is perhaps justified because it brings together a group of microbial agents which exhibit underlying scientific similarities and which present similar problems to the field-worker and particularly to government regulatory authorities.

1.2 Economics

A major factor which complicates the development of a microbiological control agent is the confused economics which are almost inevitably attendant on the exercise. By comparison the economics of a chemical pesticide are straightforward; the product will be sold at a price which produces an adequate return on investment and manufacturing costs for the producer, and the venture will be successful if its use leads to an increased crop value such as to enable the user at least to recover the costs of its application and preferably show a profit. The continued production of such an agent will depend upon the establishment of substantial and repeated sales of the pesticide. Similar considerations apply to the marketing of *B. thuringiensis* preparations and to those viruses that persist only for a short time in an insect population in the field, but they cannot be applied to persisting infectious agents.

When the application of a pesticide is followed by its persistence, possibly for years, in the target insect population and its vigorous spread beyond the territorial boundaries of the user, the situation is unlikely to be attractive either to the manufacturer or to the individual farmer and we inevitably find ourselves dealing with governmental or state contracts and integrated pest control programmes. Microbial insecticides are often expensive to produce but have the advantage of very high pathogenicity. The present effectiveness and future potential of microbiological insecticides depends on a subtle interaction between cost/effectiveness as compared with established chemical agents, safety in the sense of absence of toxic affects on man and other animal and plant species and efficacy in the field against the target insect. Selectivity of action is a feature of many insect pathogens and it is probably fair to comment that the present growing concern about the use of less specific chemical control agents will be an important factor in ensuring the continued, and indeed increasing, use of microbiological agents in the future. Nevertheless, crop damage is a serious problem for the farmer and the fact that many microbiological agents do not have an immediately visible 'knockdown' effect on the host insect must mean that chemical insecticides will often be preferred.

The effective use of microbial control agents in many cases requires a thorough understanding of the life cycle and ecology of an insect pest and careful monitoring of the time and frequency of application of the insecticide. Such sophisticated control procedures are possible only for the large agency working over substantial areas and the development of such programmes is necessarily slow. All of these factors combine to suggest reasons why microbial insecticides have been slow to grow to their present status and why their future development, although steady and persistent, is unlikely to be spectacular.

2 History

Agostino Bassi was a curious and ingenious naturalist – one of those intriguing persons that sometimes emerge from the early history of a science. He was born in Italy in 1773 and attended the University of Pavia in order to study law, apparently to satisfy his parents' wishes. But the University of Pavia was at that time vibrating with energy and enthusiasm and we soon find Bassi studying chemistry, physics, mathematics, natural history and parts of medicine for his own interests. Spallanzani was working in Pavia at the time on the problems of spontaneous generation and this work must profoundly have influenced Bassi and before long he was drawn into a study of a disease of silkworms called muscardine, or calcino, which was causing serious losses to the Italian silk industry. Bassi was not the first person to describe disease in insects; indeed, Aristotle had done that with a masterly treatise on the diseases of the honey-bee published in his *Historia Animalium.* Although Bassi was too early to take part in the momentous development of bacteriology which was to occur in the nineteenth century (it was for Pasteur to describe in detail the infectious diseases of the silkwork caused by bacteria) he was the first person to demonstrate that an insect disease could be caused by a microorganism; indeed he was the first person ever to demonstrate that diseases of any kind could arise from infection with a microscopic life-form, thereby establishing the doctrine of microbial parasitism. In 1834 Bassi showed experimentally that the fungus *Beauveria bassiana* was responsible for muscardine and he suggested that the microorganism might be used for control purposes, thus demonstrating a capacity for perception and foresight which he rapidly applied to an analysis of human diseases, advancing ideas which were later to be shown accurate by the work of Pasteur, Koch, Lister and other bacteriologists of the late nineteenth and early twentieth centuries. Today spores of Bassi's fungus are produced by fermentation and are the subject of extensive field trials, and indeed some applications, as microbiological control agents for a range of pest insects.

The silkworm played a major role in Bassi's work. It is an excellent experimental animal and has continued to play a major part in the development of insect pathology and microbiological control. About 2 000 years ago the silkworm *Bombyx mori* was introduced from China into France and became the basis of a valuable industry which reached its zenith in the first half of the nineteenth century. Then disaster struck the industry in the form of a rapidly spreading disease, pébrine, which decimated the silk farms and in 1849 Pasteur was asked to examine this devastating disease. He showed that pébrine was caused by an infectious agent which he could see under the microscope and he taught silk farmers how to avoid it by careful attention to hygiene and management.

Whilst studying pébrine Pasteur came across another disease of the silkworm, flacherie.

Pébrine was characterised by the appearance of brown or black spots on the surface of the caterpillar, flacherie was quite different; the disease spread rapidly through an insect population, caterpillars died and rapidly liquefied, and a healthy colony could be reduced overnight to a mass of blackened cadavers. It was this disease, flacherie, that struck the Japanese silk industry towards the end of the nineteenth century, epidemics sweeping through the silk farms and bringing ruin to their owners. A Japanese bacteriologist, Ishiwata, isolated from diseased caterpillars an aerobic spore-forming bacterium which he called *Bacillus sotto* and which he showed to be the cause of the disease. Japanese bacteriologists made what was to be seen later as a most significant observation; old sporulated cultures of *B. sotto* were intensely infective for healthy silkworms but young cultures harvested before spores formed could be fed to the caterpillars in relatively enormous quantities without ill effects. They concluded that it was the spores of *B. sotto* that were pathogenic for silkworms not the vegetative cells.

The isolation of *B. sotto* marks the start of the development of microbiological control of insects but progress was at first slow. The Japanese isolate attracted little attention and the bacterium was deposited in a Japanese culture collection and to all intents and purposes forgotten. In 1915 a German microbiologist, Berliner, isolated a similar bacterium from diseased caterpillars of the flour moth *Anagasta kühniella,* a worldwide pest of stored grain and flour and a similar isolation was made in 1927 by another German, Mattes, working with the same insect. Both of these bacteriologists made an important observation; as the vegetative cells grew and matured they produced, in addition to the typical refractile oval spores, a second body about the same size as the spore which they referred to as the *Restekörper* and which came to lie together with the spore inside the wall of the sporangium, subsequently to be liberated when this wall broke down. The importance of this observation, which was later seen to be crucial for the development of our understanding of the pathogenicity of the organism, was not appreciated but the availability of Mattes' isolate, which he called *B. thuringiensis,* and its wide dissemination amongst research laboratories stimulated interest in the possibility of using the bacterium for control purposes.

The idea of using a bacterium which was highly toxic for caterpillars, and which produced resistant spores which could be stored for long periods as dried powders without loss of viability, was attractive to workers interested in controlling pest insects and the 1930s saw fairly extensive field trials of spore dusts based on Mattes' bacterium against a serious pest of growing corn, the European corn borer. These were largely successful and a commercial product called Sporiene was produced in France just prior to the Second World War. It was a dust formulation containing large numbers of spores of *B. thuringiensis* suspended in an inert carrier.

The firm which manufactured Sporiene was largely destroyed during the war but the ready availability of the material for experimental work and an interesting study of the use of it for controlling the flour moth in stored flour carried out by Jacobs at Imperial College, London, during the war focused attention on *B. thuringiensis* as a potential control agent and set the scene for the rapid development of the subject which was to take place immediately after the war.

3 *Bacillus thuringiensis*

3.1 The basis of pathogenicity

Bacillus thuringiensis grows well on nutrient media and Steinhaus, working in the University of California, found little difficulty in producing spore crops and using them in a series of field trials against the Alfalfa caterpillar, a serious pest of cabbage crops in the western United States. These observations made in the late 40s and early 50s stimulated several manufacturing concerns in the United States to take up the challenge of producing *B. thuringiensis* on a semi-commercial scale. These attempts were based on the concept of the spore as the important pathogenic agent and serious difficulties were soon encountered; the product proved to be difficult to standardise and early field trials were marred by excessive batch-to-batch variation in potency. Light was thrown on this extensive variability in 1953 when a Canadian, Hannay, applied the recently developed electron microscope to a study of spore formation in *B. sotto* and immediately rediscovered the *Restekörper* (or parasporal body, as he called it), showing it to be a minute diamond-shaped crystal of pure protein. Another Canadian, Angus, in an elegant series of experiments separated the parasporal crystals from the spores in aged cultures of *B. sotto* and showed that pathogenicity for the silkworm was a function of the crystal alone; spores freed from parasporal bodies were incapable of causing disease on feeding to silkworms but the animals rapidly succumbed to minute doses of the toxic crystals.

3.2 The parasporal body (δ-endotoxin)

With the quickening interest in *B. thuringiensis* more strains were isolated and characterised. At present we recognise 12 groups on a basis of their flagella (H) antigens and all of them produce protein crystals. Indeed, the ability to produce the protein crystal is the main diagnostic feature of these organisms which in other respects are closely similar to the ubiquitous *Bacillus cereus*. Although the crystals vary from strain to strain in shape, size and antigenic composition, they appear to play similar roles in pathogenicity.

The end product of the growth of *B. thuringiensis* on culture media is a mixture of spores and crystals released by the breakdown of the residual wall of the sporangium. The crystal is soluble only with difficulty, dissolving under alkaline reducing conditions such as are found in the mid-gut of susceptible caterpillars. Solutions of the protein are highly toxic when fed to caterpillars such as the silkworm, causing gut paralysis followed by general body paralysis often in as little as 60 min. Mixtures of spores and crystals are pathogenic for a wide range of lepidopterous larvae and for a few sawflies but are without effect for other kinds of insect and are apparently non-toxic to other life forms. On ingestion the protein crystal dissolves in the mid-gut fluid and the protein is digested by the complex of proteolytic enzymes present there. Attempts to characterise the fragments released by this digestion have not been entirely successful since the smaller molecular weight fragments, which are probably important for toxicity, prove to be unstable under laboratory conditions.

The basic subunit of the protein crystal has been studied by electron microscopy and by X-ray diffraction techniques and appears to be a rod-shaped or dumb-bell structure of average length 12 nm and width 5 nm (Fig. 19.1) (Norris 1969 and 1971). Crystal synthesis and spore formation proceed at the same time in *B. thuringiensis* cells and the two processes appear to be intimately associated with one another. Asporogenous mutants of *B. thuringiensis* usually fail to produce crystals but this is not invariably the case and mutants blocked at a late stage in sporulation may still be capable of producing apparently normal crystals. Acrystalliferous mutants producing fully mature spores are much more readily isolated.

The protein of the crystal is synthesised from amino acids that result from the breakdown of cellular protein during the massive protein turnover occurring in the cell early in sporulation. Antigens typical of the protein crystal can be detected in the cell at an early stage in spore formation coincident with the appearance of the forespore membrane and the exosporium and there is some evidence to suggest that the crystal structure begins to develop on the exosporial

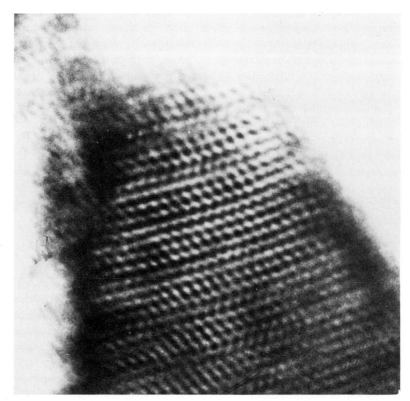

Fig. 19.1 Electron micrograph of a negatively stained parasporal body of *Bacillus thuringiensis* showing the surface arrangement of protein subunits (× 450 000). For details of preparation and other illustrations see Norris (1969 and 1971).

membrane of the developing spore itself, although this is by no means certain (Fig. 19.2). Using the presence of common antigens and other structural similarities, it has been suggested that there is a close generic relationship between the protein of the crystal and that of the protein coat of the spore. The suggestion is made that *B. thuringiensis* represents a *B. cereus*-like organism in which the mechanisms controlling the synthesis of spore coat protein have become distorted leading to an overproduction of this component. Although the circumstantial evidence is perhaps impressive, there are serious difficulties arising from contamination of one component with antigens derived from another and the evidence in support of the hypothesis is by no means complete at the present time (Bechtel and Bulla 1976).

The parasporal body is highly toxic when tested against caterpillars. Some preparations of toxic protein give LD_{50} (that is quantities killing 50 per cent of test larvae) values as low as 0·9 μg per g of larvae, a value which is comparable with some of the more widely used chemical insecticides. Larvae of some 130 lepidopterous species, including many important pests, are susceptible to spore/crystal mixtures of *B. thuringiensis*. The symptoms following ingestion of a toxic dose vary greatly from one species to another but profound changes in the permeability of the gut wall appear to be a universal early feature of pathogenicity and are attributable to the toxic products of protein digestion. In some species such initial toxicity may be followed by extensive germination of ingested spores and multiplication of vegetative cells in the body of the caterpillar resulting finally in a further round of spore and crystal synthesis and a blackened cadaver packed with spores and crystals. Often the potency of a spore/crystal mixture in the

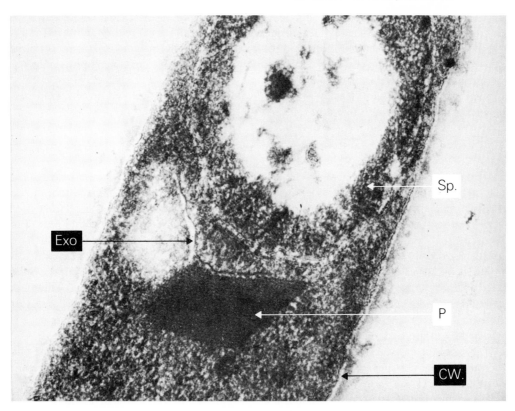

Fig. 19.2 Electron micrograph of a section through a sporulating cell of *Bacillus thuringiensis* showing the parasporal body growing in close association with the developing exosporium (× 128 000). P, parasporal body; Sp, developing spore; CW, wall of sporangium; Exo, exosporium. (Reproduced by kind permission of Dr H. J. Somerville of Shell Research Ltd.)

field is dependent on, or at least enhanced by, the presence of both components but with some species this is not so, the spore apparently playing a minor role in determining the pathogenicity of a product.

Paralysis of the gut with consequent cessation of feeding is an early feature of intoxication in most species. This is a commercially important factor since foliage damage by a leaf-eating pest will cease shortly after the insecticide is applied even though the insects may not die for several days.

As commercial development of *B. thuringiensis* insecticides led to the study of different isolates, it was soon recognised that a spore/crystal mixture derived from a particular strain could show a wide range of toxicity for different insect species and that different strains of *B. thuringiensis* showed widely different abilities to cause disease when tested against a single species of caterpillar. The reasons for these very substantial differences in performance are still largely obscure in spite of very close study but they are, of course, critical for the selection of commercially valuable products for use against a range of crop pests.

Careful control and manipulation of fermentation conditions can have profound effects on the amount of toxic protein produced by a culture and this can be further enhanced by mutation and selection of *B. thuringiensis* strains for increased toxicity. The application of such techniques following the careful selection of isolates of *B. thuringiensis* by screening against a range of test insects has led to the production of a current range of insecticides whose activity is far greater than that of earlier products. The search for better strains goes on, however, and it seems certain that as our ability to control protein synthesis by growing microorganisms increases and as our understanding of the underlying mechanisms controlling protein crystal production expands, we shall be able to produce cultures of *B. thuringiensis* with increased inherent toxicity and with species activity spectra more closely tailored to the crop pest complexes they are intended to control.

3.3 Thermostable toxin (θ-exotoxin)

The parasporal protein is the toxin which is important for the use of *B. thuringiensis* against pest species. It is active against a wide range of lepidopterous caterpillars but is without toxic effect on most other insects and forms of life. The parasporal body is present in all isolates of *B. thuringiensis* and, being protein in nature, its toxicity is destroyed by moderate heat. The discovery that autoclaved culture filtrates of some strains of *B. thuringiensis* could have insecticidal activity drew attention to a second toxin, the so-called thermostable or θ-exotoxin produced by some, but not all, strains of *B. thuringiensis*. This toxin is active against a wide range of insect species including lepidoptera, diptera, hymenoptera, coleoptera and orthoptera as well as showing marked toxicity for other animals. The effects of feeding small doses of exotoxin to larvae are only seen at moulting or during metamorphosis, when its presence is indicated by the appearance of physical deformities in the emerging insect.

The exotoxin has a low molecular weight (800–900) and is a nucleotide containing one phosphate group per adenine molecule. In addition, an unusual sugar, allomucic acid, is present in the molecule (Fig. 19.3).

The exotoxin appears to act by inhibiting the synthesis of RNA and it is a highly toxic compound quite unsuitable for use as an insecticide. Indeed, its presence in batches of *B. thuringiensis* would be sufficient to prevent their use on the market and the exclusion of exotoxin production is an important part of the selection of strains of the bacterium suitable for insecticide manufacture.

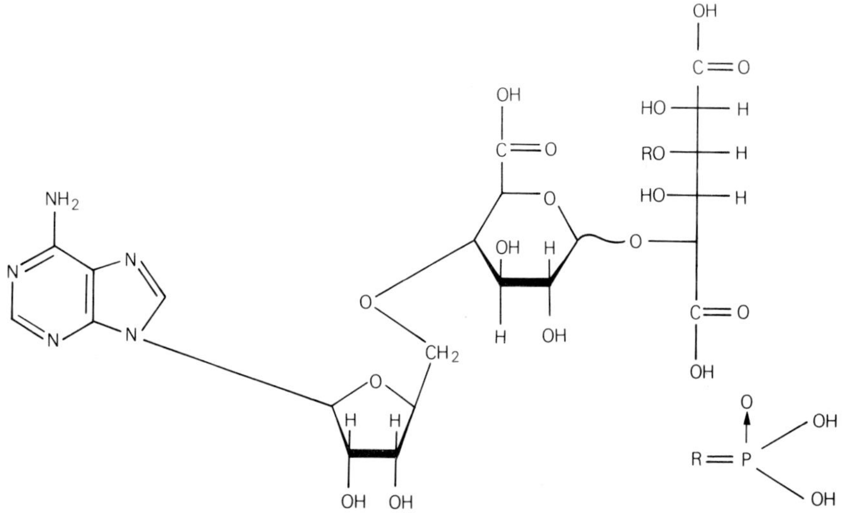

Fig. 19.3 Structural formula of *Bacillus thuringiensis* θ —exotoxin (Farkas *et al.*, 1969.)

3.4 *Bacillus thuringiensis* in the field

Commercial products containing *B. thuringiensis* have been produced by at least 12 manufacturers in 5 different countries in recent years. In the USA hundreds of tons are manufactured each year and production seems to be rising. Commercial preparations of *B. thuringiensis* are currently registered in the USA for use on more than 20 agricultural crops in addition to trees, ornamental shrubs and forests for the control of at least 23 insect pest species (Table 19.1).

Insecticides based on *B. thuringiensis* are used for short-term control. They neither persist from year to year nor do they spread extensively in the insect population. Since the insect must ingest a toxic dose of the organism and it is important to limit the destruction of foliage, efficient coverage of a leaf surface is an important feature of crop spraying. Equally important is the protection of the sprayed toxin from inactivation by sunlight since it shares with other proteins a degree of ultraviolet sensitivity. The use of emulsifying agents, stickers and ultra-violet protectants constitutes an important feature of *B. thuringiensis* insecticide formulation and can do much to enhance the value of a product.

Current uses of *B. thuringiensis* based insecticides are chiefly in the areas of forestry and on green crops where the absence of toxicity for man commends them for application right up to

Table 19.1   Some registered uses of *Bacillus thuringiensis* in the USA
(From Falcon 1971)

Pest species	*Crop*
Vegetable and field crops	
Alfalfa caterpillar (*Colias eurytheme*)	Alfalfa
Artichoke plume moth *(Platyptilia carcuidactyla)*	Artichokes
Bollworm (*Heliothis zea*)	Cotton
Cabbage looper (*Trichoplusia ni*)	Beans, broccoli, cabbage, cauliflower, celery, collards, cotton, cucumber, kale, lettuce, melons, potatoes, spinach, tobacco
Diamondback moth (*Plutella maculipennis*)	Cabbage
European corn-borer (*Ostrinia nubilalis*)	Sweet corn
Imported cabbageworm (*Pieris rapae*)	Broccoli, cabbage, cauliflower, collards, kale
Tobacco budworm (*Heliothis virescens*)	Tobacco
Tobacco hornworm (*Manduca sexta*)	Tobacco
Tomato hornworm (*M. quinquemaculata*)	Tomatoes
Fruit crops	
Fruit-tree leaf roller (*Archips argyrospilus*)	Oranges
Orange dog (*Papilio cresphontes*)	Oranges
Grape-leaf folder *(Desmia funeralis)*	Grapes
Forests, shade trees, ornamentals	
California oakworm (*Phryganidia californica*)	
Fall webworm (*Hyphantria cunea*)	
Fall cankerworm (*Alsophila pometaria*)	
Great Basin tent caterpillar (*Malacosoma fragile*)	
Gipsy moth (*Porthetria dispar*)	
Linden looper (*Erannis tiliaria*)	
Salt-marsh caterpillar (*Estigmene acrea*)	
Spring cankerworm (*Paleacrita vernata*)	
Winter moth (*Operophtera brumata*)	

the time of harvest on such crops as lettuce, tomatoes, and tobacco where toxic residue problems limit the use of other chemical insecticides.

The field-trial results presented in Table 19.2 are typical of many and serve to emphasise the effectiveness of a *B. thuringiensis* insecticide, Dipel, against various pest species, and also the considerable potential for combining the microbial insecticide with small amounts of chemical agents, a field which is rapidly being explored at the present time. Dipel is one of the new generation of insecticides based on a potent isolate of *B. thuringiensis*, strain HD-1, recovered from a diseased caterpillar by Dr. Howard Dulmage of the US Department of Agriculture in 1968.

Early *B. thuringiensis* products were standardised in terms of viable spore counts which we now know to be largely irrelevant. With the discovery of the importance of the endotoxin, bioassay has replaced spore count as the standardisation procedure. The standard is a particular preparation of strain E-61 held by the Pasteur Institute and is defined as containing 1 000 International Units per mg. Dipel is calibrated for field use at 16 000 IU mg^{-1}, some older commercial preparations have been found to contain as little as 800–1 000 IU mg^{-1}. The possible secondary importance of viable spores is recognised by setting a minimum viable spore count of $2 \cdot 5 \times 10^{10}$ viable spores g^{-1}.

4 *Bacillus popilliae/Bacillus lentimorbus*

The Japanese beetle (*Popillia japonica*) was introduced into the USA in about 1915 and soon spread to become a serious problem in the eastern seaboard and into Canada. The various larval instars of the beetle feed actively on the roots of grasses and other plants during the late summer months and cause extensive damage to lawns, pastures and shrubberies. Pupation takes place in the soil during May and adult beetles, emerging about the second week in June, live for some 40 days during which time the female will lay up to 60 eggs. Following its introduction and the recognition that a serious pest situation existed the US Department of Agriculture set up a Japanese Beetle Control Laboratory in 1929 with the object of limiting and possibly eradicating the pest. Before long workers from this laboratory found diseased larvae in field populations. These were characterised by a milky white appearance as a result of the large numbers of spore-forming bacterial rods appearing in the haemocoele and rendering the body fluid opaque. The milky diseases are caused principally by two closely similar bacteria, *Bacillus popilliae* and *Bacillus lentimorbus*. Both of these have been developed for microbial control of the pest but it is *B. popilliae* which has been studied the more extensively since it has shown most potential in the field.

Larvae ingest viable spores of the bacterium in nature and these germinate in the gut, the resulting vegetative cells penetrating the wall to multiply in the body cavity. Infected larvae usually live for a considerable time and extensive growth of the microorganisms occurs in the haemolymph. The pathogenicity of the bacterium is far from understood. Each sporangium contains, in addition to the spore, a second refractile body known as the parasporal body. Unlike the similar, though apparently unrelated, structure in *B. thuringiensis* there is no evidence that the parasporal body of *B. popilliae* plays any significant role in pathogenicity. Much of the effect of the bacterium on the host may be attributed to the removal from the blood of nutrients and essential growth factors which become locked away in the growing bacterial cells. Toxins are also involved since cell-free filtrates of cultures are lethal when injected in small amounts into larvae.

Table 19.2   *Bacillus thuringiensis* field trials.
(Reproduced by kind permission of Abbott Laboratories of North Chicago, Illinois, USA)

Crop	Location	Treatment	Dose	Insect counts at peak activity		Damage estimate or yield
				Cabbage moth	*Cabbage butterfly*	*Yield of marketable cabbages*
Cabbage var.	Queensland	Dipel	50 g/100 l	3	3	27
Early Jersey	University	DDT	50 g/100 l	2	40	3
Wakefield		Endrin	50 g/100 l	5	7	39
		Diazinon	0·05%	0	2	13
		Control	—	17	102	0
				Cabbage moth		*Formation of heads (%)*
Cauliflower	Angel Vale,	Dipel	25 g/100 l	96		63
	SA	Dipel	37·5 g/100 l	50		65
		Thuricide HP	50 g/100 l	159		40
		Methomyl	25 g/100 l	118		42
		Carbaryl	125 g/100 l	323		28
		Control	—	2 400		12
				Heliothis sp.		*Saleable fruit (%)*
Tomatoes var.	Peats Ridge,	Dipel	25 g/100 l	3		90
Grosse lisse	NSW	Dipel +	12·5 g/100 l +			
		Chlordimeform	0·01%	3		93
		Carbaryl	125 g/100 l	10		96
		Control	—	35		54
				Heliothis sp.		*Boll count*
Cotton var.	Wee Waa,	Dipel	0·25 kg/hc	6·7		94·6
Deltapine	NSW	Dipel +	0·125 kg/hc +			
		Chlordimeform	0·01%	7·0		105·9
		Chlordimeform	0·02%	7·2		98·0
		DDT	25% 4·5 l/hc	11·7		93·8
		Control	—	10·6		80·8
				Heliothis sp.		*No. damaged cobs*
Sweet corn	Bathurst,	Dipel	50 g/100 l	20		28
	NSW	Dipel	25 g/100 l	18		22
		Dipel +	25 g/100 l +			
		Chlordimeform	0·02%	12		18
		Control	—	18		46
						0 = no damage
						90 = total damage

4.1 Production of the pathogen

Shortly after the discovery of the milky disease organisms, it was shown that the presence of viable spores was essential for infection of larvae in the field. Unfortunately, unlike *B. thuringiensis,* the milky disease bacteria do not readily sporulate when grown on nutrient media away from the body of their host. Attempts to produce spores by cultivation of the micro-organisms *in vitro* have been largely unsuccessful and such success as has been achieved has been under conditions which preclude the possibility of extensive spore production on a commercial scale. Material for use in disseminating the disease in the field has been produced by the infection of larvae collected from the field using a method originally developed by the USDA Bureau of Entomology in 1939. Spores are preserved in the dried state and used to infect healthy larvae which are then stored until the disease is well developed. The moribund larvae are then disintegrated and the spores harvested to be formulated with a powder carrier as a dust containing 10^6 spores per gram for field use. Such a manufacturing process is slow, tedious and expensive and would be quite unsuitable for the production of the relatively vast quantities of material needed for an agent like *B. thuringiensis*. That it is in practice an effective process which has played an important role in the control of Japanese beetle is due to the fact that, unlike *B. thuringiensis*, *B. popilliae*, once introduced into a native population of the insect, gives rise to a persistent and spreading infection. Relatively small amounts of spore-containing material are therefore required to initiate infection and the product has been marketed in the USA under the trade name 'Doom' for many years.

4.2 Use in the field

The spore-containing powder is applied to soil in 2 g amounts at intervals of a few feet and the disease spreads in the insect population by various means, including the wandering of infected larvae, dispersal by water and wind, passive transfer by birds and animals and by human activity. Such a treatment programme is followed by spread of the disease throughout the Japanese beetle population in the treated area over a period of three seasons, leading to the reduction of larval numbers and the effective control of the pest.

The use of milky disease bacteria has been an important part of an integrated programme by the US Department of Agriculture which has resulted in a spectacular reduction in the numbers of beetles over the treated area so that it is no longer a commercially serious pest over much of its original range and complete eradication from the USA is seen by some authorities as a distinct possibility.

The milky disease bacteria are in general a promising group of pathogens whose use extends far beyond the control of the Japanese beetle. Several similar bacteria have been isolated from related larvae in other parts of the world. Members of the scarab beetle family are found in restricted areas of the world but show a marked ability to colonise fresh territories when accidentally introduced. The spread of such larvae in areas of Australia has caused concern in recent years but again milky diseases have been located in native populations of larvae and their use in controlling infestation is giving promising results. The growing awareness of the potential of this group of bacteria for controlling important pests is leading to the isolation and characterisation of a variety of new bacteria and is serving to emphasise that *B. popilliae* and *B. lentimorbus* are only representatives of a much larger and surprisingly diverse group of microorganisms. The interested reader is referred to an article by Falcon (1971) for a more detailed discussion of this subject and a key to the relevant literature.

5 Viruses

Production difficulties are also the major problem confounding the use of viruses as insecticidal agents. Nevertheless, the use of viruses for microbial control represents the most rapidly expanding aspect of the subject at the present time and is the area where we are likely to see the most significant developments in the next decade or so. Over 300 insect viruses are known to exist and many of these cause rapidly spreading epizootics amongst insect populations in nature. Thirty years ago the Cabbage White butterfly, *Pieris brassicae*, was probably the commonest of all butterflies in the UK. Then in 1955 it all but vanished. For years it was impossible to locate significant populations for experimental purposes in the field and it is only in the last few years that it has begun to reappear in significant numbers although it is still well below its former level. The reason for this drop in numbers was an explosive outbreak of granulosis virus disease which decimated the population and continued to control numbers for many years. The regular cyclic changes in population numbers of the Gipsy moth in the forests of southern Europe are largely attributable to the rapid spread of a virus disease when population numbers exceed a threshold level. When the European spruce sawfly was introduced into Canada during the 1930s it found itself free from natural enemies and spread rapidly with devastating effect on the spruce forests. The chance introduction of a nuclear polyhedrosis virus (NPV) from Europe checked the sawfly population and the species has gradually declined until it is now no longer a serious pest. In 1949 the Canadians imported another sawfly virus from Europe and this has been so effective that it has generally replaced chemical control for that particular insect.

5.1 Insect viruses

Many insect viruses are unlike plant and animal viruses in that the individual virus particles are encased, either singly or in large numbers, in protein crystals which are insoluble in water (Fig. 19.4). These crystals, which are produced in large numbers in the infected larvae and released when the cadaver disintegrates, protect the virus particles and ensure that they remain infective even if stored for years outside living tissues. It is the existence of these protective protein

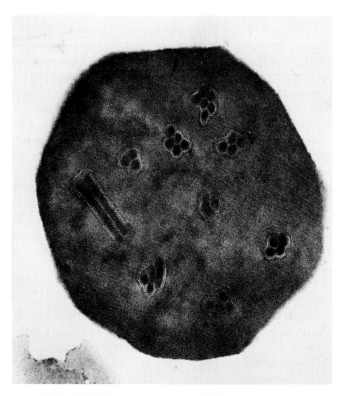

Fig. 19.4 Electron micrograph of a section through a polyhedral body of the nuclear polyhedrosis virus of the Cabbage worm (*Mamestra brassicae*) showing viral particles embedded in the protein matrix (X 76 000). (Reproduced by kind permission of Dr R. Gardner of Shell Research Ltd.)

crystals, or polyhedra as they are often called, that gives the viruses their remarkable potential as insect control agents. Aqueous suspensions of viral polyhedra or dry powder formulations remain viable under normal storage conditions for many years.

There are two main groups of insect viruses which concern us here; the polyhedroses in which many hundreds of virus particles are contained in each polyhedral crystal and the granuloses in which each crystal contains only one virus particle. Both types have some potential as insecticides since they spread rapidly in insect populations. In practice, most attention has been paid to the nuclear polyhedrosis viruses which develop within the nuclei of host cells following ingestion and affect the skin, fat body and haemolymph of larvae. Infected tissues become packed with the characteristic polyhedral bodies and millions of these polyhedra are liberated when the insect dies and its skin ruptures. Characteristically the liquefying body contents of a disintegrating cadaver leak from the ruptured skin as a highly infective fluid which is readily dispersed amongst the population by the effects of wind, rain and by the positive attraction it appears to exercise for healthy insects.

5.2 The development of viral insecticides

Since the end of the Second World War, the development of viral insecticides has fallen into two distinct phases, the transition from the first into the second occurring in the early 1960s with the development of artificial diets on which insect larvae could be reared in large numbers for experimental and production purposes.

Viruses, of course, will grow only in the infected cells of susceptible hosts and early attempts to produce viral control agents were based on the collection of diseased insects from the field, the harvesting of viral bodies from them and the introduction of the resulting preparations into healthy field populations. Such a process is necessarily expensive and inefficient but, nevertheless, substantial progress was made in some applications particularly in forestry where the actual amount of virus which must be disseminated is quite small. The NPV of the sawfly, for instance, was capable of controlling the insect following the application of virus harvested from only 15 diseased larvae to an acre of young pine trees. The virus can be administered using a low volume spraying machine or by spot treatment of individual trees. In both cases a vigorous and persisting virus epizootic follows application.

The second method of production used in the early stages of development was to rear larvae under laboratory conditions and purposely to infect them with the disease-causing agent. Most interest, naturally, attaches to leaf-eating caterpillars and attempts to rear large numbers involve the collection of substantial amounts of foliage of particular plant species, a process which is again costly and time-consuming and which carries with it the danger of introducing non-specific disease to the laboratory insect population.

In the early 1960s, however, workers in the US Department of Agriculture produced artificial rearing diets for a wide range of insect pests and it was this important development that ushered in the present vigorous phase in the exploitation of viruses for control purposes. It is now possible to rear in quantity sufficient susceptible larvae to justify commercial production of the large amounts of virus required for use against crop pest insects and the last few years have seen products of this kind registered for use in the USA and field-tested on a substantial scale in various parts of the world.

Logically the final stage in the development of production techniques would be the growth of viruses in tissue cultures derived from appropriate host insects. For many years insect tissue culture lagged behind the culture of mammalian cells but recent years have seen the establish-

(a)

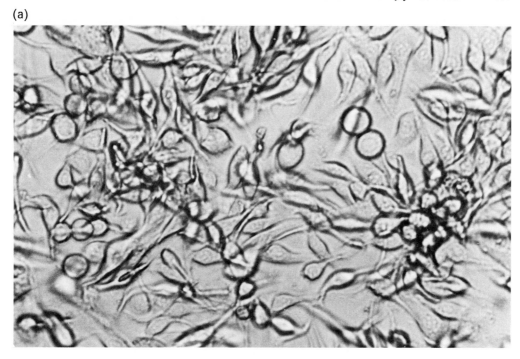

(b)

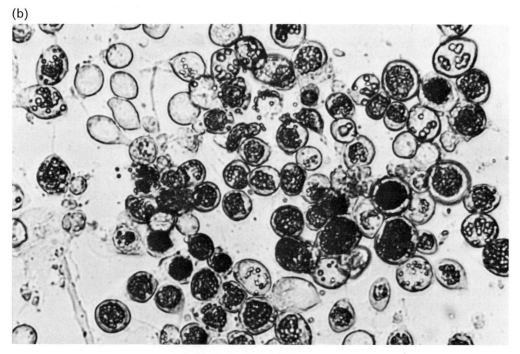

Fig. 19.5 Low-power light micrographs of tissue cultures of cells of the Cabbage looper (*Trichoplusia ni*). (a) uninfected cells. (b) 60 h after infection with a nuclear polyhedrosis virus of the Alfalfa looper (*Autographa californica*) showing development of intracellular polyhedral bodies. (Reproduced by kind permission of Dr R. Gardner of Shell Research Ltd.)

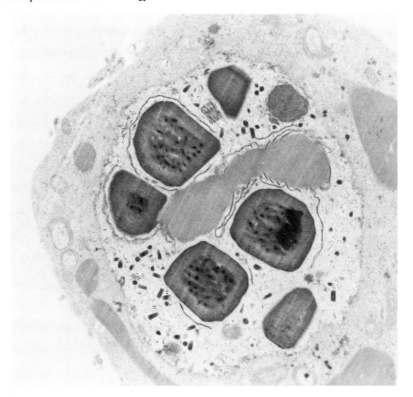

Fig. 19.6 Electron micrograph of a section through tissue culture cells of the Cabbage looper (*Trichoplusia ni*) infected with a nuclear polyhedrosis virus of the Alfalfa looper (*Autographa californica*) showing polyhedral bodies containing virus particles and free-lying virus (X 9000). (Reproduced by kind permission of Dr R. Gardner of Shell Research Ltd.)

ment of a number of stable insect tissue culture lines and a simplification of the complex growth media originally required to a point where the growth of viruses in insect tissue cultures on a large laboratory scale is an established possibility (Figs. 19.5 and 19.6). Commercial-scale tissue culture presents many technical and biological problems but the attractions of a tissue culture method for the production of viruses are substantial when compared with the existing processes using whole larvae. Several teams are now working actively to develop commercial processes and it seems likely that the future will see insect viruses produced in this way. One of the big attractions of this approach concerns the specificity of viruses; most insect viruses are highly species specific, growing only in one, or a small number of closely related, insect species. At the tissue culture level specificity is not nearly so marked and it proves possible to grow several different viruses in a tissue culture derived from one species. In nature, insect pests only rarely occur in monoculture and viral insecticides would be used as part of an integrated control programme involving control of a number of different species; the ability to make cocktail mixtures of viruses appropriate for a particular pest complex is important and would be much facilitated by the ability to produce different viruses in one production unit growing one tissue cell line.

At the present time one viral preparation (Viron H) is fully registered in the USA for efficacy in the field and for safety for use against the Cotton bollworm, *Heliothis zea,* a pest of corn, cotton and tomatoes, and the closely related *Heliothis virescens,* the Tobacco budworm.

Products are also undergoing development for use against the Cabbage worm (*Trichoplusia ni*) and the Army worm (*Prodenia*). Promising field trials results have also been obtained with a nuclear polyhedrosis virus against the Gipsy moth and for a similar virus active against the Spruce budworm. And many others are being assessed for their field potential.

The infective particle of a polyhedrosis virus is the small viral unit embedded in the protein polyhedron. The protein has a marked protective effect and the viability of a virus when stored outside the infected animal is satisfactory for insecticidal purposes. Formulation presents little difficulty, although viruses tend to lose activity when spread in the field. This is probably largely due to damage by ultraviolet radiation and ultraviolet absorbants are incorporated in the formulations. In the field disease may take some time to develop so that timing of the application is an important matter, particularly in those cases where the young instar larvae are more susceptible to the virus than older, more mature pests. This is the case with the NPV of *Heliothis.* First instar larvae will die very rapidly often within a day or two of application but it may take from four days to a week to kill a third instar larva. Viruses must be ingested by the insect and really effective coverage of foliage is an important feature of field application. Repeated spraying during the growing season is required since persisting infections are not usually established.

6 Safety of microbial insecticides

6.1 Specificity of insect pathogens

Insect pathogens differ widely in the extent of their specificity for hosts. *Bacillus popilliae* and *B. lentimorbus* are specific for larvae of the scarabaeid beetles. *Bacillus thuringiensis* attacks a wide range of lepidopterous larvae, including many pest species, with an underlying pattern of adaptation of particular biotypes of the microorganism to different insect species. It is without effect on the vast majority of other insects and on other forms of life. The insect viruses of interest for biological control are, with few exceptions, highly specific attacking only insects of one species and sometimes of one type within a species (Table 19.3). Some of the fungi, including *Beauveria bassiana,* and the nematodes of insecticidal interest on the other hand have a very low order of specificity being active in the laboratory and in the field against a very wide range of insect species.

For the insecticide manufacturer high specificity is a two-edged weapon: traditionally the grower has favoured broad-spectrum insecticides capable of killing each of the different pest species growing on his crops and sees specificity as a disadvantage in an insecticide, on the other hand specificity usually implies safety in use and the one big asset that microbial insecticides possess when seen against the whole background of efforts to control pest insects is their acceptability for wide dissemination in the environment stemming from their demonstrated safety.

6.2 Safety requirements

There is no *a priori* reason why a microbial agent should be treated in any way other than as a chemical agent for insect control. Since the agent being used is a viable one and apt to multiply in the environment, it must be proved safe for all other forms of life. The pattern of legislation of pesticidal chemicals differs widely in different countries but many regulatory bodies take a lead from the USA who require evidence of efficacy in the field for the purpose defined and

Table 19.3   Specificity of NPV of *Heliothis zea* (from Ignoffo 1968)

Organism	Method of application	Result
Plants: Cotton, Corn, Sorghum, Bean, Soybean, Snapbean, Kidney Bean, Tomato, Tobacco, Radish	External application	Negative: no apparent phytotoxicity or pathogenicity
Invertebrates: Grass Shrimp, Brown Shrimp, Oyster	*Per os,* topical	Negative
Insects: Heliothis, 5 species; Tobacco hornworm, Tomato hornworm, Wax moth, Cabbage looper, Beet armyworm, Fall armyworm, Southern armyworm, Lucerne moth, Honey-bee, Housefly	*Per os,* intrahaemocoelic	Positive only for *Heliothis.* Negative for all other listed species
Vertebrates: Man, Monkey, Dog, Rabbit, Guinea Pig, White Rat, White Mice, Chicken, Chicken egg, Quail, Sparrow, Mallard, Killifish, Spotfish, Rainbow Trout, Bluegill, Black Bullhead, White Sucker, Sheepshead Minnow	General, *per os,* inhalation, topical intradermal, intracerebral, intravenous intramuscular, intracellular, intraperitoneal, tissue cells	Negative: clinical and laboratory; cutaneous or respiratory sensitivity; toxicity, pathogenicity, teratogenicity, carcinogenicity

evidence of safety in the sense that the insecticide formulation itself and possible residues from it are subjected to stringent toxicological and metabolic studies. Only after an extensive review of all these questions will a product receive a tolerance or an exemption from a tolerance. In the case of microbial insecticides the US regulatory agencies have made it clear that they require proof of safety sufficient to allow the granting of an exemption from tolerance before a product can be fully registered for use. In addition to the normal toxicological and effectiveness data, a petitioner for registration of a microbial insecticide must also produce information about the method of production and demonstrate an adequate technique for the production of a uniform product and for the standardisation of that product. And he must also produce evidence that he can eliminate extraneous microorganisms from the product or control them to an acceptable level.

Full registration with exemption from tolerance requirement is now established in the USA for several *B. thuringiensis*-based insecticides and for products based on the nuclear poly-hedrosis virus of *Heliothis*. These are at present the only insecticidal agents which may be used without tolerance limitation right up to harvest on edible crops in the USA.

From what has been said above, it will be clear that the production of a microbial insecticide is a more difficult process than the synthesis of a chemical insecticide. To the problems of handling, harvesting, drying and standardising a biological product must be added the likelihood of contamination at some stage during manufacture or of loss of viability or potency as a result of degeneration of the agent itself. It is hardly surprising that microbial insecticides are expensive when compared to the more familar chemical agents; indeed what is surprising is that *B. thuringiensis* and virus-based agents can be marketed at anything like the price of chemical insecticides. Microbial insecticides owe their present success and future potential primarily to their safety.

7 The development of resistance to microbial control agents

One of the major problems facing the insecticide industry is the rapidity with which insects develop resistance to chemical insecticides. Experience with microbial agents extends over a much shorter period of time than is the case with chemical insecticides but a detailed study reported in 1971 revealed no records of indisputable resistance to microbial agents in field trials or control programmes. Resistant stocks of insects have certainly been produced in the laboratory. These have particularly concerned virus diseases where, for instance, cultures of the Cabbage White butterfly have shown susceptibility to a granulosis virus disease varying by as much as a factor of 1 000. There has been no report of increased resistance to spores and toxic crystals of *B. thuringiensis* although the Housefly (*Musca domestica*) has been reported to have developed a 14-fold resistance to the exotoxin of *B. thuringiensis* in 50 generations.

Biological control agents have one advantage over chemicals with regard to resistance. They are alive and are themselves subject to variation which may lead to the natural development of more effective strains or even to a capacity to infect new species. There is scant evidence on which to predict whether resistance will become of widespread importance in practice. It shows no signs of doing so at present in such well-known microbial control successes as the use of *B. popilliae* against the Japanese beetle and of the NPV of the larch sawfly for the control of that forest pest. However, microbial control is generally still in its infancy and it is not possible to predict the effects of increased selection pressure on host resistance.

When the myxomatosis virus was released into the rabbit populations of Australia and Europe in the early 1950s the effects of the disease were dramatic, the highly virulent viruses first introduced having a mortality rate usually exceeding 99 per cent. Within a year of the introduction of the virus into Australia, strains with a 90 per cent mortality rate appeared in the field. These have remained dominant ever since but still more attenuated strains, causing mortality sometimes as low as 20 per cent, have been recovered since 1955. In Europe the fully virulent strains have persisted longer and on a wider scale but here also attenuated strains have become common. In Australia the involvement of the mosquito in the summer spread of the virulent virus which results in the death of the host animal in a few days militates against the over-winter survival of the virus in the absence of the vector. Long persistence of infectious lesions is an essential condition of successful survival of the virus through the winter and it is probably this factor which has led to the rapid emergence of attenuated strains of the virus in Australia. In Britain the spread of the virulent virus was slower than in Australia and was effected by man and the rabbit flea which here continue to be the major vectors of myxomatosis. Virulent strains have persisted much longer as significant features of the population and the virulence of attenuated strains has remained somewhat higher than in Australia. Such findings suggest that the mode of transmission is of overriding importance in determining the relative frequency of emergence of attenuated strains of the virus.

Changes in resistance of the rabbit host have also been noted. Under the rigorous selection pressure of regular exposure to strains of virus killing in excess of 90 per cent of the population a rapid increase in resistance was seen in Australian rabbits. In a period of seven years of exposure to successive epizootics, resistance to a standard virus which originally killed 90 per cent of wild rabbits had developed to such an extent that it would kill only 30 per cent of newly-caught young ones.

The myxomatosis episode has taught people interested in microbial control a great deal about the use of disease processes for control of pest species in the field. It has served to emphasise that natural selection of the host and of the infective microorganism may occur in a fairly short time in such a way as to lead to a fall-off in control. This, however, will depend on a number of environmental factors and decreased effectiveness is not necessarily a feature of long-established infections in nature. Experience with myxomatosis also underlines the need for proper studies of the factors controlling the spread and persistence of an infectious disease in a natural population, topics of which our knowledge is often sadly deficient but which are fundamental to the full exploitation of the potential of biological control methods. Such considerations are no less true for the effective development and use of microbial insecticides (Burges 1973).

One possibility open to the biological control technique which is denied to the manufacturer of chemical insecticides is that of genetic manipulation of microbial agents in the laboratory, thereby supplementing any natural tendency for the pathogen to defeat developing resistance by the evolution of more effective strains. The future success of biological control programmes will depend to a large extent on the ability and ingenuity of the insect pathologist in the laboratory.

8 Future prospects

The effectiveness of *B. popilliae* is well demonstrated and its use in the USA for the control of Japanese beetle will undoubtedly continue. In other parts of the world, Australia in particular, the potential of the milky diseases for controlling scarabaeid beetle infestations is now recognised and field applications are beginning to play a significant part in control programmes. The future development of milky disease agents must be governed by the success of attempts to grow the organisms in pathogenic form in nutrient media.

The effectiveness of *B. thuringiensis* as an insecticide is also well established for certain types of crop, particularly leaf vegetable crops such as cabbage and lettuce and for forestry applications. Significant pointers for future developments are to be seen in the use of such preparations against the Cabbage looper in Arizona and California, an insect which has long been resistant to many chemical insecticides and where chemicals that are active against it may leave toxic residues on the edible plants. By using a carefully selected chemical insecticide which controls other pests such as the thrips and aphids in combination with *B. thuringiensis* an acceptable control of the pest complex can be achieved without danger of build-up of toxic residues. It is then common practice to withdraw the chemical insecticide some 30 days before harvest and apply *B. thuringiensis* from then until the crop is picked. For control of insects on tobacco *B. thuringiensis* is as effective as chemical insecticides for the Tobacco hornworm (*Manduca sexta*) and for Tobacco budworm (*Heliothis virescens*). Again the attraction of *B. thuringiensis* is the freedom from chemical residues in the crop.

Several insect viruses are known to be persistent and effective control agents for forestry pests and application in this field will undoubtedly continue to expand. The ability of a virus disease to spread and control an insect population is often a question of the host ecology. An interesting and typical example of the successful use of a virus disease in this way is the control of the Coconut rhinoceros beetle, *Oryctes rhinoceros,* on islands in Western Samoa. It is more difficult to forecast the future potential for the non-persistent virus diseases. Nevertheless, these are receiving a great deal of research and development attention at the present time since they represent difficult pest situations associated with commercially valuable crops. Here persisting infections are not established in the field following treatment and repeated application is called for during the growing season. The recent registration of the first viral insecticide in the USA establishes a landmark in the development of these agents requiring as it does convincing evidence of safety in use and of efficacy in the field.

We are seeing today a developing pattern of agriculture towards intensive large-scale, single-crop operations and this, coupled with the present interest in controlling environmental pollution and preventing interference with natural animal and plant populations, will almost certainly lead to more widespread development and adoption of integrated pest control programmes. Narrow-spectrum microbial insecticides with their high innate safety will certainly play an increasing role in such integrated programmes. The application of integrated control methods will inevitably require a much more detailed understanding of the population dynamics of insect pests than we at present possess. Falcon (1971) discussing the future potential of bacterial control agents comments that the future of these materials looks extremely bright in the dawning era of ecological pest control. Today that statement must be broadened to include the viral agents whose development since 1971 has been dramatic.

The effectiveness of many of the microbial pathogens depends primarily on their remarkable specificity for the insect as opposed to other forms of life. A study of the scientific basis for this specificity could open up whole new areas in our understanding of the way in which microorganisms bring about disease and invade host bodies. When we understand the reasons for this specificity, we may well find ourselves in a better position to exploit the knowledge by the design of highly specific insecticidal molecules. The chemical insecticide industry spends a great deal of time, money and effort looking for agents capable of exploiting the minute differences in metabolism or behaviour which serve to mark off insects from other forms of life. The search is difficult and not often successful yet many microbial pathogens have developed means of attacking these Achilles heels of the insect world. It could well be that a thorough understanding of the mechanisms of microbial pathogenicity by illuminating the molecular basis of specificity could make a fundamental contribution to man's efforts to control his insect pests.

Bibliography

BECHTEL, D. B. and BULLA, L. A. (1976) 'Electron microscope study of sporulation and parasporal crystal formation in *Bacillus thuringiensis*', *J. Bacteriol.*, **127**, 1472–81.
  A well illustrated critical study of toxic crystal biosynthesis.
BURGES, H. D. (1973) 'Enzootic diseases of insects', *Ann. N.Y. Acad. Sci.*, **217**, 31–49.
  An interesting discussion of the principles governing the origin and spread of enzootic infections.
BURGES, H. D. and HUSSEY, N. W. (Eds.) (1971) *Microbial Control of Insects and Mites.* Academic Press, London and New York.
  The best all round source of detailed information on all aspects of the subject with extensive references.
FALCON, L. A. (1971) 'Use of Bacteria for Microbial Control', in *Microbial Control of Insects and Mites* (Eds. Burges, H. D. and Hussey, N. W.), pp. 67–95. Academic Press, London and New York.
  A useful summary with numerous references.
IGNOFFO, C. M. (1968) 'Specificity of insect viruses', *Bull. Ent. Soc. Am.*, **14**, 265–76.

FARKAS, J., ŠEBESTA, K., HORSTÁ, K., SAMEK, Z., DOLIJS, L. and SORM, F. (1969) Collection of Czechoslovak Chemical Communications (English edn.), 'Structure of the exotoxin of *Bacillus thuringiensis* var. *gelechiae*', **34**, 1118–33.

NORRIS, J. R. (1969) 'Macromolecule synthesis during sporulation of *Bacillus thuringiensis*', in *Spores IV* (Ed. Campbell, L. L.), pp. 45–58. Amer. Soc. Microbiol. Washington.

NORRIS, J. R. (1971) 'The protein crystal toxin of *Bacillus thuringiensis*: biosynthesis and physical structure', in *Microbial Control of Insects and Mites* (Eds. Burges, H. D. and Hussey, N. W.), pp. 229–46. Academic Press, London and New York.

Index

A gene, in polyoma virus and SV40, coding for protein (probably T antigen), 18, 20; mapping on DNA of mutants in, 24; required for initiation of transformation, 21

accumulation body, in symbiotic dinoflagellates, 390

acetate, as repellant in chemotaxis, 335

acetyl choline: effects on release of, of botulinum toxin, 141, and of tetanospasmin, 139; mosaicism in receptor for, at neuromuscular junction, 278

acetylene: inhibits hydrogenase, 354; inhibits methane oxidation, 356; reduced to ethylene by nitrogenase, 345; test for nitrogenase with, should included test for repression by ammonia, 355

N-acetylmuramidase, in changes of form of *Arthrobacter*, 245, 246

acetyltransferase, plasmid-coded: detoxifies chloramphenicol, 87, 88, 95

Achlya (water mould), sterol sex hormones of, 211–13

Acholeplasma laidlawii: effects of growth temperature, and different fatty acid supplements, on phospholipids of, 283–4

acid phosphatase, of *S. cerevisiae* mutants, 198, 199

Acinetobacter, in methane community, 201, 202

acrasin (= cAMP), autotactic hormone of *Dictyostelium,* 207, 208–9

Acropora cervicornis (coral), calcification in, 394

actinomycin D, inhibits doubling of DNA strand in oncornaviruses, 37

acyl-CoA synthetase, in transport of fatty acids, 306

acyl transferases of phospholipid synthesis, in homeoviscous adaptation, 285

adenosine diphosphate ribose, transferred from NAD to EF-2 by diphtheria toxin, 135

adenosine phosphosulphate (adenylyl sulphate): in oxidation of sulphur compounds by some thiobacilli, 380; reductase for, 380

adenoviruses, 1, 2; human, induce tumours in newborn rodents, 2

adenyl cyclase, enterotoxins and, 89–90, 137, 138, 139

adenylate kinase, 173

ADP: formed in nitrogenase reaction, and inhibits it, 345; system for removal of, 346

aflatoxins, produced by *Aspergillus flavus*, 419

ageing, characteristic of organisms with polar growth, 236

agglutinins of cell surface: on amoebae of *Dictyostelium,* 207; induced by sex hormones on cells of both *Blepharisma* mating types, 217–18; on mating cells of *Saccharomyces,* 207, and of *Hansenula wingei,* 215; on mating zygophores of Mucorales, 215

Agrobacterium tumefaciens, plant pathogen carrying tumour-inducing plasmids, 93–4; attempts to control, by non-pathogenic strain producing a bacteriocin, 123; transfer of genes for nitrogen fixation to, causes formation of nitrogenase proteins, but no fixation, 359; unusual amino acids produced by (nopaline or octopine), 94

Aiptaisia (sea anemone), extrusion of symbionts by, 395

akinetes (resting cells, cysts) of cyanobacteria, 260

alanine, transport of, 318

alder, nitrogen-fixing symbionts in, 356

algae, marine: hydrocarbon sperm attractants of, 210–11

alimentary tract, of mammals: ecology of enterobacteria in, 119–20, 121 alkaline phosphatase: in *Paramecium,* repressed by symbiotic algae and chloroplasts, 400, 410; periplasmic protein in Gram-negative bacteria, 287−8

allomucic acid, unusual sugar in θ-endotoxin of *B. thuringiensis,* 466

Allomyces (water mould), sesquiterpene sperm attractant of, 210−11

amensalism between organisms (one is restricted, the other unaffected), 185

amino acids: analogues of, in study of transport systems, 310, 311; effect on bacterial yield of supply of, during growth on sugars, 166, 168; in flagellin of *B. subtilis,* 326, 327, and in similar protein in axial filament of spirochaetes, 331; periplasmic binding proteins for, 308, 310; transport of, 309−12, 318; unusual, produced by *Agrobacterium,* 94; zoospores and zygotes of *Allomyces* attracted by, 211

amino-glycoside antibiotics, R-plasmid enzymes set up permeability barrier against, 87−8

ammonia: ATP equivalents required for assimilation of, via glutamate dehydrogenase and via glutamine synthetase, 169, 175; concentration of, and pathway of assimilation employed, 358−9; from dinitrogen complexes of metals, treated with acid, 350; energy available from oxidation of, to nitrite, 370; energy required for oxidation of, to hydroxylamine, 370, 371; nitrogen-fixers grown with, lack nitrogenase, 345; organisms oxidising to nitrite, 364; probable mechanism of oxidation of, 379; represses expression of nitrogen-fixing genes, 358

'ammonia hydroxylase', 379

AMP, cyclic: as aggregation hormone of *Dictyostelium,* 208; in chemotaxis of Myxomycetes, 339; cholera toxin increases amount of, 137, 138; and motility of myxobacteria, 261; staphylococcal δ-toxin and, 144; and synthesis of flagella, 328; in yeast-mycelial transformation in *Histoplasma,* 423

Amphidinium (Entodinium) chattonii, dinoflagellate symbiont of pelagic marine invertebrates, 390, 395

Amphidinium klebsii, dinoflagellate symbiont of *Amphiscolops,* 390

Amphiscolops (Convoluta) langerhansi, platyhelminth with symbiotic dinoflagellate, 388, 390

amphotericin D, polyene antibiotic: as antifungal agent, 423, 424, in combination with 5-fluorocytosine, 425; toxic side effects of, 426; variation of resistance of *Candida albicans* to, with phase of growth, 425−6

amphotropic viruses (C-type murine, genetically transmitted, infect both own and other species), 45

Anabaena: cell cycle of, 239, 259−60; differentiation of filament cells of, to become cysts or heterocysts, 224−5; nitrogen fixation by, free-living and in lichens, 355

Anagasta kühniella (flour moth), susceptible to *B. thuringiensis,* 462

Ancalomicrobium sp., phenotypic variation of, 230, 231

Anemonia sulcata: factor in, stimulating release of nutrients from symbiont dinoflagellates and chloroplasts, 406

Anodonta cygnea, bivalve with symbiotic *Chlorella,* 388

antheridiol: sterol produced by female hyphae of *Achlya,* affecting male hyphae, 211−13

Anthopleura elegantissima, coelenterate with green symbionts in some strains, 388; interactions of symbionts and, 393, 394

antibiotics: in amensalism, 185; antibacterial, may encourage development of fungi, 417, 418; co-integrate and aggregate states of plasmid DNA in resistance to, 86−7; heavy selection pressure exerted on bacterial populations by, 103, 105; plasmid-carried resistance to, 83−6, 113, (mechanisms of), 87−9

antibodies: common in humans, to adenoviruses, 2, to *Candida,* 418, and to *Toxoplasma,* 437; to viral antigens, 33

antigens: assay of virus by immunological detection of, 31; changes of, on surface of *Plasmodium,* 452, and of *Trypanosoma,* 443; foetal, in tumours, 53; fungal, allergies to, 415; fungal, in diagnosis, 418, 426; produced by K plasmids on cell surface of *E. coli,* 89, 90; synthesised in cells infected with polyoma virus or SV40, 16−18; viral group-specific, type-specific, and interspecies, 33, 34; viral group-specific, transcribed as large precursor polypeptides, cleaved into virion proteins, 39

antiporters (uniporters), in chemiosmotic theory, 316

antisera to viral proteins, for determining their amount and location, 32

Aphanocapsa, as symbiont in sponges, 409

aquatic environment, bacteria of, 222−3

arabinose, systems for transport of, 217

Archaias angulatus, foraminifer with symbiotic *Chlamydomonas,* 388

arginine, transport of, 317

arsenate, transport systems inhibited by, 317

Arthrobacter crystallopoietes, nutritionally controlled morphogenesis in, 245−6

Arthrobacter spp., dimorphic cell cycle of, 233, 239, 245, 246

arthropod-borne viruses, transmission of, 40

aspartate: threshold and saturation concentrations of, for chemotaxis by *E. coli,* 334

Aspergillus flavus, aflatoxins produced by, 419

Aspergillus fumigatus, 'opportunistic pathogen', 415. 416; antigens of, 418; toxins produced by, 419; variation in resistance of, to polyene antibiotics, with stage of cell cycle, 426

Asticaccaulus sp., dimorphic cell cycle of, 237

ATP: colicin and cell content of, 117; and conformation of iron protein of nitrogenase, 349–50, 351; direct coupling of, to transport, in permease model, 314; electron flow coupled to generation of, in chemolithotrophs, 367, 368, 374, 377; electron flow driven 'uphill' by, for reduction of NAD by chemolithotrophs, 374–6; energisation of transport by, 317–18; energy required for formation of, from ADP and phosphate, 369; equivalents of, available from aerobic and anaerobic metabolism, 160, and from catabolism of different substrates, 167–8; level of, in regulation of change from aerobic to anaerobic metabolism, 172–3; level of, and respiration pathway used by *Azotobacter*, 352; level of, and sporulation in bacilli, 247; maximum possible microbial yield in terms of, 165–6; microbial yield in terms of equivalents of, expended, 158–9, 167; possible 'uncoupled' breakdown of, 173–4; required for fixation of carbon dioxide by Calvin cycle, 366; required for nitrogen fixation, 169, 345, 346, 351; requirement of, for nitrogen fixation, higher for organisms than for isolated nitrogenase, 353; requirement of, for reduction of NAD, 376

ATPase: of *E. coli*, 288, 289; inhibition of, prevents fall in ATP concentration caused by colicin K, 117; as peripheral protein of prokaryote cell membrane, 275; required for transport systems driven by glycolytically produced ATP, not for those driven by aerobic respiratory processes, 317

attractants: methods of showing chemotaxis towards, 334, 335; receptors for, in *E. coli*, 336; threshold and saturation concentrations of, 334

autoradiography: to show transfer of photosynthate from symbiont to host, 391; for study of cell wall growth, 241

autotactic hormones, 207, 208–9

avian leukaemia viruses, transmission of, 41, 42

avian leukaemia-sarcoma virus, glycoprotein of, 32

avian myeloblastic leukaemic virus: helper viruses of, 49; tumour-specific surface antigens, in cells transformed by, 54

avian oncornaviruses, host range of, 45

avian sarcoma viruses, recombinants of leukaemia viruses and defective mutants of, 50–1

Azolla, nitrogen fixation by *Anabaena* in symbiosis with, 355

Azotobacter beijerinckii: poly-β-hydroxybutyrate in, 353; symbiotic with rice, 356

Azotobacter chroococcum: ATP requirement for nitrogen fixation by, 353; hydrogen evolution by nitrogenase of, 346; nitrogenase proteins of, 347, 348

Azotobacter insolita: scavenging by, of traces of fixed nitrogen from atmosphere, 355

Azotobacter paspal: nitrogen fixation by, in symbiosis with *Paspalum*, 355

Azotobacter spp., respiratory protection of nitrogenase in, 352

Azotobacter vinelandii: ATP requirement for nitrogen fixation by, 169; branched electron-transport pathway in, 174–5; crystallisation of molybdenum-iron protein of nitrogenase of, 348; dissolved oxygen, and growth efficiency of, 169, 170; maintenance energy requirement of, 164; nitrogen-fixing ability restored to defective mutants of, by transfer of *Klebsiella* genes, 359; nitrogenase and hydrogenase of, 345–6

bacilli: bacteriocin-producing strains of, 122, growth temperature and fatty acids of, 282; monomorphic spore-forming cell cycle of, 239; nitrogen-fixing species of, 356; peptide antibiotics produced during sporulation of, 216, 249–50; sporulation in, 246–8; transcriptional control of sporulation in, 248–9

Bacillus alvei, toxin of, 145

Bacillus amyloliquefaciens, cleavage of virus DNA by endonucleases of, 6

Bacillus brevis, peptide antibiotics produced in sporulation of, 249–50

Bacillus cereus: enterotoxin of, 139; membrane-damaging toxins of, 141, 145

Bacillus laterosporus, toxin of, 145

Bacillus lentimorbus, causing disease of Japanese beetle, 468; specific for Scarabaeid beetles, 475

Bacillus licheniformis: release of exoenzyme from membrane-bound penicillinase of, 287; unsaturated fatty acids in, at different temperatures, 282

Bacillus macerans, nitrogen fixation by, 355, 356

Bacillus megaterium: lipid asymmetry in membranes of, 277, 288; transmembrane movement of phosphatidyl ethanolamine in, 276

Bacillus polymyxa: mutualism between *Proteus vulgaris* and, 183–4; nitrogen fixation by, 355, 356; nitrogenase of, 345, 348

Bacillus popilliae, causing disease of Japanese beetle, 468–9, 478; specific for Scarabaeid beetles, 475; spore production by, 469–70; use in the field, 470

Bacillus sotto, silkworm pathogen, 462

Bacillus stearothermophilus, growth temperature, and phospholipids of, 283

Bacillus subtilis: amino-acid sequence of flagellin of, 324–6, (effect on flagella of single amino-acid substitution in) 327; group transport of carbohydrates in, 303; membrane-bound transport proteins of, 309; sporulation in, 230, 232; sporulation loci on chromosome of, 250

Bacillus thuringiensis, for control of lepidopterous pests by poisoning caterpillars, 460, 462, 478; in combination with chemical insecticide, for thrips and aphids, 478; crystalline δ-endotoxin of (protein), 145, 463–6; species of pests susceptible to, 467; thermostable θ-exotoxin of some strains of (nucleotide), 466, 477

bacteria: monomorphic vegetative, polymorphic vegetative, and spore and cyst types of, 230; morphogenesis in stalked and budding species of, 221; possibility of introducing new genes into, for production of proteins or enzymes, 105; symbiotic mixture of, with higher growth efficiency on methanol than that of only methanol-utiliser present, 171; system for studying cell differentiation in, 59–60; *see also individual species and groups of bacteria*

bacteriocins, 90, 109, 121–4

Bacteroides, sphingolipids in, 268

bacteriophage β, carries gene for diphtheria toxin in *C. diphtheriae,* 234

bacteriophage λ: for genetic manipulations, 107; termed episome because of existence autonomously or integrated, 77

bacteriophage M 13: coat protein of, in cell membrane of *E. coli,* 288

bacteriophage P1, transfer of genes for nitrogen fixation by, 357

bacteriophage φ X174: assignment of genes to DNA of, 19; ssDNA of, 36; as internal molecular-weight standard in electron micrograph, 99

bacteriophage Qβ, 36

bacteriophage R17, binds specifically to F-pilus, 96

bacteriophage T1: *E. coli* cell surface receptor for, is also receptor for colicin M, 115

bacteriophage χ, flagellotropic, 327

bacteriophages: *E. coli* receptors for, are also receptors for colicins, 115; incorporation of sections of plasmid DNA into, 105; lysogenic, in some cases of bacterial toxin production, 132; parallel between bacteria carrying lysogenic, and colicinogenic bacteria, 117; sensitivity to, in cell cycle of *Caulobacter,* 253; specific for F-pilus, 95, 96

bacteriophages, T-even, dsDNA of, 36

bacteriorhodopsin, forming mosaic in cell membrane of *Halobium,* 278; in photophosphorylation mechanism, 279

Balantidium coli, parasite of humans, 435–6; epidemiology of, 440

basal body: of axial filament of spirochaetes, 331; of flagellum, 322, 323

B/C gene, in SV40, coding for major capsid protein, 18; mapping of, on DNA, 24; mutants in, 19

Bdellovibrio bacteriovorus, sheathed flagella of, 324

Beauvaria bassiana, fungus pathogenic to silkworms, 461; low specificity of, 475

Beggiatoa, 223; reproduces by release of short lengths of filament, 224

biotin, in mutualism of *B. polymyxa* and *P. vulgaris,* 183

Blakeslea trispora, production of sex hormone by, 214

Blastomyces dermatitidis, causing lung infection, 416; parasitic form yeast-like, saprobic form mycelial, 420

Blepharisma intermedium (ciliate protozoan), sex hormones released by two mating types of, 217–18

blepharismone: calcium salt of aromatic acid, sex hormone of type II *Blepharisma,* 217–18

blepharmone: glycoprotein, sex hormone of type I *Blepharisma,* 217–18

blue-green algae, *see* cyanobacteria

botulinolysin, 145

botulinum neurotoxins, 128, 129, 140–1; lysogeny in production of, 132; potency of, 131; protoxins of, 132

Brevibacterium albidum, cleavage of virus DNA by endonucleases of, 6

Brevibacterium linens, in orcinol community, 202

'budding' bacteria, 239–40: asymmetric binary fission in, 233–5; 'mushroom-shaped', 234, 235

calcium: in activity of phospholipase C of *Cl. welchii* α-toxin, 143; in release of acetylcholine, affected by botulinum toxin, 141

calcium carbonate: stimulation by symbiotic algae of deposition of, in coral reefs, 393–4

Calothrix crustacea, nitrogen-fixing cyanobacterium of coral reefs, 392

camphor, plasmid coding for degradation of, 91, 92

Candida albicans, 'opportunistic pathogen', 416; antibodies for, in practically all adult humans, 418; environmental factors affecting morphological change in, 421, 422; exocellular products of, 419; mycelial and yeast forms of, 420, 421; resistance of, to amphotericin B, varies with phase of growth, 425–6

candidicin, polyene antibiotic, anti-fungal agent, 423, 424

capsid proteins of viruses, 16; genes coding for, 18, 20, 21, 24; mutant in, defective for uncoating, 19

capsomers: of herpes virus, 2; of polyoma virus and SV40, 4

carbon: balance of, needed for calculation of growth efficiency, 160, 161; competition for source of, in commensalism of *Lb. casei* and *S. cerevisiae,* 185; source of, and growth efficiency, 167–8

carbon conversion efficiency (percentage of food carbon in carbon of cell mass), 157–8 carbon dioxide: concentration of, and cell cycle of *Rhodomicrobium,* 257; energy requirement for fixation of, by reductive pentose phosphate cycle, 366–7, 369; exposure to, causes *Convoluta* to expel symbionts, 402; as source of carbon for chemolithotrophs, 363

carbon monoxide: inhibits hydrogen evolution by hydrogenase, but not ATP-activated hydrogen evolution by nitrogenase, 346

carbonic anhydrase, in corals, 394

carboxypeptidase II of *E. coli* (periplasmic enzyme), and synthesis of septum-specific peptidoglycan, 74–5

carcinogens, cause release of infectious virus from DNA provirus, 43

carotene: trisporic acid stimulates production of, and metabolism of, to trisporic acid precursors, 215

Cassiopea (jellyfish), nutrient supply from symbionts in, 392

Casuarina, nitrogen-fixing symbionts in, 356

cats: endogenous baboon virus in domestic and Mediterranean species of, not in Asian species, 46; leukaemia viruses of, 27, 42

Caulobacter: dimorphism in cell cycle of, 221, 237, 239, 250–2; polar stalk formation and function in, 239, 252–3; swarm cells of, 250, 252, 254

cell composition: knowledge of, needed for estimation of growth efficiency, 160–1

cell cultures: for study of mechanism of toxin action, 134; transformation of cells of, by viruses, 2, 3, 4, 30–2

cell cycle: biochemical features of, in *E. coli,* 61–4; of host cell, and transcription of provirus, 39

cell density, in cell cycle of myxobacteria, 262

cell division: asymmetrical in polar growth, 236, 238, 239; by binary fission, 233–4; biological clock regulating, 59–60; of *Chlorella* symbionts in *Hydra,* 396; onset of, in *E. coli,* 240, 241; by tertiary fission, in colony formation in *Pelodictyon,* 226, 227; unregulated, of transformed cells, 55

cell lines: permissive and non-permissive, for growth of viruses, 3

cell mass, initiation of DNA replication in *E. coli* coincides with time of doubling of, 64–5, 69

cell membranes: adhesions between cell walls and, 289, 290, 291; bacterial, invaginated before septation, 71; from different groups of prokaryotes, 265, and from eukaryotes, 266; energisation of transport across, by ATP, 317–18; energised state of, and damage by colicins, 115–17, 124; as fluid mosaic, 271–9; fluid mosaic model and variability of, 286–91; fluidity in variability of, 279–85; functions of eukaryote organelles in prokaryote, 265; isolation of, 266–8, and analysis of, 268–71; M-ring of flagella embedded in, 324; replicon hypothesis of attachment of plasmids to unique sites on, which are involved in replication, 67–8, 102–3; sterols of, 423; study of transport in vesicles of, 299, 305, 306; synthesis of, distinguished from cell growth, 290; toxin segments involved in binding to, 135, 138; toxins damaging, see cytolytic toxins; toxins producing 'functional pores' in, 144–5; in transformed cells, 53; in two forms of *Candida albicans,* 421; viruses and, 32, 33, 37, 39

cell walls, of bacteria, 60; adhesions between cell membranes and, 289, 290, 291; chemically different in rod and spore forms of *Arthrobacter,* 246; as diffusion barrier between cell membrane and extracellular fluid, 302; of Gram-negative bacteria, phospholipid asymmetry in, 288–9; models for growth of, in rod-shaped organisms, 241–5; point of origin of DNA replication attached to? 67–8; in regulation of DNA initiation, 66–9, and of septation, 71–2

cell walls: of *Chlorella,* retained in symbiosis, 397; of dinoflagellates, lost in symbiosis, 389; of pathogenic fungi, change in composition with change in morphology, 420–1, in resistance to antibiotics, 426, and as target of chemotherapy, 423

cells: definition of events in (morphogenesis, differentiation, development), 229–30; definition of types of (monomorphic vegetative, polymorphic vegetative, spore, and cyst), 230; internal compartmentation in, for protection of nitrogenase from oxygen, 352; mother and daughter relation between, 237; normal, evidence for existence of endogenous virus in, 43; polarisation of, 237–8; stimulants of division of, may activate latent proviruses, 44; system for studying differentiation of, in bacteria, 59–60; transformed, 52–4

central nervous system, tetanospasmin blocks inhibitory neurones controlling motor activities of, 139

cephalexin, β-lactam antibiotic binding to *E. coli* protein involved in septation, 73

cephalosporinases, plasmid-coded, 87

ceramide (*N*-acylsphingosine), produced from sphingomyelin in erythrocytes affected by staphylococcal β-toxin, 142

chemiosmotic theory of energy coupling to transport, employing protonmotive force, 314–17, 319

chemolithotrophic bacteria, oxidising different inorganic substrates, 363, 364–6, 383; ATP generation in, 377; calculation of energy balances for, 377–8; continuous-flow culture of, for studying energy-coupling mechanisms, 381–3; electron transport in, 367–8, 373–4; energy from inorganic oxidations by, 369–71; energy requirement for carbon dioxide fixation by, 366–7, and for NAD reduction, 376; NAD reduction in, 374–6; pathways of oxidation in, 378–9, for ammonia, 379, and for sulphur compounds, 379–81; reducing potential from inorganic oxidations by, 372–3

chemostat systems, inclusion of solid particles in, 204

chemotaxis of bacteria, 333–4; chemoreceptors in, 309, 335–6; environmental conditions and, 334–5;

mutants in, 338−9; in non-flagellate bacteria, 339; of single cells, 336−8; techniques for demonstrating, 334

chemotherapy, of fungal infections, 417, 423; drug resistance in, 425−6; mechanism of action of antifungal agents in, 423−5

chitin, in walls of mycelial and yeast-like forms of pathogenic fungi, 420

chitin synthetase: polyoxin D, *in vitro* inhibitor of, does not permeate into pathogenic fungi, 423

Chlamydomonas hedleyi, symbiont in foraminifer *Archaias,* 388

chloramphenicol, detoxified by plasmid-coded acetyltransferase, 87, 88, 95

Chlorella vulgaris, symbiont in freshwater invertebrates, 388, 395−7, 409; metabolic interactions between hosts and (releases maltose), 397−8; regulation of population of, by host, 398−9; transmission of, 399−400

Chlorobium, anaerobic nitrogen fixation by, 355, 356

Chlorobium dimicola (photosynthetic sulphur bacterium), mutualism between *Desulfovibrio* and, 184

Chlorogloea fritschii (cyanobacterium): cell cycle and morphology of, in different conditions, 230, 232, 233, 261

chlorophyll: labelled, in symbiotic chloroplasts in *Elysia,* permitting estimate of chlorophyll turnover, 408

chloroplasts, in invertebrates, 388, 404, 409; autonomy of, and regulation of population of, by hosts, 406, 408; metabolic interaction between hosts and, 405−6; from Siphonales, in sacoglossan molluscs, 388, 404, 405

cholera enterotoxin, 127, 129, 136−8; potency of, 131; toxoid of, 133, 136−7

cholesterol, as main sterol of cell membranes in mammals, 423; required for growth by many species of mycoplasma, 271; toxins interacting with, 144−5

Chromatium, anaerobic nitrogen fixation by, 355

chromatoid bodies (arrays of ribosomes), in cycts of *Entamoeba histolytica,* 433

chromosome, of dinoflagellates, 389

Cladosporium carrionnii, tropical pathogenic fungus, 416

clams, giant, with symbiotic dinoflagellates, 388, 389, 390; carbon of symbiont photosynthate in, 392; have to be reinfected with symbionts in each generation, 395

Clinocardium nuttallii, cockle with green symbionts, 388

cloacin DF 13: cleaves 16S rRNA, 117; DNA of plasmid coding for, 118

clostridia: bacteriocins from, 122; cyclopropane fatty acids in phospholipids of, 270; growth temperature and fatty acids of, 282; *N*-methylated phosphatidyl ethanolamine in, 268; possible association of plasmid content with pathogenicity in, 90; *see also individual species*

Clostridium bifermentans, toxin of, 145

Clostridium botulinum, toxins of, 128, 129, 131, 132, 140−1, 145

Clostridium butyricum, nitrogen fixation by, 355

Clostridium histolyticum, toxin of, 145

Clostridium oedematiens, toxin of, 145

Clostridium pasteurianum, 355; ATP requirement of, for nitrogen fixation, 353; harboured in cultures of other organisms, 355; nitrogenase of, 344, 345; nitrogenase proteins of, 347, 348

Clostridium septicum, toxin of, 145

Clostridium sporogenes, insertion of flagellum in, 324

Clostridium tetani, toxins of, 128, 129, 131, 139−40, 145

Clostridium thermoaceticum: yield coefficient of, grown anaerobically on glucose, 166

Clostridium welchii, toxins of, 129, 130, 143, 145; α-toxin of (phospholipase C), 141, 143; enterotoxin of type A strains of, 139; protoxins of ε- and ι-toxins of, 132; θ-toxin of (reacting with sterols), 141, 144, 145

coccidia, life cycle of intestinal, 437

Coccidioides immitis, 'facultative pathogen', 415, 416; course of infection after inhalation of arthrospores of, 420

coelenterates, with photosynthetic symbionts, 388; ejection of symbionts by, 394−5; symbionts transmitted in eggs of, 395; *see also individual species*

Col plasmids (Group II larger, conjugative; Group I smaller, non-conjugative), coding for colicins, 90, 109, 112−14, 124; confer immunity to colicins on cells containing them, 118; in invasive strains of *E. coli,* 120−1

colicins, active against enteric bacteria, 90, 109, 124; in amensalism, 185; amino-acid sequences and molecular weights of, 111; classification of, 110-11; conditions for synthesis of, 111−12; E1, K, and I, affect energy-dependent processes in cell membranes, 115−17; E2, degrades cell DNA, 118; E3, cleaves 16S rRNA, 117; mutants insensitive to, 118−19; question of effects of, in mammalian alimentary tract, 119−20; receptors for, 114−15

collagenase, of *Cl. welchii,* 130

colonial microbes, aquatic, 223−8

commensalism between organisms (one derives advantage, the other is not affected), 184−5, 186 communities of microbes, 201–3

competition between microbes, 186; free, in an open environment, 189–93, (examples of) 193–8; kinetics of, 186–9; as a mechanism in enzyme evolution, 198–201

conifers: mycorrhiza of, not nitrogen-fixing, 357

conjugation of bacteria, 99–102; plasmids transferred by (conjugative), and not transferred by (non-conjugative), 79; replication of DNA in, 100, 102–3; transfer of chromosomal genes by conjugative plasmids during, 83; transfer of non-conjugative by conjugative plasmids during, 80, 82

contact inhibition: in normal cells, not in transformed cells, 21

Convoluta roscoffensis, requires infection with *Platymonas* for development, 401; digestion of symbionts by, 411; interactions between symbiont and, 402–3; specificity of association between symbiont and, 403–4; transmission of symbiont in, 403

Convoluta spp., with green symbionts, 388

corals, reef-building, with dinoflagellate symbionts, 388, 389, 390; recycling of nitrogen and phosphorus between symbionts and, 392–3; symbionts stimulate deposition of calcium carbonate by, 393–4; transfer of photosynthate to, 391

Corynebacterium autotrophicum, aerobic nitrogen fixation by, 355

Corynebacterium diphtheriae, toxin of, 127–8, 129, 131, 132, 134–5, 136

counterflux, of solutes in transport, 301–3

creatine phosphate-creatine kinase system, for removal of ADP from nitrogenase, 346

Cryptella cyanophora, protozoan with cyanellae, 408

Cryptococcus neoformans, causing lung infection, 416; acapsular variant not toxic for mice, 419–20; polysaccharide capsule of, 418, 419

Curtobacterium sp., in orcinal community, 202

Cutleria multifida, hydrocarbon sperm attractant of, 210–11

cyanellae (blue-green algae), symbiotic with aquatic invertebrates, 388, 408–9

cyanobacteria (blue-green algae): different cell forms of, 230, 232, 239, 257, 259–61; nitrogen fixation by, 224–5, 259, 260, 352, 357, 392; *see also* cyanellae, *and individual species*

Cyanophora paradoxa, protozoan with cyanellae, 408, 409

cyanophycin, polypeptide containing only aspartic acid and arginine, in akinetes of cyanobacteria, 260

cyclopropane fatty acids, in phospholipids of Gram-negative bacteria, clostridia, and lactobacilli, 270

cysteine: and morphology of *Candida albicans,* 421, 422; none in flagellin, 325; required for yeast-like but not mycelial form of *Histoplasma,* 422, and involved in yeast-mycelial transformation, 423; transport of, 317

cysts: of *Balantidium,* 435; of cyanobacteria (akinetes), 260; of *Entamoeba,* 433; of *Giardia,* 434; of myxobacteria, 262; of *Sarcocystis,* 439; of some groups of bacteria, 224; of *Toxoplasma,* 437–8; *see also* heterocysts

cytochrome b_5, cytochrome b_5 reductase: hydrophobic segments of, as sites of interaction with cell membrane, 275, 287

cytochromes: in different electron-transport pathways in *Azotobacter,* 175, 352; in electron transport of chemolithotrophs, 367–8; of *E. coli* and *K. aerogenes,* 175

cytolytic (membrane-damaging) bacterial toxins, 130, 141–4, 145–7; affect permeability properties, 134; oxygen-labile group of, 144–5

Cytophagales, gliding movement of, 332

cytosine: uptake of, by *E. coli,* 306

cytosine deaminase, cytosine permease: present in fungi, but not in mammalian tissues, 425; fungal mutants lacking, 425

D gene, in polyoma virus and SV40, coding for minor capsid protein, 18; mutants in, defective for uncoating, 19

Dalapon herbicide (2,2'-dichloropropionic acid): community growing on, 202–3; mutant of *Ps. putida* able to metabolise, 198, 200

death and autolysis of cells: 'maintenance energy requirement' may represent a constant loss of energy by, 163, 164; temperature and, 170

degradation of organic compounds by bacteria, 90–1; inducible enzymes for, 91

degradative plasmids, 90–3

dehalogenase, in mutant of *Ps. putida* acting on herbicide Dalapon, 198, 200

density-gradient separation of smallest cells, for synchronisation of bacterial cultures, 61

depurination of DNA, by treatment with formic acid: 'fingerprints' of fragments obtained by, 9, 10, 49–50

dermatophytic fungi, 415, 416; exocellular enzymes of, able to hydrolyse keratin, 418–19

Derxia gummosa: oxygen concentration, and nitrogen fixation by, 352

Desulfotomaculum, anaerobic nitrogen fixation by, 355

Desulfovibrio: mutualism between *Chlorobium* and, 184; nitrogen fixation by, 355, 356 detoxification, plasmid-coded antibiotic resistance effected by means of, 87–8, 95

development, definition of, 230

diaminopimelic acid, component of peptidoglycan, 67; labelling of, 73; transport of, 317

Dictyostelium discoideum, predator on *E. coli,* 186; acrasin in aggregation of, 207, 208–9; spacing of amoebae of, 207

differentiation, definition of, 230

diffusion of solutes into microbes, passive and facilitated, 297–8; carrier-mediated facilitated, 303; energy required for facilitated, 298, 316

digestion: of excess or moribund symbionts, 411; resistance of symbionts to, 400, 410

Digitaria (grass), nitrogen-fixing *Spirillum* symbiont in, 356

dihydrofolate reductase, bacterial: replaced by plasmid-coded enzyme not sensitive to inhibition by trimethoprim, 88–9

dihydropteroate synthetase, bacterial: replaced by R-plasmid-coded enzyme not sensitive to inhibition by sulphonamides, 88–9

dinitrogen complexes of metals (molybdenum, ruthenium, tungsten), giving ammonia on reaction with acid, 350

dinoflagellates, symbiotic with marine invertebrates, 388, 389–90; metabolic interactions between hosts and, 390–4; recycling of nitrogen and phosphorus between hosts and, 392, 393; regulation of population of, by hosts, 394–5; sites of, in hosts, 390

diphtheria toxin, 127–8, 129, 134–5; A and B sections of molecule of, 135–7; lysogenic bacteriophage carries genes for production of, 132, 134; potency of, 131

dithionate, can replace natural reducing agents of nitrogen fixation, 346, 348

DNA: of bacteria, transferred in conjugation, 99, 100; of cell, degraded by colicin E2, 118; conformational changes of, in sporulation of bacilli, 249; with histones, in nucleosomes of eukaryotes, 9; methylation of, in bacteria, as protection against own endonucleases, 5–6;
    of plasmids, 80–2; in bacterial conjugation, 77, 79–80, 99–100; determination of homology of, 97–8, 99; replication of, 100, 102–3; transfer of, between plasmids in same cell, or to chromosome or bacteriophage, 103–5; proteins attached to, 102;
    of viruses, 19, 21, 23; of adenoviruses, 2; of oncornaviruses, 27, 28, 31, 37; of polyoma virus and SV40, 4–5, (homology between) 3, 22, (maps of) 8, 14–15, 24, (replication of) 9, 11–13, (sequences in) 5–9, 10, 23, (transcription of) 13–15

DNA ligases, in introduction of new genes into DNA, 105–7

DNA polymerase: DNA-directed, reverse transcriptase acts as, 37; in maturation of viral DNA, 11, 12–13

DNA synthesis (replication): in *Caulobacter* cell cycle, 253, 254; inhibited by edeine A, peptide antibiotic produced by *B. brevis,* 249–50; inhibitors of, and septation, 69, 70; initiation of, in *E. coli,* (cell wall in) 66–9, (initiator proteins in) 66, (relation of cell division to) 61–4, (time of, correlated with doubling of cell mass) 64–5, 69; termination of, in *E. coli,* 69–71, 240, 241

DNA viruses (parvo-, papova-, adeno-, herpes-, and pox-viruses), 1; double-stranded, utilise cellular polymerases for replication and transcription, 36; single-stranded, (animal) have virus-coded RNA replicase, (bacterial) modify host polymerase to use RNA as template, 36

ecotropic viruses (C-type murine, genetically transmitted, only reinfect own species), 45

Ectocarpus siliculosus, hydrocarbon sperm attractant of, 209, 210

EDTA, in isolation of cell membranes, 267–8

edeine A, peptide antibiotic produced by *B. brevis,* reversible inhibitor of DNA synthesis, 249–50

eggs, assay of Rous sarcoma virus by growth on, 30

electron micrography: of *Caulobacter,* 251; of colony formation in *Rhodomicrobium,* 227, and swarm-cell maturation, 257; of dividing and newly divided *E. coli,* 60, 61; of flagella, 321, 322, 323; of *Hyphomicrobium* cell cycle, 229; of plasmid DNA, 82, and of heteroduplex molecule from two F plasmids, 99; of spirochaetes, 331; of toxins, 463, 464, 465;
    of viruses: cells transformed by, 30, 53; nuclear polyhedrosis, 471, 474; oncorna C-type particles, 32; polyoma particles and DNA, 5; replication of SV40 DNA, 9

electron paramagnetic resonance, for observing iron in nitrogenase proteins, 349

electron transport chain: basic, found also in chemolithotrophs, 367–8; points of entry into, for electrons from oxidations by chemolithotrophs, 373–4; in scheme for nitrogenase action, 351; 'uphill' electron flow in, in chemolithotrophs using ATP to transfer reducing equivalents to NAD, 374–6

elongation factor 2, diphtheria toxin inhibits protein synthesis by transferring ADP ribose from NAD to, 135

Elysia cauzescops, acquires chloroplasts from *Caulerpa,* 408

Elysia viridis, sacoglossan mollusc with symbiont chloroplasts, 406, 407; digestion of chloroplasts by, 411; factor in, stimulating release of nutrients from symbionts, 406; transfer of photosynthate to, 405

encapsulation, of some aquatic bacteria in adhesive polysaccharide sheath, 225, 226

Encephalitozoon, rare accidental parasite of humans, 432

endogenous metabolism (occurring in absence of growth), distinction between maintenance energy requirement and, 163–4

endogenous viruses, 41, 46; detection and expression of, 42–4

endonuclease, bacterial T: recognises guanine residues, 49; *see also* restriction enzymes

endoplasmic reticulum, protein:lipid ratio in, 267

endospores of bacilli: formation of, 246–8; germination of, 248

endotoxins, 128, 130; of *B. thuringiensis,* 145, 463–6; lipopolysaccharide, of Gram-negative bacteria, 128, 130, 147–50

energy: available from inorganic oxidations, 368–71; calculation of balance of, for chemolithotrophic growth, 377–8; calculation of coupling of, for *Thiobacillus denitrificans,* 381–2; coupling of, in transport, 298, 318, (chemiosmotic theory for) 314–17, (energisation by ATP in) 317–18, (mutants impaired in) 299, (permease and redox models for) 314; required for NAD reduction in chemolithotrophs, 376

energy charge, in terms of adenine phosphates, 173; fall in, as signal for sporulation in bacilli, 247

Entamoeba coli, E. gingivalis, E. hartmanni, non-pathogenic, 433

Entamoeba histolytica, parasite of humans, 431, 433–4

Enterobacter spp., nitrogen fixation by, 355

enterobacteria, in alimentary tract of mammals, 119–20, 121

enterochelin, cell surface receptor for colicin B is also receptor for, 115

enterotoxins: bacterial, 129, 130, 135–9; plasmids coding for, 89–90, 121

envelope of cell, *see* cell walls

envelopes: of herpes viruses (lipid-glycoprotein), 2; of mixed origin, from different viruses, 49; of oncornaviruses (lipoprotein derived from plasma membrane of host cell), 32

enzymes: competition as a mechanism in evolution of, 198–200; feedback regulation of, by ATP, ADP, AMP, 173; of halophilic, psychrophilic, and thermophilic bacteria, 279; regulation of growth by inhibition and stimulation of, or by induction and repression of, 172

epidemiology: of coccidia, 440; of intestinal amoebae and flagellates, 439–40; of *Leishmania,* 451; of *Plasmodium,* 451–2; of *Trypanosoma brucei,* 448–9, and *T. cruzi,* 449–51

epidermal growth factor (specific mitogen for epithelial and fibroblastic cells), in cells transformed by mouse sarcoma virus, 55

epidermolytic toxin (exfoliatin), of *Staphylococcus aureus,* 129, 130, 147; plasmids in production of, 132, 147; potency of, 131; varying effects of, in different species, or in same species at different ages, 133

Epidermophyton spp., dermatophytes, 416

episome, use of term, 77

Epstein-Barr (herpes-like) virus, 2

ergosterol, as main sterol of cell membranes in fungi, 423

erythrocytes: asymmetric distribution of membrane lipids in, 277; effects on, of phospholipase of *Cl. welchii,* 143, and of sphingomyelinase of staphylococci, 142; integral membrane glycoprotein of, 275–6; merozoites of *Plasmodium* in, 447, 448

erythrogenic toxin, of *Streptococcus pyogenes,* 129, 130; lysogeny in production of? 132

erythromycin: mechanism of plasmid-coded resistance to, in *S. aureus,* 88, 94

Escherichia coli: amino-acid transport in, 310, 311–12; ATPase of, 288; binary fission in (symmetrical), 223; cell cycle in, 60–4, 240–1; cell membrane of, 274, 278, 316; cell septation in, 69–74; cell wall of, 71–2, (growth of) 234, 243, (models for) 241–5; chemotaxis in, 334; colicin-producing strains of, 109–10, 120–1; colicin receptors mapped on chromosome of, 115; in competition with marine species of *Spirillum,* 197; *Dictyostelium* as predator on, 186; DNA synthesis in, (initiation) 64–9, (termination) 69–71, 240, 241; endonuclease of, for cleavage of virus DNA, 6; enterotoxins of some strains of, 129, 138–9, (plasmids in production of) 89, 132; fatty acid transport in, 306; fatty acids in phospholipids of, 282–3, 284; flagellar genes in, 327; galactose transport systems in, 308, 309, 312–13, 317; β-galactoside transport in, 299–300, (kinetics of) 300–3; group transport of carbohydrates in, 303, 305, 306, 307; hexose phosphate transport in, 314; induction of enzymes for aerobic metabolism in, 172; lipopolysaccharide endotoxin of, 148; monomorphic cell type of, 232, 239; nucleic acid bases in, 306; oxygen limitation and growth efficiency of, 175; peptidoglycan synthesis in, 72–3, 74–5; peptide transport in, 312; periplasmic binding proteins in, for sugars and amino acids, 308; pH and yield of, 170; plasmids in, 89–90, 93 (*see also* Col plasmids, F plasmid); receptors in, for attractants and repellants, 335, 336; transfer of genes for nitrogen fixation to, 357–8; transport proteins of, membrane-bound, 309

Escherichia coli mutants: colicin-resistant, 115, 118–19; with different requirements for fatty acids at different temperatures, 284, 290; in flagella (straight rather than helical), 323; galactokinase-deficient, 298, 312; in galactose transport, 317–18, 336; in galactoside transport, 300, 318; lacking component of sugar phosphotransferase system, 305; lacking plasmid coding for β-galactosidase, but with chromosomal gene coding for increased activity of this enzyme, 195; lipoprotein-deficient, 72; mutator, with competitive advantage, 196–7; septation-deficient, with filamentous growth, 71; temperature-sensitive for DNA initiation, 66; tyrosine prototroph and auxotroph, in competition in tyrosine-limited chemostat, 193–4 ethidium bromide, for separation of plasmid DNA, 82

N-ethyl maleimide, inhibitor of β-galactoside transport, 302, 307

ethylene, produced from acetylene by nitrogenase, 345

eukaryotes: DNA with histones, in nucleosomes of, 9; phosphatidyl choline in, 268; possibilities for nitrogen fixation in, 360

exogenous viruses, 41; inheritance of, 46

exotoxins, 128; of *B. thuringiensis,* 466

F plasmid (sex factor) of *E. coli,* existing autonomously or integrated into chromosome, 77–8, 83; cells containing, have pili on the surface, 95; homology of DNA of different groups of, 97–8; promotes its own conjugative transfer, 95, 98

fatty acid synthetase system, in homeoviscous adaptation, 285

fatty acids: ability to transport, induced in *E. coli* along with ability to metabolise, 306; of bacterial phospholipids, 270–1; of cell phospholipids in mammals, more unsaturated at low growth temperature, 279–80; effect of supplements of different, on growth temperature and lipoprotein in *Acholeplasma,* 283–4; growth temperature of mesophiles, and modification of, to preserve fluidity, 282; of membrane phospholipids, and membrane fluidity and transition temperature, 274, 280–2; metabolism of, in *Convoluta,* dependent on symbionts, 403

fermentative pathways, yield of ATP equivalents in, 171

ferredoxins, reducing agents in nitrogen fixation, 348, 351

ferrichrome, cell surface receptor for colicin M is also receptor for, 115

fibroblasts in culture: characteristics of, normal and transformed by sarcoma viruses, 53–4

'finger-prints' of depurination products from DNA, 9; for viruses, 9, 10, 49–50

flagella, 321–2; composition of, 324–7; eukaryotic, 329; genes for structure and operation of, 327, 329; lost from symbionts, 389, 401; mechanism of movement of, 329–31; production of, 327–8; regeneration and terminal growth of, 328–9; ultrastructure of, 322–4, 325, (hollow centre) 323, 329

flagellins, 324–7; molecules of, are self-assembling into shaft, 328

Flavobacterium: in Dalapon community, 202; in methane community, 201, 202

flavodoxins, reducing agents in nitrogen fixation, 348, 351

fluid mosaic model of membrane structure, 271–3; applied to bacterial membranes, 273–7; cause of mosaic as phase separation, 277, or phase transition, 277–8; and membrane synthesis, 286–91; visible example of, in *Halobium,* 278–9

5-fluorocytosine, antifungal agent producing malfunctioning RNA, 425; fungal mutants resistant to, 425

focus assays, of Rous sarcoma virus and lymphoma virus, 30

forests, viruses for control of insect pests in, 471, 472, 478–9

Frankia spp., nitrogen-fixing symbionts in non-legumes, 355, 356

fruiting bodies: of *Myxobacter,* 231; of *Myxococcus,* 261–2

fucoserratene, hydrocarbon sperm attractant of *Fucus* spp., 209, 210

Fucus serratus, F. vesiculosus, sperm attractant produced by eggs of, 209

fungi pathogenic to humans, 415–17, 426–7; chemotherapy for, 423–6; detection of, 417–18; dimorphism of, 419–23; exocellular products of, in infections, 418–19; factors predisposing to infection by, 417; inhibitors of change of form of, as possible future antibiotics, 427; pathogenic form of, usually unicellular, saprobic form mycelial, 420; *see also individual species*

galactokinase, *E. coli* mutants lacking, 298, 312

galactose: chemotaxis to, in *E. coli,* 334, 336; transport systems for, 308, 309, 312–13, 317

β-galactosidase of *E. coli,* 300; competition between *E. coli* with plasmid coding for, and mutant without plasmid, but with gene coding for increased activity of, 195; hydrolysis of nitrophenol-β-galactoside by, 298–9; synthesised on cytoplasmic ribosomes, 287

β-galactoside permease, as envelope marker in study of cell-wall growth in *E. coli,* 241, 242

β-galactosides: transport of, in *E. coli,* 290, 299–300, (kinetics of) 300–3

gangliosides (glycosphingolipids), of cell membrane, 137, 140; as receptors for enterotoxins, 138, 139, and tetanospasmin, 140

genetic complementation, 48

genetic manipulations: dangers of, 105; plasmids in, 105–7, 110

genetic recombinants, of related oncornaviruses, 28, 50–1

genetic transmission of virus, as DNA provirus, 40–1

genetics, of polyoma virus and SV40, 18–20

Geodermatophilus, cell cycle of, 228, 230, 239

germinal virus, *see* provirus

germination: of myxospores (cysts), 262; of spores of bacilli, 248

Giardia lamblia (G. intestinalis), parasite of humans, 434; epidemiology of, 440 gliding movement in bacteria, 332–3, 339

Gloeocapsa: internal membrane structure of, and nitrogenase, 352

glucans: changes in relative amounts of α- and β-linked, in cell walls of two forms of some pathogenic fungi, 420

glucocorticosteroid hormones, enhance production of viruses from proviruses, 44

glucose: ATP equivalents available from, aerobically, and anaerobically by different metabolic pathways, 168; microbial growth on, in terms of ATP equivalents obtained, indicates metabolic pathway in operation, 159; as principal product released to host by symbiont chloroplasts, 405–6; represses aerobic metabolism in *Saccharomyces,* 172, 174; represses production of flagella in *E. coli,* 328; respiration in trypomastigotes of *Trypanosoma* based on conversion of, to pyruvate, 442

glutamate: transport of, driven by pH gradient, 316

glutamate dehydrogenase, assimilation of ammonia via, 169, 358

glutamate synthase, 169, 175; as regulator of nitrogen metabolism, 359

glutamine: reaction of, with 2-oxoglutarate to give glutamate, 359; transport of, 308, 317

glutamine synthetase, assimilation of ammonia via, 169, 175; 358–9

glycerol: auxotrophs for, cease phospholipid synthesis when deprived of, 285, 290; ether and vinyl ether linkages to, in phospholipids, 271; induction of cyst formation in myxobacteria by, 262; L configuration of, in phospholipids of halophilic bacteria, 271; in lipids of host, appearance of carbon from symbiont photosynthate in, 391

glycine, transport of, 317

glycogen: formation of, introduces error into calculation of growth efficiency, 160–1

glycolic acid, product of symbiont chloroplasts, 406

glycolipids, glycophospholipids, in cell membrane of Gram-positive bacteria, 268

glycophorin: integral glycoprotein of erythrocyte membrane, exposed on both surfaces, 275–6, 287, 289

glycoproteins: antibodies to, in fowls infected with avian leukaemia viruses, 42; in envelope of herpes virus, 2; of cell membrane, in transformed cells, 53; gene for, in Rous sarcoma virus, 50; as helper factor for defective Rous sarcoma virus, 43; of mouse leukaemia virus, 34; receptor for colicin E composed of, 114; receptors for, in host specificity for oncornaviruses, 45, 48; Rous sarcoma virus mutant lacking, 48; of sex hormones of *Blepharisma* type I, 217, and *Volvox,* 217; on *Trypanosoma brucei,* 443; two, protruding from oncornaviruses, 32, and promoting adsorption to host cell, 33

GMP, cyclic: increases at initiation of sporulation in bacilli, 248; and stalk length in *Caulobacter,* 253

Golgi bodies, in free-living and symbiont *Platymonas,* 401

gonidia: cells released from filaments of *Leucothrix,* capable of gliding movement, 225

gorgonian coelenterates: infection of larvae of, with *Gymnodinium* symbiont, 395

Gram-negative bacteria: adhesions between cell wall and cell membrane in, 289, 290; cell membrane of, 265; fatty acids in cell membrane of, 270; insertion of flagellum in, 322, 324; lipopolysaccharide endotoxin of, 148; phospholipid asymmetry in outer membrane of, 288–9

Gram-positive bacteria: cell membrane of, 265; constituents of cell membrane of, 268–9, 270; insertion of flagellum in, 322, 324

granulosis virus, of *Pieris brassicae,* 470–1

griseofulvin, anti-dermatophytic agent, affecting microtubular function, and accumulating in keratinised tissues, 424–5

growth: intercalary, of rod-shaped organisms, 235–6; polar, 236–7, (gives potential for morphogenesis) 238, 239

growth efficiency, 165–7; of chemolithotrophs, 371, (maximum theoretical) 381; factors affecting, (environmental) 170–1, (organism) 166–7, (substrate) 167–9; importance of, ecologically, 175, and in industrial processes, 175–7; maintenance energy requirement, and 'true yield' in, 161–5; regulation of, 171–2, (by feedback) 172–3, (mechanisms for changing) 173–5; sources of error in estimating, 160–1; in terms of yield coefficient, 157–8, of yield expressed in terms of ATP equivalents, 158–9, and of maximum yield in terms of ATP, 165–6

growth factors, mimicked by transforming viruses? 55

growth rate: growth efficiency usually increases with, 381; portion of substrate utilisation independent of (maintenance energy requirement, or loss of cell mass, through death and autolysis?) 163, 164

gums of aerobic nitrogen-fixing bacteria, obstruct diffusion of oxygen to nitrogenase? 353

Gymnodinium microadriaticum, dinoflagellate symbiont of benthic marine invertebrates, 389–90, 409; transmitted in eggs of coelenterates (except gorgonians), 395

haemagglutinin, in botulinum toxin complexes, 132, 140

α-haemolysins, plasmids coding for, 89

haemolytic bacterial toxins, 130, 142, 143, 146; assay of, 131

Haemophilus spp., cleavage of virus DNA by endonucleases of, 6, 7, 8

Halobacterium cutirubrum, phytyl ether lipids in membrane of, 285

Halobacterium halobium, bacteriorhodopsin in membrane mosaic of, 278 halophilic bacteria, phospholipids of cell membrane of, 270–1

Hansenula anomala; growth coefficient of, on methanol, 167

Hansenula wingei, agglutinins on mating cells of, 215

Hartmannella, soil amoeba occasionally pathogenic to humans, 434; epidemiology of, 440, 441

Heliothis virescens, H. zea, insect pests: commercial production of nuclear polyhedrosis virus for control of, 474, 475, 476; specificity of virus for, 476

helper viruses, for defective strains of oncornaviruses, 28, 43, 48–9

herpes viruses, 1, 2; lytic infection by, 28; sometimes latent in humans, 57

heterocysts of cyanobacteria, nitrogen-fixing, 224–5, 259, 260; lack of photosystem II in, 352

heteroduplex plasmids, 97–8, 99, 113

hexose phosphates, transport of, 314

histidine, transport of, 308, 309, 311

histin, inhibitor of RNA polymerase produced in mycelial form of *Histoplasma capsulatum,* 423

histones: associated with DNA in virion, 18; in maturation of viral DNA? 13; in nucleosomes of eukaryotes, 9

Histoplasma capsulatum, 'facultative pathogen', 415; biochemistry of morphological changes in, 420, 422–3; parasitic form yeast-like, saprobic form mycelial, 420

holdfasts, of aquatic microbes, 222–3

homeoviscous adaptation to temperature change (keeping membrane fluidity constant), 282, 285

hook: of axial filament of spirochaetes, 331; of flagellum, 322, 327

hormogonia (short lengths of filament), in reproduction of *Beggiatoa,* 224

hormones, microbial, 208, 218–19; autotactic, 207, 208–9; sex, 209–18

hospitals: antibiotic-resistant bacteria in, 84, 86; bacteraemias associated with, 148

host range: changed in recombinant viruses, 51; mutants of viruses for, 20; of oncornaviruses, 44–6

human cells in culture: semi-permissive for SV40, non-permissive for polyoma virus, 3

humans: fungi pathogenic to, *see* fungi; infection of, by *E. coli* carrying plasmids coding for enterotoxins, 89; papova and herpes viruses sometimes latent in, 57; plasmid-mediated antibiotic resistance in, 84–6; protozoans pathogenic to, 431–2 (*see also individual species*); question of RNA tumour viruses in, 56–7

Hydra, with symbiotic *Chlorella,* 388; *Chlorella* in buds and eggs of, 399; expels dead *Chlorella,* 410; regulation of symbiont population in, 398–9; symbiont-free strains of, 396, (not found in nature) 398, (reinfection of) 399–400

Hydra viridis, site of symbionts in, 396

hydrocarbons; plasmids coding for degradation of, 91; sperm attractants of marine algae as, 210–11

hydrogen: evolution of, by hydrogenase (inhibited by carbon monoxide) 344–5, 346, and by nitrogenase (not inhibited by carbon monoxide) 346, 354; fate of, in nitrogen fixers, 354; organisms oxidising, to methane, 366, and to water, 365; reducing potential from oxidation of, 372

hydrogenase: associated with nitrogenase in nitrogen-fixers, but present also in non-nitrogen-fixers, 344–5, 346, 353–4; in hydrogen-oxidising bacteria, 367

hydroxylamine: energy available from oxidation of, to nitrite, 370; as intermediate in oxidation of ammonia to nitrite, 379; organisms oxidising, 364; oxidation of ammonia to, requires energy, 370, 371

hydroxylamine dehydrogenase, 379

Hyphomicrobium, methanol-oxidising budding bacterium, 233; in methane community, 201; polarisation of cell of, 238; polymorphic cell cycle of, 228, 229, 230, 238, 239, (binary fission in) 234, (polar growth in) 236–7

imidazole derivatives, antifungal agents, 424; variation in resistance to, with phase of growth, 425

immune response in humans: impairment of, by disease or treatment, favours development of fungal infections, 417, 418

immunofluorescence methods, 32; for study of cell-wall growth, 241

industrial processes using microbes, growth efficiency in: production of single-cell protein, 175–6; waste-processing, 176–7

infection with virus, congenital, 40, 41

influenza virus, segmented ssRNA of, 36

insect pests, microbial control of, 459–60; artificial rearing diets for, 472; development of resistance in, 477–8; economics of, 460–1; history of, 461–2; prospects of, 478–9; safety of, 475–7, 479; *see also Bacillus popilliae, B. thuringiensis,* viruses

insect vectors: *Anopheles* mosquitoes, of *Plasmodium,* 447, 451–2, 453; mosquitoes and rabbit fleas, of myxomatosis virus, 477; *Phlebotomus* sandflies, of *Leishmania,* 445, 451, 453; Reduviid bugs, of *Trypanosoma cruzi,* 444, 449–50; tsetse fly, of *Trypanosoma brucei,* 441, 448–9, 452

interactions: between microbial populations, 181–2, 203–4 (*see also* amensalism, commensalism, competition, mutualism, neutralism, prey-predator relations); between symbionts and hosts, 392–4, 397–9, 402–4, 405–6

interfering units, assay of leukaemia virus by, 31

interferon, and synthesis of T antigen in cells infected with polyoma virus or SV40, 16

invasiveness, of tumour cells, 54

ionising radiations, cause release of infectious virus from DNA provirus, 43, 44

ionophores, and transport across membrane, 316, 317

iron: cell-surface receptors for colicins are also receptors for, 115; energy available from oxidation of ferrous to ferric, 370, 371, 378; as ferric hydroxide, adsorbed to sheath of colonial *Sphaerotilis*, 223; lack of, affects growth efficiency of aerobic yeast, 171; in proteins of nitrogenase, 349–50; required for nitrogen fixation, 344

iron protein of nitrogenase, 347, 348, 351; biophysical probes of, 349; gene for, 359; interchangeable between species, 348

iron-molybdenum protein of nitrogenase, *see* molybdenum-iron

isoelectric focusing, for separation of proteins, 132, 133

isoleucine, transport of, 317

isoprenoid alcohol phosphokinase, as integral protein of prokaryote cell membrane, 275, 276

isoschizomers, endonucleases from different sources recognising same nucleotide sequences in DNA, 5

Isospora belli, parasite of humans, 436–7

Isospora hominis, as phase of bovine and porcine *Sarcocystis*, 439

K plasmids, producing antigens (large filamentous proteins) on cell surface of *E. coli*, 89, 90

kanamycin B, plasmid-coded resistance to, 88, 94–5, 98

kappa particles (endosymbionts), in killer strains of *Paramecium*, 109

keratin: breakdown of, by keratinases of dermatophytes, 418–19

keratinised tissue, griseofulvin accumulates in, 425

killer strains, of yeasts and *Paramecium*, 109

kinetics: of competition of mutants with parent population, 190–3; of transport of β-galactosides in *E. coli*, 300–3

Klebsiella aerogenes: aerobic metabolism in, 160, 172; energy charge of, with glucose in excess or limited, 173; maintenance energy requirement of, 164; mutants of, for ribitol metabolism, 195–6, 197; yield of, in terms of ATP equivalents expended, 159; yield of, while fixing nitrogen, compared with that on fixed nitrogen sources, 169; yield coefficient of, 166, 170

Klebsiella pneumoniae: ATP requirement of, for fixing nitrogen, 353; nitrogenase of, 345, 347, 348; oxygen concentration, and nitrogen fixation by, 352; plasmids from, carrying genes for nitrogen fixation, 357–9

Klebsiella spp.: antibiotic-resistant strains of, 84; lipopolysaccharide endotoxin of, 148; nitrogen fixation by, 355, 356

β-lactam antibiotics: detoxified by R-plasmid-coded β-lactamases, 87, 88, 94; inhibit septation in *E. coli*, 72

lactate: affinity for, in competition between *E. coli* and *Spirillum*, 197

lactate oxidase, membrane enzyme of *E. coli*; lipid fluidity and, 278

lactobacilli, cyclopropane fatty acids in phospholipids of, 270

Lactobacillus casei: commensalism between *S. cerevisiae* and, 185; yield coefficient of, grown anaerobically on glucose, 166

Lactobacillus plantarum: yield coefficient of, grown anaerobically on glucose, 166

Lactobacillus sp., not affected by presence of *Streptococcus* sp. in chemostat culture, 183

lactose, plasmids coding for fermentation of, 93

lactose permease, integral membrane protein, 287

Lampropedia, formation of colony by, 227

lattice formation, by aquatic bacteria, 226, 227

leaf nodules, bacteria in, 357

lectins, plant: activation of latent oncornaviruses by, 44; transformed cells sensitive to agglutination by, 53

leghaemoglobin, as oxygen buffer in legume nodules, 353

legume nodules, 344, 353

Leptospirillum ferrooxidans, 365

Leishmania spp., parasites of humans, transmitted by sandflies, 445–7; control of, 453; cutaneous and visceral forms of disease, caused by strains of, with different temperature tolerances, 446; epidemiology of, 451; primitive method of immunisation against cutaneous, 453

leucine, transport of, 317

leucocidin, of *Staphylococcus aureus*, 130

Leucothrix, cell cycle of, 223, 224, 225

leukaemia viruses: acting as helper viruses, 48; of cats and cattle (oncornaviruses), 27, 42; in cell cultures, 29, 30–1; defective, requiring helper viruses, 48; titration of, as units preventing entry of sarcoma virus into cell, 30–1; *see also* avian leukaemia viruses, mouse leukaemia viruses leukoencephalopathy, progressive multifocal: virus related to SV40 isolated from brain tissue of patients with, 1

lichens, nitrogen fixation by cyanobacteria in, 355, 357

light: high intensity of, causes expulsion of *Chlorella* from *Hydra*, 399

Limnaea peregra, gastropod with symbiotic *Chlorella*, 388

limulus test, for lipopolysaccharide endotoxin of Gram-negative bacteria, 148–9

lincomycin, mechanism of plasmid-coded resistance to, 88

lipases, destroy envelope and infectivity of herpes viruses, 2

lipid: ratio of protein to, in cell membrane of prokaryotes, resembles that in mitochondrial membranes and endoplasmic reticulum, 267

lipid A component (glucosamine phosphate, carrying long-chain fatty acids), responsible for toxicity of lipopolysaccharide endotoxin, 148, 149

lipid bilayers, artificial, 266; fluidity of, 272; pure phospholipids dispersed in saline spontaneously assemble into, 272

lipid-glycoprotein, of envelope of herpes viruses, 1, 2

lipids, not present in polyoma virus and SV40, 4

lipopolysaccharide, of cell envelope of Gram-negative bacteria, 265, 270; as endotoxin, 128, 130, 147–50; mostly confined to outer half of bilayer, 288–9; in *Salmonella typhimurium*, 131, 277

lipoprotein, of cell membrane of Gram-negative bacteria, 67, 288; envelope of oncornaviruses derived from, 32; mutants deficient in, 72; protein of, 71

lipoteichoic acids, in cell membrane of Gram-positive bacteria, 269

Listeria monocytogenes, toxin of, 145

lupus erythematosis, systemic, in humans: associated with xenotropic virus of mice, 45, 57

lymphoma (lymphocytic leukaemia) viruses, difficulties of assaying, 30

lysine: ϵ-N-methyl, in flagellin of some serotypes of *Salmonella*, 327; transport of, driven by membrane potential, 316

lytic infection of cells, 3, 28; cells susceptible to, by either polyoma virus or SV40, are usually not susceptible to the other, 22

M protein, membrane-bound, for β-galactoside transport in *E. coli*, 307

macrophages, human: multiplication of *Leishmania* in, 446

magnesium: and ATP, in conformation of iron protein of nitrogenase, 349–50, 351; colicin K causes lethal loss of, from cell, 116, 117; in haemolysis by staphylococcal β-toxin, 142; required for nitrogen fixation, 348

maintenance energy requirement for microbes, 161–3; nature of, 163–4; and 'true yield', 164–5

malate dehydrogenase, peripheral protein of prokaryote cell membrane, 275

maltose: periplasmic binding protein as receptor for chemotaxis to, 336; synthesised by symbiotic *Chlorella*, and released to host, 397, 412

mannitol, produced by symbiotic *Platymonas*, 403

marker rescue method, for mapping viral mutations, 19

mating pairs of bacteria, 99, 100

maturation (encapsulation) of viral DNA, 11, 13, 19; mutants affected in, 47; occurs at cell surface, 37

membrane elution method, for synchronisation of bacterial cultures, 60–1, 62

mercury; plasmids conferring resistance to compounds of, by conversion to metallic mercury, which evaporates, 92, 93

Mesodinium rubrum, marine ciliate with symbiotic algae, 388, 404

mesosomes, 268; conditions of formation of, 279, 291

metabolic pathways: and growth efficiency, 159, 167, 168, 171; oxygen supply and, 172

methane: microbial community utilising, 201–2; organisms oxidising hydrogen to, 366

methane-oxidising bacteria: in methane community, 201; methane oxidation by, inhibited by acetylene, 356; nitrogen fixation by, 355

methane-producing bacteria (anaerobic), nitrogen fixation by, 355

Methanobacterium, 366

methanol: potentially toxic, 175; yield of different organisms grown on, 167, and of symbiotic mixture of bacteria grown on, 171

methionine (as S-adenosyl methionine): methylation of a membrane protein by, in generation of tumbling behaviour in motile bacteria, 338, 339–40

7–methyl guanosine, on 5' end of mRNAs of SV40 and probably polyoma virus, 15

methyl thio-β-galactoside, non-metabolised analogue of lactose, 298, 301–3

methylation: of a membrane protein, in generation of tumbling behaviour in motile bacteria, 338, 339–40; modification of bacterial DNA by, as protection against own endonucleases, 5–6

Methylosinus sp., spore formation in, 230, 232

micro-injection technique, for insertion of nucleic acids into cells in culture, 16–17

Microsporum sp., dermatophyte, 416

microtubules, griseofulvin interferes with function of, 424

mitochondria: in free-living and symbiont *Platymonas,* 401; non-functional, or partly functional, in trypo-mastigotes of *Trypanosoma,* 442, 443; protein:lipid ratio in inner membranes of, 267; in transition from yeast-like to mycelial form of pathogenic fungi, 420

mitomycin C, induces colicin production, 112, 124

molybdenum, required for nitrogen fixation, 344

molybdenum-iron protein of nitrogenase, 347, 348, 351; biophysical probes of, 349; genes for, 359; interchangeable between species, 348

monkey oncornaviruses, some associated with humans? 56–7

morphogenesis: definition of, 229; potential for, in obligate polar growth, 238

Mossbauer spectroscopy, for detecting oxidation state of iron in molybdenum-iron protein of nitrogenase, 349

motility, bacterial, 339–40; chemotaxis in, 333–9; flagellar, 321–31; gliding, 332–3; spirochaetal, 331–2

mouse erythroleukaemia (Friend) virus, 52; has helper virus itself causing lymphoid leukaemia, 49

mouse leukaemia (hereditary lymphoma) virus, 27, 28; assay of, 30; glycoprotein of, 32; proteins of, 33, 34, 39; transmission of, 40, 42, 46

mouse mammary tumour virus, 27, 29; factors promoting expression of latent, 44; glycoprotein of, 32–3; transmission of (viral genome in host DNA, activated in lactating mammary gland), 40–1

mouse oncornaviruses, host range of, 45–6

mouse sarcoma virus, 55, 56

Mucor sp., mutualism between *Rhodotorula* and, 184

Mucorales, joint production of trisporic acid by two mating types of, leading to formation of zygospores, 214–15

multicellular prokaryotes, 232, 239; cooperation between cells of (*Anabaena*), 260

multifidene, hydrocarbon sperm attractant of *Cutleria multifida,* 210–11

murein, *see* peptidoglycan

'mushroom-shaped' bacterium, 235, 236; monomeric spore-forming cell cycle of, 239

mutagens: agents for 'curing' cells of plasmids may also act as, 78; cause release of infectious virus from provirus, 43, 44

mutants: in chemotaxis, 338–9; kinetics of competition of, with parent population, 190–3; in membrane lipids and proteins, 279; mutator, of *E. coli,* 196–7; in study of transport systems, 298, 299; *see also under individual species of bacteria*

mutualism between organisms (both derive advantage), 183–4

Mycobacterium flavum, aerobic nitrogen fixation by, 352, 355

mycology, medical, 415–17, 426–7; chemotherapy in, 423–6; detection of fungi in, 417–18

Mycoplasma hominis, wriggling movement of, 333

mycoplasmas: aminoacyl derivatives of phosphatidyl glycerol in, 268; cell membrane of, 265; many species require cholesterol for growth, 271

mycorrhiza of conifers, 357

mycoses, 415

Myrica, nitrogen-fixing symbionts in, 356

Myrionema amboinense, coelenterate with green symbionts, 388

Myxobacter; cellular interdependence in, 232; chemotaxis in, 339; fruiting body formation in, 231

Myxobacterales, gliding movement of, 332

myxobacteria, cell cycle of, 239, 261–2

Myxococcus xanthus: cell cycle of, 261–2; gliding movement of, 333

myxomatosis virus of rabbits, 477

Myxomycetes, chemotaxis in, 339

Myxophyceae, gliding movement of, 333

myxospores (cysts), 262

myxoviruses, negative-strand RNA of, 36

NAD: diphtheria toxin and, 135; energy required for reduction of, 376; hydrogen oxidation by bacteria coupled to reduction of, 367; indirect reduction of, in most chemolithotrophs, by 'uphill' electron flow, 373, 374–6

NADH, NADPH: ATP equivalence of, 159, 166; required for carbon dioxide fixation by Calvin cycle, 366

NADH dehydrogenase, peripheral protein of prokaryotic cell membrane, 275

NADH oxidase, membrane enzyme of *E. coli,* lipid fluidity and, 278

Naegleria, soil amoeba occasionally pathogenic to humans, 434; epidemiology of, 440, 441

'negative growth' (death of organisms), in calculation of growth efficiency, 161, 163

Neisseria meningitidis, bacteriocin-producing strains of (in a few instances inhibiting *N. gonorrhoeae*), 122

nematodes: insecticidal, with low order of specificity, 475

neomycin, plasmid-coded resistance to, 95, 98

nerve cells: destruction of, by *Trypanosoma cruzi,* 445

neurotoxins, bacterial, 129, 139–41; in botulinum type A toxin, 132; enterotoxin of *Staphylococcus aureus* acts also as, 139; nerve-muscle preparations for study of, 134

neutralism between organisms (no interaction), 183

Nevskia ramosa, polysaccharide matrix of floating colonies of, 225, 226

nicotinic acid, in mutualism of *B. polymyxa* and *P. vulgaris,* 183

nitrate: acting as electron acceptor in anaerobic culture, introduces error into calculation of growth efficiency, 160; ATP equivalent required for conversion of, to ammonia, 168–9; conversion of ammonia of fertilisers to, 344; energy available from oxidation of nitrite to, 370; organisms oxidising nitrite to, 364; reduced by *Thiobacillus denitrificans,* 365; uptake of, by *Convoluta,* 403

nitrite: energy available from oxidation of ammonia and hydroxylamine to, and from oxidation of, to nitrate, 370, 371; growth efficiency of organisms oxidising, 371; organisms oxidising, to nitrate, 364, 378; organisms oxidising ammonia and hydroxylamine to, 364

Nitrobacter, 364, 375; ATP required for reduction of NAD by, 376; calculation of energy balance for, oxidising nitrite and fixing carbon dioxide, 378; P/O ratios for, oxidising nitrite, 377

Nitrococcus, 364

nitrogen: cycle of, 343; glutamate synthase as regulator of metabolism of, 359; heavy, in study of nitrogen fixation, 344; recycling of, by symbionts of coral reefs, 392; scavenging from atmosphere of traces of fixed, by *Azotomonas* and *Pseudomonas* spp., 355; source of, and growth efficiency, 168–9

nitrogen fixation: ATP required for, 169, 345, 346; biochemistry of, 345–51; ecology of, 354–7; genetics of, 357–9; organisms effecting, 354, 355; physiology of, 352–4; *see also under* cyanobacteria

nitrogenase, 344, 345–6; acetylene test for, 345, 355–6; components of, 347–8; inactivated by oxygen, 345, 352, 356; protection of, against oxygen, 169, 352–3; scheme for action of, 351; triple-bonded-nitrogen substrates reduced by, 348–9

Nitrosococcus, 364

Nitrosolobus, 364

Nitrosomonas, 364; P/O ratios for, 377

Nitrosospira, 364

Nitrosovibrio, 364

Nitrospira, 364

Noctiluca, protozoan with green symbionts, 388

nopaline (unusual amino acid), in plant galls produced by *Agrobacterium* with tumour-inducing plasmid, 94

Nostoc: nitrogen fixation by, free-living and in lichens, 355

nuclear polyhedrosis viruses of insects, 471–2, 473, 474; commercial production of, 475

nucleoids: central in C-type oncornaviruses, eccentric in B-type, 32, 39

nucleoside phosphorylases, in transport of nucleosides, 306

nucleosides: cell-suface receptor for colicin K is also receptor for, 114; purines and pyrimidines converted to, during transport into cell? 306; transport of, 319

nucleosomes, structures of DNA and histones in eukaryotic cells, 9

nucleotides: accumulation of highly phosphorylated, at onset of sporulation in bacilli, 248

nystatin (polyene antibiotic), antifungal agent, 423, 424

octopine (unusual amino acid), in plant galls produced by *Agrobacterium* with tumour-inducing plasmid, 94

Oedigonium, sex hormone of, 207

oncogenes, viral, 27, 33, 47; carried by acutely oncogenic, but not by slowly oncogenic viruses, 52; concept of activation of, 44; derived from host cell genes? 51, 55, 56

oncogenesis: slow, usually clonal, rapid, usually by spread of virus into neighbouring cells, 52

oncoprotein, virus-coded: question of existence of, 54–5

oncornaviruses (RNA tumour viruses), 27, 32–4; assay methods for, 29–32; defective mutants of, 47–8, and helper viruses for, 28, 48–9; exist as infectious particles containing RNA genes, and as DNA genes integrated into host DNA (provirus), 27–8; genetic recombinants of related species of, 28, 50–1; hosts of, 44–6; infections with, productive, 28, or non-productive, 29; inheritance of exogenously acquired, 46; mapping genome of, 49–50; oncogene in, 27, 55–6; oncogenesis induced by, 52; oncoprotein coded by? 54–5; properties of cells transformed by, 52–4; provirus of, 34–7; (detection and expression) 42–4; question of occurrence of, in humans, 56–7; reverse transcription of RNA of, and integration of resulting DNA into host genome, 37–9; transcription and maturation of, 39; transmission of, horizontal and vertical, 40–2

oocysts: of *Isospora,* 436; of *Toxoplasma,* 438

oogoniol: sterol produced by male hyphae of *Achlya,* affecting female hyphae, 213 orcinol, community of three bacteria growing on, 202

Orychtes rhinoceros (coconut rhinoceros beetle), virus for control of, 478

Oscillatoria princeps, gliding movement of, 333

osmotic shock, transport systems sensitive to, and resistant to, 317–18

oxidative phosphorylation: ATP generation by, in chemolithotrophs, 374, 377; partial uncoupling of, in bacteria? 174; transport systems sensitive to uncouplers of, 317

oxygen: dissolved, and growth efficiency, 170–1; induction by, of enzymes for aerobic metabolism, 172; nitrogenase inactivated by, 345, 352, 356; nitrogenase proteins and, 347, 348; particulate preparations of *Azotobacter* nitrogenase fairly tolerant to, 346, 348; protection of nitrogenase against, 169, 352–3; yield in terms of consumption of, 158

oxygenase reaction: question of occurrence of, in oxidation of sulphur compounds, 379–80

papilloma (wart) viruses, 3

papova DNA viruses, 1, 3; induce tumours in newborn rodents, 2; lytic infection by, 28; sometimes latent in humans, 57; *see also* polyoma virus, SV40

Paracoccidioides brasiliensis, causing lung infection, 416; parasitic form yeast-like, saprobic form mycelial, 420

Paramecium, protozoan with *Chlorella* symbionts, 396; alkaline phosphatase of, repressed by symbionts, 400, 410; population of *Chlorella* in, 399; symbiont-free, not found in nature, 398; wide range of *Chlorella* strains in, 400

paramyxoviruses, with negative-strand RNA, 36

parasites: protozoan, of humans, acquired congenitally, 453–5, orally, 432–41, venereally, 455–6, or via the blood, 441–53; use of term, 431

parasporal bodies (protein crystals): in *B. popilliae,* not involved in pathogenicity, 468; in *B. sotto,* toxic, 462, 463; in *B. thuringiensis,* toxic, 463

Paspalum (grass), with nitrogen-fixing *Azotobacter* symbiont, 356

Pasteur effect (stimulation of glycolysis in anaerobic conditions), in *Saccharomyces,* 172

pathogenicity of bacteria for mammals: plasmids conferring, 89–90; toxins in, 127–8

Paulinella chromatophora, testate amoeba with symbiotic cyanellae, 408, 411

Pelodictyon, colony formation by, 226, 227

Peliaina cyanea, protozoan with symbiotic cyanellae, 408, 409

penicillin-binding sites in *E. coli,* equated with potential division sites, 73–4, 243

penicillinases: membrane-bound, of *B. licheniformis,* from which exoenzyme is released, 287; plasmid-coded, 87, 94

penicillins, inhibition of peptidoglycan synthesis by, 72

peptide antibiotics, produced by spore-formers at sporulation, 249

peptides, transport of, 312

peptidoglycan (murein): enzymes involved in synthesis and degradation of, in cell cycle of *E. coli,* 74–5; of *E. coli* cell envelope, 66, 67; layer of, invaginated before septation, 71; layers of, and insertion of flagella, 324; septum-specific and basal syntheses of, 72–3; from spirochaetes, is helical, 331

permeability barrier, plasmid-coded restistance to antibiotics acting by means of, 88

permeases: model employing, for energy-coupling to transport, 314; use of term, 300

pH: affects growth rate more than growth efficiency, 170; gradient of, across membrane, in chemiosmotic theory of transport, 315, 316; and release of maltose by *Chlorella,* 397; separation of plasmid DNA by means of its compact state at high values of, 82

Phaenocora, platyhelminth with symbiotic *Chlorella,* 388; reinfected each generation, 399, by ingestion of *Chlorella,* 400

Phaenocora typhlops, in loose facultative association with free-living *Chlorella,* 396, 399

phagocytosis, entry of symbionts to host cells by, 399, 410

phenotypic mixing, 48–9

phenyl alanine, transport of, 317

Phialophora spp., tropical pathogenic fungi, 416

phosphate: affinity for, in competition between *Spirillum* and rod-shaped bacterium, 197–8; concentration of, and length of stalk in *Caulobacter,* 252–3; as germinant for myxospores, 262; periplasmic binding protein for, 308; recycling of, by symbionts of coral reefs, 393; shortage of, and phospholipids of *Pseudomonas,* 279

phosphatidyl choline, 269: dioleyl and distearyl, transition temperatures of, 274; less common in bacteria than in eukaryotes, 268

phosphatidyl ethanolamine, 269; *N*-methylated, in *Thiobacillus,* clostridia, and rhizobia, 268; transmembrane movement of, in *B. megaterium,* 276

phosphatidyl glycerol, 269; aminoacyl derivatives of, in Gram-positive bacteria and mycoplasmas, 268, 269; dihydrophytyl ether form of, in *Sulfolobus,* 271; glucosaminyl derivative of, 268, 269; in phosphoenol pyruvate-sugar phosphotransferase system, 304 phosphatidyl serine decarboxylase, temperature-sensitive bacterial mutants unable to synthesise, 285

phosphodiesterase of cell surface, in response of *Dictyostelium* amoeba to cAMP, 209

phosphoenol pyruvate: carbon dioxide fixation by carboxylation of, in *Thiobacillus denitrificans,* 382

phosphoenol pyruvate-sugar phosphotransferase system, 303–5; enzyme II of, as integral protein with phospholipid requirement, 276; genetics of, 305; glucose and mannose receptors as components of, 336; isolation of enzymes of, 306, 309

phosphofructokinase: in changes of morphology in *Candida albicans,* 422; as key enzyme in regulation of change between aerobic and anaerobic metabolism, 172

phospho-β-galactosidase of *Staphylococcus aureus,* gene for, 305

phospholipase C, α-toxin of *Cl. welchii,* 141, 143

phospholipases, in isolation of cell membranes, 267

phospholipids: of cell envelope, attachment of *E. coli* DNA to, 67; of cell membranes, major types of, 268–71; fluidity of, and orientation of bacteriophage coat protein in cell membrane of *E. coli,* 288; fluidity gradient within molecules of, 276; interaction between proteins and, 274; lateral diffusion of, within membrane bilayer, 276, 289; liquid crystal form of, 272–3; phase transition temperature of, 273–4, 278; phase transition temperature and viscosity of, from *E. coli* grown at different temperatures, 283; phytyl ether, of halobacteria, 270–1, 279, 285; pure, dispersed in saline, spontaneously assemble into stable bilayers, 272; replaced by glycolipids in phosphate-limited *Pseudomonas,* 279; restricted to inner layer of outer membrane of Gram-negative bacteria, 289

phosphoribosyl transferase, in transport of nucleic acid bases and nucleosides, 306

photolithotrophs, 363

photosynthate of symbionts: factors in host stimulating release of, 393, 403, 406, 412; transfer of, to host, from *Chlorella,* 397, from chloroplasts, 405–6, from dinoflagellates, 390–2, and from *Platymonas,* 403

photosystem 2 (oxygen-evolving), lacking from heterocysts of cyanobacteria, 352

phototaxis, in photosynthetic bacteria, 339

Phytophthora (plant parasite), requires sterols for formation of oospores, 213

Pichia pinus: growth coefficient of, on methanol, 167

Pieris brassicae: granulosis virus disease of, 470–1; varying resistance of, to virus, 477

pigs, young: infections of, by *E. coli* carrying plasmids coding for enterotoxins, 89, 90

pili: F-, on surfaces of cells containing F plasmids, 95, 96, (in conjugation) 99; I-like, shorter than F-, with different antigen, 95, 113; plasmid genes for formation of, 99; on surfaces of cells containing some varieties of Col plasmid, 113

Pityrosporum furfur, obligate parasite of humans, 416

placenta, human: parasites able to cross, 438, 454

Placopecten magellanicus, bivalve with green symbionts, 388

Planctomyces: dimorphic cell cycle of, 239; holdfast and fimbriae of, 222

plaque titrations: assay of lytic viruses by, 29; syncytial, assay of some lymphoma strains by, 30

plasmalogens (phospholipids with alkyl groups linked to glycerol by vinyl ether bond), 271

plasmid chimeras, 106

plasmids, 77–8; 'curing' of cells by loss of, 78–9; classes of, 82–3; cryptic, 94; DNA of, 80–2, (homologies of) 97–8, 99, (replication of) 100, 102–3, (transfer of, to other plasmids in same cell, or to chromosome or bacteriophage) 103–5; incompatibility groups of, 95, 97, 103; introduction of new genes into, 105–7; responsible for pathogenicity in mammals, 89–90, for resistance to antibiotics, 83–6, 113, and to mercury, 92, 93, and for toxin production by bacteria, 132; in staphylococci, 94–5; transfer of, in bacterial conjugation, 77, 79–80, 99–100; transfer of genes for nitrogen fixation by means of, 357–9; tumour-inducing (in plants), 93–4; *see also* Col, degradative, F, K, *and* R plasmids

plasminogen: protease cleaving, to plasmin, 54

Plasmodium spp., parasites of humans, transmitted by mosquitoes, 447–8, 449; control of infection by, 453; epidemiology of, 451–2

Platymonas convoluta, symbiont in *Convoluta,* 388, 401; interactions between host and, 402–3; specificity of association, 403–4; transmission of, 403

Platymonas tetrathele, 403

Plectonema, aerobic nitrogen fixation by, 352, 355

pneumolysin, toxin of *Streptococcus pneumoniae,* 145

polio virus, 36

polyene antibiotics: act on cell membrane of pathogenic fungi (which differs in sterol from mammalian membranes), 423–4; variation in resistance to, with phase of growth, 425–6

poly-β-hydroxybutyrate, in *Azotobacter beijerinckii,* 353

polynucleotide ligase, in maturation of viral DNA, 11, 12–13

polyoma virus, 1, 3–5; antigens in cells infected by, 16–18; DNA of, 4–5, (cleavage of, by bacterial endonucleases) 6–8, (map of) 8, (replication of) 9, 11–13, (transcription of) 13–15; nucleoprotein complexes in, 15–16; proteins of capsid of, 15–16; SV40 and, compared, 20, 22, 23, 24; temperature-sensitive mutants of, 18–20 polyoxin D, inhibitor of chitin synthetase *in vitro,* does not permeate into pathogenic fungi, 423, 426

polysaccharides, capsular bacterial: adhesion to surfaces by, 223, 225; floating colonies in, 225, 226

Polysphondylium spp.: aggregation hormone of, 209; not responsive to hormone of *Dictyostelium,* 208

Popillia japonica (Japanese beetle), pest in eastern USA and Canada: *B. popilliae* for control of, 468–70

potassium: colicin K causes lethal loss of, from cell, 116, 117; concentration of, in medium, and growth efficiency of different organisms, 171; efflux of, induced by valinomycin, establishing a membrane potential with inward transport of solutes, 316; membrane-damaging toxins affect permeability of cell to, 145

potential difference across membrane, in chemiosmotic theory of transport, 315, 316

Prasinocladus: can replace *Platymonas* as symbiont in *Convoluta,* but is ejected if *Platymonas* becomes available, 403–4

prey-predator relations, between microbes, 186

primers: for DNA polymerase of DNA tumour viruses, 12, 13; for reverse transcriptase of RNA tumour viruses, 33, 37, 38–9

prokaryotes, levels of organisation in (single-celled, occasionally multicellular, multicellular), 231–2, 239

proline, transport of, 310, 317

prosthecate linking, colony formation by, 227

Protaminobacter ruber: growth coefficient of, on methanol, 167

proteases: of cell surfaces, increased in transformed cells (affecting invasiveness?), 54; during isolation of cell membranes, 267; release toxins from protoxins, 132, 141

protein disulphide reductase, of *Candida albicans* (lacking in mycelial form), 422

protein-lipopolysaccharide, *Sphaerotilis* sheath composed of, 223

protein synthesis: induced by antheridiol in male hyphae of *Achlya,* 212–13; inhibited by diphtheria toxin, 134–5; inhibition of, in study of amino-acid uptake, 298; inhibitors of, activate latent proviruses, 44, and prevent new rounds of DNA replication in *E. coli,* 66; involved in replication of DNA of polyoma virus and SV40, 12, and in responses to *Blepharisma* sex hormones, 217, and to trisporic acid, 215

proteins

attached to plasmid DNA, 102; DNA-binding, in cell envelope of *E. coli,* 68–9; DNA-binding, specific to spores, 249; in DNA replication in *E. coli,* (initiator) 66, (terminator) 240, 241;

binding: involved in chemotaxis, 336; involved in transport, 297, 300, 302–3, 306–9; for penicillin, at septation sites in *E. coli,* 73–4, 243; transport systems dependent on, not coupled to proton movement, 317–18;

of cell membrane: effects of isolation procedure on, 267–8; fluidity of, 277; integral (intrinsic, hydrophobic), and peripheral (extrinsic, water-soluble), in fluid mosaic model, 274–6, 286, 287, 288; interaction between phospholipids and, 274; lipid fluidity and orientation of, 288; ratio of, to lipids, in prokaryotes, resembles that in eukaryote mitochondrial membranes and endoplasmic reticulum, 267;

on cell surface: colicin receptors, 114; LETS (large external transformation-sensitive), not present in transformed cells, 53; required for gliding motion, 333;

of viruses: of mouse leukaemia virus, 33, 34; mutants in genes for, 18–20; of oncornaviruses, 31, 32–3; of polyoma virus and SV40, (of virions) 15–16, (virus-induced) 16–18; of Rous sarcoma virus, genes for, 50

Proteus mirabilis: amino acids of flagellin of, 326; peritrichous flagella of, 322, 324

Proteus spp.: antibiotic-resistant strains of, 84; dissociation of co-integrate plasmids in, 87

Proteus vulgaris, mutualism between *B. polymyxa* and, 183–4

protonmotive force, in chemiosmotic theory of energy coupling to transport, 314–17; transport systems dependent on binding proteins, not coupled to, 317–18

protoplasts, 291

protoxins, activated by proteases, 132, 141

protozoans: with cyanellae, 408–9; pathogenic to humans, 431–2 (*see also individual species*); with photosynthetic symbionts, 388 (*see also Paramecium*)

provirus (viral DNA bound to host DNA), 21, 34–7, 40–1; detected in cell by use of radioactive viral RNA, 31; of oncornaviruses, copied from their RNA, 27, 28; of Rous sarcoma virus, 43, 44; synthesis of, *in vitro,* 38; transcribed by host RNA polymerase to give RNA for progeny viruses, and mRNA for viral proteins, 39

Pseudomonas aeruginosa: antibiotic-resistant strains of, 84, 85; enterotoxin of, 139; lipopolysaccharide endotoxin of, 148; R plasmids isolated from, found also in many other genera, 97; strains of, producing bacteriocins, 122; transport of amino acids in, 310; use of bacteriocins in typing strains of, 123

Pseudomonas azotocolligans: scavenging by, of traces of fixed nitrogen from atmosphere, 355

Pseudomonas diminuta, effect of phosphate shortage on phospholipids of, 279, 285

Pseudomonas extorquens: growth coefficient of, on methanol, 167

Pseudomonas fluorescens: motile outgrows non-motile mutant, in oxygen-limiting conditions, 321; polar flagellum of, 322

Pseudomonas lindneri: yield of, in terms of ATP equivalents expended, 159

Pseudomonas putida: in community growing on Dalapon, 202–3; mutant of, with dehalogenase acting on Dalapon, 198, 200

Pseudomonas spp.: in Dalapon community, 202–3; degradative plasmids in, 91, 92; plasmids in, conferring resistance to mercury, 93

Pseudomonas stutzeri: able to metabolise orcinol, 202; sodium concentration and growth efficiency of, 171

pseudotypes of viruses, 48

psychrophilic microbes, phospholipids of, 279, 283

purines: removal of, from DNA, in study of nucleotide sequences, 9; transport of, 306, 319

Purshia, nitrogen-fixing symbionts in, 356

pyrimidines: halogenated, cause release of infectious virus from DNA provirus, 43; transport of, 306, 319

pyruvate: ATP equivalents available from, aerobically and anaerobically, with different residual products, 168; as substrate for nitrogen fixation by nitrogenase, 345; trypomastigotes of *Trypanosoma* oxidise glucose to, 442

Pythium (plant parasite), requires sterols for production of oospores, 213

R (antibiotic-resistant) plasmids, 83; co-integrate and aggregate forms of, 86–7; DNA homology among, 98; ecology and epidemiology of, 83–6; mechanisms of resistance coded by, 87–9; transfer of nitrogen-fixing genes by, 357

radioimmunoassays, 32

receptor sites on cells: for colicins, 114–15, lost in some colicin-resistant mutants, 110, 119; for glycoproteins of oncornaviruses, 33, 37; in host specificity of oncornaviruses, 45; lack of, does not protect against genetic transmission of virus, 41

redox model, for coupling of energy to transport, 314

reducing potential, from inorganic oxidations, 371, 372–3

refractory period: in response of *Allomyces* male gametes to sirenin, 211; in response of *Dictyostelium* to aggregation hormone, 209

reoviruses, with genome segmented into genes, 36

repellants: methods of showing chemotaxis away from, 334, 335; receptors for, in *E. coli,* 336; threshold concentrations of, 335

replicon hypothesis, of DNA attachment to membrane in *E. coli,* 67–8, 102–3

reservoirs, mammalian, for protozoan infections of humans, 431, 456; for *Balantidium,* 436, 440; for *Entamoeba,* 433; for *Giardia,* 434; for *Leishmania,* 451, 453; for *Sarcoplasma,* 439; for *Toxoplasma,* 438, 440; for *Trypanosoma brucei,* 441, 443, 448–9; for *Trypanosoma cruzi,* 444, 449–50, 452

resistance markers, in plasmids: transposition of, 87

respiratory protection, of nitrogenase against oxygen, 352

restriction enzymes (endonucleases) of bacteria, cleaving both strands of dsDNA at specific nucleotide sequences, 5; creation of mutants by, 20; in introduction of new genes into DNA, 105–7; in mapping of Rous sarcoma virus genome, 49–50; as protection against invading species, 5–6; in study of provirus in host DNA, 39

retroviruses, *see* oncornaviruses

reverse transcriptase (RNA-directed DNA polymerase) of oncornaviruses, synthesises DNA on RNA, hydrolyses RNA, and synthesises second DNA strand, 28, 37; assay of, by use of radioactive precursors, 31; gene for, in Rous sarcoma virus, 50; host-derived tRNAs as primers for, 33, 37; of mouse leukaemia virus, 34, 35; mutants defective for, 47, 48; mutant of Rous sarcoma virus with temperature-sensitive, 47; mRNA for, 39; treatment of virus particles to induce activity of, 37; used to prepare viral DNA, 31

rhabdoviruses, with negative-strand RNA, 36

rhizobia: evolution of hydrogen by, 354; nitrogen fixation by, free-living and in association with non-legumes, 356, 360, and in roots of legumes, 344; *N*-methylated phosphatidyl ethanolamine in, 268

Rhizobium meliloti: transfer of nitrogen-fixing genes to, causes formation of nitrogenase proteins, but no nitrogen fixation, 359

Rhizopus, and trisporic acid, 215

Rhodomicrobium vannielii (photoheterotroph), 233; binary fission in, and polar growth, 234; cell forms and spore cycle in, 230, 238, 239, 254–5, 258; colony formation by, 227; sporulation in, 256, 259; swarm cells of, 255–7

Rhodopseudomonas acidophila, cell cycle of, 238, 239

Rhodopseudomonas palustris: asymmetric membrane formation in, 236, 237; dimorphic cell cycle of, 237, 238, 239; grows from one pole only, 234; polarised cell of, 237–8

Rhodopseudomonas spp., nitrogen fixation by, 355

Rhodospirillum, nitrogen fixation by, 355, 356

Rhodotorula, mutualism between *Mucor* and, 184

ribitol dehydrogenase: activity of, towards xylitol, in *K. aerogenes* and mutants, 195–6, 197

riboflavin, in commensalism between *Lb. casei* and *S. cerevisiae,* 185 ribose: periplasmic binding protein as receptor for chemotaxis to, 336; transport of, 317

ribosomes: arrays of, in chromatoid bodies of *Entamoeba* cysts, 433; modified in plasmid-induced resistance to erythromycin and lincomycin, 88; modified in streptomycin-resistant mutants, 83

rice: *Azotobacter beijerinckii* as symbiont of, 356, cyanobacteria in fields of, 357

RNA: early and late, formed in transcription of viral DNA, 13–14; malfunctioning, produced in pathogenic fungi by 5-fluorocytosine, 425; of oncornaviruses, 27, 28, 31, 33, 37; of Rous sarcoma virus, inheritance of, 43; short lengths of, as primer for DNA polymerase in replication of viral DNA, 12, 13

mRNAs: for flagellin, 328; specific, in sporulating bacilli, 248;
    of viruses: coding for capsid proteins, 16; direction of transcription of, 15; mapping of, on DNA of polyoma virus and SV40, 24; production of, from various viral genomes, 36; three specific, isolated from cells infected with polyoma virus or SV40, 14

rRNA: methylation of specific guanine on 23S by plasmid prevents binding of erythromycin and lincomycin to, 88; 16S, cleaved by colicin E3-CA38, and by cloacin DF13, 117

tRNAs: derived from host cell, acting as primers for reverse transcriptase of oncornaviruses, 33, 37

RNA polymerase: of host, transcribes provirus, 39; inhibited by tyrothricin, 249; of polio virus, 36; RNA-directed (in negative ssRNA viruses), incorporated into virion ready to synthesise mRNA, 36; in sporulation of bacilli, 248–9; in yeast-like and mycelial forms of *Histoplasma,* 423

RNA synthesis: inhibited by θ-exotoxin of *B. thuringiensis,* 466; involved in production of antheridiol, 212, and trisporic acid, 215; on an RNA template, in viruses, 35–6

RNA tumour viruses, *see* oncornaviruses

RNA viruses, double-stranded, and single-stranded positive and negative, 36

rodents: tumours induced in newborn, by adeno- and papova viruses, 2

Rous chicken sarcoma virus, 27; assay of, 30–1; as endogenous virus, 42–3; gene maps of, and of transformation-defective mutant, 50; mutants of, 47, 48–9; rapid development of, 52; recombinants of, with inherited viral genome of chicken cells, 51; transformation of cells by, 47

rubella virus, transmission of, 40

rumen organisms, plasmalogens in, 271

Saccharomyces, agglutinins on mating cells of, 207

Saccharomyces cerevisiae: acid phosphatase of, 197, (mutants in) 198, 199; commensalism between *Lb. casei* and, 185; competition between *Schizosaccharomyces kephir* and, 188–9; glucose repression of aerobic metabolism in, 172, 174; maintenance energy requirement of, 164; peptide sex hormones of mating types of, 216

Saccharomyces rosei: yield coefficient of, grown anaerobically on glucose, 166

sacoglossan molluscs, chloroplasts from Siphonales in, 388, 404, 405; factors in, stimulating nutrient release from chloroplasts, 406; metabolic interactions between chloroplasts and, 405–8; transmission of chloroplasts to, 408

Salmonella spp.: plasmids in, conferring resistance to metals, 93

Salmonella typhimurium: antibiotic-resistant strains of, 84, 85; doubling time of, in intestine, 119; enterotoxin of, 139; flagellar genes in, 327–8; lipopolysaccharide endotoxin of, 131, 277; mutants of, with straight instead of helical flagella, 323; partial homology between flagellins of *B. subtilis* and, 327; plasmid DNAs in, 86–7; transport in, of aromatic amino acids, 311, of group of carbohydrates, 303, 305, 306, of hexosephosphates, 314, and of histidine, 308, 309, 311

Sarcocystis lindemanni, parasite of humans, 439; epidemiology of, 440

sarcoma viruses: in cell cultures, 29, 30–1; many defective, requiring helper viruses, 48; oncogenes carried by, 27

Schizosaccharomyces kephir, competition between *S. cerevisiae* and, 188–9

septation, in *E. coli,* 60, 61; cell envelope and, 71–2; penicillin-binding proteins and, 73–4; peptidoglycan synthesis and, 72–3; primordium of, 240; sites of initiation of, 290; termination of DNA replication and, 69–71

serine, transport of, 317

serological tests, for fungal infections, 418

sex factor plasmids, 83; *see also* F plasmids

sex hormones, 207–8; control of production of, 218; of fungi, 211–17; of protozoans, 217–18; receptors for, 218; sperm attractants, of algae, 209–10, and of water moulds, 210–11

shaft of flagellum, helix except when cell is helical, 322, 323

Shigella flexneri: antibiotic-resistant strains of, 83–4, 85; co-integrate plasmid isolated from, 86

Shigella sonnei: colicin-producing strains of, 109; enterotoxin of, 139; use of colicins in typing strains of, 123–4

silkworms: bacterial diseases of, 461–2; fungal disease of, 461

single-cell protein, growth efficiency in production costs of, 175–6

siphonein, siphonoxanthin: pigments of Siphonales, found in sacoglossan molluscs with symbiont chloroplasts, 405 sirenin, sesquiterpene sperm attractant of *Allomyces,* 210–11

sodium: concentration of, in medium, and growth efficiency of different organisms, 171

sperm attractants, *see under* sex hormones

Sphaerotilis natans, flagellated rod in nutrient-rich media, colonial in dilute media, 223, 224; cell cycle of, 239

sphaeroplasts, liberation of periplasmic proteins during formation of, 307–8

sphingolipids, in *Bacteroides,* 268

sphingomyelinase C, β-toxin of *Staphylococcus aureus,* 130, 132, 141, 142

Spirillum: marine species of, in competition for lactate with *E. coli,* 197, and for phosphate with rod-shaped organism, 197–8; rotation of flagella of, 330

Spirillum lipoferrum: aerobic nitrogen fixation by, 355; in symbiosis with *Digitaria,* 356

spirochaetes: axial filament in periplasmic space, and movement of, 331–2; chemotaxis in, 339

sponges, marine: with cyanellae, 409

spores: of bacilli used for control of insect pests, 459; of pathogenic fungi, infection by inhalation of, 420

sporopollenin, in cell walls of *Chlorella* symbiotic in *Paramecium,* 396, 410

Sporothrix schenkii, tropical pathogenic fungus, 416; parasitic form yeast-like, saprobic form mycelial, 420

sporulation, 247–8; induced by substrate deficiency (removal of catabolite repression?), 246–7; loci for, on chromosome of *B. subtilis,* 250; protein turnover during, in *B. thuringiensis,* 463–4; in *Rhodomicrobium,* 256, 259; sites of initiation of spore membranes in, 290

squalene and derivatives, in cell membrane of obligate halophiles, 271

staphylococcal toxins, 129: α-toxin, membrane-damaging, 130, 131, 132, 145–6; β-toxin, sphingomyelinase C, 130, 132, 141, 142; γ-toxin, membrane-damaging, 130; δ-toxin, membrane-damaging, 130, 141, 143–4, 146; enterotoxin B, 131; epidermolytic toxin, 129, 130, 147; leucocidin, 130

staphylococci, possible correlation of plasmid content and pathogenicity in, 90

staphylococcins, 109, 122–3

Staphylococcus aureus: mechanism of plasmid-mediated antibiotic resistance in, 88; non-conjugative plasmids in, conferring resistance to antibiotics and to toxic anions and cations, 93, 94; strain of, preventing colonisation of skin by other strains, 123; toxins of, *see* staphylococcal toxins; transport in, of group of carbohydrates, 303, 305, 306, and of hexose phosphates, 314

sterols: of cell membranes, mainly ergosterol in fungi and cholesterol in mammals, 423; of *Convoluta,* are of plant origin (from symbionts), 403; required for production of oospores by *Phytophthora* and *Pythium,* 213; sex hormones of *Achlya* as, 212, 213; toxins interacting with, 141, 144–5

streptococci, high potassium content of, 316

streptococcins, 109, 123

Streptococcus faecalis: mechanism of plasmid-mediated antibiotic resistance in, 88; yield of, in terms of ATP equivalents expended, 159; yield coefficient of, grown anaerobically on glucose, 166

Streptococcus pneumoniae, toxin of, 145

Streptococcus pyogenes, toxins of, 129, 130

Streptococcus sp., not affected by presence of *Lactobacillus* in chemostat culture, 183

streptolysin O, 130, 131, 145

streptolysin S, 130

Streptomyces sp.: cell cycle of, 239; commensalism of, with bacterium metabolising trichloroacetic acid, 185, 186

streptomycin: bacterial mutants resistant to, with modified ribosomes, 83; plasmid-coded resistance to, 95

substrate, affinity for: in competition between organisms, 197–8; in microbial community, 202

succinic dehydrogenase, integral protein of prokaryotic cell membrane, 275

sugar phosphotransferase system, *see* phosphoenol pyruvate-sugar phosphotransferase system

sugars: periplasmic binding protein for, in *E. coli,* 308; uptake of protons by symport of protons and, in *E. coli* mutants, 317–18

sulphate: acting as electron acceptor in anaerobic culture, introduces error into calculation of growth efficiency, 160; periplasmic binding protein for, 308

sulphate/hydrogen sulphide, in mutualism between *Chlorobium* and *Desulfovibrio,* 184

sulphite: in oxidation of sulphur and sulphur compounds, 379, 380; product of dermatophytes, effecting breakdown of disulphide bridges of keratin, 419

sulphite oxidase, linked to cytochome *c,* 380

Sulfolobus, oxidises iron, 365

Sulfolobus acidocaldarius, thermophile: ether-linked lipids in membrane of, 271

sulphonamides, mechanism of bacterial resistance to, 89

sulphur and sulphur compounds, organisms oxidising, 364–5; ATP generation in, 377; energy obtained by, 370, 371; growth efficiencies of, 371; mechanisms of oxidations by, 379–81; reducing potentials available to, 372, 373

surfactin, membrane-damaging toxin of *B. cereus,* 141

SV40 (papova virus), 1, 3–4; antigens in cells infected by, 16–18; deletion mutants of, 20; DNA of, 4–5, (cleavage of, by bacterial endonucleases) 6, 8, (map of) 8, (replication of), 9, 11–13; nucleoprotein complexes in, 9; polyoma virus and, compared, 20, 22, 23, 24; proteins of capsid of, 15–16; temperature-sensitive mutants of, 18–20

swarm cells: binary fission in formation of, 234; of *Caulobacter,* 250, 252, 254; of myxobacteria, 261; of *Rhodomicrobium,* 255–6, 257; of *Sphaerotilis,* 224

symbionts, photosynthetic: entry of, into host cells, and resistance to digestion, 410; host factors stimulating release of nutrients by, 393, 403, 406, 412; metabolic interactions between host and, (*Chlorella*) 397–8 (dinoflagellates) 390–4; multiplication of, in host, 411; recognition of, by host, 395, 400, 410–11; regulation of population of, by host, 394–5, 398–9, 411–12; transmission of, 395, 399–400, 412

symbiosis, 387; between photosynthetic organisms and invertebrates, 387–9, 409–12

symporters, in chemiosmotic theory, 315–16, 317

synchronisation: of cell division in cultures of *E. coli,* 60–1, 62; of *Rhodomicrobium* swarm cells, 255

Synechococcus (cyanobacterium), symbiont in *Paulinella,* 408

T (tumour) antigen, in cells infected with polyoma virus or SV40, 16–17, 54; gene coding for, 18, 20, 21

teichoic acids, in cell walls of Gram-positive bacteria, 269

temperature: chemotaxis more sensitive than motility to, 335; and growth efficiency, 170; homeoviscous adaptation to change of, 282; in isolation of cell membranes, 267, 268; membrane-bound functions dependent on, 273; permissive and non-permissive, for temperature-sensitive viral mutants, 18, 53, 54; and transformation between mycelial and yeast-like forms of pathogenic fungi, 421; transition, from crystalline to liquid crystal, in phospholipids, 272–3, (in biological membranes) 273–4, (with different fatty acids) 280

tetanolysin, 145

tetanospasmin, 128, 129, 139–40; potency of, 131

tetracyclines, plasmid-coded resistance to, 87–8, 95, 106

Tetrahymena pyriformis (predator on bacteria): effect of, on competition between two species of bacteria, 203

thermodynamics, of growth efficiency, 155–7, 158

thermophilic microbes, phospholipids of, 279, 283

thiamine: synthesis of, in mutualism between *Mucor* and *Rhodotorula,* 184

thiobacilli, oxidising thiosulphate: ATP requirement for NAD reduction in, 376; P/O ratios in, 377; scheme for energy coupling in, 381; 'uphill' electron flow in, 374–5

Thiobacillus, N-methylated phosphatidyl ethanolamine in, 268

Thiobacillus A2, 365

Thiobacillus denitrificans, oxidising sulphur and sulphide anaerobically, 365, 374, 379; adenosine phospho-sulphate in, 380; energy coupling in, calculated from growth-linked carbon dioxide fixation, 381–2; energy yield estimations for, based on carbon assimilation, 382; theoretical 'true' growth yield of, 381; thermodynamic efficiency of, 383

Thiobacillus ferrooxidans, oxidising iron and sulphur compounds, 365; aerobic nitrogen fixation by, 355; cytochrome *c* reductase in, 374; energetics of, on iron, 382; growth efficiency of, 371; rapid growth of, 369; thermodynamic efficiency of, 383

Thiobacillus neapolitanus, 365, 375–6; adenosine phosphosulphate in, 380; oxidative phosphorylation in, 377; theoretical 'true' growth yield of, 381

Thiobacillus novellus, ATP-dependent NAD reduction at expense of succinate in, 375

Thiobacillus thiooxidans, 365

Thiobacillus thioparus, 365

Thiobacterium, 365

Thiocapsa, Thiocystis, encapsulated colonies of, 226

β-thiogalactoside transacetylase, of *E. coli,* 300

Thiomicrospira, 365

Thiothrix (cyanobacterium), cell cycle of, 223, 224

tissue cultures, from insects, for production of control viruses, 472–4

toxins, bacterial, 128–30: assay of, 131; conditions of production of, 131–2; may lose toxicity without losing antigenicity, 133; mechanisms of action of, 133–4; and pathogenicity, 127–8; as proteins, 128, 132–3; purification of, 132; *see also individual toxins and groups of toxins*

toxins, fungal, 415

toxoids, 133

Toxoplasma: in mouse macrophages, prevents fusion of lysosomes with vacuole containing it, 410

Toxoplasma gondii, parasite of humans, 437–8; congenital infection with, 438, 454; epidemiology of, 440

transacetylase, of lactose operon, 287

transduction, 77; exclusion of, in detection of conjugational transfer of plasmids, 79

transfection, of normal chick cells by DNA from cells transformed by Rous sarcoma virus, 35 transfer factor, conjugative plasmids acting as, 80, 82–3

transformation of cells by viruses, 77; abortive, 21; assay of cells affected by, 30; of cells in culture, 2, 3, 4, 30–2; cells susceptible to, by either polyoma virus or SV40, are usually not susceptible to the other, 22; exclusion of, in detection of conjugational transfer of plasmids, 79; malignant, by oncornaviruses, 29–30; mutants temperature-sensitive for maintaining cells in state of, 18, 19, 47; mutants unable to effect, 47, 49; properties of cells after, 52–4; Rous sarcoma virus gene for, 50, (in host DNA) 55

transmission of oncornaviruses, horizontal and vertical, 40–2

transport, of organic solutes by bacteria: of amino acids and peptides, 309–12; energy coupling to, 314–18; group translocation in, 303–5, 306; of hexose phosphates, 314; mechanism of uptake in, 297–8; methods of studying, in intact organisms, 298–9, and in membrane vesicles, 299; proteins involved in, 306–9; *see also under E. coli*

transposons (fragments of plasmid DNA, transferred to another plasmid, or to a chromosome or bacteriophage), 103–5

Trema, non-legume with symbiotic nitrogen-fixing rhizobia, 356

Tremella (fungus), sex hormone of, 207

Treponema pallidum, movement of, 331

trichloroacetic acid: commensalism between bacterium metabolising, and *Streptomyces,* 185, 186

Trichoderma viride, fungus able to grow on Dalapon herbicide, in Dalapon community, 202, 203

Trichodesmium, nitrogen fixation in centres of clusters of, 353

trichomes (filaments of cyanobacteria containing different kinds of cell), 260

Trichomonas vaginalis, parasite of humans, acquired venereally, 455–6

Trichophyton spp., dermatophytes, 416

Tridachia crispata, Tridachiella diomedea, sacoglossan molluscs with symbiont chloroplasts, 408; transfer of photosynthate to, 405

trimethoprim: bacterial resistance to, dependent on replacement of sensitive enzyme by plasmid-coded resistant enzyme, 88–9

trisporic acid (mixture of interconvertible metabolites formed from β-carotene), sex hormone of Mucorales, 214–15

Trypanosoma brucei, parasite of humans, transmitted by tsetse flies, 441–3; control of infection by, 452; epidemiology of, 448–9, 450

Trypanosoma brucei gambiense, 441, 444, 449, 450

Trypanosoma brucei rhodesiense, 431, 441, 444, 448, 449, 450

Trypanosoma (Schizotrypanum) cruzi, cause of Chagas's disease in humans, transmitted by Reduviid bugs, 444–5; congenital infection by, 454, 455; control of infection by, 452–3; epidemiology of, 449–51

trypomastigotes: of *Trypanosoma brucei,* with abnormal respiratory system, 441–3; of *T. cruzi,* 444, 445

tryptophan: blepharismone (sex hormone of type II *Blepharisma*) synthesised from, and inhibited by, 217, 218; none in flagellin, 325

tryptose, in cell cycle of *Geodermatophilus,* 228

tumbling behaviour in motile bacteria, 337–8, 339–40; genes for, 339; methionine in generation of, 338

tumour-specific surface antigen (TSSA), associated with transformation by oncornaviruses, 34

tumour-specific transplantation antigen (TSTA), produced in cells infected with polyoma virus or SV40, 18

tumours: caused by oncornaviruses, 27–9; induced in plants by *Agrobacterium,* 93–4; induced in susceptible animals by transformed cells, 21; production of (oncogenesis), 52; viruses associated with, 1–3

tyrothricin (peptide antibiotic produced by *B. brevis*), inhibits RNA polymerase, 249

ultraviolet irradiation: δ-endotoxin of *B. thuringiensis* sensitive to, 467; induces colicinogeny and lysogeny, 112, 124

ultraviolet/visible spectroscopy, for detecting oxidation state of iron in iron protein of nitrogenase, 349

uncoating of virion, mutants defective for, 19

Unio pictorum, bivalve with symbiotic *Chlorella,* 388

uniporters (antiporters), in chemiosmotic theory, 316

Urceolaria viridis, protozoan with green symbionts, 388

uric acid, accumulated by *Convoluta* and utilised by *Platymonas,* 402

uricase, in *Platymonas,* 402

valinomycin, ionophore, 316

vesicular stomatits virus (rhabdovirus): phenotypic mixing of, with defective oncornaviruses, 49; separate mRNAs transcribed for each protein of, 39

Vibrio cholerae, enterotoxin of, *see* cholera enterotoxin

Vibrio metchnikovii, sheathed flagella of, 324

Vibrio parahaemolyticus, possible correlation of plasmid content and pathogenicity in, 90

virogenes, 41

viruses, *see individual viruses and groups of viruses*

viruses for control of insect pests, 470–2; development of insecticides containing, 472–5; specificity of, less marked in tissue cultures than in whole insect larvae, 474

vitamin B_{12}: cell surface receptor for colicin E is also receptor for, 114; in commensalism between *Streptomyces* and trichloroacetic-acid-metabolising bacterium, 185, 186

Volvox spp.: glycoprotein hormones of, inducing sexuality, 216–17

waste treatment, growth efficiency of microbes in costs of, 176–7

X-rays, induce lymphoma in mice by activation of provirus, 44

Xenopus laevis: genes for rRNA of, transferred into plasmids, 106

xenotropic viruses (C-type murine, genetically transmitted, only reinfect other than own species), 28, 45

yeast: killer strains of, 109; pink budding, in Dalapon community, 202; *see also individual species of yeast*

yield: per average electron, 158; based on heat production, 156–7, 158; effect on, of carbon source, 167–8, and of nitrogen source, 168–9; maximum, in terms of ATP, 165–6; in terms of ATP requirement, 158–9; in terms of oxygen consumed, 158; 'true', 163, 164–5

yield coefficient (weight of cells produced per unit weight of substrate consumed), 157; for different organisms grown anaerobically on glucose, 166–7; for different organisms grown on methanol, 167; molar, 157

zygophores of Mucorales, produced in response to trisporic acid, 214–15; of opposite mating types, grow towards each other in response to volatile effectors, 215

Zygorhynchus moelleri, trisporic acid and fertility of, 215